Biochemical and Immunological Taxonomy of Animals

Biochemical and Immunological Taxonomy of Animals

Edited by

C. A. WRIGHT

British Museum (Natural History),
Department of Zoology,
London, England

1974

ACADEMIC PRESS · LONDON · NEW YORK
A Subsidiary of Harcourt Brace Jovanovich, Publishers

ACADEMIC PRESS INC. (LONDON) LTD.
24–28 Oval Road
London NW1

United States Edition published by
ACADEMIC PRESS INC.
111 Fifth Avenue
New York, New York 10003

Library of Congress Catalog Card Number: 73–19025
ISBN: 0–12–765350–3

PRINTED IN GREAT BRITAIN BY
WILLIAM CLOWES & SONS, LIMITED
LONDON, BECCLES AND COLCHESTER

List of Contributors

J. E. Ahlquist, *Peabody Museum of Natural History, Yale University, New Haven, Connecticut, U.S.A.*

T. K. R. Bourns, *Department of Zoology, University of Western Ontario, London, Canada.*

A. B. Champion, *Department of Biochemistry, University of California, Berkeley, California, U.S.A.*

K. W. Corbin, *Peabody Museum of Natural History, Yale University, New Haven, Connecticut, U.S.A.*

H. C. Dessauer, *Louisiana State University Medical Center, New Orleans, Louisiana, U.S.A.*

A. Ferguson, *Peabody Museum of Natural History, Yale University, New Haven, Connecticut, U.S.A.*

F. J. O'Rourke, *Department of Zoology, University College, Cork, Ireland.*

M. L. Johnson, *Puget Sound Museum of Natural History, University of Puget Sound, Tacoma, Washington, U.S.A.*

E. M. Prager, *Department of Biochemistry, University of California, Berkeley, California, U.S.A.*

C. G. Sibley, *Peabody Museum of Natural History, Yale University, New Haven, Connecticut, U.S.A.*

*W. P. Stephen, *Department of Entomology, Oregon State University, Corvalis, Oregon, U.S.A.*

D. Wachter, *Department of Biochemistry, University of California, Berkeley, California, U.S.A.*

A. C. Wilson, *Department of Biochemistry, University of California, Berkeley, California, U.S.A.*

C. A. Wright, *British Museum (Natural History), London, England.*

* Present address: *Naciones Unidas, Provecto Alfalfa FAO–INTA, Casilla de Correo 2257, Buenos Aires, Argentina.*

Preface

As specialized branches of the biological sciences become more complex so communication between divergent disciplines becomes more difficult. As these divisions separate more widely so the need for integrative activities, drawing together the results from separate sources, becomes greater. Taxonomy is one of the most strongly integrative parts of biology because, as Simpson has said, it gathers together and utilizes everything that is known about organisms. Furthermore, never has the need for sound taxonomic work been greater than it is now. Current awareness of acute environmental problems has accelerated intensive ecological studies and these place heavy demands upon existing taxonomic services. Not only are the demands for 'traditional' taxonomy great but there is an increasing need for methods useful in characterizing infra-specific categories such as stocks of food-resource animals, disease-carrying organisms and pesticide- or disease-resistant strains. The biochemical and biophysical techniques which have been developed to fulfil the requirements of molecular biology have made available a bewildering array of data of potential taxonomic value at all systematic levels and it is the purpose of this book to review the progress in animal taxonomy based upon these methods.

The extent to which the techniques have been developed for taxonomic purposes varies greatly in different animal groups, also there have been relatively few workers who have been consistently active in this rapidly developing field. As a result a certain element of personal bias, either in the choice of methods or in the systematic level of their application, may be apparent in some of the chapters. It is a matter of regret that the whole of the animal kingdom has not been covered in this volume. There are many excellent studies in some invertebrate groups which merit report but it proved impossible to find an author prepared to draw these together for review. It is to be hoped that this book will stimulate further activity in these groups and that some future synthesis of the subject will be able to accord them appropriate treatment. Meanwhile, however, a most useful bibliography of the immunotaxonomic literature of the animal kingdom by Charles A. Leone can be found in Bulletin No. 39 of The Serological Museum (1968).

With no previous general review of the subject as a baseline some of the contributors found the task of assessing the material available to them more time-consuming than had been anticipated. In consequence several of the chapters were delayed in their preparation and others have had to be updated either by partial revision or addenda. It is with pleasure that I acknowledge the co-operation of both the contributors and the publishers and I hope that the results of this co-operation will be of use to taxonomists in all groups of animals.

C. A. Wright *London, May 1974.*

Contents

Appendix: Techniques

1 Mammals

MURRAY L. JOHNSON

Puget Sound Museum of Natural History,
University of Puget Sound,
Tacoma, Washington, U.S.A.

I. Introduction

The purpose of this chapter is direct and simple: to bring together data provided by biochemistry and immunology and show application of these data to the science of taxonomy of mammals. Underlying this simplicity, however, is the disturbing feature that we are dealing with several separate sciences.

We must therefore provide a common ground for judging the worth of information. On the one hand we have accurate and reproducible techniques giving data that fulfill the basic tenets of scientific approach. There should be no question of their worth. On the other hand, we have taxonomic problems that absorb such information like sponges, but still remain biological questions.

Let us then view the subject as a whole. After this we may begin, taking pieces and fitting them into the jigsaw puzzle of the problem. Fortunately we are at a point of experience in mammalian taxonomy where new techniques may be integrated and tested with reasonable confidence. Most new types of data, if worthwhile, find a ready spot to plug in and fit. Such is the case with information derived from biochemistry and immunology. If we may use the term 'molecular taxonomy' for this type of study, both the need and the benefits have been amply demonstrated. This we will review.

In order to gain a common ground of understanding with persons of diverse backgrounds, we must clearly state definitions and concepts. With such statements the mammalogist may sit at the same table with the biochemist and the immunologist. Each may be intellectually nourished by the same material. And the person in the most specialized of related fields may make contributions and partake of the same fare.

HISTORY OF MAMMALIA

In the various classes of animals, histories and antiquities vary. Methods of taxonomic approach vary, imposed by the organisms themselves. Theoretically one would think that a method of study would have equal significance in each biological division. It is not so. This is important to remember, not only in dealing with various classes of organisms, but also with subdivisions down into the species and population level.

The class Mammalia is distinct in many ways and this is reflected in the systematic study of the class. Mammalia shares with the class Aves the distinction of being the most recent in evolution (Romer, 1966).

Compared with unicellular organisms, which possibly arose two billion years ago, the first primitive mammals are found toward the end of the Triassic period of the Mesozoic era, something more than 180 million years ago. Related to proteins, however, simple organisms, even as man, contain complicated and homologous protein molecules. Molecular affinity of all animal life is, therefore, greater than the cellular aggregate that makes up the external form.

Other classes such as Pisces and Amphibia go back to the Paleozoic era, over 400 million years ago. The so-called 'Age of Mammals,' the Cenozoic, began some 63 million years ago and progressed through the Paleocene epoch (63 million years ago), Eocene (58 million), Oligocene (36 million), Miocene (25 million), and Pliocene (13 million). Mammal species were at a maximum during the Miocene. During the last million years, the Pleistocene, there was a great reduction in the numbers of species of mammals. This is all relative to molecular studies, for the molecule must change and adjust as the needs of the evolved species change.

Paleontologic information concerning mammals is aided by the fact that the extremely hard teeth, as well as bones, are available to show relationships. It is now generally accepted that mammals arose from an ancient lineage of mammal-like reptiles, the therapsids.

Although evolution in mammals has produced great diversity of external form, with specialization, as for flying and marine life, the proteins have maintained a surprising general similarity. And to carry the concept back even further, protein similarity and homology can be traced into other classes of animals.

It is as well to bear in mind the time element stated above for there are theoretical considerations concerned with the rate of evolutionary change of proteins. In order to judge and digest some of the recent publications, one must understand the total time table of evolution.

II. Taxonomy of Mammals

Taxonomy has proceeded through several phases since the time of Linnaeus. Among other classes of animals, mammalian taxonomy has long been on the most solid basis, simply because of the parts of the specimen available for study. The traditional mammal specimen presented more, and more varied, data for analysis. Hard structure of bone and teeth, because it has endured and is therefore available, has provided maximum material for investigating phylogeny of mammals. This undoubtedly has produced a more realistic and stable phylogenetic appraisal than is possible in other classes. Darlington (1970) supports this generalization.

In a modern and conservative sense, a specimen for proper taxonomic evaluation should be one of a series of specimens of skins and matching skulls. Standard external measurements of body, tail, hind foot and ear give immediate recognition to general anatomic proportions. The skull, being a composite of many bones, may be subjected to an almost infinite number of analyses. In a high proportion of taxonomic evaluations the combination of skins and skulls is sufficient for proper definition.

Many other anatomic features have been used and have been found helpful. Each of these parts are at times quite good: hair, nails, horns and hooves, skeletons, including special bones such as the baculum, hyoid, and ear ossicles, soft structures, including viscera and muscles. Microscopic to ultramicroscopic evaluations are the natural outgrowth of this background; these are proceeding with sporadic success in taxonomy.

Because good criteria are already available, taxonomists in mammalogy have been less sanguine to accept radically new techniques. By the same token, a trial of new methods is probably best evaluated in situations where a maximum background of information is already available. Relative to molecular taxonomy, it may be stated that this newest of taxonomic data fits well with traditional evaluation, as it should.

A. DEFINITIONS: TAXONOMY, GENETICS AND PHYLOGENETICS

Misunderstanding of the working definition of these three sciences in relation to mammals (as indeed in other groups, also) has led to considerable useless conflict from the podium, from the pen, and in conversation. In cases of agreement of data much energy has been expended in explaining a point of view.

Let us then define in practical and working terms, each of these as used in this chapter, so that arguments may be properly directed and more telling.

Taxonomy (from Gr. *taxo* = future of *tasso*, to put in order, arrange) is the science of placing a biological form in order. This *implies* that in a systematic way, one form is most closely related to the next. This is the crux of many arguments by the purists: on one hand, there are those who wish to have an absolute relationship shown, and on the other those who merely wish a name by which to identify an organism at hand. Most taxonomists of experience accept the fact that unequal treatment must be accorded different taxa because of practical considerations. At the same time, no one denies the desirability to show relationships truly and accurately. Somewhere between the two extremes lies the broad and well-trodden path of the working taxonomist. In practice it cannot be as accurate as the data that support it.

Systematics has been used as a synonym for taxonomy by many, but is probably best used in a more restrictive sense as 'the scientific study of the kinds and diversity of organisms and of any and all relationships among them' (Simpson, 1961).

Genetics (from Gr. *genos*, a race, kind, descent) is more pure, more definite and is a more reproducible, analyzable science. The basic concepts of genetics related to hereditary characters carried by the genes of the chromosomes are well documented, though there is some evidence that other pathways are effective.

Phylogenetics (from Gr. *phyle*, a tribe, race, plus genetics) is also a pure science which shows derivation of the 'family tree' of an organism. Because of limitation in source material, much of phylogenetics is theoretical. And because certain parts of molecular taxonomy can be subjected to mathematical analysis, a number of thought provoking considerations have been applied to mammalian phylogenetics.

B. THE SPECIES PROBLEM IN MAMMALS

Species (from L. *species*, a shape, sort, a particular kind) is essentially a biological concept as I consider it. It is the most important taxonomic category. 'Species are groups of actually (or potentially) interbreeding populations which are reproductively isolated from other such groups,' (Mayr, *et al.*, 1953; Mayr, 1969). Practical limitations exist in obtaining interbreeding, thus morphological characters are generally described and used as determining criteria.

The species definition has been further refined (Mayr, 1963): (1) 'Species are defined by distinctness rather than difference;' (2) 'Species consist of populations rather than unconnected individuals;' (3) species are more certainly defined by comparison with populations of other species than by comparison of conspecific individuals; (4) a species may be characterized as possessing a large intercommunicating gene pool, with an individual having a small portion of that pool for a short time.

The typological species concept must also be mentioned for here is a base of criticism, often justified, levied at publications of molecular taxonomy. In essence this concept states that if two individuals or groups of individuals appear sufficiently different, they are different species. The purely morphologically defined species is in this category. A tendency has existed to ascribe undue significance to difference found in molecular studies. For this reason, series rather than individual specimens should be used, and new data should be integrated with, rather than proposed to supersede, prior information.

Mammalian, and most other, specialists in every-day taxonomy work largely at the species level. The category of 'compleat taxonomist' of mammalogy includes well known investigators from over the world, and a continual crop of rising young mammalogists. These persons have developed the ability to recognize, in the particular groups they study, characters that define a species. 'Numerical taxonomy,' with its increasingly important contributions to mammalian systematics, and into which molecular data can be injected, does not at all supplant the need for a balanced and experienced judgement for taxonomic decisions.

While many characters are clearly adaptive, this attribute is frequently lacking, at least in light of present knowledge. In any event many characters cannot be related to the question at hand: reproductive isolation. A sharp parallel may be drawn here: the characters discussed in this chapter which are analyzed by biochemical or immunologic means are, in our present state of intelligence: (1) *not* proven to be adaptive, and (2) *not* related to reproductive isolation, i.e. innate species determination. These characters are in general more subject to judgement (than are the usual morphologic characters) before being used in mammalian taxonomy.

C. TAXONOMY OF GROUPS OTHER THAN SPECIES

1. Subspecies and Populations

These are shades of the same phenomenon: geographically assignable groups of animals which have characters in common that are different from neighboring groups. They are usually capable of interbreeding freely with other populations or subspecies. When they cannot freely interbreed on contact, they then must be evaluated as a full species. Determination of intergradation is generally acceptable from characters of field caught specimens, but there is an increasing tendency to bring these problems into the laboratory for definition.

It is this general category of taxonomic study where some of the most interesting problems lie and where there is most need for help from other than traditional methods. Chromosome studies and molecular studies each have much to offer in re-studying some of the puzzling biological arrangement with which we now live. It is also quite likely that our complacency will be shaken in some fully accepted situations, when thoroughly evaluated.

2. Higher Taxonomic Categories of Mammals—Phylogeny

In the rank of genus and above—family, order and intermediate classifications, it must be recognized that the separations are arbitrary. There is no biological principle involved except assessed lineage. Intermediary forms may be found living in the lower categories such as genus, but in most cases even here taxonomists have relied greatly upon paleontologic data to help seek out and assess degrees of relationship in time and in lineage.

The use of biochemical and immunologic data in each rank is a situation unto itself, comparable to the morphologic characters. Not only is each rank a unique situation, but in each mammal group the usability of a given character or set of characters may be quite different from the situation in the next group.

It is in these higher categories where phylogenetic assessment by many means other than traditional morphology is likely to make contributions. There are, however, two cautions that I believe are valid: (1) the best judgement will be made by the taxonomist with data from many fields at his disposal, not by the specialist in some other field, even though it is he who may discover exciting new data; (2) evaluation of living forms, especially those more widely separated on the phylogenetic tree, will continue to be plagued by errors of theory and judgement; decisions will not be to everyone's satisfaction. The wider the separation, the more we need solid fossil evidence to support our assertions.

In dealing with the phylogeny of these higher categories, the questions of the relationships of one order to another, what genera to properly include into what families, these and many similar ones, will have improved answers as works in biochemical and immunological comparisons progress.

Already theoretical considerations of the rate of evolution of proteins are being tested. These will be elaborated on later.

Attractive to the mammalian taxonomist is the theoretical probability that new methods will help to establish placement on the phylogenetic tree. This is particularly promising in the use of macromolecules. Through the past century a solid base of facts has been built to support the present contentions of relationships. These facts are both morphologic and physiologic. They are from many disciplines. They interweave to reinforce one another and molecular data is expected to do the same.

From fossil bones and teeth logical evolutionary schemes have been provided. Gaps between past and present day structures have often been filled and homology verified.

From embryologic and developmental studies the derivation of structures of modern animals has been noted. Relationships have been clearly shown.

By comparative anatomic analyses logical sequential relationships of modern forms have been produced.

By detailed physiologic analyses of metabolic cycles, of hormones, of enzymes and literally hundreds of inter-related processes, good phylogenetic information has been obtained.

In each of the above sciences, no matter how worked over, a great deal remains to be done, and much is there to re-evaluate. The point is that we have this great wealth of factual material that has given us a good basic understanding of phylogeny of mammals, but innumerable questions remain to be answered.

It is tempting at this point to project our techniques and look to the time when complete amino acid sequences and other internal anatomy of the protein molecules will be known. In molecular weights less than 50,000 to over a million, the 'comparative anatomy' must be revealing. There must be homologous chains, homologous peptide groups, as well as identical parts, down to the exact connections of each amino acid. Some of this information is already available from biochemical laboratories. Certain proteins such as hemoglobin appear most suitable for this analysis in the present point of time, however, with automated equipment and computer analysis rapidly becoming improved, all molecular forms, even the most complex, are becoming accessible.

This kind of information becomes easily transposable into the taxonomic and phylogenetic milieu. The problems of convergent, parallel and divergent evolution exist as in comparative anatomy or physiology. The definition of analogy and homology come back into focus. And hopefully, as this kind of information becomes available, some of the questions of adaptability and reasons for evolution will be answered.

D. VARIATIONS AND POLYMORPHISMS

Polymorphism is an important enough phenomenon, in dealing with taxonomy, to clearly define as to occurrence and significance in mammals. 'Variation' is often used interchangeably in a broad sense, but in a stricter sense polymorphism has come to designate, in protein studies, the presence of nearly identical forms of a functional molecule in members of a population or species, which are genetically controlled and may be single or multiple within an individual. It will be so used here. 'Balanced polymorphism' describes the stable condition (in an evolutionary sense) where the presence of a number of forms of a protein are apparently favored by selective pressures.

The difference in form in polymorphism may be a substitution of one or more amino acids so that different molecular forms of a macromolecule are produced. It may be an alteration in combination of parts (chain) of a molecule. It may be a combining (polymerizing) of several molecules which are identical.

Polymorphisms have been shown to be under strict genetic control, and as such may be useful markers for populations. Their practical significance will be reviewed later. In most cases, however, little taxonomic use has been made of them, at and above the species level. As with other biological variations, the usual rule of examining adequate biological series must be followed. On the other hand, if the proteins being examined have been found to have stability of character, a single specimen may be reliable for making a preliminary determination. To repeat, in a taxonomic sense, there is no essential difference between a protein character and another morphologic character. Each requires determination as to whether it is stable and conservative or is variable and either must be examined in series or discarded as worthless for the purpose at hand.

Some authors have used polymorphism to define the presence of more than one molecular form within a taxonomic unit, and essentially without variation within the unit. Such use should be strictly defined.

E. BACKGROUND—MOLECULAR TAXONOMY OF MAMMALS

We are dealing with two essentially different types of investigations—biochemical and immunologic. It is quite proper that we state the systematic position of each of these in relation to each other and to mammalian taxonomy. Immunology is discussed first in this section because of historical priority.

1. Immunological Taxonomy in Mammals

Immunology is concerned with the interaction of antigens and antibodies. These interactions present the following phenomena (Boyden, 1949): (a) agglutination of particulate antigens; (b) precipitation of soluble antigens; (c) neutralization of toxins or inhibition of enzymes; or (d) lysis of cellular antigens. The principle is that an antigen from a species of animal injected into another animal stimulates antibody production. The general method is to give a series of injections of serum, or a particular fraction of serum, or other protein into a rabbit or horse. After a few weeks the serum of the injected animal contains circulating antibodies to the injected protein or proteins (antigens) in the immunoglobulin fractions. This can then be titrated against antigens from the same and other species of animals. The relative amount of reaction, usually indicated by precipitation zones, is then used to assess the closeness of immunologic relationship and, presumably, taxonomic position. Refinements of techniques and evaluation have produced significant and interesting results.

These technical refinements in general allow examination of particular components. An example is immunoelectrophoresis which first separates fractions by electrophoresis; on the same preparation antibody containing serum challenges the strip of separated proteins; precipitation zone reactions indicate relationships. Specific protein fractions may be obtained to which immunological manipulations may be done. Another approach is comparative studies on strictly homologous molecules such as albumin (Sarich, 1969a).

Blood grouping involves other immunologic reactions: (1) agglutination of the intact red blood cells, and (2) hemolysis of the red blood cells. Blood groupings have been devised for many mammals in addition to man. Their use has largely been related to genetic studies; taxonomic use has potential and should be explored.

2. Immunologic Taxonomy—History

Immunologic advances began with Kraus's discovery of the precipitin reaction in 1897, related to micro-organisms. Landsteiner in 1900 noted the agglutination of human erythrocytes mixed with the serum of other individuals; the study of blood groups and the genetics of blood proteins were begun. Nuttall in 1901 applied the precipitin reaction to many organisms and noted correlation to their general taxonomic position. Mammals, especially rabbits and horses, have largely been (and still are) used for the production of antibodies, but many other animals may be used, including those in other classes.

Some of the earlier assertions regarding mammalian relationships, based on immunologic studies were not in conformity with traditional concepts. This tended to keep mammalogists from accepting the results of immunologic studies. However, continuing work has placed immunology into a respected and honored category, with accent upon phylogeny.

A remark on the term 'serology' is in order here. This is sometimes used loosely to refer to any studies involving blood serum. However, by virtue of tradition 'serology' is equated with immunology, and 'serologic' with 'immunologic'.

3. Biochemical Taxonomy in Mammals

In analyzing the structure and reactions of proteins there was a tendency to over-state, as with immunologic techniques, the significance of the findings. This is now settling down. Basically the mammalian taxonomist can integrate biochemical data more easily than immunologic. The anatomy of the molecule is more like a traditional character carried to the submicroscopic range. It is possible to find characteristics which may be assigned to any taxonomic unit from populations through subspecies and species into the higher categories.

As with immunological information, the biological concept of species cannot be equated with similarities and differences of protein molecule structure or reaction until thoroughly evaluated. For instance, it has long been considered that the hemoglobin molecules of each species are characteristic and different. Thus far, when amino acid sequences are established, this has proven to be the case. Other complex molecules generally have assignable differences to follow this pattern of species specificity. However, associated in most macromolecules there are homologies and identities of segments that allow comparison between even remotely related groups.

There is no biological reason why related mammals may not share identical macromolecules through complicated linkages into ranges of 60,000 molecular size, or more. Such cases will very likely be found as data accumulate. Recent publication by Seal (1969) presents some evidence that this may be true in hemoglobins of carnivores; his information, however, must be checked by other methods before acceptance.

As evolution progresses it is certain that mutations will occur within the complexity of the molecular systems. In a hypothetical situation, if a protein or any other molecule functions just as well with certain amino acid substitutions, and if these substitutions are made at the right time and place to go along with other changes of adaptive significance, then there is produced a difference in a molecule proceeding with evolution. The relative importance of natural selection versus neutral mutations and random genetic drift is controversial (King and Jukes, 1969; Clarke, 1970). Mechanistic concepts of evolution of proteins are well presented by Ohno (1967, 1970), Manwell and Baker (1970) and Kimura and Ohta (1971).

That molecular evolution occurs at a steady rate is the basis of theoretical consideration by certain modern workers, such as Sarich (1969a). Relationship in time as related to the position on the phylogenetic tree may then be calculated. This is an attractive and well conceived theory, and logically should have greater chance of accuracy when applied to a conservative protein such as mammalian albumin. On the other hand, there is little in the documented history of evolution to suggest that a constant rate of change is ever maintained in a character of any group. Conversely, fossil evidence indicates greatly varied rates of evolution (Darlington, 1970). Also, analysis of statistical models indicate that better data is needed for substantiation (Read and Lestrel, 1970). Thus this theoretical basis for calculating phylogenetic relationships must remain open to challenge.

In summary, regarding protein molecular evolution in mammals, we can say that it may or may not be adaptive, in light of our present theoretical considerations and abilities to analyze the term, 'adaptive.'

4. Biochemical Taxonomy—History

The origins of molecular biology are reviewed by Hess (1970), and mention was made of work of Reichert and Brown (1909) who published crystallographic data on the hemoglobins of more than 100 species of mammals, birds, amphibians and reptiles. The significance of this structural data was not really appreciated for nearly 50 years, when the associated concept arose that proteins were composed of long chains of amino acids held together by covalent bonds.

In 1945 (Deutsch and Goodloe, 1945; Moore, 1945) it was shown that difference between mammalian species could be demonstrated by electrophoresis of serum proteins. No taxonomic evaluations were attempted, but the presence of a new character for investigation was established. There followed a decade and a half of in-depth investigation of mammalian proteins. These were not related to taxonomy, but to biochemistry and especially genetics.

Because of the great impetus in basic clinical investigations (and the money available for such) many excellent studies relating to human proteins and laboratory animals were published. The numbers of proteins which could be separated or demonstrated by differential staining methods grew enormously. Undisputed evidence for strict genetic control of protein synthesis was presented.

The presence of biological variations of molecular forms, comparable with characters long used in taxonomy were proven. Applied to this the term 'polymorphism' was used by many workers. This of course was nothing new to taxonomists, whose very basis of science consists of the study of variations and applying this to the matter of segregating and classifying forms.

Thus in the late 1950s the stage was set for applying a wealth of sophisticated and diverse methods and data to taxonomy. Much of the original research and information was with the proteins of man and continues to be. So it has happened that *Homo sapiens* became the control animal. Taxonomic and other studies concerning proteins were clustered around the basic information derived from humans and domesticated animals.

It is of some interest that taxonomic application of electrophoretic and allied techniques on mammalian proteins lagged behind some excellent studies on other classes of animals.

Electrophoresis was found to provide sensitive, direct data regarding polymorphisms, quantity and presence or absence of certain proteins. It provided indirect evidence of molecular changes by differences in migration. This amounted to a large amount of information not previously available.

I have personally been involved in an organized taxonomic survey of mammalian proteins since 1957, proceeding at a lower level of intensity than many of my colleagues. During this time I have seen the interest grow to its present status where evaluation is being done by many men in many groups of mammals, and from different points of view. In the following part of this chapter I will summarize much of published work, adding some original data of my own, and will attempt to provide a perspective of the work done up to this time. Hopefully, this will allow evaluation and give a jumping off place for mammalogists, taxonomic biochemists and others with special interests in allied fields of genetics and phylogeny.

5. *Taxonomic Use of Mammalian Proteins—Done by Whom?*

A word must be said regarding the usability of the techniques by which proteins are evaluated.

In general, the complicated procedures such as amino acid sequence, analyzing of the more elusive protein fractions, determinations of the physicochemical properties of the macromolecules and relationships within metabolic pathways—these and many others are the province of specialists in biochemistry, biophysics or physiology. There are few taxonomists who have either the background or time to productively engage in these fields. Here, the approach is multi-disciplinary. It is the taxonomist's duty to have sufficient expertise to communicate with his confreres and to evaluate their opinions and results.

There are, however, several of the evaluative procedures that can be performed by a single individual. These involve electrophoresis and chromatography and some of the immunologic techniques. One might say that these are suitable for graduate students in taxonomy; or, conversely a student with a biochemical or immunological background might profitably be invited into the taxonomist's realm. In our university undergraduate students in advanced genetics satisfactorily perform several of these laboratory maneuvers.

However, to repeat, it is in the best interests of taxonomy to have decisions of significance made by an experienced taxonomist, *not* by the data gatherer, unless he is one and the same.

III. Systematic Review—Biochemical and Immunological Taxonomy

There is a burgeoning literature available for research and study, related (usually not directly) to biochemical and immunological taxonomy of mammals. I have reviewed this in depth and offer here samples of what I have judged to be important. That I have missed some important works I do not doubt. The authors and readers of experience with this sort of review will understand and be tolerant. (Literature search extended into 1971.)

It is not too difficult to encompass the major works until about 1960; in 1957 when I began work in this field a single page of references might be considered a reasonably good bibliography. Now, however, a complete list of the literature of biochemical and immunological publications related to comparisons of mammal species would fill books.

Despite the multiple works, it is prudent to include many references that are not taxonomic studies *per se*. As one reviews the literature over the years, one is struck by the numbers of characters which have been well studied in different contexts, before being 'discovered' as traits for clarifying relationships. With a little experience and imagination, one can look ahead to the use of many now untried characters suitable for mammalian taxonomy.

Thus I have attempted to include, in addition to numbers of solid taxonomic contributions, many descriptive articles. The characters described, as I see it, appear to have a potential. They deserve to be looked at, evaluated, then possibly injected into the taxonomic stream. There may be answers to some of our questions as we add these bits of information.

In order to make this review usable, it has been organized by subject matter and by taxonomic position. Most references I have evaluated myself. Some have been obtained from abstracts or other authors; these have not all been traced to the original article, depending on the relative importance, the information already in hand, and in some cases the availability. References so treated are specified. When there are multiple publications on a single subject, I have arbitrarily chosen representative ones; my choice does not necessarily reflect upon the worth of those omitted. Regarding the particular groups of species, man and his fellow primates, and laboratory and domestic animals, it has been necessary to omit a great wealth of information. This may be the raw material for tomorrow's mammalian taxonomy. As a poor substitute, references are made to some of the monographic treatments now available. These are with increasing regularity presenting up-to-date summaries of data. They include pronouncements related to phylogeny and taxonomy; these should continue to have critical appraisal rather than blind acceptance. At the same time taxonomists should maintain an open-minded appreciation for the fine opportunities made available.

Scientific names have been used for all wild-type animals. Common names are not given for many of the lesser known mammals; their taxonomic position is indicated, however. In cases of domestic and laboratory animals, the scientific names are given in Table I. It is to be noted even today that an author not biologically oriented may specify no more than 'rat' or 'mouse' or 'rabbit', in his appropriate language.

Table I
Scientific and Common Names of Laboratory and Domestic Mammals

Oryctolagus cuniculus	Laboratory rabbit, domestic rabbit, Old World rabbit, rabbit, lapin (Fr.), Kaninchen (Ger.)
Mesocricetus auratus	Hamster, Golden hamster, hamster (Fr.), Hamster (Ger.)
Rattus norvegicus	Laboratory rat, white rat, rat, rat (Fr.), Ratte (Ger.)
Mus musculus	Laboratory mouse, white mouse, mouse, souris (Fr.), Maus (Ger.)
Cavia porcellus	Guinea pig, cobaye, cochon d'inde (Fr.), Meerschweinchen (Ger.)
Canis familiaris	Domestic dog, dog, chien (Fr.), Hund (Ger.)
Mustela putorius	Domestic ferret, domestic polecat, furet (Fr.), Frettchen (Ger.)
Felis catus	Domestic cat, cat, chat (Fr.), Katze (Ger.)
Equus caballus	Horse, pony, cheval (Fr.), Pferd (Ger.)
Equus asinus	Ass, burro, donkey, âne (Fr.), Esel (Ger.)
Sus scrofa	Domestic pig, pig, wild boar, cochan, porc (Fr.), Schwein (Ger.)
Bos taurus	Domestic cattle, cow, ox, beef, 'bovine,' 'milk,' bétail, bovin, boeuf (Fr.), Vieh, Rind, Kuh (Ger.)
Ovis aries	Domestic sheep, sheep, lamb, mouton (Fr.), Schaf (Ger.)
Capra hircus	Domestic goat, goat, chèvre (Fr.), Ziege (Ger.)

A. TAXONOMIC MATERIALS—SPECIES AND MOLECULES

1. Taxonomic Units

There are in the world some four thousand different species of mammals, in nineteen orders (Table II). Intertwined biologically, and altered by evolutionary pressures is the internal milieu of the mammalian body. Here there are literally thousands upon thousands of different molecules ranging from the simplest chemical compounds to the most complex. These in turn are involved in complicated biochemical–metabolic pathways that require much elucidation.

2. Simple Molecules

Taxonomic use of the simpler chemical compounds is little, but in terms of comparative physiology or composition of body parts must be considered.

Table II
Tabulation: Living Mammals Of The World
(Modified from Anderson and Jones, 1967)

	Number of families	Number of genera	Number of species
1. Monotremata	2	3	6
2. Marsupialia	8	81	242
3. Insectivora	7	72	391
4. Dermoptera	1	1	2
5. Chiroptera	16	173	875
6. Primates	11	52	181
7. Edentata	3	14	31
8. Pholidota	1	2	8
9. Lagomorpha	2	9	63
10. Rodentia	34	355	1,687
11. Cetacea	10	38	84
12. Carnivora	7	96	253
13. Pinnipedia	3	20	31
14. Tubulidentata	1	1	1
15. Proboscidea	1	2	2
16. Hyracoidea	1	3	11
17. Sirenia	2	3	5
18. Perissodactyla	3	6	16
19. Artiodactyla	9	75	171
TOTAL	122	1,005	4,060

The chemical composition is, in general, stable and identical, one species to the next.

3. Intermediary Molecules

As we proceed into the complexities of larger molecules, it has been known for many years that a specific pattern exists, but deviations in chemical composition or metabolic use may be found between taxonomic units. Good examples are the bile salts of various species (Haslewood, 1964, 1968).

4. Large Molecules

As we reach the macro–molecular size, demonstrable differences are the rule, from species to species. Even here, however, the changes may be infinitesimal. Whole molecules may be identical in makeup except for one or two substitutions.

The relationship is that the larger the molecule, the more is the chance of difference at the lower taxonomic levels. Thus it is noted that taxonomic uses and potential in mammalogy are in the area of macromolecules (primarily protein).

5. Protein classifications

Most of living tissue is protein. Proteins have the greatest diversity of function and structure. They may be classified broadly into four types (Steiner, 1965), according to the function within the organism: (1) structural, as collagen; (2) contractile, as muscle; (3) catalytic, as enzymes; (4) transport, as hemoglobin. It is obvious that each of these is related to adaptive biological conditions. It is postulated that relationships between phenotypic expression of proteins and adaptation will eventually be explained in many situations.

Because this book is concerned with interdisciplinary matters, it is important that we recognize other names and classifications for proteins. Various names may be used in reports. These may be synonymous with others or may refer to a particular property of the protein. The following list of definitions and synonyms may help clarify the 'types of proteins', particularly those involved in taxonomic studies of mammals (after Sober, 1968).

a. Physical definitions (categories may overlap).

i. Globular proteins—approximately egg-shaped with maximal frictional coefficient of 1·5.

ii. Fibrous—'cigar shaped' or 'threadlike' with frictional coefficient above 1·5.

Little, if any, taxonomic use is now possible within this classification. Electron microscopy of protein molecules (Höglund and Levin, 1965) reported a simple external morphology of serum proteins of various classes. Svehag and Bloth (1970), however, found Y-shaped molecules (125 by 140 angstroms) of IgA from human and rabbit colostrum. Expanded techniques, such as use of the scanning electron microscope or field ion microscope allows visualization of atoms and small biomolecules (Müller, 1970), and may change the physical classification drastically.

b. Chemical definitions.

i. Heme proteins—compounds in which a heme group is bound to polypeptide chains. In myoglobin a single chain is present; in hemoglobin a tetrameric form, with four heme groups and four polypeptide chains exists as a single molecule. Cytochromes are intracellular respiratory pigments. Cytochrome C is present in the mitochondria of all aerobic cells; it contains the same heme group as hemoglobin and myoglobin and functions in oxidative phosphorylation within the cells.

ii. Lipoproteins—globulins combined with lipids. In man these occur principally in the alpha-1 range (high density lipoproteins) and alpha-2 range (low density lipoproteins).

iii. Glycoproteins (mucoproteins)—globulins combined with sugars or polysaccharides. These occur in all segments of the globulin spectrum in mammalian blood and include many of the specially-named fractions such as haptoglobin, hemopexin, transferrin, immunoglobulin, orosomucoid and prothrombin. Hormones and enzymes are largely glycoprotein. 'Seromucoid' sometimes designates serum glycoproteins not precipitated by heat. Several of the proteins listed here lack acceptable definition.

iv. Haptoglobins (Hp). Serum globulins that combine with free hemoglobin (but not with free heme).

v. Hemopexins. Serum globulins that combine with free heme, and have also been called seromucoid, cytochromophilin and beta-1-haptoglobin.

vi. Transferrins (Tf.). Serum globulins that combine with iron. Also called siderophilins.

vii. Immunoglobulins (gamma globulins). These are classified as gamma G (IgG), gamma M (IgM), gamma A (IgA), gamma D (IgD) and gamma E (IgE) immunoglobulins in humans. IgG is the most important. These are involved primarily in antibody production.

c. Electrophoretic definition (based upon migration rates of human plasma fractions).

i. Albumin. In mammals the fastest moving (referring to distance travelled in unit time) major band, toward the anode. It is usually single. It may be double typically in some of the Sciuridae or aberrantly as a polymorphic trait. Pre-albumins and post-albumins are lesser fractions, designated according to their mobilities and irregularly present according to taxons.

ii. Alpha globulins. These are included in the next fastest moving fractions on paper electrophoresis, and are divided into 'alpha-1,' the faster and 'alpha-2' bands. These are further broken down by the better resolving gel techniques and may be termed 'fast' and 'slow' alpha globulins with numbers −1, −2, −3, indicating successively slower migration. With starch-gel, the 'slow' human alpha bands are slower than the beta globulins, converse to the paper technique; also human pre-albumin-2 on starch-gel is an alpha-1 globulin by paper technique.

iii. Beta globulins. These are slow moving fractions. By paper technique this is in between alpha-2 and gamma globulins. By starch-gel in humans these lie between the 'fast' and 'slow' alpha globulins.

iv. Gamma globulins. These are the fractions that are nearly isoelectric or cathodally migrating, frequently in a broad, diffuse band. These are immunoglobulins, which term also includes some other globulin fractions in the alpha and beta range.

The designation of the alpha and beta globulins, in order to prevent confusion, should be properly qualified as to type of technique being used and should be related to mobility, not specific protein type.

6. Biosynthesis of Protein-relevance to Taxonomy

The major argument that proteins should be plugged into the taxonomic circuit is the knowledge (e.g. Steiner, 1965; Ohno, 1970) of the tight control of their formation: Genetic control is principally, if not entirely, by the system of genes within chromosomes. The active genetic portion of the chromosome is deoxyribonucleic acid (DNA), a high molecular weight polymer of deoxyribose nucleotides. There are only four nucleotides in appreciable amounts:

deoxyadenylic acid, deoxyguanylic acid, deoxycytidylic acid, and thymidilic acid. The only internucleotide present is of the phosphodiester type. Genetic control is achieved by the nucleotide sequence within the DNA, transferred by way of ribonucleic acids (RNA).

The genetic code is accomplished by triplets of 64 possible ribotrinucleotides, specified in DNA and replicated in RNA.

During synthesis of protein, occurring from messenger RNA on the cellular ribosomes, each of 20 amino acids used by modern organisms has specific triplet 'codons'. Protein synthesis is regulated by the messenger RNA with integral involvement of smaller transfer RNA from the soluble pool (Khorana, 1968) and a multi-enzyme system of many kinds of enzymes (Nirenberg, 1968). The beginning, end, and sequence of amino acids in each peptide is controlled by the RNA nucleotide sequences, thus refering back to the DNA of the gene and chromosome.

The mammalian (human) chromosomes contain sufficient information to synthesize 3 to 6×10^6 different kinds of proteins, though much information stored is probably redundant (Nirenberg, 1968).

It is obvious that the distinctness of taxonomic forms exists within the DNA, but thus far little can be done to demonstrate constant variation at this level. Some differences between species in cellular DNA content of certain parts of the body have been found (Sober, 1968).

Almost identical translations of nucleotide sequences to amino acids have been found in such diverse organisms as bacterial, amphibian and mammalian amino-acyl transfer RNA (Nirenberg, 1968). Many 'universal' kinds of aminacyl-tRNA were found. However, seven kinds of mammalian tRNA were not detected in bacterial preparations. Conversely, five kinds of tRNA from a bacterium were not found in mammalian preparations.

More than 100 different macromolecules must interact orderly to translate coded information from messenger RNA into the amino acid sequences of proteins (Kurland, 1970). This gives some indication of the complexities of study, but also of the promises of rewards from the many investigations now in progress.

7. *Evolutionary Time Scale*

Inasmuch as taxonomic concepts involve evolution and phylogeny, it is well to include some theoretical concepts that have been published in recent years relative to molecular biology.

Williams (1964) mentions the 'irresistible temptation' to compute time rates of evolution, using molecular data. Several imaginative workers have been bold enough to proceed down this street.

Nirenberg (1968), on the basis of his extensive work and pertaining to all life, suggests that the genetic code probably evolved more than 5×10^8 years ago, and that 'some species-dependent differences in codon recognition apparently serve as regulators in protein synthesis'.

Zuckerkandl (1965) reviewed the principles of 'chemical paleogenetics'. He stated that this science would eventually have taxonomic value in providing fundamental measurements, but that the prime objective was to understand natural selection related to different types of mutations. By comparing hemoglobins of several species, humans, horse, pig, cattle, and rabbit, and computing the number of amino acid substitutions (about 22 or 11 pairs) that have taken place since their common divergence (estimated 80 million years), a figure of one change per seven million years was obtained. Single chains could be used with statistically satisfactory results. Figuring may be retrograde, to obtain a rough estimation of the time elapsed since a common ancestor of a given polypeptide chain. In human beta and delta chains, for example, which differ at ten sites, only five changes in the gene line are necessary, or 35 million years.

McLaughlin and Dayhoff (1969) estimated the evolutionary rate of change of proteins, indicating that each protein accepts mutations at a different rate. Rates of different proteins may vary by a factor of 1400: Fibrinopeptides 90 PAM (accepted point mutations per 100 residues estimated to occur in 100×10^6 years of evolution) Hb, beta chain 12 and alpha chain 13, myoglobin 9, histone 0·06. They agreed with other workers that tetraploidation may be an important mechanism in protein evolution. It must be pointed out, however, that the uniformity of the genome of all mammals (as determined by DNA content) indicates that ploidy of the genes must have occurred before mammals as a class evolved (Ohno, 1970).

Fitch and Margoliash (1969) constructed a model of evolution based on details of cytochrome C and its presumed evolution. Eleven mammals are included in their model, including 'ancestral' mammal, primate and ungulate. The theoretically constructed model allowed placement of amino acid sequences with great accuracy. The more recent the event, the more accurate was the model placement with the proven sequence. It was suggested that differences between the model case and evolution of cytochrome C might provide clues related to natural selection (not built in the model case). Parallel mutations appeared to occur more frequently in reality than expected by chance. Double mutations also occurred more frequently than expected by chance, related to the apparent greater possibility of fixation of the next mutation in a codon which had a recently fixed mutation. Margoliash *et al.*, (1969) stated that the molecule cytochrome C has been 'sheltered' because there was no need for further perfection throughout evolution. Therefore, most changes that have occurred are the result of random processes; any control by natural selection was only in 'the sense of keeping them within rather elastic functionally permissible boundaries'. The resulting effect is related to the tertiary structure of the protein.

Available data for comparative protein structure is at this time principally mammalian. Of 282 sequence studies presented, 201 related to mammals (Dayhoff, 1969).

Sarich and Wilson (1966, 1967a, 1967b) and Sarich (1969a, 1969b) have staunchly promoted a theory of constant rate of evolution. This is based largely upon their studies with antigenic reactions of the albumin molecules, particularly of primates, carnivores and pinnipeds. Rates of evolution of species and and times of divergence have been proposed for a substantial number of species (see also the sections of this chapter dealing with specific taxonomic groups).

Hafleigh and Williams (1966) analyzed immunochemical correspondence among primate albumins. They did not assume that the modifications were introduced at a constant rate, but stated that as a longer evolutionary time is considered, the more regular is the rate function.

These and other analyses and proposals are most exciting and thought-provoking. The data of relationships of taxonomic units pertaining to evolution and phylogeny deserves continued evaluation, with other factors such as generation time, total gene pool, and other interactive factors considered. The present analyses of rates of evolution need more seasoning and correlated data, and should not yet be accepted as fact.

8. Proteins of the Ancient Dead

It has been possible to apply some immunological and biochemical techniques to prehistoric and fossil material. These show some promise of taxonomic use. In any event, they are of interest.

Paleoserology and the application of forensic medical principles in examination of human 'ancient dead' was reviewed by Smith (1960). Specifically, blood grouping could be applied, with varied success to mummified tissue and old cancellous bone. This was based on the principles that: (1) blood group substances are sometimes in greater concentration in organs and body fluids; and (2) inhibition of the actions of corresponding agglutinins (red cells) can be determined by examination of such tissues. A highly purified product was not essential. Harrison *et al.* (1969) demonstrated serological kinship of Egyptian mummies, 3,000-year-old Smenkhare and Tutankhamen; both were probably blood group A_2 and MN, and thus could well have been brothers.

Mammoth (*Mammuthus primigenius*) hair, age about 32,000 years, was found by Gillespie (1970) to have preserved its gross structure. Decrease in molecular size of the protein fractions was found; this was assumed to be degeneration related to age.

The amino acid content of collagen and dentine of bones and teeth in Pleistocene fossils was determined by Ho (1966 and 1967). About 70% of demineralized proteinaceous material from compact bones and 13% from spongy bones was recovered as pure collagen. Similarities with modern forms were marked, but with minor differences. Fossil animals were ground sloth (*Northrotherium shastense*), dire wolf (*Canis dirus*), coyote (*C. orcutti*), shortfaced bear (*Tremarctotherium simum*), saber-tooth cat (*Smilodon californicus*), cat

(*Panthera atrox*), horse (*Equus occidentalis*), camel (*Camelops hesternus*), and bison (*Bison antiquus*).

B. REVIEW BY BIOCHEMICAL GROUP (INCLUDING REPORTS OF GENERAL NATURE)

Reports of general nature are defined as those regarding species of diverse taxonomic groups, and are included in this section (B). Reports limited to taxonomic categories of order and below are included in Section C.

In this section and the following, (C), unless otherwise noted, the unqualified terms of 'band', 'zone', and 'migration' refer to electrophoretic determinations, which may be called 'electropherograms'; 'fast' and 'slow' and 'speed of migration' refer to the distances travelled during the standard time of the procedure.

Many pieces of data are provided that should facilitate correlation between taxonomy and other special fields.

1. *Serum Proteins, General and Multiple Species Examinations*

The old-fashioned nomenclature of human serum globulins is generally of little use except to locate the relative position in the electropherogram. Within the total spread of 'alpha' to 'gamma' globulins exist many proteins with special structure and function, such as haptoglobins, transferrins, enzymes and others.

In 1945 there were two reports (Deutsch and Goodloe; Moore) of electrophoretic examination of serum proteins in a number of different animals: man, monkey, rabbit, cotton rat (*Sigmodon*), hamster, rat, guinea pig, dog, fox (*Vulpes*), cat, mink (*Mustela vison*), swine, cow, sheep, goat, and chicken, pheasant, pigeon, duck and carp. Reproducible characteristics and distinct species-specific patterns were observed. Tiselius' moving boundary technique was used at this time (Tiselius and Flodin, 1953).

In 1952, Ganzin *et al.*, reported the use of paper electrophoresis on six common species. They reported qualitative and quantitative differences similar to those observed in moving boundary electrophoresis. Gleason and Friedberg (1953) compared several healthy and diseased mammals and amphibians and demonstrated abnormalities in disease.

Latner and Zaki (1957) used starch-gel electrophoresis and illustrated differences between common species.

During the later 1950s numerous reports appeared that indicated strict genetic control of polymorphisms of several mammalian serum proteins. At this time also evaluation of the taxonomic value of electrophoresis was being made in reptiles and amphibians (Dessauer and Fox, 1956; Zweig and Crenshaw, 1957) and in birds, using egg albumin (Sibley, 1960). Many excellent papers related to human and laboratory and domestic mammals were appearing and have continued to be produced into the present.

In 1959 our laboratory published two reports related to taxonomy (Johnson *et al.*; Johnson and Wicks). The need for considering factors of age, physiologic states, disease and natural variations was mentioned. A number of taxonomic situations were compared with the patterns of serum proteins and hemoglobin.

Riou *et al.* (1962) used starch-gel to compare plasma proteins of different species, including several strains of mice, the rat, rabbit, pullet and guinea pig. They also used bidimensional electrophoresis and identified ceruloplasm, haptoglobins, lipoproteins, and transferrins. Species specificity was observed.

We further reported our experience (Johnson and Wicks, 1964) with a wide spectrum of Mammalia. Serum proteins including glycoproteins and lipoproteins were examined. Reference was especially made at the higher taxonomic levels, but it was concluded that the primary taxonomic usefulness was at the level of genus and species.

In 1968 Johnson described 25 taxonomic situations correlated with electrophoretic data. In 19 instances clarification of relationships was obtained; in 6 situations the protein studies were of no help.

In recent years increasing numbers of studies of particular species have been published. These are reviewed in Section C below.

2. *Albumin (Alb) and Prealbumin*

Albumin has long been distinguished in a rough way by biochemical means as an important and constant fraction of blood plasma. With refined techniques it has proved to be a stable and workable molecule. Human albumin has a molecular weight of 68,460, a Svedburg coefficient at 4·60 and a diffusion coefficient of 6·10 (Sober, 1968). Albumins of other mammals have similar physical characteristics. This molecule is quite stable and maintains its identity under adverse circumstances and when other parts of the electropherogram have become denatured. We have kept several serum samples for over a year at refrigerator temperature without denaturation of the albumin.

Methemalbumin forms in some animals when the extracellular concentration of hemoglobin exceeds the binding capacity of haptoglobin. Baur (1969) found a remarkable distribution in the ability to form methemalbumin: Methemalbumin formed easily in higher primates (19 species examined), reptiles and amphibians, but not in other mammals (65 species of 8 orders examined), birds or fishes. Catalase preparation (Fig. 3) also reveals methemalbumin.

Boyden (1964) stated on the basis of his long experience in immunology that 'certain kinds of serum proteins, albumins, for example, are conservatively inherited'. The albumin molecule has been effectively used in immunologic studies (Sarich and Wilson, 1969; Seal, 1969).

Polymorphism has been found in several species, including man (Melartin *et al.*, 1967), horse (Osterhoff, 1966), and cattle (Ashton, 1964). 'Double albumin' regularly occurs in certain species of the rodent family Sciuridae as a specific pattern (Johnson and Wicks, 1959; Johnson, 1968).

Prealbumins and postalbumins occur in many species. In humans two prealbumins have been identified—prealbumin-1, the faster, and prealbumin-2. Prealbumin-2 is an acidic glycoprotein of low molecular weight which on paper electrophoresis migrates as an alpha-1 globulin; it is sometimes called orosomucoid (Barnicott *et al.*, 1965).

3. Immunoglobulins (Ovary, 1966; Sober, 1968; Putnam, 1969)

a. General information. Immunoglobulins (Ig or gamma globulins) are proteins with antibody activity produced by the lymphoid cell system of vertebrates. Also included in this category are some other proteins which are structurally similar, but for which antibody activity has not been demonstrated. The basic chemical structure is a pair of heavy and a pair of light polypeptide chains. Immunoglobulins occur in other body tissues besides blood plasma, and may have widely differing physico-chemical properties.

Classification has been a problem. With the great increase in information of biological behavior and chemical structure, a disorganized and confusing nomenclature is gradually being clarified, based upon World Health Organization recommendations of human immunoglobulins. Similar recommendations have been made by the WHO regarding notation of genes, genotypes and allotypes of human immunoglobulins.

A brief review of the characteristics of immunoglobulins is necessary equally for an understanding of past contributions, and for a forward look at the taxonomic potential. This knowledge becomes doubly important in taxonomy because immunologic phenomena are related. The subject is full of complicated data and plagued with past biochemical designations based on partial understanding. Taxonomy before Linnaeus was much like this.

b. Classification (Ovary, 1966; Sober, 1968; Aalund, 1968; and Kohler *et al.*, 1970a). This refers to human molecules unless specified otherwise. Most mammalian species probably have similarly endowed molecules.

i. IgM or gamma M (also referred to as beta 2M, gamma 1M, macroglobulin or 19S gamma globulin). Electrophoretic migration in the beta globulin region; molecular weight about 1,000,000 (polymer of five to six disulphide-bonded monomer units); sedimentation constant (Svedburg constant) 19S; it may combine with five antigenic determinants (pentavalent); the carbohydrate content is hexose 5·4%, acetyl hexosamine 4·4%, sialic acid 1·3% and fucose 0·7%.

ii. IgG or gamma G (also referred to as gamma, gamma 2, 7S). Electrophoretic migration in gamma region; molecular weight about 156,000; there are two sites for antigenic determinants on each molecule; carbohydrate content is hexose 1·1%, acetyl hexosamine 1·3%, sialic acid 0·3% and fucose 0·2%. Four subclasses present in humans; identified in many species.

iii. IgD or gamma D. Electrophoretic migration fast; sedimentation constant 7S; recently described, only in humans.

iv. IgA or gamma A (also referred to as beta 2A or gamma 1A). Electrophoretic migration in beta region; sedimentation constant 7S (or variable, related to polymerization to 13S); carbohydrate content is hexose 3·2%, acetyl hexosamine 2·3%, sialic acid 1·8% and fucose 0·22%. Has been found in rabbits and mice, but not guinea pigs.

v. IgE or gamma E. Recently described in humans and related to allergies; electrophoretic migration similar to IgA. Correlated by biological behavior to guinea pigs and mice; similar antibodies in rats and rabbits present in very small quantities only.

c. Polypeptide chains. The chains of all immunoglobulins are divided into a constant COOH-terminal region and a variable NH_2-terminal region. The antigen-combining site is thought to be at the variable region (Köhler *et al.*, 1970a).

Two major groups of polypeptides occur: light and heavy chains.

i. Light chains (also designated L chains, B chains). Molecular weight of about 23,000. Two forms of light chains are distinguished according to the C terminal amino acid: *Kappa chains* with cysteine; important for this provides the disulphide bond between light and heavy chains; and *lambda chains* with serine (preceded by cysteine) as the C terminal amino acid.

Usually both kappa and lambda chains are present in every Ig molecule, but this varies with the species. Each molecule is symmetrical and made up of either two kappa or two lambda chains. The chains are different in amino acid composition and antigenic determinants. Antigenic factors are the Oz and the Inv serologic systems. Oz (+) has arginine, and Oz (−) lysine, at position 190 of the lambda chain (Putnam, 1969). Inv (a+) has leucine, and Inv (b+) valine, at position 191 of the kappa chain.

The kappa chain has also been referred to as Type I and the notation for the immunoglobulin containing this is Type K. The lambda chain has also been referred to as type II and the Ig containing this, Type L.

ii. Heavy chains (also designated H chains, A chains). Molecular weight about 50,000. These are quite different in amino acid content and antigenicity. They are designated by the Greek letters according to the immunoglobulin class, i.e. IgG contains gamma heavy chains, IgA, alpha heavy chains, and IgM, mu heavy chains. The isoantigens present on human gamma chains are called Gm factors. The gamma chains are best known and different 'classes' are found in humans, guinea pigs and mice.

There are other characteristics of the Ig molecules that further complicate their classification and relationship. Subclasses of the molecules exist and more are being described as this is written. There may be polymerization of molecules within an immunoglobulin class. 'Fragments' of molecules resulting from cleavage of peptide bonds are being studied and subjected to additional systems of naming ('F'). These may be distinguished by immunological and chemical properties, but may not be homogeneous according to present naming. Pertaining to biological activity, antibodies of the same specificity may be in different classes of Ig molecules; antibodies within the same class may be heterogeneous, with different electrophoretic mobility; they may bind identical antigenic determinants in a different manner. Part, but not all, of the specific reactions of an antibody may be retained as immunoglobulins are fragmented.

All of these findings indicate a most fascinating complex of structure and function. The relationships to taxonomy are many, but the intricacies of the immunoglobulins dictate care and judgement in their use, and indicate that much investigation remains to be done.

d. Immunoglobulins of other species of animals related to taxonomy and phylogeny. Students of immunoglobulins have suggested that the same principles of nomenclature be applied to other species. Criteria used, such as molecular weight, carbohydrate content, antigenic relationships and amino acid sequences are potentially valuable for taxonomic use.

Two subclasses of IgG immunoglobulins have been found in guinea pigs, three in mice. IgA has been found in mice and rabbits, but not in guinea pigs. Only one type of light polypeptide chain has been described in rabbits, two types in guinea pigs. Regarding the heavy polypeptide chains, not much is known except related to the gamma type. In humans, four different gamma chains have been found, determining the four subclasses of the IgG molecule. In guinea pigs and mice different heavy chains have been delineated that produce the subclasses of IgG. Relative carbohydrate fraction of various immunoglobulins of the horse, cow and rabbit have been analyzed. Horse IgA is quite different from cow IgA; both are dissimilar to human IgA; horse and rabbit IgG carbohydrate fractions are fairly similar to each other and human (Sober, 1968).

According to Wikler *et al.* (1969) and Putman (1969), evolutionary significance may be attached to the similarities of sequence of the three heavy and two light chains in humans. Normal kappa and lambda chains of all species have variability of the amino acid sequence of the NH_2-terminal half of the polypeptide chain of about 110 residues; the COOH-terminal half, however, is invariant except for single amino acid substitutions in its 105 to 107 residue chain. The variations thus far analyzed are genetically controlled. Identity of portions of Ig molecules of different mammals indicate relationship, but thus far information is fragmentary: There is great similarity of the NH-terminals of the human mu chain and the rabbit gamma chain (closer than the gamma chains of man and rabbit). In light chains, the kappa chain of mouse and man have 60% identity, whereas the kappa and lambda chains of man have only about 40% identity.

Hill *et al.* (1966) suggest that the genes for light and heavy chains possessed a common ancestor. This ancestor could have been a gene controlling the sequence of a chain of 110 to 120 residues; by duplication a light chain of 212 to 215 residues could be produced; again, by duplication a heavy chain could result. Such heavy chains are recognized in each major immunoglobulin class.

They postulate this as a mechanism of origin of other proteins which are structurally analogous to the immunoglobulins.

Oudin (1966) stated that in man the evidence is that genetic differences of the immunoglobulins are essentially the same as those among hemoglobins. Köhler, *et al.* (1970b) found that three independent variable-gene pools existed which were common to the IgM, IgG and IgA immunoglobulins.

In summary, the comparative anatomy of the portions of the Ig molecules thus far known suggests a common origin of heavy and light chains of all species examined. Related to immunology, the concepts of antibody formation are controversial. These cannot be expected to agree until the complexities of immunoglobulin heterogeneity has been resolved (Aalund, 1968).

4. Haptoglobins (Hp)

Haptoglobins were first reported by Polonovski and Jayle in 1938. They occur in most species of animals and are related to the transport of free hemoglobin in the serum. Analysis is by adding specific quantities of free hemoglobin to serum before electrophoresis, and then staining for hemoglobin afterward.

Haptoglobins are glycoproteins composed of alpha and beta chains which may be polymorphic and which are determined by autosomal genes (Sober, 1968); type 2–2 of man is characterized as a fibrous protein with a Svedburg coefficient of 4·3, diffusion coefficient of 4·7, and molecular weight of 79,230; as Hp-methemoglobin complex, this changes to 6·1, 3·4, and 168,000, respectively; type 1–1 has a Svedburg constant of 4·2 and a molecular weight of 85,000; haptoglobin occurs in normal human plasma in the amount of 30 to 190 mg/100 ml; carbohydrate content is hexose 7·8% acetyl hexosamine, 5·3%, sialic acid 5·3% and fucose 0·2%. The Hp–Hb binding is so firm that there is practically no dissociation (Adams and Weiss, 1969).

Smithies (1955a, 1955b) first noted an inherited variation. He found migration in the alpha-2 globulin position in filter-paper electrophoresis, but a slower migration when starch was used. He was able to distinguish three groups on this basis. Kirk (1968) reviewed the genetics structure and function of human haptoglobins.

Polymorphism occurs in several mammals as an intraspecific genetic character. There is considerably less polymorphism than found in the iron binding globulin, transferrin. A quantitative difference existed between haptoglobins of various domestic animals, according to the kind of hemoglobin (horse, etc.) added to the serum; the horse, ass, mule and pig haptoglobin migrated as human Hp 1–1; no haptoglobin band was found in the goat, sheep or bull (Navarro *et al*., 1964).

Haptoglobins of several species of primates (*Macaca mulatta*, *Pan troglodytes*, and *Papio papio*); ground squirrel (*Spermophilus undulatus*), Marmot (*Marmota broweri*), rabbit, seal (*Callorhinus ursinus*) and cow were similar to human type 1–1 (Blumberg *et al*., 1960).

Patterns characterizing each of four rodent species of Capromyidae and

Myocastoridae were found by Kimura and Johnson (1970). Three phenotypes were present in the species *Capromys pilorides*.

5. *Transferrins (Tf)*

Transferrins are globulins (in the human, beta-globulins) which have the property of binding two atoms of ferric iron per molecule. They are glycoproteins which are physiologically important in the transport of iron. A molecular weight of 68,000 and Svedburg coefficient of 5·1 to 5·25 in man, rhesus monkey and rat was indicated by ultracentrifugation studies (Charlwood, 1963). They are identified by (a) adding Fe^{59}, a radioactive iron isotope, to the serum and performing radioautography (Giblett *et al.*, 1959), or (b) specific iron strain of the electropherogram. Because transferrins comprise a large proportion of the globulin fraction (200–320 mg/100 ml) some authors have reported transferrins as the prominent globulin bands without specifically identifying them. Polymorphism is common in mammals at the infraspecific level. Some 15 human variants have been reported (Wang and Sutton, 1965).

Transferrin passes readily from the maternal to the fetal circulation of the rat, near term. The reverse was true in the rabbit. This explains the differences in iron binding capacity of late fetal rats and rabbits (Morgan, 1964).

Other tissues than serum contain transferrins, or proteins with similar functions, related to iron transport. These are imperfectly known but appear to vary between taxonomic groups. Best studied are those in milk; there is a bovine 'red protein' and a colorless apoprotein, both of these are chemically different from bovine serum transferrin (Gordon *et al.*, 1963). In humans the iron in milk is bound principally by 'lactotransferrin' and transferrin (Blanc and Isliker, 1963).

Because of polymorphism and high concentration in serum, this group of molecules has been particularly attractive for analyzing populations. Allele systems, relationships of populations, and presumptive evolutionary or zoogeographic history of a species may be inferred.

There has been a tendency to equate Tf bands and genetic relationships between species on the basis of electrophoretic similarity. Until more exact biochemical identity of the transferrins is determined, this may be misleading.

6. *Enzymes* (Vennesland, 1966; Sober, 1968)

a. Characteristics. Enzymes are catalytic proteins produced by living cells. Most chemical changes in the body are initiated and controlled by enzymes. There are probably thousands of mammalian enzymes. There were over 650 enzymes of all types in 1961 when the International Union of Biochemistry issued recommendations for standardization and classification. New ones have been identified continually since then. An up-to-date version is to be published (Hoffman-Ostenhof, 1969). The Enzyme Handbook (Barman, 1969) listed 1,300 enzymes.

Enzymes are characterized by their large molecular weight 10,000 to several million) and may be composed of amino acids only or may have prosthetic groups of non-protein material attached. They are usually present in low concentration relative to the substrate they act upon and are usually much larger in molecular weight; but due to recycling they may react over and over. Recycling is usually over 1,000 times a minute; it may be as high as 18×10^6 a minute in choline esterase and catalase—the highest known. Some proteins of known function may also possess enzymatic activity, such as human hemoglobin A and its separated alpha and beta subunits which have peroxidase activity that is affected by interactions between unlike polypeptide chains (Smith and Beck, 1967).

b. Classification. Six main classes of enzymes are named. An enzyme is classified by the type of reaction it catalyzes and the substrate upon which it works (both, however, may be more than one). The classes are: (1) oxidoreductases, (2) transferases, (3) hydrolases, (4) lyases, (5) isomerases, and (6) ligases (synthetases). A number key for classifying all enzymes organizes this most complex group of proteins and is under the jurisdiction of International Union of Biochemistry, but will not be referred to further in this chapter.

c. Taxonomic implications. There is a growing number of mammalian enzymes which have been characterized by one means or another and tied into the genetic scheme of many species. Enzymes from different species with the same substrate specificity and catalyzing the same reaction may nevertheless differ markedly in other properties (Gregory, 1961). Because enzymes are functional, extremely specific, and may well be adaptive, it is inevitable that they will become deeply involved in future taxonomic and phylogenetic thinking. It is to be noted that although a large number of enzymes can be identified and classified biochemically, the exact function within metabolic pathways is frequently unknown or poorly understood. The amount of data needed to apply this information remains a great challenge to present and future workers.

Examples of enzyme data dealing with multiple taxonomic groups are reviewed in the following section. Other publications dealing with specific taxons are included in Section C. These references may point the way to taxonomic use.

Abbreviation of LDH (lactate dehydrogenase) and G6PD (glucose-6-phosphate dehydrogenase) are used below. All other enzymes are spelled out, in conformity with names used in reports.

d. Variants. Large numbers of variants are being reported. In G6PD, for example, more than 50 variants in man had been reported by early 1970, all compatible with X-linkage on genetic studies (Yoshida *et al.*, 1970). The amount of difference between variants and normal enzyme molecules may be extremely slight; for instance, a single amino acid substitution occurs in the common negro variant (A+) of human G6PD.

e. Occurrence in various organs. Another facet in the complexity of enzyme

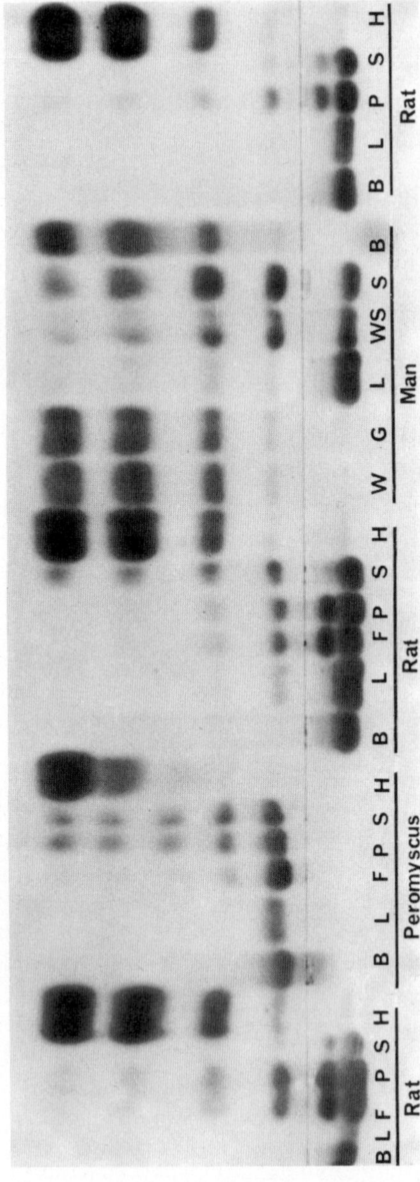

Fig. 1. Tissue lactate dehydrogenase of the laboratory rat (3 specimens), deer mouse (*Peromyscus maniculatus*) and man. B-whole blood; F-marrow of femur; G-gray matter of brain; H-heart; L-liver; P-marrow of pelvis; S-marrow of sternum; WS-washed marrow of sternum; W-white matter of brain.

study is the occurrence of identical or isomeric enzymes in various tissues. The relative amounts, and presence or absence of isozymes between tissues gives us another potential source of analysis. Figure 1 shows the differential distribution of lactate dehydrogenase within organs and between species. Many reports indicate this type of distribution for enzymes; there is no reason to doubt the strict genetic control nor the species specificity of the occurrences.

Bernstein *et al.* (1966) described the occurrence of LDH in the lens and cornea of cattle and rabbits; isozyme 1 was predominant in the bovine lens, isozyme 3–5 in the rabbit; the cornea of the cattle had primarily isozymes 3–5, rabbit 1–4; there was a progressive loss of enzyme activity with age, more in the cattle.

Burton and Waley (1968) isolated triose phosphate isomerase (TMI) from bovine lens and compared it with rabbit muscle TMI; one major band of each was present, possessing different mobilities. Cuatrecasas and Segal (1966) were able to identify five isozyme bands of galactose dehydrogenase in various tissues of the rat. Penhoet *et al.* (1966) isolated a new fructose diphosphate aldolase from rabbit brain with five isoenzymes; kidney and liver, four isoenzymes; spleen, heart and muscle, one; they did not find cross-reactivity between species. Davis *et al.* (1967) separated isoenzymes of glycogen phosphorylase from rabbit heart extracts; three fractions were found; fraction III was identical to the muscle form; comparisons were made with other species, with skeletal muscle and heart extracts; in all, a single form was found in muscle, and one to three forms in heart muscle; migration rates were different. Jacyszyn and Laursen (1968) reported heterogeneity of gamma-glutamyl-transpeptidase enzyme in serum, urine and tissue. Akhtar *et al.* (1968) found three types of alkaline phosphatase isoenzymes in rat serum which they classified as liver, kidney and bone types.

f. Comparative studies, multiple enzymes. There are numerous reports which compare enzymes of various species. By and large these are in the 'man–mouse–cow' category, not taxonomic in concept, but placed at the periphery and pointing discerning arrows towards the problems of systematics and adaptive phenomena.

Filov (1963) examined the blood esterase activity related to splitting butyl ether of acetic acid in frog, pigeon, man, mouse, rat, guinea pig, dog, cat, swine, sheep and cattle; values were relatively constant within species, but varied greatly between some species. Cholinesterase, esterase, and proteolytic activity were examined in man, monkey, rabbit, rat and cat (Hess *et al.*, 1963); as many as seven esterases were found; except in the monkey, some activity related to the albumin fraction; the enzymes varied with the species. Hyyppä *et al.* (1966) studied the carbonic anhydrase and naphthyl esterase distribution in erythrocyte hemolysates of the human, rat, guinea pig, pig, sheep and goat, and in rat kidney. Carbonic anhydrase activity was different for each species studied.

Berk *et al.* (1963) studied normal rabbit serum to elucidate amylase dis-

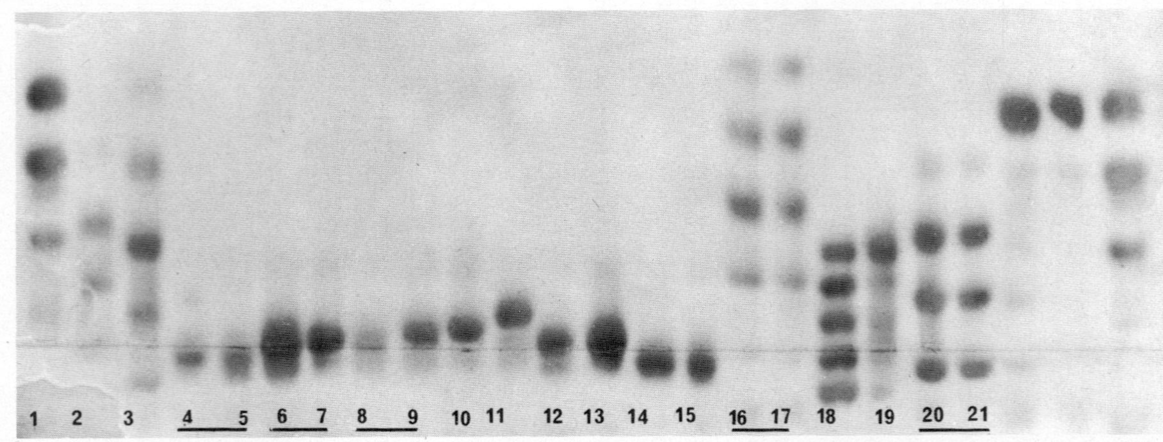

Fig. 2. Serum lactate dehydrogenase. 1 Man; 2 Chipmunk (*Eutamias amoenus*); 3 Chipmunk (*E. quadrivittatus*); 4–5 Pocket gopher (*Thomonys talpoides*); 6–7 Red-backed mouse (*Clethrionomys gapperi*); 8–9 Vole (*Microtus longicaudus*); 10 Vole (*M. oregoni*); 11 Deer mouse (*Peromyscus maniculatus*); 12 Lemming (*Lemmus sibiricus*); 13 Collared lemming (*Dicrostonyx groenlandicus*); 14 Jumping mouse (*Zapus trinotatus*); 15 Jumping mouse (*Z. princeps*); 16–17 Nutria (*Myocastor coypus*) 18 Shrew (*Sorex bendirii*); 19 Shrew (*S. nanus*); 20–21 Skunk (*Mephitis mephitis*).

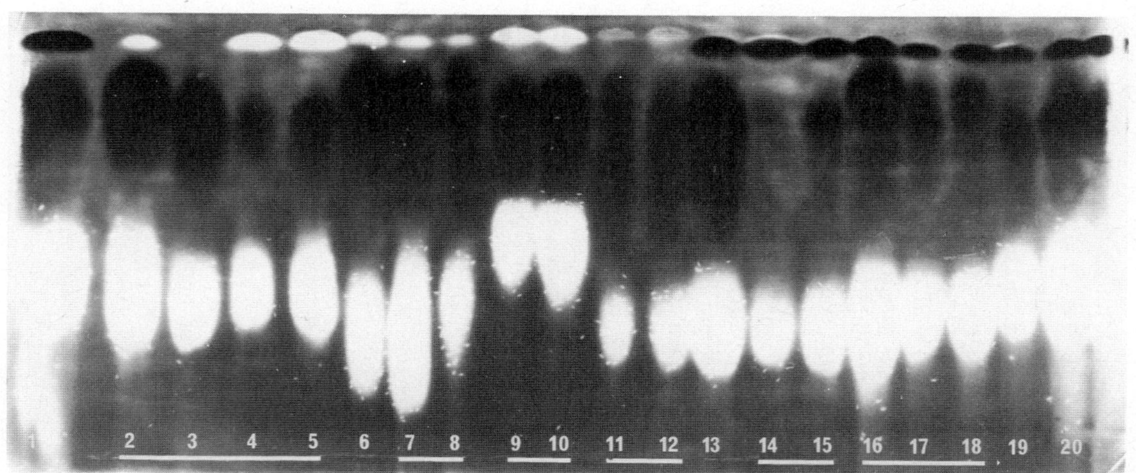

Fig. 3. Erythrocyte catalase, using filter paper specimens. 1 Man; 2–5 Pangolin (*Manis pentadactyla*); 2 lysate; 3 washed lysate; 6 Pika (*Ochotona princeps*); 7–8 Rabbit (*Sylvilagus nuttalli*); 9–10 Lemur (*L. variegatus*); 11–12 Galago (*G. crassicaudatus*); 13 Baboon (*Papio anubis*); 14–15 Grivet (*Cercopithecus aethiops*); 16–18 Macaques (*Macaca nemestrina* and *M. silenus*); 19 Gibbon (*Mylobates lar*); 20 Chimpanzee (*Pan troglodytes*). Note also the differential formation of methaemalbumin in the higher primates.

tribution. They stated that serum amylase is heterogeneous in origin and is species specific.

Searcy *et al.* (1966) found a single band of amylase in human, rabbit, guinea pig, cat, horse, and sheep. Multiple forms occurred in the rat, dog, pig, goat and cow.

Khanolkar *et al.* (1963) reported that 60 sheep and 40 goats showed complete absence of G6PD; it was found to be present in the rabbit, dog, horse, cattle and buffalo (*Bubalus bubalis*).

Smith and Holdridge (1967) compared erythrocyte G6PD activitity in man, horse, swine, sheep, goats and cows; Michaelis constants of the triphosphopyridine nucleotide and the pH optimum (9·0) were similar in all species; the sheep and goat were found to have markedly decreased activity. Hori and Kamada (1967) showed species and organ differences of G6PD in several laboratory species.

Genetic polymorphism of tetrazolium oxidase in dogs was found by Baur and Schorr (1969). Two types of patterns, each with multiple isozymes (up to 12) were detected and named A and B. Two coyotes (*Canis latrans*) and one hybrid had type A. Gene frequencies differed in different breeds of dogs. A general survey of 83 mammalian species indicated that isozymes of several species within a genus were identical by electrophoresis: five species of *Macaca* and five species of *Microtus* each appeared identical. *Felis catus* and *F. concolor* were alike, as were *Pan troglodytes* and *Homo*.

Rivello *et al.* (1969) report five isoenzymes of adipose tissue esterase (alkaline lipolytic activity) without genetic variation in humans. Rabbit, rat, mouse and lamb patterns were similar to man, with rabbit showing genetic variants.

Figure 2 illustrates differences of serum lactate dehydrogenase between many species, using thin layer starch-gel. Figure 3 shows erythrocyte catalase differences between species, using whole blood specimens. Methemalbumin formation is also identified by this technique.

Of all protein groups, the enzymes must eventually be the most revealing in terms of physiologic and adaptive significance.

7. Hemoglobins and Other Heme Proteins

a. Characteristics and structure. The hemoglobin molecule is found enclosed in cells in all vertebrates. In humans some 280 million molecules of hemoglobin occur in each red cell (Perutz, 1964). Hemoglobin is also found in lower forms of life (Scott, 1966). In invertebrates the heme moiety is the same as in higher forms, but the molecular size varies from 24,000 in an echinoderm to 3 million in some annelid worms. The giant molecules are found only dissolved in plasma. Hemoglobin has been found in legume root nodules, but only in the presence of bacteria; and it has been identified in yeasts and molds. It is found in a number of unicellular primitive animals. The ability to produce hemoglobin is therefore an ancient character. The hemoglobin in the primitive hag fish and lamprey consists of a single globin chain of a molecular weight of about 17,000. All other vertebrates contain a usual 4 globin chain per molecule.

In mammals, hemoglobin consists of a protein, globin, and four heme groups (Ingram, 1963; Lehmann and Huntsman, 1966). The four identical heme groups are protoporphyrin rings containing single iron atoms in the centers, normally in the ferrous state; this iron is capable of combining reversibly with molecular oxygen. One molecule of hemoglobin may carry four molecules of 0_2, one for each heme group. Globin is made up of four polypeptide chains; each chain is composed of a linear arrangement of 19 kinds of amino acids joined by peptide linkages. Normally two identical 'alpha' chains and two identical 'beta' chains are present ($alpha_2/beta_2$), each composed of about 140 amino acids, in specific sequence according to chain and according to the species of animal. In humans, $alpha_2/epsilon_2$ (Gower II) and less well-studied $epsilon_4$ (Gower I) occur in the very early embryo of 7–12 weeks (Huehns *et al.*, 1964). $Alpha_2/gamma_2$ is the fetal form (HbF). $Alpha_2/delta_2$ is hemoglobin A_2, a minor fraction (about 2·5%) in normal humans. Also in humans many abnormal hemoglobins have been detected, usually in small amounts, and genetically determined. These molecules may have three or even four different chains. A single amino acid may be deleted, without substitution (Jones *et al.*, 1966). In general, however, abnormal hemoglobin in man has a single (or two) amino acid substitutions in the beta chain or less commonly in the alpha chain. Dessauer (1969) cautioned regarding over-interpretation of the significance of amino acid differences, citing the seven amino acid difference in two beta chain variants in sheep.

Portions of the hemoglobin chains are remarkably constant and can be traced phylogenetically throughout Mammalia. The molecular weights of mammalian hemoglobins are about 68,000.

b. Other red cell proteins. Over 98% of the soluble proteins in the red cells is hemoglobin. The remainder is largely enzymes, including methemoglobin reductase (Lonn and Motulsky, 1957; Ingram, 1961), phosphatase (Hopkinson *et al.*, 1963), carbonic anhydrase (Barnicott *et al.*, 1964), catalase (Baur, 1964), G6PD, aldolase, phospho-hexoseisomerase, glyceraldehyde-3-phosphate dehydrogenase, lactate dehydrogenase (Smith *et al.*, 1965) and others. Peroxidase activity is possessed by the hemoglobin molecule (Smith and Beck, 1967).

c. Adaptive significance. It is not known just how various molecular forms of hemoglobin act in an adaptive capacity. However, it has been assumed that the differences between taxonomic units somehow fits each unit for better survival in its own ecological niche. The conclusion of Taketa *et al.* (1967) that the functional properties of a hemoglobin relate to both the amino acid structure and specific interactions between the alpha and beta chains points in the direction of better understanding. The fact that hemoglobin is enclosed in the erythrocytes would seem to simplify the oxygen transport interreaction, but in practice the species (or individual) blood grouping factors limit this.

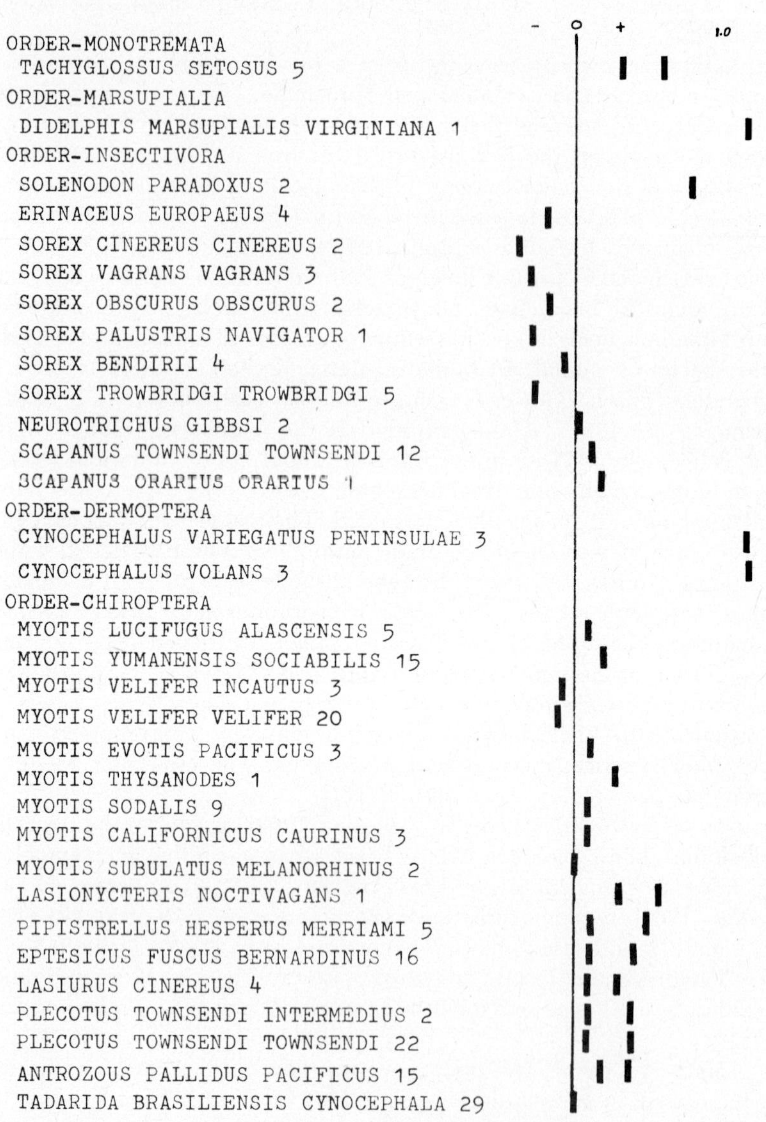

FIG. 4a–i. (pp. 34–42) Relative electrophoretic mobilities of major mammalian hemoglobins. Each band illustrated can be plainly identified in the original electropherogram by the color of the hemoglobin, without additional staining. The number of specimens examined is specified after each name. The migration is determined from the point of application, 'O' and the migration of the control (human hemoglobin A) set at an arbitrary figure of 1·0, toward the anode. Diffuse areas are indicated by closely placed bands. Absolute mobilities in different runs may vary 10% or more from that illustrated, due to limitations of technique; side-by-side comparison of fresh specimens of a single run are advisable for taxonomic purposes.

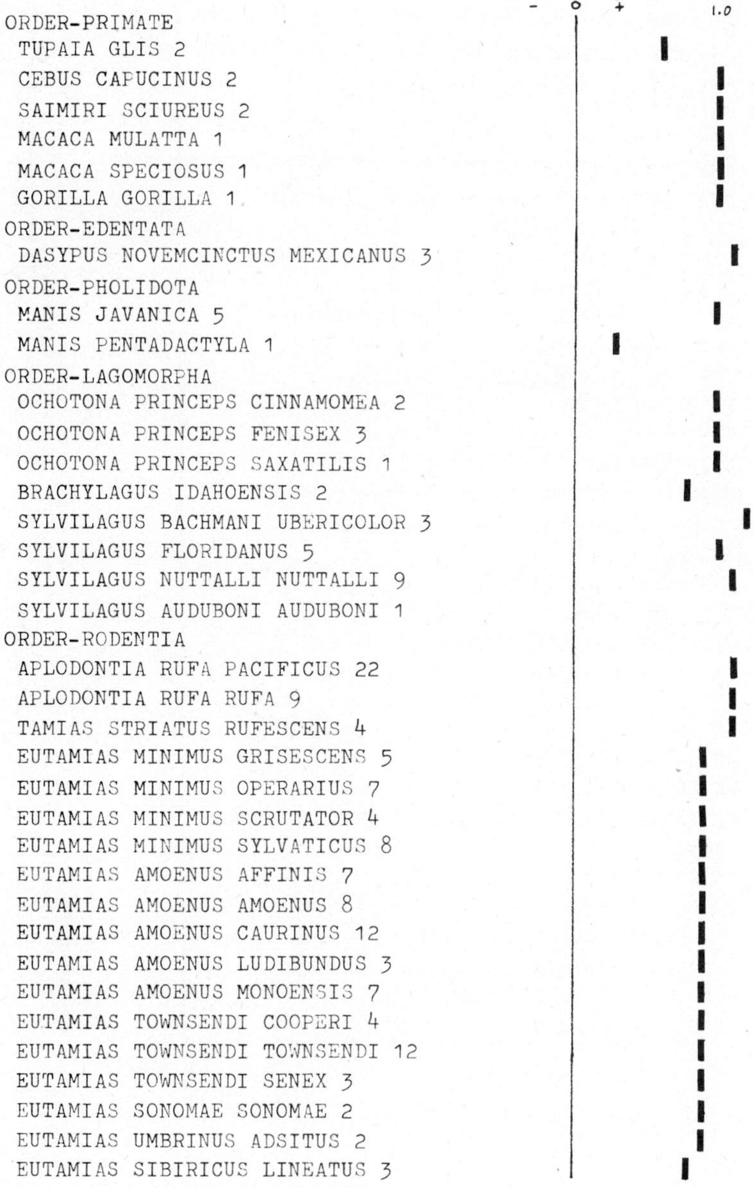
- 0 + 1.0
ORDER-PRIMATE
TUPAIA GLIS 2
CEBUS CAPUCINUS 2
SAIMIRI SCIUREUS 2
MACACA MULATTA 1
MACACA SPECIOSUS 1
GORILLA GORILLA 1
ORDER-EDENTATA
DASYPUS NOVEMCINCTUS MEXICANUS 3
ORDER-PHOLIDOTA
MANIS JAVANICA 5
MANIS PENTADACTYLA 1
ORDER-LAGOMORPHA
OCHOTONA PRINCEPS CINNAMOMEA 2
OCHOTONA PRINCEPS FENISEX 3
OCHOTONA PRINCEPS SAXATILIS 1
BRACHYLAGUS IDAHOENSIS 2
SYLVILAGUS BACHMANI UBERICOLOR 3
SYLVILAGUS FLORIDANUS 5
SYLVILAGUS NUTTALLI NUTTALLI 9
SYLVILAGUS AUDUBONI AUDUBONI 1
ORDER-RODENTIA
APLODONTIA RUFA PACIFICUS 22
APLODONTIA RUFA RUFA 9
TAMIAS STRIATUS RUFESCENS 4
EUTAMIAS MINIMUS GRISESCENS 5
EUTAMIAS MINIMUS OPERARIUS 7
EUTAMIAS MINIMUS SCRUTATOR 4
EUTAMIAS MINIMUS SYLVATICUS 8
EUTAMIAS AMOENUS AFFINIS 7
EUTAMIAS AMOENUS AMOENUS 8
EUTAMIAS AMOENUS CAURINUS 12
EUTAMIAS AMOENUS LUDIBUNDUS 3
EUTAMIAS AMOENUS MONOENSIS 7
EUTAMIAS TOWNSENDI COOPERI 4
EUTAMIAS TOWNSENDI TOWNSENDI 12
EUTAMIAS TOWNSENDI SENEX 3
EUTAMIAS SONOMAE SONOMAE 2
EUTAMIAS UMBRINUS ADSITUS 2
EUTAMIAS SIBIRICUS LINEATUS 3

ORDER-RODENTIA (CONT.)   - 0 +   1.0

MARMOTA FLAVIVENTRIS 4
MARMOTA BROWERI 2
MARMOTA OLYMPUS 2
AMMOSPERMOPHILUS LEUCURUS LEUCURUS 7
AMMOSPERMOPHILUS NELSONI 9
SPERMOPHILUS TOWNSENDI TOWNSENDI 17
SPERMOPHILUS BELDINGI 2
SPERMOPHILUS COLUMBIANUS COLUMBIANUS 13
SPERMOPHILUS TRIDECEMLINEATUS 3
SPERMOPHILUS SPILOSOMA ANNECTENS 9
SPERMOPHILUS BEECHEYI DOUGLASI 5
SPERMOPHILUS BEECHEYI SIERRAE 1
SPERMOPHILUS TERETICAUDUS NEGLECTUS 3
SPERMOPHILUS LATERALIS CHRYSODEIRUS 5
SPERMOPHILUS LATERALIS CINERASCENS 6
SPERMOPHILUS LATERALIS LATERALIS 4
SPERMOPHILUS LATERALIS TESCORUM 5
SPERMOPHILUS SATURATUS 11
CYNOMYS LUDOVIANUS LUDOVICIANUS 8
CYNOMYS LEUCURUS LEUCURUS 8
CYNOMYS GUNNISONI GUNNISONI 8
CYNOMYS GUNNISONI ZUNIENSIS 4
CYNOMYS LEUCURUS X GUNNISONI 8
SCIURUS CAROLINENSIS CAROLINENSIS 12
SCIURUS ABERTI FERREUS 1
SCIURUS IGNITUS 1
TAMIASCIURUS HUDSONICUS FREMONTI 2
TAMIASCIURUS HUDSONICUS PREBLEI 1
TAMIASCIURUS HUDSONICUS STREATORI 6
PTEROMYS MOMONGA 1
GLAUCOMYS VOLANS SATURATUS 6
GLAUCOMYS SABRINUS FULIGINOSUS 1
GLAUCOMYS SABRINUS LATIPES 1
GLAUCOMYS SABRINUS OREGONENSIS 5

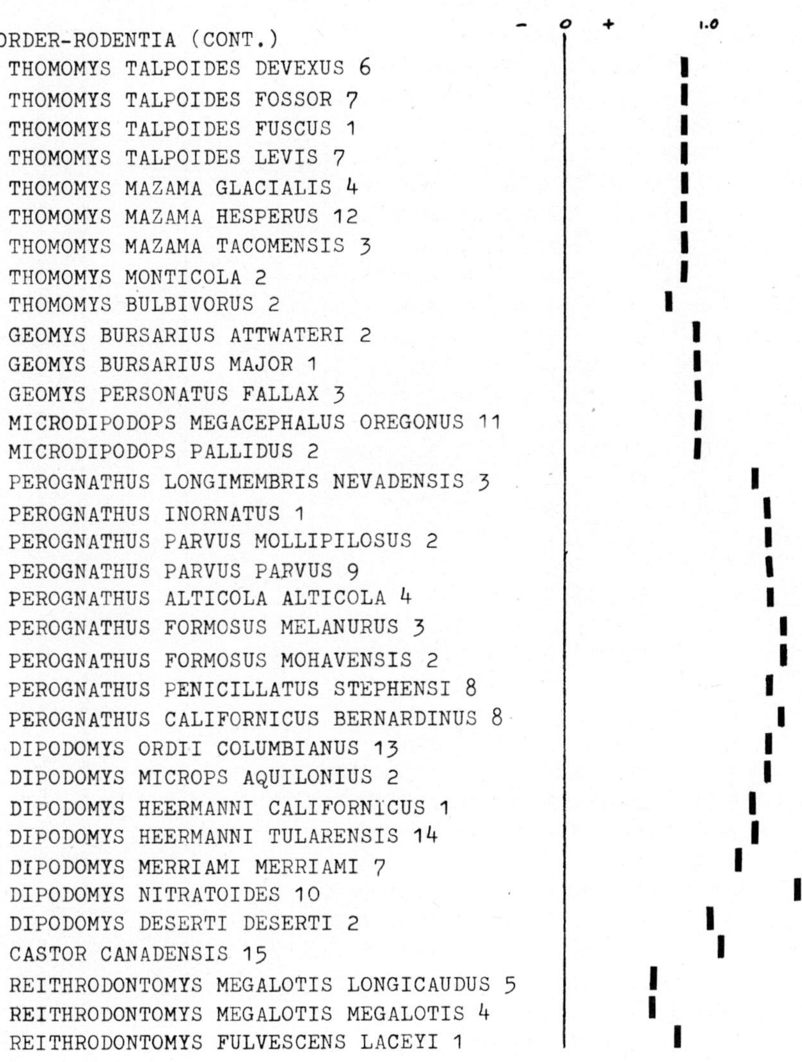

- 0 + 1.0
ORDER-RODENTIA (CONT.)
THOMOMYS TALPOIDES DEVEXUS 6
THOMOMYS TALPOIDES FOSSOR 7
THOMOMYS TALPOIDES FUSCUS 1
THOMOMYS TALPOIDES LEVIS 7
THOMOMYS MAZAMA GLACIALIS 4
THOMOMYS MAZAMA HESPERUS 12
THOMOMYS MAZAMA TACOMENSIS 3
THOMOMYS MONTICOLA 2
THOMOMYS BULBIVORUS 2
GEOMYS BURSARIUS ATTWATERI 2
GEOMYS BURSARIUS MAJOR 1
GEOMYS PERSONATUS FALLAX 3
MICRODIPODOPS MEGACEPHALUS OREGONUS 11
MICRODIPODOPS PALLIDUS 2
PEROGNATHUS LONGIMEMBRIS NEVADENSIS 3
PEROGNATHUS INORNATUS 1
PEROGNATHUS PARVUS MOLLIPILOSUS 2
PEROGNATHUS PARVUS PARVUS 9
PEROGNATHUS ALTICOLA ALTICOLA 4
PEROGNATHUS FORMOSUS MELANURUS 3
PEROGNATHUS FORMOSUS MOHAVENSIS 2
PEROGNATHUS PENICILLATUS STEPHENSI 8
PEROGNATHUS CALIFORNICUS BERNARDINUS 8
DIPODOMYS ORDII COLUMBIANUS 13
DIPODOMYS MICROPS AQUILONIUS 2
DIPODOMYS HEERMANNI CALIFORNICUS 1
DIPODOMYS HEERMANNI TULARENSIS 14
DIPODOMYS MERRIAMI MERRIAMI 7
DIPODOMYS NITRATOIDES 10
DIPODOMYS DESERTI DESERTI 2
CASTOR CANADENSIS 15
REITHRODONTOMYS MEGALOTIS LONGICAUDUS 5
REITHRODONTOMYS MEGALOTIS MEGALOTIS 4
REITHRODONTOMYS FULVESCENS LACEYI 1

- 0 + 1.0

ORDER-RODENTIA (CONT.)
PEROMYSCUS CRINITUS PERGRACILIS 13
PEROMYSCUS CRINITUS STEPHENSI 12
PEROMYSCUS CALIFORNICUS INSIGNIS 5
PEROMYSCUS CALIFORNICUS PARASITICUS 5
PEROMYSCUS MANICULATUS ARTEMISIAE 26
PEROMYSCUS MANICULATUS AUSTERUS 27
PEROMYSCUS MANICULATUS BAIRDII 6
PEROMYSCUS MANICULATUS BOREALIS 11
PEROMYSCUS MANICULATUS GAMBELII 76
PEROMYSCUS MANICULATUS GRACILIS 27
PEROMYSCUS MANICULATUS KEENI 13
PEROMYSCUS MANICULATUS NEBRASCENSIS 4
PEROMYSCUS MANICULATUS OREAS 87
PEROMYSCUS MANICULATUS OSGOODI 8
PEROMYSCUS MANICULATUS RUBIDUS 23
PEROMYSCUS MANICULATUS RUFINUS 29
PEROMYSCUS MANICULATUS SONORIENSIS 11
PEROMYSCUS SITKENSIS PREVOSTENSIS 7
PEROMYSCUS POLIONOTUS POLIONOTUS 5
PEROMYSCUS LEUCOPUS LEUCOPUS 2
PEROMYSCUS LEUCOPUS NOVEBORACENSIS 3
PEROMYSCUS LEUCOPUS TEXANUS 4
PEROMYSCUS LEUCOPUS TORNILLO 1
PEROMYSCUS BOYLII BOYLII 8
PEROMYSCUS TRUEI CHLORUS 1
PEROMYSCUS TRUEI GILBERTI 6
PEROMYSCUS TRUEI MONTIPINORIS 1
PEROMYSCUS TRUEI TRUEI 1
OCHROTOMYS NUTTALLI AUREOLUS 7
ONYCHOMYS LEUCOGASTER UTAHENSIS 8
ONYCHOMYS TORRIDUS PULCHER 6
NEOTOMA LEPIDA LEPIDA 8
NEOTOMA LEPIDA NEVADENSIS 4
NEOTOMA LEPIDA SANRAFAELI 1
NEOTOMA FUSCIPES FUSCIPES 5
NEOTOMA FUSCIPES MONOCHROURA 2
NEOTOMA FUSCIPES SIMPLEX 2
NEOTOMA CINEREA ALTICOLA 2
NEOTOMA CINEREA FUSCA 8
NEOTOMA CINEREA OCCIDENTALIS 15

ORDER-RODENTIA (CONT.)	- 0 + 1.0
CLETHRIONOMYS RUTILUS DAWSONI 6	
CLETHRIONOMYS RUTILUS MIKADO 1	
CLETHRIONOMYS GAPPERI BREVICAUDUS 4	
CLETHRIONOMYS GAPPERI CASCADENSIS 1	
CLETHRIONOMYS GAPPERI GALEI 2	
CLETHRIONOMYS GAPPERI OCCIDENTALIS 1	
CLETHRIONOMYS GAPPERI SATURATUS 35	
CLETHRIONOMYS GLAREOLUS 2	
CLETHRIONOMYS CALIFORNICUS CALIFORNICUS 2	
CLETHRIONOMYS RUFOCANUS BEDFORDIAE 6	
CLETHRIONOMYS (N.S.) 4	
EOTHENOMYS KAGEUS 5	
EOTHENOMYS SMITHI 3	
ASCHIZOMYS ANDERSONI 2	
ASCHIZOMYS IMAIZUMI 1	
ASCHIZOMYS NIIGATAE 3	
PHENACOMYS INTERMEDIUS INTERMEDIUS 1	
PHENACOMYS INTERMEDIUS ORAMONTIS 1	
ARBORIMUS LONGICAUDUS LONGICAUDUS 11	
ARBORIMUS LONGICAUDUS SILVICOLA 8	
ARBORIMUS (N.S.)4	
MICROTUS PENNSYLVANICUS DRUMMONDI 2	
MICROTUS PENNSYLVANICUS FUNEBRIS 4	
MICROTUS PENNSYLVANICUS INSPERATUS 5	
MICROTUS PENNSYLVANICUS PENNSYLVANICUS 7	
MICROTUS AGRESTIS 1	
MICROTUS MONTANUS CANESCENS 14	
MICROTUS MONTANUS FUSUS 2	
MICROTUS CANICAUDUS 4	
MICROTUS CALIFORNICUS CALIFORNICUS 7	
MICROTUS CALIFORNICUS PALUDICOLA 2	
MICROTUS TOWNSENDI TOWNSENDI 17	
MICROTUS TOWNSENDI SSP. 3	
MICROTUS OECONOMUS INNUITUS 3	
MICROTUS LONGICAUDUS ABDITUS 2	
MICROTUS LONGICAUDUS LITTORALIS 3	
MICROTUS LONGICAUDUS MACRURUS 5	
MICROTUS LONGICAUDUS MORDAX 14	
MICROTUS OREGONI OREGONI 7	
MICROTUS MONTEBELLI 3	
MICROTUS OCHROGASTER OCHROGASTER 15	

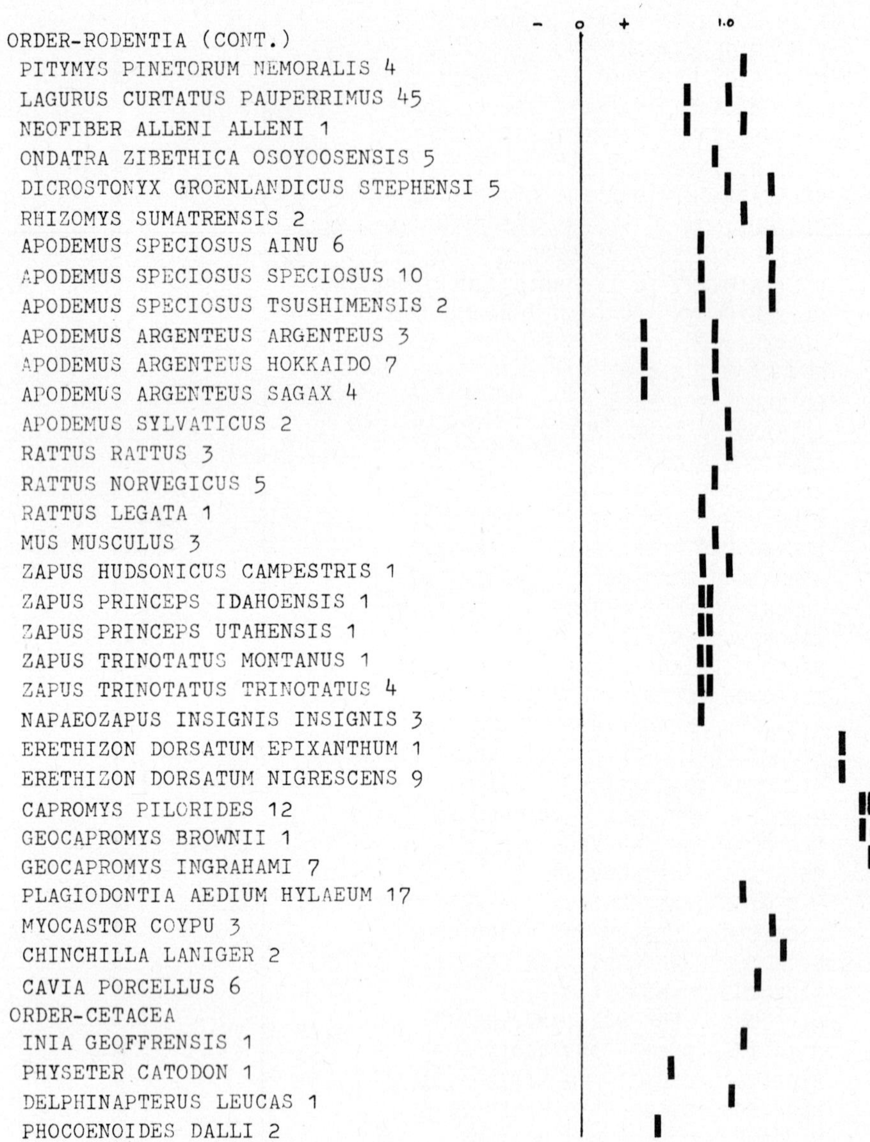
- 0 + 1.0
ORDER-RODENTIA (CONT.)
PITYMYS PINETORUM NEMORALIS 4
LAGURUS CURTATUS PAUPERRIMUS 45
NEOFIBER ALLENI ALLENI 1
ONDATRA ZIBETHICA OSOYOOSENSIS 5
DICROSTONYX GROENLANDICUS STEPHENSI 5
RHIZOMYS SUMATRENSIS 2
APODEMUS SPECIOSUS AINU 6
APODEMUS SPECIOSUS SPECIOSUS 10
APODEMUS SPECIOSUS TSUSHIMENSIS 2
APODEMUS ARGENTEUS ARGENTEUS 3
APODEMUS ARGENTEUS HOKKAIDO 7
APODEMUS ARGENTEUS SAGAX 4
APODEMUS SYLVATICUS 2
RATTUS RATTUS 3
RATTUS NORVEGICUS 5
RATTUS LEGATA 1
MUS MUSCULUS 3
ZAPUS HUDSONICUS CAMPESTRIS 1
ZAPUS PRINCEPS IDAHOENSIS 1
ZAPUS PRINCEPS UTAHENSIS 1
ZAPUS TRINOTATUS MONTANUS 1
ZAPUS TRINOTATUS TRINOTATUS 4
NAPAEOZAPUS INSIGNIS INSIGNIS 3
ERETHIZON DORSATUM EPIXANTHUM 1
ERETHIZON DORSATUM NIGRESCENS 9
CAPROMYS PILORIDES 12
GEOCAPROMYS BROWNII 1
GEOCAPROMYS INGRAHAMI 7
PLAGIODONTIA AEDIUM HYLAEUM 17
MYOCASTOR COYPU 3
CHINCHILLA LANIGER 2
CAVIA PORCELLUS 6
ORDER-CETACEA
INIA GEOFFRENSIS 1
PHYSETER CATODON 1
DELPHINAPTERUS LEUCAS 1
PHOCOENOIDES DALLI 2

\- 0 + 1.0

ORDER-CETACEA (CONT.)

ESCHRICHTIUS GIBBOSUS 4
BALAENOPTERA PHYSALUS 6
BALAENOPTERA BOREALIS 3
MEGAPTERA NOVAEANGLIAE 2
BALAENA MYSTICETUS 2

ORDER-CARNIVORA

CANIS LATRANS UMPQUENSIS 4
CANIS LUPUS 2
CANIS FAMILIARIS 48
CANIS LATRANS X FAMILIARIS 7
CANIS HALLSTROMI 2
VULPES FULVA 4
CHRYSOCYON BRACHIURUS 1
URSUS AMERICANUS ALTIFRONTALIS 4
URSUS AMERICANUS CINNAMOMUM 1
URSUS AMERICANUS PERNIGER 1
URSUS ARCTOS HORRIBILIS 10
BASSARICUS ASTUTUS FLAVUS 2
PROCYON LOTOR EXCELSUS 1
PROCYON LOTOR PACIFICUS 4
MARTES AMERICANA ABIETINOIDES 2
MARTES AMERICANA ACTUOSA 1
MARTES PENNANTI COLUMBIANA 1
MUSTELA ERMINEA MURICA 1
MUSTELA ERMINEA OLYMPICA 10
MUSTELA ERMINEA RICHARDSONII 1
MUSTELA FRENATA ALTIFRONTALIS 3
MUSTELA VISON ENERGUMENOS 2
MUSTELA VISON 5
TAXIDEA TAXUS NEGLECTA 1
TAXIDEA TAXUS TAXUS 1
SPILOGALE PUTORIUS AMBARVALIS 8
SPILOGALE PUTORIUS INTERRUPTA 3
SPILOGALE PUTORIUS LATIFRONS 8
SPILOGALE PUTORIUS PHENAX 3
MEPHITIS MEPHITIS SPISSIGRADA 5
ENHYDRA LUTRIS 2
AMBLONYX CINEREA 2
FELIS PARDALIS 4
LYNX CANADENSIS 3
LYNX RUFUS FASCIATUS 1

– 0 + 1.0

ORDER-PINNIPEDIA
EUMETOPIAS JUBATA 7
ZALOPHUS CALIFORNIANUS 1
ODOBENUS ROSMARUS DIVERGENS 1
PHOCA VITULINA RICHARDSONI 3
PUSA HISPIDA BEAUFORTIANA 2
PUSA SIBIRICA 1
ERIGNATHUS BARBATUS NAUTICUS 2
CYSTOPHORA CRISTATA 7
ORDER-TUBULIDENTATA
ORYCTOPERUS AFER 2
ORDER-PROBOSCIDEA
ELEPHAS MAXIMAS 3
LOXODONTA AFRICANA 2
ORDER-HYRACOIDEA
PROCAVIA CAPENSIS 1
ORDER-SIRENIA
TRICHECHUS MANATUS 1
ORDER-PERISSODACTYLA
EQUUS ASINUS 2
EQUUS CABALLUS 1
EQUUS CABALLUS X ASINUS 3
ORDER-ARTIODACTYLA
SUS SCROFA 10
TRAGULUS MEMINNA 2
CERVUS CANADENSIS ROOSEVELTI 65
CERVUS ELAPHUS 1
ODOCOILEUS HEMIONUS FULIGINATA 1
ODOCOILEUS HEMIONUS HEMIONUS 1
ODOCOILEUS VIRGINIANUS DACOTENSIS 2
ODOCOILEUS VIRGINIANUS OCHROUROUS 1
ALCES ALCES GIGAS 3
ANTILOCAPRA AMERICANA 2
BISON BISON 13
BOS TAURUS 5
BOS TAURUS 12
OVIS CANADENSIS CANADENSIS 7
CAPRA HIRCUS 9

By agglutinating or rupturing the erythrocytes, these factors demand that the erythrocytes and their hemoglobins remain part of the genetically predestined biological system.

Multiple hemoglobins, some specific for a particular stage of development, occur naturally in many species (Braun *et al.*, 1967). The maximum number of eight has been demonstrated in an insect (*Chironomus thummi*) and additional components have been indicated. Within a single species four adult hemoglobins are found in *Rana catesbeiana*, five adult hemoglobin types are known in the white Leghorn chicken. In the human A_2 comprises about 2·5% of the total hemoglobin.

The occurrence of multiple (usually two) hemoglobins in domestic species is well known and genetically important. Of interest to taxonomists is the frequent occurrence of multiple hemoglobins in wild species; these may be population, species or genus characteristics (Fig. 4).

Polymorphisms within species are being found as more indepth studies are done. Kimura and Ohta (1971), referring to proteins in general, supported the theory that polymorphism is a phase of molecular evolution, as a 'random genetic drift of neutral mutations in finite populations.'

d. Technical note. Lehmann and Huntsman (1966) advocated the use of paper or cellulose acetate techniques as a reliable and simple first-look at the hemoglobin molecule. They spoke mainly for its use in looking for abnormal human hemoglobin; the same principle exists for taxonomic use where major fractions are important. However, starch-gel is indispensable for examination of minor fractions, hybridization and many other uses. The point to be made is that reliance on any one technique may at times be misleading.

e. Review of general reports. Herner and Riggs (1963) dissociated the alpha and beta chains of various hemoglobin molecules with dilute acid. They produced molecular hybrids by restoring a neutral pH in mixtures of various animal hemoglobins. The results indicated that hemoglobin subunits of distantly related animals (bullfrog and mammal species) are similar. Chiancone *et al.* (1966) produced dissociation of hemoglobin of various animal species, including eight mammals; dissociation of hemoglobin molecules into subunits is a general phenomenon; all mammalian hemoglobins behaved in a similar fashion, different from other vertebrates. Antonini *et al.* (1964) examined the oxygen equilibrium of hybrids of canine and human hemoglobins; a basic similarity of the hemoglobins of different species was found. Antonini *et al.* (1963) studied relationships of molecular and functional properties of hemoglobin. Measurement of the Bohr effect relative to proton binding of Hb and HbO_2 or $HbCO_2$ show that dog, human and horse hemoglobins are closely similar.

Braunitzer (1966) reviewed the phylogenetic variation in the primary structure of hemoglobins. The mechanisms involved included base exchanges, deletions, insertions and gene duplications. In vertebrate hemoglobins, the number of amino acids per chain varied from 127 to 156 in fish, and 141 to 146 in mammals. In all of the hemoglobins compared, including both fish and mammals, eight identical positions of amino acids were found.

Beetlestone and Irvine (1964) found differences in the thermodynamics of ionization of 15 methemoglobins studied. They conclude these are related to the charge configuration on the globins, arising from amino acid differences.

Ramakrishnan and Barnabas (1962) examined a number of different forms of animals, including 11 mammals. Among these were the fruit bat (*Pteropus*), wild rabbit (*Lepus nigricollis*), black buck (*Antilope cervicapra*), and buffalo (*Bubalus bubalis*). Double hemoglobins were found in the rat, goat and buffalo. All six artiodactyls had high alkali resistance. The wild rabbit exhibited slower migration than the domestic. The fruit bat showed the slowest migration, but only slightly less than the rat.

Other reports restricted to taxonomic rank of order and below are included in Section C.

f. Fetal hemoglobin. This has taxonomic potential related to its presence or absence and its biochemical characteristics in various species. Fetal hemoglobin (HbF) is the major hemoglobin during embryonic life in man and many other animals. It has not been found in the marsupial (*Marmosa*), dog, or horse; it has been found in cats, sheep and cattle (Shifrine and Stormont, 1970). Other reports have indicated the presence of fetal hemoglobin in the rabbit (Tyuma *et al.*, 1964); rat (Brdička, 1966); mouse (Craig and Russell, 1963); elephant, *Loxodonta* (Kleihauer *et al.*, 1965); deer, *Odocoileus virginianus* in two forms (Kitchen *et al.*, 1967); roe deer, *Capreolus* (Maughan and Williams, 1967) and sheep and goats (Breathnach, 1964). This listing is only partial.

At the time of birth in man there is a shift to HbA. Both types may be formed within a single erythrocyte at various ratios in the blood of a newborn. This has been dramatically shown by Dan and Hagiwara (1967) by the use of fluorescent antibodies. The beta chain and the gamma chain are remarkably homologus, more than the alpha and beta chains (Ingram, 1963). Each consists of 146 residues and two-thirds of the amino acid sequences are identical. The amino acid isoleucine occurs in human HbF, but not in adult hemoglobin. The oxygen affinity of HbF in man and other species studied is generally considerably higher than adult hemoglobin of the same species.

g. Myoglobin. This is a heme protein similar to hemoglobin (Ingram, 1961). Its molecular size is only one fourth that of hemoglobin; it is composed of a single polypeptide chain to which a heme group is attached. It is found in muscles in varying amounts and provides the red color. In high concentration it gives the extreme dark color of the muscle of such mammals as the pinnipeds and cetaceans when it must act as a reserve supply of oxygen during prolonged periods of apnoea. Its function is the storage and conveying of oxygen within the muscle cells. Myoglobin concentration of skeletal muscle is reported to be 'close to 1%' (Nauman, 1970) and in Cetacea 9 to 44 mg/kg (Rice, 1967).

Present concepts of molecular evolution state that myoglobin was the primitive form. Chemical structure of human myoglobin is more like sperm whale myoglobin than human hemoglobin (Ingram, 1963).

Dautrevaux *et al.* (1969) found 18 differences (17 substitutions and one inversion) in the 153 amino acid chain between horse and sperm whale (*Physeter*) myoglobin. Bradshaw and Gurd (1969) compared amino acid sequences of the seal (*Phoca vitulina*), the sperm whale and porpoise (*Phocoena phocoena*); all replacements were confined to 29 positions, leaving 123 invariant. The seal differed from the sperm whale in 26 positions, the porpoise in only 15.

h. Cytochrome C. This is present in the mitochondria of all aerobic cells. It is a globular protein, molecular weight in mammals, in the range of 12,500 to 15,500 (Sober, 1968); it contains a heme group and functions with other cytochromes in electron transport for the basic metabolic process of oxidative phosphorylation, by which most energy is obtained from food and storage materials. Cytochrome C is readily available and is a suitable molecule for sequence studies. McLaughlin (1969) indicated amino acid chains of 104 residues for all vertebrates. Man and chimpanzee (*Pan*) appeared identical; the patas monkey (*Erythrocebus patas*) varied in one residue out of 76 identified. Pig, sheep and bovine cytochrome C appeared identical. This group of molecules should be able to supply good phylogenetic data.

8. Miscellaneous

a. Glycoproteins. Many species have been surveyed by us. Glycoprotein content of the serum increases with various pathologic states, as cancer, infection and trauma (Hudson *et al.*, 1970). Thus it may be less reliable as a taxonomic indicator. However, in view of the large numbers of functional proteins in this general group, there may be more usefulness for taxonomic purposes than now appears.

b. Lipoproteins. These have been extensively sampled in our laboratory. Because of considerable variability we found it of little use as a taxonomic character. Immunoassay of low density lipoproteins by Lees (1970) and disc electrophoresis of various animal sera by Narayan (1967) also indicate variable concentrations of this type of protein. To date I am not aware of any taxonomic contribution made by lipoprotein analysis.

c. Hemopexins. These have not been extensively studied, but show promise similar to haptoglobin and transferrin. Hemopexins are glycoproteins which have been identified in the serum of several mammals (Bremner, 1964). In distinction to haptoglobins which are serum proteins binding hemoglobins, hemopexins bind free heme. The heme-hemopexin complexes are identified on the electropherograms by the peroxidase reaction, much as the hemoglobin–haptoglobin complex. In man, hemopexin has a molecular weight of 80,000, a Svedburg constant of 4·8 and contains 9% hexose, 7·4% acetyl hexosamine, 5·8% sialic acid and 0·4% fucose. It is present in the amount of 80–100 mg% in man (Sober, 1968).

Lush (1967) mentioned a polymorphism of hemopexin in pigs which was confused with haptoblobin. Shaughnessy (1970) found two heme-binding protein phenotypes in the Southern fur seal, *Arctocephalus*, which fit the criteria for hemopexin, having been derived experimentally from denatured ('old') hemoglobin. Population differences were indicated.

d. Fibrinogen. This is a glycoprotein found in the plasma of all vertebrates and is composed of three types of peptide chains, A, B, and C, linked together by disulfide bonds; the molecular weight in man is 341,000 (Sober, 1968; Dayhoff, 1969).

The primary structure of fibrinogen was investigated by Blombäck and Blombäck (1968). They analyzed the amino acid sequences of fibrinopeptides split off the fibrinogen molecule by the action of thrombin in 34 species of animals, most of which were mammals. The fibrinopeptides were considered quite suitable for comparative studies. The following species were analyzed by grouping: one lizard, two marsupials, four Primates, one Lagomorpha, two Rodentia, seven Carnivora, four Perissodactyla, and thirteen Artiodactyla. In general the classification agreed with existing taxonomy.

Bovine fibrinogen has an electrophoretic mobility on paper similar to haptoglobin (Bremner, 1964). Fibrinopeptides A and B of the gibbon (*Hylobates lar*) and other old world monkeys were examined by Mross *et al.* (1970); relationships were appraised; a double allele system of glycine-serine at position B–3 was found in the gibbon. Fibrinopeptides A and B of the mandrill (*Mandrillus leucophaeus*) were characterized by Doolittle *et al.* (1969) and compared with other primates and mammals of other orders; they termed this procedure 'fine structure taxonomy'.

Dayhoff (1969) presented amino acid sequence data of many mammalian fibrinogen and fibrinopeptide chains.

e. Collagen. This is found in most animals, but not in micro-organisms or plants. It is therefore not as old phylogenetically as some other molecules; in vertebrates about one-third of the total body protein is collagen, composed largely of glycine, proline, hydroxyproline and alanine (McLaughlin, 1969). Collagen molecules have a molecular weight of about 300,000 and are composed of two types of amino acid chains of different amino acid composition, but with the same molecular weight (Nold *et al.*, 1970); these are termed 'alpha-1' and 'alpha-2' chains, and there is a ratio of two alpha-1 chains to one alpha-2 chain within the molecule; the guinea pig was used by these authors. Little comparative data is available at this time.

f. Ceruloplasmin. Ceruloplasmin in human serum was reported by Hirschman *et al.* (1961). It is present in the amount of 30 mg/100 ml of plasma. Almost all copper in the plasma is contained therein. Quantitative variations occur in some diseased states, but its significance is not known. Two fractions of ceruloplasmin were found in the serum of normal adults. Imlah (1965) reported the presence of three phenotypes of ceruloplasmin in pigs, with two alleles (Cp^a and Cp^b).

g. Hageman factor (Factor XII). This has been found to be related to two basic physiologic defence mechanisms of the mammalian body blood clotting and the kallikrein-kinin system (Sherry, 1970). What little data is now available indicates that the Hageman factor could become important taxonomically, related as it is to two adaptive phenomena. In the absence of the Hageman factor neither the first step of coagulation nor kinin production can be accomplished. Kinins are polypeptides cleaved from globulins by the enzyme action of kallikrein. Kallikreins and kinins apparently are part of a whole 'family' of substances having slightly different actions and control mechanisms; they may be the chemical keys to the repair function in pathologic states, as well as being an integral part of the systemic physiologic mechanisms.

Hageman factor was absent in three cetacean species, *Tursiops truncatus*, *Orcinus orca*, and *Lagenorhynchus obliquidens* (Robinson *et al.*, 1969; Lewis *et al.*, 1969). This factor is reported to be present in 17 species, including the man Hageman, of 7 different orders by Ratnoff (1966) and in 21 species of 10 orders by Lewis *et al.* (1969).

h. Lens proteins. These have been studied by immunologic techniques for several decades. In comparison to serum proteins, they are synthesized by only one cell type which is present in a unicellular layer in the eye (Manski *et al.*, 1964). A peculiarity is that an antiserum against the lens of any vertebrate species gives precipitin reactions with lens of all vertebrates; serum proteins are much more limited in this ability. There are soluble (crystalline) and insoluble (albuminoid) proteins present. Most investigations have been concerned with phylogenetics, in view of the wide range of reactivity. There are as yet too few repeated studies to evaluate ultimate significance, but considerable promise is shown.

i. Hair protein. Keratin is the principle protein of hair, nails and the stratum corneum. Examination has revealed distinctness between unrelated species (Shechter *et al.*, 1969). Nine bands were found in human hair extract, 14 in monkey, 10 in rabbit, 9 in guinea pig and 13 in dog. Taxonomic usefulness is indicated.

Mammoth (*Mammuthus primigenius*) hair (about 32,000 years old) was compared with the Asian elephant (*Elephas maximus*) and sheep. Differences in both low-sulfur proteins and high-sulfur proteins were found by electrophoresis (Gillespie, 1970). There were some differences ascribed to degradation of proteins of the mammoth hair.

j. Milk. This characteristic substance of the class Mammalia presents possibility of taxonomic use at many levels. It is a complex mixture containing proteins (Feeney and Allison, 1969), carbohydrates, lipids, and includes enzymes, transferrins and many other types of biochemical materials. It has adaptive significance. Polymorphisms of various constituents are genetically determined in domestic animals. There is increasing information concerning milk of many species (Jenness and Sloan, 1970). The extremely high fat content in Pinnipedia is a widely known ordinal character. Many studies are being reported: Lipase in primate milk (Freudenberg, 1966; Buss, 1969), transferrin types in cattle (Ashton *et al.*, 1964), electrophoretic examination of milk protein of the polar bear (Baker *et al.*, 1967) and others. We may expect a growing amount of data that will better allow us to characterize taxonomic units or understand phylogenetic relationship within this compartment of study.

k. Potassium level. Rasmusen and Hall (1966) reported a high and low potassium level in red cells of goats, similar to that previously described for sheep. They reported that the alleles influencing phenotypes in the M system of red cell antigens influence also the potassium levels in both sheep and goats. Levels of red cell potassium below 40 mEq/l were classified low, those over that figure high. This trait has not been used taxonomically and little comparative data exists.

l. Bile salts. Haslewood (1964, 1968) has examined the bile salts of many animals. Monotremes are similar to other mammals. Possible help in rodent classification is offered by finding differences of glycine conjugation in such forms as the South American *Myocastor coypus* (Myocastoridae) and the African *Thryonomys swinderianus* (Thryonomyidae). Unique bile acids are found in each Suidae and Pinnipedia. In general these findings have phylogenetic rather than taxonomic implications.

9. *Immunology*

a. General review. There have been intermittent immunologic studies in taxonomy of mammals for several decades. Most of these are related to particular groups and are reviewed under the taxonomic unit involved.

Williams (1964) reviewed immunologic principles, stressing that the immunochemical similarity was one direct measurement of phylogenetic relationship.

Bauer (1969) investigated heterologous reactions of mammalian serum proteins with monospecific anti human sera, using the Ouchterlony plate. Eighteen monospecific antisera were used, against various serum fractions. Serum for reaction was from several species of monkeys, rats, guinea pig, pig, sheep, cattle, chick and eel (which did not cross react). Animals reacted with various antisera in general related to their phylogenetic position. Most reacted against albumin, ceruloplasmin, and alpha-2 macroglobulin.

b. Histocompatibility. The phenomenon of 'immunologic memory' of proteins is obviously in a different category from that of the biology of speciation. However, one cannot refrain from considering the theories regarding the recognition of 'self' by the immunologic system. Does this go back in time as a basic mechanism of survival of an individual and therefore the species, to the time when the first aggregations of proteinaceous material were formed? Is this, therefore, a primordial species instrument?

In the modern context this recognition of self is now being intensively studied in relation to tissue transplants. Within the web of multiple systems in the mammalian body the genetics of histocompatibility is being meticulously worked out by immunologic techniques. The immediate purpose is tissue-preserving of transplanted parts, and, therefore, life-preserving of the host. It is likely that this type of complex analysis will become useful in a taxonomic sense. The population, species, genus, or whatever category needing characterization may well be identified in ways not now considered possible.

10. Immunology—Blood Groups

There is a wealth of material published in recent years regarding the principles and practice of blood grouping. It is in two general categories.

(i) Related to man in antigens and systems of naming, with publications in the several specialty journals on blood, blood banking and transfusions, and books on the subject. Much of this is directly applicable to taxonomic studies of primates but some usage spills over into other more distantly related orders.

(ii) Related to domestic animals, especially horses, pigs, sheep and cattle, and using special antigens and special nomenclature for each species group. This is applicable to taxonomic studies of Perissodactyla and Artiodactyla, and has been applied to diverse species. Much of this activity centers about the European Society for Animal Blood Group Research (E.S.A.B.R.). Reports of recent conferences of this group have been published in 'Proceedings' (9th in 1965, 10th in 1966, and the 11th in 1970). In 1970 the journal *Animal Blood Groups and Biochemical Genetics* was instituted. Much of the material is basic genetics, related to identifying and refining the breeds of domestic animals. A trend is seen, however, to include much data concerning comparative studies of wild species, other groups of vertebrates as fowl and fish, and information about biochemical polymorphisms.

Taxonomic studies of mammals have not often used blood group studies as an adjunct. This kind of information, however, is the definite type to be used in systematic studies. A brief review is demanded in order that perspective may be obtained.

Blood group determinations in lower mammals is a direct outgrowth of the ABO system. Many additional factors such as MN, Rh, Lewis, Lutheran and Duffy are found in humans. The reactions of the red cells in blood grouping studies may be either hemolysis or agglutination. The source of the antibody used varies considerably.

Coombs (1966) reviewed recent advances in blood grouping and stressed the genetic importance of the Xg human blood group antigen and the molecular activity of the red cell membrane. He stated that little is known of the chemistry of most red cell antigens, but in the ABH and Lewis antigens (mucopeptides) the specificity is in the carbohydrate portion; he considered it paradoxical that we know less about the antigens 'if such exist' which characterize species while knowing much about antigens that vary within the species.

For primates, reagents for testing human red cells are used (Moor-Jankowski *et al.*, 1965; Moor-Jankowski and Wiener, 1968, 1969). These reagents must be further processed to remove nonspecific antisimian heteroagglutinins which most human blood grouping sera contain; many other maneuvers may be applied.

Podliachouk (1970) found 18 blood factors in Equidae which belong to at least 8 genetic systems. Sandberg (1970) mentions 20 blood factors, some of which are controlled by multiple alleles.

In cattle, there was by 1965 (Bouw) an imposing list of over 70 internationally accepted reagents. These factors have been used in investigating relationships of wild type artiodactyls with considerable success.

Studies on sheep are less advanced. Some 7 systems have been discovered within the 15 blood groups studied (Lipecka, 1970).

Pigs have also been extensively studied. Buschmann (1965) investigated large numbers of the Bavarian breed by use of blood grouping. Nineteen different reagents were available for study. Also checked were 283 blood samples of wild boars (*Sus scrofa ferus*). Some blood group systems existed without polymorphism in the wild boar; some did not appear; in other systems the frequency distribution was 'completely different' from that of domestic pigs.

Varga *et al.* (1965) examined blood groups of rabbits; Frenzl *et al.* (1965) and Spiteri and Eyquem (1965) examined rats. There were evident problems of nomenclature.

In summary the following statements may be made regarding blood groups of mammals and their taxonomic use.

1. Individual blood groups characterize individuals or genetically similar populations.
2. Combinations of blood group characteristics may be used to identify populations of interbreeding animals.
3. Reaction to particular blood group reagents may occur in diverse taxonomic groups, but in general the reaction is less, and less likely, the more distant the taxonomic placement.
4. The possibility, or difficulty, of typing blood appears to vary with the taxonomic unit. (This may be modified as more investigation is done.)
5. In those species which can react to blood group reagents, it is possible to determine combinations of these reactions that will characterize the species and show relationships in the immunologic sense.

C. REVIEW BY TAXONOMIC GROUP

For purposes of organization, articles pertaining to diverse taxonomic groups of animals were included under Section B above. The reports reviewed below in the main are restricted to taxonomic ranks of order and below.

1. Monotremata

This most primitive of mammalian orders probably evolved during the Mesozoic (Dawson, 1967), but has no fossil record beyond the Pleistocene. The serum protein pattern on two specimens of the echidna (*Tachyglossus setosus*) containing a fat globulin band close behind the albumin which stained well for glycoprotein and was assumed to be transferrin. A second band, slow in migration contained less glycoprotein. There was more cathodally migrating protein in these specimens than in the human control. Lipoprotein stains indicated low levels. No further special stains were done. There were two well defined major hemoglobins, each migrating at a slow rate. Lewis *et al.* (1969) reported that coagulation factor XII (Hageman Factor) was present in the echidna, as in other mammals thus far examined, except Cetacea.

Tachyglossus aculeatus had levels of intestinal maltase, isomaltase and trehalase similar to man and other placentals, but only trace levels of lactase and cellobiase; sucrase was completely absent (Kerry, 1969). The milk whey of the same species was rich in total solids (37% to 47%), protein (11·3% to 13·0%), fat (4·8% to 19·6%); iron binding was high: 24 to 36 μg/ml as compared to 3 to 5 μg/ml for serum (Jordan and Morgan, 1969).

Kirsch (1968a) stated that the monotremes were serologically more distant from the eutherians than the marsupials.

2. *Marsupialia*

This order is of particular interest for two reasons. It is a separate infraclass of Mammalia and it has principal centers of evolutionary distribution in Australia and in South America with some fossil records dating back to late Cretaceous (Dawson, 1967). Analysis of this situation from the molecular viewpoint should be rewarding.

A single hemoglobin migrating slightly faster than human HbA (Fig. 4) was found in the opossum (*Didelphis marsupialis*); glycoprotein stains revealed a larger amount in the albumin fraction as compared to the intermediate globulin fraction; lipoprotein was found primarily in the intermediate area. A single type of transferrin was found in the pigmy opossum (*Marmosa mitis*) (Shifrine and Stormant, 1970); two types of haptoglobin were present, one a single zone, the other double; a single zone of hemoglobin was present; no fetal hemoglobin was found in three 36 day old fetuses. Two transferrin types consisting of three and four bands were found in the kangaroo, *Megaleia rufa* (*Macropus rufus*) (Cooper and Sharman, 1964). Polymorphism of transferrins was also found in the grey kangaroo (*Macropus*) and applied to taxonomic studies by Kirsch and Poole (1967); they correlated this with characteristic antigens and other findings, and were able to separate the eastern greys (*M. giganteus*) from the western greys (*M. fuliginosus*).

Red cell acid phosphatase was studied in two species of kangaroo (*M. rufus* and *M. giganteus*) by Lai (1966); a difference in the pattern of the two species was found, and two distinct patterns were found in *M. giganteus*. Interpopulation differences in red cell enzyme levels were found in the wallaroo (*M. robustus*) by Richardson and Czuppon (1970). A high acid phosphatase and blood urea were found by Parsons *et al.* (1970) in the Tasmanian devil (*Sarcophilus harrisii*).

Polymorphism of hemoglobin in *M. giganteus* was reported by Thompson *et al.* (1969). Air *et al.* (1971) calculated the time of divergence of eutherians and marsupials to be about 130 million years ago, from differences in the amino acid sequences of hemoglobins; the potoroo (*Potorous tridactylus*), with at least 16 differences in the beta chain probably diverged from *M. giganteus* about 57 million years ago. A constant rate of evolutionary change is accepted in these calculations.

Information regarding amino acid sequences was reviewed by Dayhoff (1969): Cytochrome C and fibrinopeptide A and B of the kangaroo (*Macropus*); and fibrinopeptide A of the wombat (Phascolomyidae).

Lewis *et al.* (1969) found Hageman Factor to be present in the opossum, quokka and wallaby, as in most other mammals except Cetacea.

Blood and milk proteins of the quokka (*Setonix brachyurus*) were compared with those of the rat (Ezekial *et al.*, 1963); transferrins in milk of the quokka were more numerous than in the serum; no phenotypic variation was observed in 25 samples of milk. Activities of intestinal enzymes were reported by Kerry (1969). He found that most marsupials possessed disaccharidase activity similar to placentals. Exceptions were: absence of sucrase in *M. giganteus* and the bandicoot (*Isoodon obeselus*) and extremely low trehalase activity in the koala (*Phascolarctos*).

Weymyss (1953) compared five Australian species with the opossum (*Didelphis*) by immunologic methods. Compared with placental mammals, the correspondence between families was slight; this corresponds with the concept that Marsupialia contains more diverse groups of animals than do the placental orders.

Kirsch (1968a, 1968b and 1969) presented considerable data and theoretical considerations regarding the serological affinities of marsupials. He agreed with concepts that differentiation within this order actually represents several separate orders (three were named). His analysis of marsupial proteins indicates varying rather than constant rates of evolution.

3. *Insectivora*

This is a catch-all order of diverse groups that are put together for convenience. There has been much argument about taxonomic relationship and phylogeny; this assemblage is the most primitive of the placental mammals. Table II indicates the richness of the taxonomic units available for investigation. The evolutionary history, beginning in the Cenozoic, has many gaps; it is likely that other orders, including Primates, arose from common ancestors. Problems of study are indicated by difficulties of capture and shipping and by poor tolerance to captivity of most insectivores. Molecular studies to date are most fragmentary and preliminary.

Picard *et al.* (1966) found that the alpha-2 macroglobulin of the hedgehog

(*Erinaceus europaeus*) is a homologue of human and rabbit alpha-2 macroglobulin; no antigenic relationship was found with several primate species, including man. Larsen and Tönder (1967) found albumin of the hedgehog to migrate about as human albumin; globulins were quite different from human; transferrin had about the same migration as human beta-2 globulin; haptoglobin banded in the area of human beta-1, and with a frequent secondary band between alpha-2 and beta-1; a broad lipoprotein band extended from the alpha-1 into the albumin region (we have found definite banding in the alpha-1 and prealbumin areas); hemoglobin was nearly isoelectric with human HbA.

Dabrowski and Skoczeń (1962) found, in the European mole (*Talpa europaea*), six bands; albumin, gamma globulin and one in the position of human beta-2 were the most marked, in decreasing order; a very high total protein content was found: 14·6 g/100 ml of serum.

Figure 4 shows the interesting biophysical behaviour of major hemoglobins of the Insectivora. No cases of multiple major forms were found in eleven species examined. The cathodal grouping within the families Erinaceidae and Soricidae is unique within Mammalia (Johnson, 1968). Talpidae of three species has a very slow anodal migration; Dabrowski and Skoczeń (1962) found *T. europaea* hemoglobin to migrate toward the anode, but less than human HbA. Two examples of *Solenodon* migrate well toward the anode, another indication of the diversity of the families included under Insectivora. Buettner-Janusch and Buettner-Janusch (1963) in two species of Macroscelididae, the elephant shrews, found migrations well toward the anode as in human HbA.

We have found well defined lipoprotein bands in the hedgehog, *Erinaceus* but little evidence in the shrew, *Sorex* and mole, *Scapanus*. Less glycoprotein was found in the albumin area in *Erinaceus* and *Sorex* than in *Scapanus*.

Hirsch and Junqueira (1965) examined amylase, protease, lysozyme and deoxyribonuclease of the submaxillary and parotid glands of European insectivores (*Crocidura russula*, *Sorex araneus*, *Erinaceus europaeus* and *Talpa caeca*). Protease was the only digestive enzyme found in all glands, except the *Sorex* submaxillary gland. Deoxyribonuclease activity was constantly present in high levels only in the *Sorex* parotid gland.

Hyvärinen and Oikarinen (1970) examined tissue alkaline phosphatase isoenzyme in the shrew (*Sorex araneus*) and the mouse. Differences between tissues were found.

4. Dermoptera

The colugos or 'flying lemurs' are a very distinct group consisting of one genus and two species in Southeast Asia and the Philippine Islands. There is little fossil record in areas where they now occur; a group known only from jaws and teeth in late Paleocene from North America (*Plagiomenidae*) is tentatively assigned to this order (Dawson, 1967). Survey samples by us showed a slower rate of migration of albumin compared to the human control, glycoprotein in small amount in the alpha-1 globulin area, lipoprotein near the alpha-2 area. Hemoglobin migrates as a single band slightly faster than human HbA.

Some solid molecular taxonomic information compared to other orders, in particular Chiroptera, will be welcome. A difficult part of the study of this order is the near impossibility of shipping and maintaining in captivity with our present knowledge of the species.

5. *Chiroptera*

The bats comprise a highly specialized order. There is poor fossil background; the ancient history is shrouded with lack of information. Thus this order presents a fertile field for applying other types of data. Study of bats is confounded with difficulty of collecting, problems of maintaining animals alive, and, frequently, small size. However, they present a fascinating diversity of specializations, and next after Rodentia are most numerous (Table II). With mounting interest in this order over the past decade, we may expect clarification of many taxonomic relationships in the near future.

Manwell and Kerst (1966) studied bat hemoglobins, lactate dehydrogenases, and esterases in four species of vespertilionid bats. They suggested that this group of bats 'are in the beginning of a period of extensive speciation and that they have not had sufficient time to accumulate conspicuous genetic differences'.

Glycoproteins (probably transferrins) have been examined by us in three vespertilionid species. Well defined bands are present in the alpha-1 to alpha-2 region of *Eptesicus fuscus*, *Lasiurus cinereus*, and *Plecotus townsendi*. An interesting pattern of lipoprotein was found in *E. fuscus* and *L. cinereus* with the strongest band midway between the alpha-1 and alpha-2 region and a second band in the albumin- to post-albumin region. *Plecotus* showed the lipro-protein to concentrate in the alpha-2 region, with little in the albumin range.

Hemoglobins of Vespertilionidae and Molossidae have a uniquely slow migration under most conditions of electrophoresis (Fig. 4). Double major hemoglobins appear to be common in Vespertilionidae (Mitchell, 1966), but were not found in the number of species of Neotropic bats by Tamsitt and Valdivieso (1969) and Valdivieso *et al.* (1969). It is to be noted that Mitchell (1966), Manwell and Kerst (1966) and Mitchell (1970) described the major hemoglobin for *Eptesicus fuscus* as single, whereas we (Fig. 4) and Valdivieso *et al.* found well defined double bands. This could be related to differences in technique or a polymorphism within the species.

Valdivieso *et al.* (1968) compared tissue isozymes of LDH in three species of Porto Rican bats, *Artibeus jamaicensis*, *Erophylla bombifrons*, and *Molossus fortis*. The heart and liver profiles were similar for all three species. *A. jamaicensis* and *E. bombifrons* were similar in muscle LDH, but much different from *M. fortis*. Other tissues were also examined for this enzyme complex.

Forman *et al.* (1968) used immunologic and electrophoretic tests to study relationships of the vampire bats (Desmodontidae), as part of a more comprehensive study. *Desmodus* was found to be related to the Phylostomatidae, through the subfamilies Phylostomatinae and Glossophaginae. Four other families showed little affinity by these examinations.

6. Primates

This order has a long fossil record, dating back to late Cretaceous (Anderson, 1967). Prosimians and primate-like insectivores have been found in the Paleocene and Eocene of North America and Europe, but later records are poor, as the order became more tropical in distribution. The separate areas of later evolution in both the New and Old Worlds have provided an attractive taxonomic complex for investigation. Man is fascinated by his own origins. Molecular scientists have not been averse to adding conjecture based upon this different kind of computation. Early hominids are known from late Miocene or early Pliocene, but *Homo sapiens* himself cannot be designated as a species back more than about 200,000 years, by fossil evidence (Clark, 1966). This is the kind of information into which modern data must integrate.

There is much molecular data regarding primates. Most of this has been gathered in recent years and is related to medical research and the availability of large numbers of animals in Primate Centers. Taxonomists will note that the general orientation is toward physiology, genetics and phylogenetics. There are many major publications which contain basic information and which review prior studies. These should be consulted by anyone seriously interested in this particular subject. Among these are Washburn (1963), Vagtborg (1965), Buettner-Janusch (1963, 1964), Napier and Napier (1968), Dayhoff (1969), and Bourne (1970).

a. General. Numerous articles explore biochemical and zoological relationships of the primates. The few cited below indicate the diversity of activity within research laboratories. Methemalbumin formation was found to be markedly increased in primates of 19 species compared to 65 species of 8 other orders examined by Baur (1969). Esterases of chimpanzee (*Pan*), gorilla (*Gorilla*), three species of *Macaca* and man and mouse were found to be qualitatively different by Arfors *et al.* (1963). Transferrins have been investigated with the finding of considerable heterogeneity (Buettner-Janusch, 1962; Barnicott *et al.*, 1965; Nute and Buettner-Janusch, 1969) and many others.

Hill *et al.* (1963), Hill and Buettner-Janusch (1964) and Buettner-Janusch and Hill (1965) analyzed the evolution of primates related to the molecular structure of their hemoglobins. In general, the results were in conformity with existing taxonomy. Computations referable to rates of evolution were done with several species as a 'favorite pastime of evolutionary biologists'. In so doing, however, they mentioned that they did not have enough information to put these rates into a 'meaningful content'. Boyer *et al.* (1969) examined hemoglobins of the New World genera, *Saguinus* of the family Callithricidae and *Aotus*, *Ateles*, *Callicebus* and *Saimiri* of the family Cebidae; by an ingenious method of comparing human beta and delta chains with selective amino acid positions, they proposed an archetype to which the amino acid positions and differences in these positions could be compared; they concluded that polymorphism and evolutionary changes can occur in an 'effectively functionless and thus selectively nearly neutral gene'.

Hudgins *et al.* (1966) compared the molecular structure of the myoglobin of 14 primates, including great apes, New World and Old World monkeys (superfamilies Hominoidea, Ceboidea and Cercopithecoidea). They determined consistency within, and definite variations of molecular structure between, superfamilies. Compared to hemoglobin, the myoglobin was more consistent in structure within superfamilies and thus tended to correlate well with the traditional taxonomy at this level.

Goodman (1962, 1965) proposed that in the evolution of primates, 'as the union of the fetus to the maternal blood became more intimate, maternal isoimmunizations decreased the heterozygosity of the prenatally acting genes and thereby increased the S (slow evolving) properties of proteins such as albumins which are synthesized early in fetal life'. He theorized that this also favored selection of genes which delayed synthesis of rapidly evolving proteins until later stages of development, thus permitting these proteins to evolve more rapidly.

Williams (1964) considered that proteins as antigens evolve in a random manner, unaffected by direct selective pressures in contradiction to Goodman's theory of antigenic evolution. He further stated that 'The general structure of protein homologues is probably more resistant to evolutionary changes than quantitative immunochemical cross reactions would indicate. The usefulness of antibodies in systematic studies is due to their ability to amplify and codify small differences by non-reaction. The data recorded (e.g. per cent cross reaction) are not different measurements, however; they are abstract estimates of similarities preserved through descent from a common ancestor in spite of evolutionary change'. He suggests time rates of evolution and phylogeny can be computed (as an 'irresistible temptation').

Mohaghegpour and Leone (1969) reported studies using antisera to several serum proteins. Computer drawn clusters of genera produced results in conformity to present concepts of family relationships.

Wilson and Sarich (1969) reviewed their own and published molecular data regarding the evolution of man and monkey. Man is genetically more similar to African apes, according to data from serum albumins, transferrins, hemoglobins and DNA. Comparing data derived from amino acid sequence of hemoglobins and immunologic study of albumins, the human lineage diverged from that leading to the African apes about 4·5 million years ago. Their theories have been attacked with enthusiasm by Darlington (1970), Read and Lestrel (1970) and Leakey (1970).

Wiener and Moor-Jankowski (1963), Moor-Jankowski *et al.* (1965) and

Moor-Jankowski and Wiener (1968, 1969) review blood grouping studies in Primates. They concluded that there is little information on simian blood group specificities and that many experiments are not yet done or completed. Relationship to human blood groups have been established for many of the better known species and many other maneuvers have been performed. Certain subfractions of serum were found to be recognized by antibodies within each species and by cross-reacting antibodies within closely related species.

b. Tupaiidae. The relationship of this family—whether a member of Insectivora or Primates has long been a question. Lange and Schmitt (1963) reported two forms of transferrins ('A' and 'B') to be present in *Tupaia glis*. Buettner-Janusch and Buettner-Janusch (1964) compared the hemoglobin with that of insectivores and primates and found it migrated slower than in all other primates, but did not resemble that of the insectivores.

The amino acid sequences of the alpha and beta chains of hemoglobins of *Tupaia glis* are listed in Sober (1968). There are about 24 alpha chain differences and 30 beta chain differences from man.

Goodman (1962) included *Tupaia* and *Urogale* with Primates by immunological data. Hafleigh and Williams (1966) found a high antigenic correspondence of *Tupaia* to human serum albumin (similar to prosimians) as compared to the hedgehog. Sarich (1969c) stated that the tupaiids are 'as much primate as any of the prosimians', in phyletic terms.

c. The prosimians and lower forms of the anthropoid families. These have considerable, though scattered, molecular information available. In general this must be considered as background data, but it provides a beginning for many kinds of serious taxonomic investigations. Napier and Napier (1968) provided a summary of the available information on each genus. The cercopithecoid genus, *Macaca*, because of its extensive use as an experimental animal, is especially well known.

d. Members of the family Pongidae, the chimpanzee (Pan), *gorilla* (Gorilla) *and orangutan* (Pongo). These, because of their close relationship to man have much information regarding biochemical and immunological relationships to man and to each other.

A number of sophisticated biochemical and immunological studies indicate the close relationship of *Gorilla* to *Homo*. The amino acid sequence of the hemoglobin molecule reveals an amazingly close identity: The alpha and beta chains are identical to human HbA except that in position 23 of the 141 amino acid alpha chain, the gorilla has aspartic acid instead of glutamic acid; in position 104 of the 146 amino acid beta chain, the gorilla has lysine instead of arginine (Zuckerkandl, 1965).

Goodman (1963) found a closer correspondence of *Gorilla* to *Homo* and *Pan* than to *Pongo* and *Hylobates*, using immuno diffusion techniques.

The antigenic correspondence of *Gorilla* to human serum albumins, by the quantitative precipitation reaction was extremely high as was that of the next ranking primate *Pan* (Hafleigh and Williams, 1966). Sarich and Wilson (1966, 1967a) found a similar situation using the quantitative microcomplement fixation technique.

Blood grouping, tested by human reagents proved the ABO groups to be more readily detected in secretions than in blood (Wiener and Moor-Jankowski, 1963). Gorillas were among the apes most different from man and most similar to other monkeys. There was some indication that the lowland gorilla, *G. g. gorilla*, was group B and the mountain gorilla (*G. g. beringei*), group A.

e. Hominidae. Man, *Homo sapiens*, continues to be subjected to innumerable biochemical, immunological and genetic studies. The information amassed is greater than for any other species; whole treatises are available for many particular molecules or groups of molecules. These data are medically rather than taxonomically directed, but they may become extremely useful in future systematic work. Sober (1968) provided lists of genetic variations in man, including structural variation of 41 proteins, 15 red cell antigen systems, quantitative variations in 10 blood proteins, and quantitative variations in 10 plasma clotting factors. Dayhoff (1969) provided additional comparative data on amino acid sequences of many molecules.

Application of these genetically controlled variations may be used to correlate populations and races of man. Relative occurrences of the ABO blood group system and sickle cell trait (HbS) have been subjects of statistical examination. Other proteins such as haptoglobins (Kirk, 1968) have been similarly studied. Statistical analyses of such intensively studied traits may well become models for study at various taxonomic levels (Boyd, 1964).

7. *Edentata*

Evolution of this order was primarily in South America beginning in the Paleocene (Dawson, 1967). It has been little studied in the present context. Its taxonomic relationships are incompletely known. Three modern families each with several genera offer an attractive though difficult group to investigate.

In the armadillo (*Dasypus*) we found that albumin migration was about as in the human control. There was a major fast-moving globulin band staining well for glycoprotein, likely transferrin, with a second band near the alpha-2 globulin area. A sharply defined lipoprotein fraction was found in the beta globulin area.

Marvin and Shook (1963) found an increased blood urea nitrogen and cholesterol and lower values for glucose, chloride, sodium and protein bound iodine in the sloth (*Choloepus didactylus*). The possibility of a low metabolic rate was suggested.

Roig (1964) used the precipitin test to clarify critical points of systematics in the armadillos of Argentina.

8. Pholidata

The pangolins consist of a single family found in Africa and Asia. They tentatively date from the Oligocene, but are poorly known. Because of differences in the albumin and especially the hemoglobin migration (Fig. 4) evaluation of *Manis* and *Paramanis* as separate genera was recommended by Johnson (1968). Three glycoprotein bands were found in the range of alpha-1 and alpha-2 globulin, and a lipoprotein band in the range of beta globulin area. Erythrocyte catalase was present, similar in migration to the human control; methemalbumin was not formed, as in most lower forms of Mammalia (Fig. 3).

9. Lagomorpha

This is a relatively small order of some 63 species world-wide. Fossil records do not indicate a special relationship with Rodentia; early lagomorphs date from the Paleocene (Dawson, 1967). In the immunologic field the domestic rabbit has been the main antibody producer for years. As a result, there is voluminous literature in this category; this includes investigation into the immunoglobulin molecules, searching into the anatomy of the act of the immunologic phenomenon. As a common laboratory animal, other parameters of this species' biochemistry have been widely recorded. The entries listed here are principally those applicable to taxonomic studies.

Tyuma *et al.* (1964) found differences in the alkali denaturation and ultraviolet absorption spectrum of fetal versus adult rabbit hemoglobin. They concluded that fetal hemoglobin was present. Grunder (1966) reported a greater amount of hemopexin in adult rabbits than in young; six phenotypes in five zones were controlled by three codominant alleles. Braunitzer *et al.* (1966) found 14 amino acid differences in the beta chain of hemoglobin, compared to man.

Stormant and Suzuki (1970) found multiple esterase isozymes in rabbit serum; specific atropinesterase activity and cocainesterase activity were each limited to a single isozyme and occurred in certain phenotypes only. In this study the relationship was to differences in drug responses, but it pointed the way towards finding similar functional differences within isozyme systems, as a definitive adaptive mechanism.

Ohno *et al.* (1965) examined G6PD in the hares *Lepus europaeus* and *L. timidus* and in hybrids of reciprocal crosses. The male hybrids possessed a single band identical with that of the mother. Both parental types were found in the female hybrids; sex linkage was indicated.

Cohen (1962) reported that both lysins and agglutinins were found during blood grouping studies; a genetically based system was present.

Moody *et al.* (1949) verified by immunologic techniques that lagomorphs and rodents are not more closely related than mammals of other orders. Most close similarities were with Artiodactyla. A firm conclusion on this relationship was considered premature.

Dayhoff (1969) provided molecular information regarding cytochrome C, fibrinopeptides A and B, hemoglobin alpha and beta chains, heavy chain of IgG, kappa and lambda Ig chains, myosin, and active site peptides of several enzymes. This correlated much information of recent authors.

10. Rodentia

Within this largest of orders (Table II) there are forms with all degrees of specialization. There are living now some ancient groups, and some that appear to be evolving rapidly. The oldest rodent fossils are known from the Paleocene (Dawson, 1967).

Because of their availability and diversity, rodents have long furnished mammalogists with problems for study. The laboratory mouse, rat and guinea pig have provided masses of information basic for understanding mammals. In the following treatment references to studies of laboratory animals are largely omitted—this would require a book unto itself. For the sake of organization, the subordinal classification Sciuromorpha, Myomorpha, and Hystricomorpha are utilized.

a. General works. Auerheimer *et al.* (1960) examined the serum proteins of eight rodent species; marked differences between species were detected, except in the genus *Perognathus.*

Hori and Kamada (1967) demonstrated two major forms of G6PD in the organs of mice, rats, hamsters, and guinea pigs; species differences were found. Baur and Pattie (1968) analyzed rodent erythrocytic LDH and discussed the phylogenetic distribution found in many species of 25 genera.

Levine and Moody (1939) published a report on the serological relationship of rodents, using rodent-immune rabbit sera. Relationships at higher categories could be defined: families and subfamilies were separable; in many instances it was possible to differentiate closely related genera within the same subfamily.

b. Sciuromorpha. Wild (1965) examined serum proteins of the gray squirrel (*Sciurus carolinensis*); 4·3 to 9·5% of total protein was prealbumin; at least 17 antigenetically different proteins were defined by immunoelectrophoresis. Blumberg *et al.* (1960) studied the marmot (*Marmota broweri*) and ground squirrel (*Spermophilus undulatus*). A similar double banded haptoglobin pattern was found except in one marmot where a single, different, band of slower migration was present.

Nadler and Hughes (1966) analyzed serum proteins of the ground squirrels *Spermophilus undulatus*, *S. columbianus*, and *S. beldingi*, by two dimensional electrophoresis. Variations of similar patterns occurred in nine fractions. Transferrins were also examined. They concluded that protein characters in this genus 'appear to offer great promise as a method for systematic investigation'. Nadler (1968b) reported a study of the serum protein and transferrins of ground squirrels *Spermophilus townsendi*, *S. richardsoni*, *S. armatus*. *S. undulatus*, *S. columbianus*, and *S. beldingi*; polymorphism of the transferrins was found in most but not all of the species; each species was easily identifiable by examination of the proteins, except for *S. beldingi* and certain subspecies of *S. townsendi*; the evolutionary significance was discussed. Nadler and Youngman (1969; 1970) suggested that transferrin examination was useful in evaluating evolution within populations as well as in clarifying taxonomy and investigating origins of *S. undulatus*.

Marsh *et al.* (1969) examined serum proteins of four species of ground squirrels from California (*Spermophilus beecheyi*, *S. beldingi*, *S. lateralis*, and *S. tereticaudus*). Transferrin patterns in each species were different; four infraspecific variants were found in *S. beecheyi* and five in *S. lateralis*; no variants were found in *S. beldingi* or *S. tereticaudus*. Double albumins were present in all the *S. lateralis*. Zymograms showed a variety of phenotypes of esterase. Hybrids of subspecies of *S. beecheyi* showed no evidence of 'hybrid substance' (proteins not present in the parents).

Hight *et al.* (1969) used two-dimensional starch-gel electrophoresis to demonstrate constant differences in the mobilities of a number of fractions of serum proteins of *Sciurus carolinensis*, *S. niger*, *Tamiasciurus hudsonicus*, *Glaucomys volans* and *G. sabrinus*. They suggested usefulness of this technique in studying relationships down to the population level.

Hall (1965) studied the oxygen dissociation of hemoglobin of seven species of North American Sciuridae and suggested that the function reflects adaptive evolutionary changes correlated with different ways of life.

Gerber and Birney (1968) used immunological precipitin tests to determine relationship of four subgenera of the ground squirrels, *Spermophilus* (*Otospermophilus*, *Callospermophilus*, *Spermophilus*, and *Ictidomys*). These were also compared with the tree squirrel, *Sciurus*. They concluded that the tree squirrels diverged from the ground squirrels long before the subgenera of ground squirrels diverged from each other. Degrees of relationships within the genus *Spermophilus* were suggested.

'Double albumins' are found regularly in certain species and species groups of Sciuridae (Johnson and Wicks, 1959; Johnson, 1968). This phenomenon can be demonstrated in some other mammals by special processing techniques, and has been found rarely as a polymorphism in other species. However, it is only in the sciurids where it can be used as taxonomic character, in our present state of knowledge.

Bongardt *et al.* (1968) examined the serum of pocket gophers (*Thomomys umbrinus*, *T. townsendi* and *T. bulbivorus*). Differences were found in transferrins amongst species and subspecies: one subspecies of *T. umbrinus* varied from three other subspecies which in turn resembled *T. townsendi*; *T. bulbivorus* was distinct. They concluded that serum protein patterns may prove to be a valuable taxonomic tool.

Studier (1967) found a drop in the serum albumin during lactation in the kangaroo rat (*Dipodomys merriami*).

c. Myomorpha, general. Dalby and Lillevik (1969) demonstrated up to five transferrin bands in the cotton rat (*Sigmodon hispidus*) which could be allocated on a regional basis. In one population, polymorphism of the phenotype was found, indicating a 'limited gene exchange'. Dolyak and Truffelli (1952) reported no isoagglutinins in 60 *S. h. hispidus*.

Marchuwska-Koj (1966) found single hemoglobins with identical migrations in three microtines, *Clethrionomys glareolus*, *Microtus arvalis*, and *M. agrestis*. Neilsen (1969) reported genetic studies on the isoenzymes of amylase in *C. glareolus*, from saliva, serum and pancreas.

Johnson (1968) reported comparisons of several related microtine species: differences of serum proteins and hemoglobins were found in *Microtus agrestis* compared with *M. pennsylvanicus*, and in *M. montanus* compared with *M. canicaudus*. Various species of the genus *Clethrionomys* and close allies *Aschizomys* and *Eothenomys* were found to have a remarkable homogeneity, in specimens over a wide range from Europe, Asia and North America; slight differences, however, separated *C. gapperi* and *C. glareolus*, and *C. gapperi* from *C. californicus*. The subgenus *Arborimus* could be separated from the genus *Phenacomys*. These indications have been supported also by evidence in other systems.

Russell and Semeonoff (1967) examined 254 Scottish field voles (*Microtus agrestis*). Two genetic loci, each with two alleles were found to control a system of four esterase enzymes. Engel *et al.* (1970) reported 6-phosphogluconate dehydrogenase polymorphism in *Microtus oeconomus* and *M. ochrogaster*.

Rosler (1965) in the serum of *Gerbillus pyramidum* found and identified five protein fractions by paper, 17 by starch-gel electrophoresis.

The murids, particularly *Rattus* and *Mus*, have been well studied for immunological and biochemical genetic factors (Lush, 1967). Most of these are related to laboratory animals, but the information can be transposed for taxonomic use. Malecha and Tamarin (1969) used transferrin polymorphism to investigate three species of *Rattus* in widely isolated Pacific islands.

Selander *et al.* (1969) analyzed protein polymorphism and genic heterozygosity in two European subspecies of *Mus musculus*. Selander and Yang (1969) reported a study involving 36 proteins controlled by 41 genetic loci in populations of *Mus*, *Peromycus polionotus* and several other vertebrates. Forty per cent of loci segregated for two or more alleles. When taxonomic units were compared in over-all genetic character, subspecies differed at 10% of their loci on the average, sibling species at 50% and a full species at 70% of their loci. Two forms of hemoglobin were reported by Ranney and Gluecksohn-Waelsch (1955); these have also been found in wild populations (Heinecke and Wagner, 1964). Amino acid sequences of the alpha chain (6 strains) and the beta chain (4 strains) of mouse hemoglobin were reported by Dayhoff (1969).

The rat has been investigated in its laboratory and feral states by Vokáč (1961), who reported 'paracrystalline' form of hemoglobin, and by French and Roberts (1965), Enoki *et al.* (1966) and others. There are five or six forms of hemoglobin in each animal; the amino acid sequences are not yet reported, probably because of this. The amino acid sequences of fibrinopeptide A and

B, the IgG kappa chain and ribonuclease of the rat are reported in Dayhoff (1969).

Arnason and Pantelouris (1966) report 23 serum esterases in the murine rodent *Apodemus sylvaticus* and 19 in *Mus musculus*; according to their concept, 'a marked evolutionary divergence' has occurred. Johnson (1968) found a single hemoglobin in *Apodemus sylvaticus* from Germany, two in *A. argenteus* from Japan; *A. specious* from Japan also had a double hemoglobin (Fig. 4). Avrech and Kalebuchov (1937) were able to distinguish between species and subspecies in the genus *Apodemus* by serological tests.

d. Myomorpha-Peromyscus. This genus has presented multiple challenges to North American taxonomists. Throughout the nomenclatural refining—*Mus*, *Hesperomys*, *Vesperimus*, *Sitomys*—and now *Peromyscus*, students have cut their eye teeth on this delightful and available group; many have remained entranced through a life time of productive work. The molecular phase of study of this genus is now under way on several fronts. Needed clarifications of its complexities are being accomplished by reports such as that of Rasmussen (1968), who reviewed the genetic studies of the genus.

Peterson (1968) examined 18 species of *Peromyscus* from the United States and Mexico. He demonstrated significant differences at the species-group level. These generally supported existing classifications.

Brown and Welser (1968) examined the albumins of 14 species of laboratory and wild stock *Peromyscus*. Polymorphisms were found in six species (two forms in each). They made several conclusions: electrophoretic migration of albumin is useful in characterizing and identifying species; aside from some polymorphism, albumins are invariable within a species; there is considerable potential for corroborating difficult species identifications. Evidence pointed to codominant, autosomal, allelic inheritance of albumin types. In areas of sympatry of *P. leucopus and P. maniculatus*, each mouse possessed the albumin predicted. The differences of allelic frequencies within polymorphism of albumins in wild populations of *P. leucopus* suggested a very limited gene exchange between local populations.

Rasmussen and Koehn (1966) analyzed serum transferrin polymorphisms. Three fractions were identified in *P. maniculatus*. Polymorphism was found in Arizona specimens of *P. crinitus*, *P. eremicus* and *P. boylii*. Ahl (1968) found several transferrin types within the subspecies *P. m. nebrascensis*.

Foreman (1960, 1964, 1966, 1968) described the occurrence of double hemoglobins in nine species of *Peromyscus*, and a single hemoglobin in *P. eremicus* and *Ochrotomys* (*Peromyscus*) *nuttalli*. Rasmussen *et al.* (1968) found three phenotypes in *P. maniculatus*. A single major hemoglobin in *P. maniculatus* occurs sporadically in certain populations. This was described by Foreman, by Rasmussen *et al.*, and was found by us (Fig. 4) in *P. m. gracilis*, by Gough and Kilgore (1964), Thompson *et al.* (1966), and Ahl (1968).

Shaw and Barto (1965) reported two forms of G6PD to occur in most tissues of *P. maniculatus*. This enzyme appeared to be controlled by an autosomal gene. A hybrid molecule occurred in the heterozygote, suggesting a dimeric structure.

Moody (1941), by use of an immunologic adsorption technique, was able to distinguish between *P. maniculatus* and *P. leucopus* living in a single area. Dolyak and Waters (1962), and Waters (1963) found no isoagglutinins; intergeneric heteroagglutinins could not differentiate one subspecies of *P. leucopus* and two subspecies of *P. maniculatus*; close relationship was indicated. Species-specific agglutinins distinguishing *P. leucopus* from *P. maniculatus*, and population-specific agglutinins distinguishing *P. m. bairdi* from *P. m. gracilis* were found by Rasmussen (1961); antigenic polymorphism was found in the subspecies *P. m. gracilis*. Rasmussen (1964) analyzed the allelic and zygotic frequencies in natural populations of *P. m. gracilis*. Genetic markers were two red cell antigens inherited as a two-allele, one locus group system, with three phenotypes demonstrable. Inbreeding within localized populations was found, with a shortage of heterozygotes. The finding supported a conclusion of less gene flow between populations than was usually thought to exist.

Selander *et al.* (1971) reported variations in 30 proteins which were encoded by 32 structural genetic loci in a population of *P. polionotus*. Seventeen of the loci were polymorphic in one or more populations. By this means the genic heterozygosity within named subspecies and additional populations was defined.

Smith and Selander (1971) and co-workers have been involved in a comprehensive survey of blood proteins of the genus *Peromyscus*. Systems analyzed included hemoglobin, plasma proteins, transferrins, and some 13 different enzyme types. Problems investigated included 'the extent of local subdivision within and between populations, clinal variation in gene frequency, and the degree of genetic similarity and divergence among *Peromyscus*'. They have been able to identify all of the species in the United States by their electrophoretic patterns, and have accumulated data of the amount of allelic variation occurring within species. This is along the taxonomic path becoming well marked during the past decade or more; a better understanding of this polymorphic genus will result (personal communication, University of Texas at Austin).

e. Hystricomorpha. This group of rodents has not received much attention with the outstanding exception of the guinea pig, as a favorite laboratory animal.

Brown (1966) studied the sera of six adult and two fetal nutria (*Myocastor coypus*). One band was found in adults, not present in fetuses.

Baur and Pattie (1968) found six species of this suborder to contain subunit LDH AB type in the erythrocytes, whereas ten myomorph species showed an LDH AB- type. This suggested a separate evolutionary line at an early stage of evolution. Radden (1968) studied the blood of four species of the Caribbean Capromyidae (*Capromys pilorides*, *Geocapromys browni*, *G. ingrahami*, and *Plagiodontia aedium*) and compared them with the *Myocastor coypus* (Myocastoridae). The two families were distinct by this study, and *Plagiodontia* was unique within Capromyidae, justifying its subfamily status. Studies included fingerprinting of hemoglobin and examination of catalase, LDH and G6PD.

Moody and Doniger (1956) showed that the African and North American porcupines are not more closely related to each other serologically than to other South American hystricomorphs.

Szynkiewicz (1970) examined the antigenic properties of red cells of the coypu (*M. coypus*); out of 710 animals only four sera were found agglutinating the cells of unrelated coypus; test sera of man, bison, cattle, and horse were used for lytic and agglutination tests of coypu red cells; only the bison serum identified antigens of the coypu red cells. Roig and Reig (1969) studied species of the Argentinian *Ctenomys* by precipitin tests; this method was helpful in clarifying relationships.

11. Cetacea

This order is the most highly specialized for aquatic life and thus in an evolutionary sense is judged to be the oldest group of living mammals to have returned to the sea. They probably evolved from some ancient eutherian stock and are first known from the Eocene (Dawson, 1967). Some cetologists are of the opinion that two orders (Mysticeti and Odontoceti) should be recognized (Rice, 1967). In the absence of paleontologic data of connecting forms, it is desirable to obtain molecular and immunologic data to see how these correspond.

Medway and Gerachi (1965) examined the blood proteins of six bottlenose dolphins (*Tursiops truncatus*). The albumin, and alpha and beta globulin levels appear to be lower than in dogs and horses. Gallien *et al.* (1967) analyzed the serum of five porpoises (*Delphinus delphis*), and identified transferrins. De Monte and Pilleri (1968) described the serum protein electropherogram of *Stenella styx*, *D. delphis*, and *Phocaena phocaena*. Each pattern was different. The hemoglobin of *D. delphis* migrated less than human HbA.

Clotting Factor XII (Hageman Factor) was absent in three odontocetes (*Tursiops truncatus*, *Orcinus orca*, and *Lagenorhynchus obliquidens*) in distinction to species of seven other orders (Robinson *et al.*, 1969). Lewis *et al.* (1969) also reported absence of Factor XII, in 12 specimens of *T. truncatus*, in distinction to 21 species of 10 other mammalian orders. It is important to find out if this trait also occurs in the mysticetes.

A number of amino acid sequences were reported (Dayhoff, 1969): Cytochrome C of the gray whale (*Eschrichtius gibbosus*); myoglobin of the sperm whale (*Physeter catodon*); myoglobin fragments of the humpback whale (*Megaptera novaeangliae*) and dolphin (species not listed); and insulins A and B of the sperm whale, finback whale (*Balaenoptera physalus*) and sei whale (*B. borealis*). It is of interest that the human, rabbit, dog, sperm whale, finback whale, and pig insulins are identical in amino acid sequence of insulin A. The sei whale has two replacements in the 21 amino acid chain. The three whales are identical in the insulin B chain (24 amino acids), in common with dog, horse, pig, sheep, goat and cow. Konagaya (1963) reported the N-terminal amino acid sequences of hemoglobin of the dolphin (*Stenella coeruleo-albus*). Bradshaw and Gurd (1969) reported 15 differences between myoglobin amino acid sequences of the sperm whale and porpoise (*P. phocoena*).

Boyden and Gemeroy (1950) studied serological relationships of Cetacea to other orders of mammals. They concluded that 'the Cetacea should be granted a greater degree of systematic relationship to the Artiodactyla than to any other mammalian order tested'. Boyden (1964) made a qualifying statement, '. . . studies on the relationship of the Cetacea to other orders of mammals . . . revealed small amounts of correspondence in most cases. Then we may say that interordinal reactions are weak;'.

Fujino (1962) characterized finback whale populations of South African and Antarctic waters; he indicated intermixing of these populations and stated that maternal-fetal incompatibilities in blood types were isolating mechanisms.

12. Carnivora

This is an interesting order composed of several well known and easily recognized families. The order originated during the Paleocene and several of the modern families have been diagnosed in early Oligoene material; two, Ursidae and Hyaenidae, were derived later (Dawson, 1967). There is an ongoing battle of opinions regarding placement of Pinnipedia as a suborder of Carnivora; molecular and immunologic information is being added to the data available.

a. Families other than Canidae and Felidae, and general references. Hemoglobins of carnivores and pinnipeds were investigated in depth by Seal *et al.* (1967) and Seal (1969). They concluded that the major hemoglobin found in Canidae, Ursidae, Procyonidae, Mustelidae, Otariidae and Phocidae was identical, and that the hemoglobins in Feloidea were diversified. Final judgement on these conclusions regarding identity of hemoglobins between species must await further and more diversified examination beyond acrylamide gel. If verified within such a wide spectrum of species, it will be a unique and unexpected phenomenon. In any event the molecular homogeneity of the Carnivora is indicated by these published results.

Comparative immunologic studies (Leone and Wiens, 1956) indicated that the superfamilies Canoidea and Feloidea were divergent enough to be considered separate suborders, but the Pinnipedia were not sufficiently different to be considered a suborder within Canoidea; the feloids, including Felidae, Viverridae and Hyaenidae showed closest serological relationship one to another; the giant panda was closest to Ursidae, compared to Procyonidae; the dog and the coyote were more similar to each other than to the wolf; the dog was closer to the wolf than the coyote.

Pauly (1962) examined seven species of Felidae by immunological reactions, and compared these with species of six other carnivore families, three pinniped families, the bison (*Bison*) and the macaque (*Macaca*). No significant differences were found amongst the cat species and none between the carnivores and pinnipeds. He suggested a revised classification.

Sarich (1969a) suggested that albumin evolution has occurred in a regular fashion in carnivores and pinnipeds. According to his analysis, it was possible to develop a quantitative phylogeny for a group of species based on an immunological study of a single protein. Sarich (1969b) measured differences amongst various carnivore and pinniped taxa by immunologic means. A quantitative phylogeny was established.

Serum albumins of bears and other carnivores were examined by Seal *et al.* (1970), using micro-complement fixation studies. *Tremarctos* could be distinguished from the remainder of Ursidae. Divergent values for *Canis*, *Mephitis* and *Panthera* spoke against a constant rate of evolution.

Genetic studies on mink (*Mustela vison*) have been considerable because of its domestication. Porter and Dixon (1966) reported examination of serum proteins of mink compared with man. Four protein bands were found by paper technique, five by cellulose acetate technique and 25 by immunoelectrophoresis on agar. Rapacz *et al.* (1965) examined 2,200 domestic mink. At least 12 blood antigenic specificities were found.

Hemoglobin and serum proteins of the badger (*Meles meles*) were described and compared with unrelated species (Mäsiar and Babják, 1968; Mäsiar and Sova, 1969).

b. Canidae. The dog family is best known from the genus *Canis* and its widespread use as an experimental animal. Canine hemoglobin has been effectively used in hybridization experiments in which alpha and beta chains of the hemoglobin molecules are disassociated and recombined with those of another species, by means of manipulating the pH of the solution. The resulting hybrid molecules are then examined for their characteristics (Maggioni *et al.*, 1963; Antonini *et al.*, 1964).

Braend (1966) examined dogs of 25 breeds. He found three different transferrin phenotypes. Kaminski *et al.* (1966) found eight transferrin types composed of four to five variable bands in 109 foxes (*Vulpes*).

Baur and Schorr (1969) described a genetic polymorphism of erythrocyte tetrazolium oxidase in dogs. Two alleles 'A' and 'B' determined the polymorphism. The gene frequencies of the German Shepherd differed from those of other breeds. Two specimens of the coyote, *Canis latrans* closely resembled the 'A' pattern of the dog.

Anthony (1966) studied registered breeds of dogs by various immunological and biochemical techniques. There was considerable molecular heterogeneity of analogous serum proteins; some correlation existed between breed and molecular variation of several enzyme molecules.

Swisher *et al.* (1962) found evidence of mild isoantibodies and a blood group system in dogs. Kaminski *et al.* (1966) reported no natural antibodies (agglutinins or hemolysins) in 49 fox (*Vulpes*) sera. They found natural hetero-antibodies in the fox serum against erythrocytes of many species except the dog. All of the animals except the dog and rabbit possessed antibodies against the fox erythrocytes.

Afanasieva (1963) found no differences in the immunochemical behavior as dog albumin was synthesized at all ages from fetuses through adults.

Of practical taxonomic interest is the attempt to find some characteristics by which the American species of *Canis* may be identified. These species are the dog, (*C. familiaris*), wolf, (*C. lupus*), red wolf (*C. rufus*), and coyote (*C. latrans*). The chromosomes are identical. Seal (1971) has been investigating these species in depth and stated, 'immunological comparisons and electrophoretic analyses of serum and red blood cell proteins and enzymes have not differentiated between North American species of the genus *Canis*. . . . Tissue esterase zymograms have demonstrated several differences which are being examined in population samples'. Some characters will certainly be found, but at this writing we must still rely upon traditional morphologic characters to distinguish these species (personal communication, Vet. Admin. Hospital, Minneapolis, Minn.).

c. Felidae. This family appears biochemically most diverse of all Carnivora Hemoglobins are unique within the order (Seal, 1969). The domestic cat was found to have a lower oxygen affinity than other mammals (Taketa and Morell, 1966; Taketa *et al.*, 1967); two major hemoglobins were found, both of which contained eight sulfhydryl groups, the highest number found in normal hemoglobins, including man, rabbit, mouse, rat, dog, horse, sheep and ox. Two very early heme proteins ('embryonal') were found in the fetus between 26 and 35 days' gestation; the differences between the two adult hemoglobins were found in the beta chains (Lessard and Taketa, 1969).

Thuline *et al.* (1967) found either a single fast band, a single slow band, or three bands of 6-phosphogluconic dehydrogenase in the cat; they proposed a dimeric structure for the enzyme. Kramer and Sleight (1968) found a single band of alkaline phosphatase in kitten and cat sera, migrating in the beta-1 globulin fraction.

13. Pinnipedia

This order is highly specialized, next after Cetacea, among the mammals, for aquatic existence. It has three families and 31 species. There continues to be resounding and stimulating controversy regarding its origin and relationship of the species and genera. The earliest known fossils are in the early Miocene, and they were already specialized (Dawson, 1967). Biochemical and immunological studies are beginning to provide fuel for the fires of controversy.

One telling argument that Pinnipedia arose from a common, rather than diverse ancestor was supplied by Haslewood (1964, 1968): only members of this order synthesize phocaecolic acid in the formation of bile salts.

Polymorphism in the serum proteins has been found in several seals. Naevdal (1966a) in the harp seal (*Pagophilus groenlandicus*) found three transferrin bands, manifested as six phenotypes which are probably under the control of three autosomal co-dominant alleles; a difference was found between populations of harp seal in the eastern and western North Atlantic; hemoglobin specimens had two fractions: the major band migrated toward the anode, the minor toward the cathode. Individual variations in hemoglobins and serum proteins of the harp seal were found by Møller *et al.* (1966); significant difference in gene frequencies of transferrins were noted between populations from Newfoundland and other samples, but not between populations from the eastern breeding areas of Jan Mayen and White Sea.

Shaughnessy (1969) reported three transferrin types, a single heme-binding zone, and two hemoglobin zones in the Weddell seal (*Leptonychotes weddelli*). In 1970 the same author reported a study of the southern fur seals (*Arctocephalus forsteri*) of southern Australia, New Zealand and Macquarie Island. Six transferrin types and two phenotypes of a heme-binding protein (considered to be hemopexin) were found. Two haptoglobin zones were noted. Populations could be characterized by these criteria. No transferrin polymorphism was found in southern elephant seals (*Mirounga leonina*).

Polymorphism in the serum protein, probably transferrin, was found in the ringed seal (*Pusa hispida*) but no transferrin polymorphism was demonstrated in the hooded seal (*Cystophora cristata*) nor the bearded seal (*Erignathus barbatus*) by Naevdal (1966b).

Blumberg *et al.* (1960) demonstrated two haptoglobin phenotypes in the northern fur seal (*Callorhinus*). They also described two major hemoglobin bands and a slower diffuse component. Bradshaw and Gurd (1969) reported the amino acid sequence of myoglobin of the harbor seal (*Phoca vitulina*). Nauman (1970) studied the primary structure of myoglobins of six pinnipeds.

Leone and Wiens (1956), with the precipitin test found that Pinnipedia showed closest affinities to the canoids, including Canidae, Ursidae, Procyonidae and Mustelidae. Borissov (1969) used the precipitin reaction to study the phylogenesis and systematics of the pinnipeds. The microcomplement fixation technique using albumin, indicated that Pinnipedia share 'more recent common ancestry' with Canoidea than either do with Feloidea (Sarich, 1969a), and further, that Pinnipedia is monophyletic in origin and that the walrus (*Odobenus*) was recently derived from otariid stock (Sarich, 1969b).

Fujino and Cushing (1960) found that individual variations existed in the erythrocyte antigens of fur seals (*C. ursinus*). At least four different types occurred in 234 samples collected off Honshu Island in 1958. They concluded that blood type antigens could be used in research on fur seal populations.

14. Tubulidentata

This is a unique order with possible early relationship to condylarths (primitive ungulates). The earliest fossil evidence is in the Miocene in Africa where the order is now confined. Fossil specimens indicate a greater original range including Europe and India (Dawson, 1967). Essentially no molecular data have been recorded about the single species which now lives.

On preliminary survey of a single specimen, all erythrocytes were found to be of target cell form. The electropherogram was not remarkable, with the albumin migrating at the rate of human Alb. Glycoprotein staining was most marked in the albumin zone, lesser in the beta globulin area; essentially no lipoprotein was found. A single hemoglobin was nearly isoelectric with human HbA (Fig. 4).

Data regarding the amino acid structure of hemoglobin and other molecules should allow us to relate this order to other modern orders in a better way than has been possible before now.

15. Proboscidea

This order is distinct and has no close relatives. Early representatives have been found in late Eocene (Dawson, 1967). It has not been extensively studied by molecular methods.

Buettner-Janusch *et al.* (1961) compared the African elephant, *Loxodonta* to the hyrax, *Procavia habessinica* and man. Two slowly moving bands of transferrin were found in *Loxodonta*. A single haptoglobin band was found, migrating slightly faster than that of man. A single hemoglobin moved very fast compared to the human, not quite as fast as that of the hyrax. The hemoglobin was found to have a high value (86·9%) of alkali resistance.

Kleihauer *et al.* (1965) found fetal hemoglobin only in the five-month-old and not in the 12-month-old fetus of *Loxodonta* (gestation period is 21 months). Both fetal and adult elephants had alkali resistance figures of between 95% and 97%.

McCullagh *et al.* (1969) reported a unique distribution of capric acid (65% of total fatty acids) in milk of *Loxodonta*.

Having two genera (*Elephas*, *Loxodonta*) within a unique order, there is an interesting time ahead as molecular data accumulate and can be compared and correlated with other information.

16. Hyracoidea

This group is not closely related to any living order, but has been compared with Proboscidea, Rodentia, Sirenia and Perissodactyla (Simpson, 1945; Weitz, 1953; Grassé, 1955). Early ancestors which do not clarify its relationships have been found only in Africa, dating back to early Oligocene (Dawson, 1967). It has been little studied.

We have found albumin in a single specimen of *Procavia capensis* to migrate slightly faster than that of the human. There is a pronounced glycoprotein-staining band near the alpha-2 position and secondary slower banding. No lipoprotein is seen in the albumin area; a moderate band is present near the beta globulin position. There is a single hemoglobin (Fig. 4), which migrates rapidly, between the elephants *Elephas* and *Loxodonta*.

Buettner-Janusch *et al.* (1964) found multiple haptoglobins and two transferrin bands in a specimen of *Procavia habessinica*. The hemoglobin of this animal migrated slightly faster than that of the elephant, *Loxodonta*.

Weitz (1953) found a close serologic relationship between the hyrax and the elephant.

The taxonomic relationships of this order have not been clarified by past studies. However, at this point in time, by use of comparative amino acid sequence of the hemoglobin and the albumin molecules (among others), we have the ability to advance this understanding. Comparison of these types of data within the three genera, (*Procavia*, *Heterohyrax* and *Dendrohydrax*), and projecting this into similar information from the orders Proboscidea, Sirenia and Perissodactyla will provide a most stimulating series of exercises.

17. Sirenia

This is a restricted order without close modern relatives and with an unknown phylogeny, 'perhaps with some special, although distant, connection with the proboscideans' (Simpson, 1945).

A survey specimen showed an unremarkable electropherogram and a single hemoglobin that migrated as the elephant (*Elephas*). Glycoprotein was principally in the region of alpha-2 globulin, none in the albumin area. A strong lipoprotein band was present in the region of beta globulin, very similar to the human.

A great stride in establishing taxonomic relationships of this order is now possible as never before. Once we have the complete architecture of a few macromolecules of this and other orders, we will be able to fit in homologies for concepts of relationships and some time-estimates of evolution.

18. Perissodactyla

This is a small (Table II), but well-studied order which dates back to the early Eocene (Baker, 1967). There are many genetic studies related to the horse; the reader is referred to the E.S.A.B.R. publications as noted in the section on blood groups. As an example, Gahne (1966) examined the serums of 147 Saleritana horses; he demonstrated the following phenotypes: 16 transferrin, 3 albumin, 8 prealbumin, and 6 esterase. Some 20 blood group factors were mentioned by Sandberg (1970). These are good control systems for taxonomic work in and about this order.

a. General studies. Osterhoff (1966) examined horses, donkeys, mules, and zebras (*Equus burchelli*). Three hemoglobin phenotypes were identified in the horse. A combination of a 'major band migrating slower than a faint secondary band' was found in horses and all mules. Only the slower band was found in donkeys. Three hemoglobin phenotypes in the zebras were similar to the horse. New types of transferrins were found and named in mules, donkeys and zebras. Three bands of albumin were demonstrated, in six phenotypes; the donkey and zebra showed closest relationship.

Kitchen and Easley (1969) compared hemoglobins of the equids, *Equus c. caballus*, *E. c. tarpan*, *E. przewalskii*, four subspecies of *E. burchelli*, and hybrids between horse and donkeys. Limited differences were found, primarily in the alpha chains. The genetic control of horse hemoglobins can be based upon three pairs of alleles for the two different alpha chains and one beta chain. Horses and most zebras (*E. burchelli*) have two hemoglobins. This was explained simply by control of two structurally different alpha chains by two separate pairs of cistrons. The combination of the products of these two cistrons combine with the respective common beta chain which are unique to the zebra and horse. The presence of all hemoglobins of each parent in interspecific hybrids indicated the hemoglobins were 'multiple rather than polymorphic'.

b. Hybrids. Studies of hybrids have been attractive to investigators. The mule has been a fine subject, and has given results as expected.

Trujillo *et al.* (1967) found that hybrids contained three hemoglobins—one from the donkey and two from the horse. Kilmartin and Clegg (1967) found in the two electrophoretically-distinct horse hemoglobins four different alpha chains; no changes were found in the beta chain; the single donkey hemoglobin had a single amino acid difference in the alpha chain and they concluded that this chain in both species was probably under the control of a single structural gene.

Trujillo *et al.* (1965) reported different electrophoretic patterns of G6PD in the horse and donkey. Additional findings suggested an X-linkage in male reciprocal hybrids, and the existence of isoenzymes.

Bonadonna *et al.* (1967) examined serum proteins of the horse, donkey, mule and hinny. In six hinnies a splitting of 'beta globulin' was seen, much as in the horse. In seven mules, no such splitting occurred, similar to the donkey.

Boyden (1942) illustrated the potential usefulness of the immunological method by placing the mule between the horse and the ass. Podliachouk (1965) found blood group factors specific for the horse and for the donkey were all present in the mule and hinny.

19. Artiodactyla

The earliest forms lived during the Eocene in North America and Europe (Dawson, 1967). This order of large and well-known hoofed animals has had considerable biochemical and immunological investigation. Especially regarding the domestic animals—pigs, cattle and sheep there is continuing large-scale research into the genetic factors of blood proteins, hemoglobins, and blood groupings. The interested reader is referred to the scientific productions of the E.S.A.B.R. as noted in the section on blood groups. These data are useful for comparative studies.

a. Miscellaneous families, excluding Cervidae and Bovidae. The domestic pig has much genetic data related to biochemical polymorphisms and blood grouping. The European wild boar, considered conspecific with the domestic pig, has nevertheless some demonstrable differences. Wiatroszak (1970) found distinct differences between wild boars and domestic pigs, using 36 pig test sera belonging to 12 blood group systems. Buschmann (1964) found two hemoglobin types in the wild pig. One, consisting of a single zone, corresponds to that of the domestic pig; type 2 is characterized by at least two zones, including that of type 1.

Weinbren (1960) examined the serum proteins of *Hippopotamus amphibius*. A 40% globulin content was found, with a great excess of gamma globulin. This was not affected by age or sex.

Information regarding amino acid sequences of several proteins of many artiodactyls was reviewed by Dayhoff (1969). Wolfe (1939) studied relationships between five species of Bovidae and three species of Cervidae, using 19 antisera. Baier and Wolfe (1942) reported quantitative serologic relationships within families in 13 species of Artiodactyla. Marable and Glenn (1964) compared sera of eight artiodactyls and two perissodactyls with 15 homologous or heterologous antisera, using densitometry of single diffusion columns; relationships determined by this method were expressed in percentages. In general in each of these studies accepted relationships were verified.

b. Cervidae. Van Tets and Cowan (1966) examined two subspecies of the white-tailed deer (*Odocoileus virginianus*) and three of the black-tailed deer (*O. hemionus*); they were unable to determine taxonomically significant characters; individual differences in relative proportions of 14 serum fractions were determined, and were related to age and sex. Miller *et al.* (1965) reported a study on the blood of 200 white-tailed deer (*O. virginianus*); two serum globulin patterns were distinguished: two bands and three bands; these were presumed to be transferrin. Seal and Erickson (1969) examined 25 substances in *O. virginianus* of Minnesota; polymorphisms were delineated.

Gahne and Rendel (1961) described six transferrin types in the European reindeer, *Rangifer tarandus* and explained them on the basis of three alleles. Braend (1964a; 1964b) found 27 transferrin phenotypes; an 8 allele system at a single locus was proposed; population analysis showed that a wild herd differed in the frequency of one or more of the alleles. Shubin (1969) postulated 5 codominant alleles determining 15 Tf phenotypes.

Nadler *et al.* (1967) examined serum proteins of North American *Alces alces*, *Rangifer tarandus* and *Ovis dalli*. Two transferrins were found in seven specimens of *Alces*. Thirty-seven specimens of *tarandus* were similar to those reported by Braend (1962, 1964b) from Norway. A specimen of *O. dalli* contained a single transferrin band; serum proteins were the most dissimilar of the three species examined. Shubin (1969) found 12 specimens of *A. alces* monomorphic for three zones of transferrin.

Hemoglobins of Cervidae have been investigated by several workers. Maughan and Williams (1967) examined nine species of deer. Single bands of identical mobility and no variation were found in *Cervus elaphus*, *C. nippon*, *C. duvanceli*, *Elaphurus davidianus*, and *Axis axis*. The hemoglobin of *Dama dama* moved somewhat faster, with that of *Capreolus capreolus* in between. In 40 Chinese muntjac deer (*Muntiacus reevesi*) three phenotypes were found. Two had double bands, the third, three, appearing to be a heterozygous condition of the other two. In 18 Chinese water deer (*Hydropotes inermis*) a single hemoglobin was found in all but three, which showed a second slower minor band. All deer hemoglobins were more than 50% resistant to alkali denaturation. Naik *et al.* (1964) found two hemoglobin variants (single bands each) in the Indian axis deer (*Axis*) and a single band in the muntjac *Muntiacus*; in three axis deer, G6PD linked with the coenzyme nicotinamid-adenine dinucleotide (NAD), not previously reported in mammals. The muntjac linked with NAD phosphate. Kitchen *et al.* (1967) reported seven adult and two fetal hemoglobins in 700 *Odocoileus virginianus*. Huisman *et al.* (1968) proposed a genetic basis for the hemoglobin heterogeneity found in this species.

The sickling phenomenon of deer erythrocytes was reported first by Gulliver in 1840. Since the discovery of the different molecular forms, HbS, by Pauling *et al.* in 1949, and with the association of this to sickling as a pathologic state, several papers have appeared trying to link this to sickling in deer. Because sickling is reported to be a trait within a taxonomic unit, it is well to review the information on this condition. There appear to be definite differences between sickling in man and in deer. In man there is a specific abnormal molecular form of the hemoglobin molecule that produces sickling. HbS differs from HbA only in the substitution of valine for glutamic acid in position 6 of the beta chain (Lehmann and Huntsman, 1966). In man sickling is produced by 'close ordering of the *deoxygenated* molecule into linear aggregates' (Bertles *et al.*, 1970), as shown nicely by the electron microscope. The bizarre distortion of the sickled cells in deer and in man are identical in form. However, in deer, sickling can be produced by a variety of abnormal conditions *in vitro*: Maughan and Williams (1967) produced sickling in five species of deer, but not in *Hydropotes* in 0·9% saline solution. Sickling in deer occurs at very high oxygen tensions and alkaline pH. Reversal to normal erythrocyte contours in deer can be produced by passing 95% N and 5% CO_2 through the blood (Kitchen *et al.*, 1964); this is the exact experimental procedure to produce sickling in man (Bertles *et al.*, 1970). In man sickled erythrocytes may be restored by oxygenating the blood (Lehmann and Huntsman, 1966). Kitchen *et al.* (1964, 1967, and 1968) indicated that sickling in *Odocoileus* was not associated with a single hemoglobin type, but rather sickling was precluded by the presence of hemoglobin 'V' or 'VII', even when combined with hemoglobin types normally found with sickling. The presence of sickling in man is pathologic and in the homozygous state crippling or fatal by means of intravascular obstruction and hemolytic anemia. In deer, no pathologic states are associated (Kitchen *et al.*, 1967).

Braend (1962) found in the European elk (*Alces*), a system similar to the J-anti J system in cattle; other cattle reagents gave reactions, but of a different order than cattle. Naik *et al.* (1964) were unable to produce reactions with available cattle blood group reagents upon the erythrocytes of *Axis* and *Muntiacus*.

c. Bovidae. Braend and Stormant (1963) found a single transferrin type consisting of three bands in bison (*Bison bison*), similar to the AA type found in cattle. Osterhoff and Young (1966) found an identical transferrin pattern in 67 samples of African buffalo (*Syncerus caffer*), apparently identical to cattle type AA. A single specimen of the gnu (*Connochaetes taurinus*) was quite different.

Sartore *et al.* (1969) found multiple forms of carbonic anhydrase in the red cells of cattle and American buffalo (*Bison*); each species could be readily distinguished. Naik and Anderson (1970) described the occurrence of 6-phosphate dehydrogenase and 6 phosphogluconate dehydrogenase in *B. bison*.

The hemoglobin of the bison (*Bison*) was examined by Braend and Stormant (1963); no polymorphism of the two bands was encountered in 112 specimens. Balani and Barnabas (1965) found in the buffalo (*Bubalus*) that the beta chains of two major hemoglobins were similar; the alpha chains were different.

Serologic relationships of the musk ox (*Ovibos*) were explored by Moody (1958), employing antigens from the goat, sheep, cattle, bison (*Bison*) and mountain goat (*Oreamnos*) and antisera against the musk ox, cattle and goat. The musk ox was found to be serologically similar to sheep and goats and 'markedly dissimilar' to cattle and bison. Their results supported the taxonomic placement of musk ox into the subfamily Caprinae, separating the tribes Ovibovini and Caprini, with *Oreamnos* in the tribe Rupicaprini.

Gasparski (1965) and Gasparski *et al.* (1966) investigated the wisent (*Bison bonasus*), cattle and cattle-wisent hybrids. Cattle blood-typing reagents were used and testing of the sera against the red cells of several species of animals was done. Reaction to antigens A, W, X_1, V_2, J and Z were identical, and others were similar in wisent and cattle; several new antigenic factors were determined. The wisent and hybrid sera gave characteristic results, with few exceptions.

Datta and Stone (1963) stated that more than 80 heritable antigenic factors had been detected on the erythrocytes of cattle by their reaction with blood grouping reagents. These reagents had also been tested on American bison (*B. bison*) and the wisent (*B. bonasus*), finding at least nine homologues. Erythrocytes of the water buffalo (*Bubalus bubalis*) have been classified into four types by using natural agglutinins of cattle; an antigenic system similar to J in cattle was found; little reactivity was found with F, V, and Z reagents. They concluded that the serological relationship of cattle and water buffaloes was less close than that of bison and cattle.

Osterhoff and Young (1966) studied the African buffalo (*Syncerus caffer*) with cattle blood group reagents. Reactions were observed in the A and B systems principally. No identical blood types were found. In 40 cattle reagents tested, 15 reacted with buffalo cells, similar to findings of Chet Ram *et al.* (1964) on the Indian water buffalo in which 16 of 35 reagents showed activity.

IV. Summary

What's to be said has already been written. I wish simply to add a quote from Alan Boyden. Dr Boyden worked for many years and with techniques that became more refined as the years progressed. He has been able to see his persistent interest in molecular and immunologic taxonomy blossom into the driving, controversial and productive force it should be. He said (1958) 'This serologic placement series does seem to be correlated with many other fundamental characters of the organisms and can serve, *not as the sole basis for a systematic arrangement*, but to lend support to some one of the possible arrangements where other data are conflicting'.

May each of us temper our enthusiasm with this reflection.

Acknowledgements

The background preparation for this chapter has been interdisciplinary and interpersonal. National Science Foundation provided Grant G 10831, GB 1738, GB 4617, and GF 157. The list of persons involved includes untrained collectors, technicians, students, highly specialized research scientists and those who fit in between, or in several categories. They have each made for a happy and appreciated relationship. Special thanks must be given several persons who have helped on a recurring basis. Dr M. J. Wicks, co-worker, of the Tacoma General Hospital; Mrs Fumiko Kimura, co-worker biochemist; Dr Robert Freeman, electronic consultant; Dr Robert Sprenger (deceased), chemist, and Dr Gordon Alcorn, biologist, co-workers at the University of Puget Sound; Dr Richard T. Jones, molecular biologist *par excellence* of the University of Oregon Medical School; Dr Ernst Baur, consultant and provider of refined techniques and the photographs used in this chapter. A last mention must be made of my many friends in the American Society of Mammalogists; within this group for a decade I prodded and was prodded back at the annual sessions as systems of molecular taxonomy of mammals evolved.

References and Bibliography

Aalund, O. (1968). 'Heterogeneity of Ruminant Immunoglobulins'. Munksgaard, Copenhagen.

Adams, E. C. and Weiss, M. R. (1969). *Biochem. J.*, **115**, 441–447.

Afanasieva, A. V. (1963). *Bull. Biol. Méd. exp. USSR*, **8**, 54–57.

Ahl, A. S. (1968). *Comp. Biochem. Physiol.* **24**, 427–435.

Air, G. M., Thompson, E. O. P., Richardson, B. J. and Sharman, G. B. (1971). *Nature, Lond.* **229**, 391–394.

Akhtar, A., Hanson, A. and Kärcher, K. H. (1968). *Z. klin. Chem. klin. Biochem.* **6**, 334–337.

Anderson, S. (1967). *In* 'Recent Mammals of the World'. (S. Anderson and J. K. Jones, eds.) pp. 151–177. The Ronald Press Co., New York.

Anderson S. and Jones J. K. (eds.). (1967). 'Recent Mammals of the World'. The Ronald Press Co., New York.
'Animal Blood Groups and Biochemical Genetics' (Vol 1, 1970). Centre for Agricultural Publishing and Documentation, Wageningen, the Netherlands.
Anthony, R. L. (1966). *Bull. serol. Mus., New Brunsw.* **35**, 7.
Antonini, E., Wyman, J., Brunori, M., Bucci, E., Fronticelli, C. and Rossi-Fanelli, A. (1963). *J. biol. Chem.* **239**, 2950–2957.
Antonini, E., Wyman, J., Bucci, E., Fronticelli, C., Brunori, M., Reichlin, M. and Fanelli, A. (1964). *Biochim. Biophys. Acta* **104**, 160–166.
Arfors, K.-E., Beckman, L. and Lundin, L.-G. (1963). *Acta genet. Statist. med.* **13**, 226–230.
Arnason, A. and Pantelouris, E. M. (1966). *Comp. Biochem. Physiol.* **19**, 53–61.
Ashton, G. C. (1964). *Genetics, Princeton* **50**, 1421–1426.
Ashton, G. C., Fallon, G. R. and Sutherland, D. N. (1964). *J. Agric. Sci.* **62**, 27–34.
Auernheimer, A. H., Cutter, W. and Atchley, F. O. (1960). *J. Mammal.* **41**, 405–407.
Avrech, V. V. and Kalabuchov, N. E. (1937). *Zool. Zh.* **16**, 135–148.
Baier, J. G., Jr. and Wolfe, H. R. (1942). *Zoologica, N.Y.* **27**, 17–23.
Baker, B. E., Hatcher, V. B. and Harrington, C. R. (1967). *Can. J. Zool.* **45**, 1205–1210.
Baker, R. H. (1967). *In* 'Recent Mammals of the World'. (S. Anderson and J. K. Jones, eds.) pp. 374–384. The Ronald Press Co., New York.
Balani, A. S. and Barnabas, J. (1965). *Nature, Lond.* **205**, 1019–1021.
Barman, Thomas E. (1969). 'Enzyme Handbook'. Springer-Verlag, New York.
Barnicot, N. A., Jolly, C., Huehns, E. R., and Moor-Jankowski, J. (1964). *Nature, Lond.* **202**, 198–199.
Barnicot, N. A., Jolly, C. J., Huehns, E. R. and Dance, N. (1965). *In* 'The Baboon in Medical Research' (H. Vagtborg, ed.) pp. 323–338, University of Texas Press, Austin.
Bauer, K. (1969). *Hum. Genet.* **7**, 76–90.
Baur, E. W. (1963). *J. Lab. clin. Med.* **61**, 166–173.
Baur, E. W. (1964). *Clin. chim. Acta* **9**, 252–253.
Baur, E. W. (1969). *Comp. Biochem. Physiol.* **30**, 657–664.
Baur, E. W. and Pattie, D. L. (1968). *Nature, Lond.* **218**, 341–343.
Baur, E. W. and Schorr, R. T. (1969). *Science, N.Y.* **166**, 1524–1525.
Beetlestone, J. G. and Irvine, D. H. (1964). *J. chem. Soc.* **977**, 5090–5095.
Berk, J. E., Kawaguchi, M., Zeineh, R., Ujihari, I. and Searcy, R. (1963). *Nature, Lond.* **200**, 572–573.
Bernstein, L., Kerrigan, M. and Maisal, H. (1966). *Exp. Eye Res.* **5**, 309–314.
Bertles, J. F., Rabinowitz, R. and Döbler, J. (1970). *Science, N.Y.* **169**, 375–377.
Blanc, B. and Isliker, H. (1963). *Helv. physiol. pharmac. Acta* **21**, 259–275.
Blombäck, B. and Blombäck, M. (1968). *In* 'Chemotaxonomy and Serotaxonomy' (J. G. Hawkes, ed.) pp. 3–20. Academic Press, London and New York.
Blumberg, B. S., Allison, A. C. and Garry, B. (1960. *J. cell. comp. Physiol.* **55**, 61–71.
Bonadonna, T., Fornaroli, D. and Succi, G. (1967). *Zentbl. Vet. Med.* **14**, 845–848.
Bongardt, H., Richens, V. B. and Howard, W. E. (1968). *J. Mammal.* **49**, 544–546.
Borissov, V. I. (1969). *Zool. Zh.* **48**, 248–255, from *J. Mammal.* **50**, 855.
Bourne, G. H. (ed.). (1970). 'The Chimpanzee' Vol 2, S. Karger, Basel.
Bouw, J. (1965). Proc. Ninth Europ. Animal Blood Group Conf., pp. 25–38.
Boyd, W. C. (1964). *In* 'Taxonomic Biochemistry and Serology' (C. A. Leone, ed.) pp. 119–169. The Ronald Press Co., New York.
Boyden, A. (1942). *Physiol. Zool.* **15**, 109–145.

Boyden, A. (1949). *Bull. serol. Mus., New Brunswick* **3**, 2.

Boyden, A. (1958). *In* 'Serological and Biochemical Comparison of Proteins' (W. H. Cole, ed.) pp. 3–24. Rutgers University Press, New Brunswick, N.J.

Boyden, A. (1964). *In* 'Taxonomic Biochemistry and Serology' (C. A. Leone, ed.) pp. 75–99. The Ronald Press Co., New York.

Boyden, A. and Gemeroy, D. (1950). *Zoologica, N.Y.* **35**, 145–151.

Boyer, S. H., Crosby, E. F., Thurmon, T. F., Noyes, A. N., Fuller, G. F., Leslie, S. E., Shepard, M. K. and Herndon, C. N. (1969). *Science, N.Y.* **166**, 1428–1431.

Bradshaw, R. A. and Gurd, F. R. N. (1969). *J. biol. Chem.* **244**, 2167–2181.

Braend, M. (1962). *Ann. N.Y. Acad. Sci.* **97**, 296–305.

Braend, M. (1964a). *Nature, Lond.* **203**, 674.

Braend, M. (1964b). *Hereditas* **52**, 181–188.

Braend, M. (1966). Proc. Tenth Europ. Conf. on Animal Blood Groups and Biochem. Polymorph. pp. 317–322.

Braend, M. and Stormant, C. (1963). *Nature, Lond.* **197**, 910–911.

Braun, V., Hilse, K., Best, J., Flamm, U. and Braunitzer, G. (1967). *Bull. Soc. Chim. biol.* **49**, 935–948.

Braunitzer, G. (1966). *J. cell. Physiol.* **67**, Sup. 1, 1–20.

Braunitzer, G., Best, J. S., Flamm, U. and Shrank, B. (1966). *Z. phys. Chem.* **347**, 207–211.

Brdička, R. (1966). *Acta biol. med. germ.* **16**, 617.

Breathnach, C. S. (1964). *Q. J. exp. Physiol.* **49**, 277–289.

Bremner, K. C. (1964). *Aust. J. exp. Biol. med. Sci.* **42**, 643–656.

Brown, J. H. and Welser, C. F. (1968). *J. Mammal.* **49**, 420–426.

Brown, L. (1966). *Comp. Biochem. Physiol.* **19**, 479–481.

Buettner-Janusch, J. (1962). *Ann. N.Y. Acad. Sci.* **102**, 235–248.

Buettner-Janusch, J. (ed.). (1963, 1964). 'Evolutionary and Genetic Biology of Primates'. Vol 1, 1963, Vol 2, 1964. Academic Press, Inc., New York and London.

Buettner-Janusch, J. and Buettner-Janusch, V. (1963). *Nature, Lond.* **199**, 918–919.

Buettner-Janusch, J. and Buettner-Janusch, V. (1964). *In* 'Evolutionary and Genetic Biology of Primates' (J. Buettner-Janusch, ed.) Vol 2, pp. 75–90. Academic Press, New York and London.

Buettner-Janusch, J. and Hill, R. L. (1965). *Science, N.Y.* **147**, 836–842.

Buettner-Janusch, J., Buettner-Janusch, V. and Sale, J. B. (1964). *Nature, Lond.* **201**, 510–511.

Burton, P. M. and Waley, S. G. (1968). *Exp. Eye Res.* **7**, 189–195.

Buschmann, H. (1964). *Zentbl. Vetmed.* **11**, 445–447.

Buschmann, H. (1965). Proc. Ninth Europ. Animal Blood Group Conf. pp. 129–136.

Buss, D. H. (1969). *Comp. Biochem. Physiol.* **29**, 313–318.

Charlwood, P. A. (1963). *Biochem. J.* **88**, 394–398.

Chet Ram, Khanna, N. D. and Prabhu, S. S. (1964). *Ind. J. vet. Sci.* **34**, 84–88. (From Osterhoff and Young, 1966.)

Chiancone, E., Vecchini, P., Forlani, L., Antonini, E. and Wyman, J. (1966). *Biochim. Biophys. Acta* **127**, 549–552.

Clark, W. E. L. G. (1966). *In* 'Encyclopedia Britannica', Vol 14, pp. 731–643. Wm. Benton Publisher, Chicago, London, etc.

Clarke, B. (1970). *Science, N.Y.* **168**, 1009–1011.

Cohen, C. (1962). *Ann. N.Y. Acad. Sci.* **97**, 26–36.

Coombs, R. R. A. (1966). Proc. Tenth Europ. Conf. on Animal Blood Groups and Biochem. Polymorph. pp. 43–51.
Cooper, D. and Sharman, G. B. (1964). *Nature, Lond.* **203**, 1094.
Craig, M. L. and Russell, E. S. (1963). *Science, N.Y.* **142**, 398–399.
Cuatrecasas, P. and Segal, S. (1966). *Science, N.Y.* **154**, 533–535.
Dabrowski, Z. and Skoczeń, S. (1962). *Acta biol. cracov.* **5**, 207–213.
Dalby, P. L. and Lillevik, H. A. (1969). *Publs Mich. St. Univ. Mus. Biol Ser.* **4**, 67–101.
Dan, M. and Hagiwara, A. (1967). *Jap. J. Human Genet.* **12**, 55–61.
Darlington, P. J., Jr. (1970). *Syst. Zool.* **19**, 1–18.
Datta, S. P. and Stone, W. H. (1963). *Nature, Lond.* **199**, 1208–1209.
Dautrevaux, M., Boulanger, Y., Han, K. and Biserte, G. (1969). *Eur. J. Biochem.* **11**, 267–277.
Davis, C. H., Schliselfield, L. H., Wolf, D. P., Leavitt, C. A. and Krebs, E. C. (1967). *J. biol. Chem.* **242**, 4824–4833.
Dawson, R. M. (1967). *In* 'Recent Mammals of the World' (S. Anderson and J. K. Jones, eds.), pp. 12–53. The Ronald Press Co., New York.
Dayhoff, M. O. (ed.). (1969). 'Atlas of Protein Sequence and Structure', Vol **4**, National Biomedical Research Foundation, Silver Spring, Maryland.
DeMonte, T. and Pilleri, G. (1968). *Blut. (Munich)* **17**, 25–30.
Dessauer, H. C. (1969). *In* 'Systematic Biology', pp. 325–365. Publ. 1692, National Acad. Sci., Washington, D.C.
Dessauer, H. C. and Fox, W. (1956). *Science, N.Y.* **124**, 225–226.
Deutsch, H. F. and Goodloe, M. B. (1945). *J. biol. Chem.* **161**, 1–20.
Dolyak, F. and Truffelli, G. T. (1952). *Bull. serol. Mus., New Brunsw.* **8**, 2.
Dolyak, F. and Waters, J. H. (1962). *Bull. serol. Mus., New Brunsw.* **27**, 2–5.
Doolittle, R. F., Glasgow, C. and Mross, G. A. (1969). *Biochim. Biophys. Acta* **175**, 217–219.
Engel, W., Bender, K., Kadir, S., Op't Hoff, J. and Wolf, U. (1970). *Hum. Genet.* **10**, 151–157.
Enoki, Y., Tomita, S. and Sato, M. (1966). *Jap. J. Physiol.* **16**, 702–709.
Ezekiel, E., Lai, L. Y. C. and Kaldor, I. (1963). *Comp. Biochem. Biophys.* **10**, 69–75.
Feeney, R. E. and Allison, R. G. (1969). 'Evolutionary Biochemistry of Proteins'. Wiley-Interscience, New York, London.
Filov, V. A. (1963). *Byul. Eksptl. Biol. Med.* 55 (4), 45–46.
Fitch, W. M. and Margoliash, E. (1969). *In* 'Structure, Function and Evolution in Proteins', Vol **1**, pp. 217–242. Brookhaven National Laboratory, Upton, New York.
Foreman, C. W. (1960). *Am. Midl. Nat.* **64**, 177–186.
Foreman, C. W. (1964). *J. cell. comp. Physiol.* **63**, 1–6.
Foreman, C. W. (1966). *Genetics, Princeton* **54**, 1007–1012.
Foreman, C. W. (1968). *Comp. Biochem. Physiol.* **25**, 727–731.
Forman, G. L., Baker, R. J. and Gerber, J. D. (1968). *Syst. Zool.* **17**, 417–425.
French, E. A. and Roberts, K. B. (1965). *J. Physiol., Lond.* **180**, 16–17.
Frenzl, B., Brdička, R., Křen, V. and Štark, O. (1965). Proc. Ninth Europ. Animal Blood Group Conf., pp. 197–203.
Freudenberg, E. (1966). *Experientia* **22**, 317.
Fujino, K. (1962). *Am. Nat.* **96**, 205–210.
Fujino, K. and Cushing, J. (1960). *Science, N.Y.* **131**, 1310–1311.
Gahne, B. (1966). *Genetics, Princeton* **53**, 681–694.

Gahne, B. and Rendel, J. (1961). *Nature, Lond.* **192**, 529.
Gallien, M. C.-L., Foulgoc, M.-T. C. and Fine, J. M. (1967). *C. r. hebd. Séanc. Acad. Sci., Paris* **264**, 1359–1362.
Ganzin, M., Macheboeuf, M. and Rebeyrotte, P. (1952). *Bull. Soc. Chim. biol.* **34**, 26–31.
Gasparski, J. M. (1965). Proc. Ninth Europ. Animal Blood Group Conf. pp. 93–97.
Gasparski, J., Gasparska, J., Kasperska, E., Krasinska, M., Seyfried, H. and Szynkiewicz, E. (1966). Proc. Tenth Europ. Conf. on Animal Blood Groups and Biochem. Polymorph. pp. 491–495.
Gerber, J. D. and Birney, E. C. (1968). *Syst. Zool.* **17**, 413–416.
Giblett, E. R., Hickman, C. G. and Smithies, O. (1959). *Nature, Lond.* **183**, 1589–1590.
Gillespie, J. M. (1970). *Science, N.Y.* **170**, 1100–1102.
Gleason, T. and Friedberg, F. (1953). *Physiol. Zool.* **26**, 95–100.
Goodman, M. (1962). *Hum. Biol.* **34**, 104–150.
Goodman, M. (1963). *Hum. Biol.* **35**, 377–436.
Goodman, M. (1965). *In* 'Taxonomic Biochemistry and Serology' (C. Leone, ed.), pp. 467–486. The Ronald Press Co., New York.
Gordon, W. G., Groves, M. L. and Basch, J. J. (1963). *Biochemistry, N.Y.* **2**, 817–820.
Gough, B. J. and Kilgore, S. S (1964). *J. Mammal.* **45**, 421–428.
Grassé, P. (1955). *In* 'Traité de Zoologie' (P. Grassé, ed.), Vol **17**, pp. 078 898, Masson et Cie, Paris.
Gregory, K. F. (1961). *Ann. N.Y. Acad. Sci.* **94**, 657–658.
Grunder, A. A. (1966). *Vox Sang.* **14**, 218–223.
Gulliver, O. (1840). *Phil. Mag.* **17**, 325–331.
Haflleigh, A. S. and Williams, C. A., Jr. (1966). *Science, N.Y.* **151**, 1530–1535.
Hall, F. G. (1965). *Science, N.Y.* **148**, 1350–1351.
Harrison, R. G., Connolly, R. C. and Abdalla, A. (1969). *Nature, Lond.* **224**, 325–326.
Haslewood, G. A. D. (1964). *Biol. Rev.* **39**, 537–574.
Haslewood, G. A. D. (1968). *In* 'Chemotaxonomy and Serotaxonomy' (J. G. Hawkes, ed.), pp. 159–172. Academic Press, London and New York.
Hawkes, J. G. (ed.). (1968). 'Chemotaxonomy and Serotaxonomy'. Systematics Association Special Vol 2, Academic Press, London and New York.
Haywood, B. J. (1969). 'Electrophoresis—Technical Applications. A Bibliography of Abstracts'. Humphrey Science Publishers, Inc., Ann Arbor.
Heinecke, H. and Wagner, M. (1964). *Nature, Lond.* **204**, 1099–1100.
Herner, A. E. and Riggs, A. (1963). *Nature, Lond.* **198**, 35–36.
Hess, A. R., Angel, R. W., Barron, K. D. and Bernsohn, J. (1963). *Clin. Chim. Acta* **8**, 656–667.
Hess, E. L. (1970). *Science, N.Y.* **168**, 664–668.
Hight, M. E., Prychodko, W. and Goodman, M. (1969). *J. Mammal.* **50**, 906 (Title only).
Hill, R. L. and Buettner-Janusch, J. (1964). *Fedn. Proc. Fedn. Am. Socs exp. Biol.* **23**, 1237–1242.
Hill, R. L., Buettner-Janusch, J. and Buettner-Janusch, V. (1963). *Proc. natn. Acad. Sci. U.S.A.* **50**, 885–893.
Hill, R. L., Lebovitz, H. E., Fellows, R. E. and Delaney, R. (1966). *Science, N.Y.* **154**, 420–421.
Hirsch, G. C. and Junqueira, L. C. U. (1965). *Z. Naturw.* **8**, 1–2.
Hirschman, S. Z., Morell, A. G. and Scheinberg, I. H. (1961). *Ann N.Y. Acad. Sci.* **94**, 960–969.

Ho, T.-Y. (1966). *Comp. Biochem. Physiol.* **18**, 353–358.
Ho. T.-Y. (1967). *Biochim. Biophys. Acta* **133**, 568–573.
Hoffman-Ostenhof, O. (1969). *Science, N.Y.* **166**, 1658.
Höglund, S. and Levin, Ö. (1965). *J. molec. Biol.* **12**, 866–871.
Hopkinson, D. A., Spencer, N. and Harris, H. (1963). *Nature, Lond.* **199**, 969–971.
Hori, S. H. and Kamada, T. (1967). *Jap. J. Genet.* **42**, 367–374.
Hudgins, P. C., Whorton, C. M., Tomoyoshi, T. and Riopelle, A. J. (1966). *Nature, Lond.* **212**, 693–695.
Hudson, R., Kitts, W. D. and Bandy, P. J. (1970). *J. Wildl. Dis.* **6**, 104–106.
Huehns, E. R., Dance, N., Beaven, G. H., Keil, J. V., Hecht, F. and Motulsky, A. G. (1964). *Nature, Lond.* **201**, 1095–1097.
Huisman, T. H. J., Dozy, A. M., Blunt, M. H. and Hayes, F. A. (1968). *Archs Biochem. Biophys.* **127**, 711–717.
Hyyppä, M., Korhaven, L. K. and Kornhonen, E. (1966). *Annls Med. exp. Biol. Fenn.* **44**, 63–66.
Hyvärinen, H. and Oikarinen, A. (1970). *Aquilo, Ser. Zool.* **9**, 65–67.
Imlah, P. (1965). Proc. Ninth Europ. Animal Blood Group Conf. pp. 109–122.
Ingram, V. M. (1961). 'Hemoglobin and its Abnormalities'. Charles C. Thomas, Springfield, Illinois.
Ingram, V. M. (1963). 'The Hemoglobins in Genetics and Evolution'. Columbia University Press, New York.
Jacyszyn, L. and Laursen, T. (1968). *Clinica chim. Acta* **19**, 345–352.
Jenness, R. and Sloan, R. E. (1970). *Dairy Sci. Abstr.* **32**, 599–612.
Johnson, M. L. (1968). *Syst. Zool.* **17**, 23–30.
Johnson, M. L. and Wicks, M. J. (1959). *Syst. Zool.* **8**, 88–95.
Johnson, M. L. and Wicks, M. J. (1964). *In* 'Taxonomic Biochemistry and Serology' (C. A. Leone, ed.), pp. 681–694. The Ronald Press Co., New York.
Johnson, M. L., Wicks, M. J. and Brenneman, J. (1959). *Murrelet (Seattle)* **39**, 32–36.
Jones, R. T., Brinhall, B., Huisman, T. H. J., Kleihauer, E. and Betke, K. (1966). *Science, N.Y.* **154**, 1024–1027.
Jordan, S. M. and Morgan, E. H. (1969). *Comp. Biochem. Physiol.* **29**, 383–391.
Kaminski, M., Podliachouk, L., Nikołajczuk, M. and Balbierz, H. (1966). Proc. Tenth Europ. Conf. on Animal Blood Groups and Biochem. Polymorph. pp. 315–318.
Kerry, K. R. (1969). *Comp. Biochem. Physiol.* **29**, 1015–1022.
Khanolkar, V. R., Naik, S. N., Baxi, A. J. and Bhatia, H. M. (1963). *Experientia* **19**, 472–474.
Khorana, H. G. (1968). *J. Am. med. Ass.* **206**, 1978–1982.
Kilmartin, J. V. and Clegg, J. B. (1967). *Nature, Lond.* **213**, 269–271.
Kimura, F. T. and Johnson, M. L. (1970). *Comp. Biochem. Physiol.* **37**, 277–280.
Kimura, M. and Ohta, T. (1971). *Nature, Lond.* **229**, 467–469.
King, J. L. and Jukes, T. H. (1969). *Science, N.Y.* **164**, 788–798.
Kirk, R. L. (1968). 'The Haptoglobin Groups in Man'. S. Karger, Basel and New York.
Kirsch, J. A. W. (1968a). *Nature, Lond.* **217**, 418–420.
Kirsch, J. A. W. (1968b). *Aust. J. Sci.* **31**, 43–45.
Kirsch, J. A. W. (1969). *Syst. Zool.* **18**, 296–311.
Kirsch, J. A. W. and Poole, W. E. (1967). *Nature, Lond.* **215**, 1097–1098.
Kitchen, H. and Easley, C. (1969). *J. biol. Chem.* **244**, 6533–6542.
Kitchen, H., Putnam, F. W. and Taylor, W. J. (1964). *Science, N.Y.* **144**, 1234–1239.

Kitchen, H., Putnam, F. W. and Taylor, W. J. (1967). *Blood* **29**, 867–877.

Kitchen, H., Easley, C. W., Putnam, F. W. and Taylor, W. J. (1968). *J. biol. Chem.* **243**, 1204–1211.

Kleihauer, E., Buss, I. O., Luck, C. P. and Wright, P. G. (1965). *Nature, Lond.* **207**, 424–425.

Köhler, H., Shimizu, A., Paul, C. and Putnam, F. W. (1970a). *Science, N.Y.* **169**, 56–59.

Köhler, H., Shimizu, A., Paul, C., Moore, V. and Putnam, F. W. (1970b). *Nature, Lond.* **227**, 1318–1320.

Konagaya, S. (1963). *J. Biochem.* **54**, 189–190.

Kramer, J. W. and Sleight, S. D. (1968). *Am. J. Vet. Clin. Pathol.* **2**, 87–91.

Kraus, R. (1897). *Wien. klin. Wschr.* **32**, 736.

Kurland, C. G. (1970). *Science, N.Y.* **169**, 1171–1177.

Lai, L. Y. C. (1966). *Nature, Lond.* **210**, 643.

Landsteiner, K. (1900). *Zentbl. Bakt. Parasitkde.* **27**, 357–362.

Lange, V. and Schmitt, J. (1963). *Folia primatol.* **1**, 208–250.

Larsen, B. and Tönder, O. (1967). *Acta physiol. scand.* **69**, 262–269.

Latner, A. L. and Zaki, A. H. (1957). *Nature, Lond.* **180**, 1366–1367.

Leakey, L. S. B. (1970). *Proc. natn. Acad. Sci. U.S.A.* **67**, 746–748.

Lees, R. S. (1970). *Science, N.Y.* **169**, 493–495.

Lehmann, H. and Huntsman, R. D. (1966). 'Man's Hemoglobins'. J. B. Lippincott Co., Philadelphia and Montreal.

Leone, C. A. (ed.). (1964). 'Taxonomic Biochemistry and Serology'. The Ronald Press Co., New York.

Leone, C. A. and Wiens, A. L. (1956). *J. Mammal.* **37**, 11–23.

Lessard, J. L. and Taketa, F. (1969). *Biochim. Biophys. Acta* **175**, 441–449.

Levine, H. P. and Moody, P. A. (1939). *Physiol. Zool.* **12**, 400–411.

Lewis, J. H., Bayer, W. L. and Szeto, I. L. F. (1969). *Comp. Biochem. Physiol.* **31**, 667–670.

Lipecka, C. (1970). Proc. Eleventh Europ. Conf. on Animal Blood Groups and Biochem. Polymorph. pp. 489–492.

Lonn, L. and Motulsky, A. G. (1957). *Clin. Res. Proc.* **5**, 157.

Lush, I. E. (1967). 'The Biochemical Genetics of Vertebrates Except Man'. North Holland Publishing Co., Amsterdam.

McCullagh, K. G., Lincoln, H. G. and Southgate, D. A. T. (1969). *Nature, Lond.* **222**, 493–494.

McLaughlin, P. J. (1969). *In* 'Atlas of Protein Sequence and Structure' (M. O. Dayhoff, ed.), Vol 4, pp. D-7 to D-20; D-179 to D-183. National Biomedical Research Foundation, Silver Spring, Maryland.

McLaughlin, P. J. and Dayhoff, M. O. (1969). *In* 'Atlas of Protein Sequence and Structure' (M. O. Dayhoff, ed.), Vol 4, pp. 39–46. National Biomedical Research Foundation, Silver Spring, Maryland.

Maggioni, G., Bottini, E., Ioppolo, C., Signoretti, A. and Benedettelli, L. (1963). *Acta haemat.* **29**, 321–335.

Malecha, S. R. and Tamarin, R. H. (1969). *Genetics, Princeton* **61**, S38–S39.

Manski, W., Halbert, S. P. and Auerbach, T. P. (1964). *In* 'Taxonomic Biochemistry and Serology' (C. Leone, ed.), pp. 545–562. The Ronald Press Co., New York.

Manwell, C. and Baker, C. M. A. (1970). 'Molecular Biology and the Origin of the Species'. Sidgwick and Jackson, London, and University of Washington Press, Seattle.

Manwell, C. and Kerst, K. V. (1966). *Comp. Biochem. Physiol.* **17**, 741–754.

Marable, I. W. and Glenn, W. G. (1964). *In* 'Taxonomic Biochemistry and Serology' (C. Leone, ed.), pp. 527–534. The Ronald Press Co., New York.
Marchuwska-Koj, A. (1966). *Folia Biol. Kraków* **14**, 177–181.
Margoliash, E., Fitch, W. M. and Dickerson, R. E. (1969). *In* 'Structure Function and Evolution in Proteins'. Vol 2, pp. 259–305. Brookhaven National Laboratory, Upton, New York.
Marsh, R. E., Howard, W. E. and Bongardt, H. (1969). *J. Mammal.* **50**, 649–652.
Marvin, H. V. and Shook, B. R. (1963). *Comp. Biochem. Physiol.* **8**, 187–189.
Mäsiar, P. and Babják, M. (1968). Biológia Bratisl. **23**, 917–920.
Mäsiar, P. and Sova, O. (1969). *Life Sci.* **8**, 919–924.
Maughan, E. and Williams, J. R. B. (1967). *Nature, Lond.* **215**, 404–405.
Mayr, E. (1963). 'Animal Species and Evolution'. Harvard University Press, Cambridge, Massachusetts.
Mayr, E. (1969). 'Principles of Systematic Zoology'. McGraw-Hill Book Co., New York and London.
Mayr, E., Linsley, E. G. and Usinger, R. L. (1953). 'Methods and Principles of Systematic Zoology'. McGraw-Hill Book Co., New York and London.
Medway, W. and Gerachi, J. R. (1965). *Am. J. Physiol.* **209**, 169–172.
Melartin, L., Blumberg, B. and Lisker, R. (1967). *Nature, Lond.* **215**, 1288–1289.
Miller, W. J., Haugen, A. O. and Roslien, D. J. (1965). *J. Wildl. Mgmt.* **29**, 717–723.
Mitchell, G. C. (1970). *Comp. Biochem. Physiol.* **35**, 667–677.
Mitchell, H. A. (1966). *Nature, Lond.* **210**, 1067–1068.
Mohaghegpour, N. and Leone, C. A. (1969). *Comp. Biochem. Physiol.* **31**, 437–452.
Møller, D., Naevdal, G. and Valen, A. (1966). *Fisken Hav.* nr. 2, pp. 1–18.
Moody, P. A. (1941). *J. Mammal.* **22**, 40–47.
Moody, P. A. (1958). *J. Mammal.* **39**, 554–559.
Moody, P. A. and Doniger, D. E. (1956). *Evolution, Lancaster, Pa.* **10**, 47–55.
Moody, P. A., Cochran, V. A. and Drugg, H. (1949). *Evolution, Lancaster, Pa.* **3**, 25–33.
Moore, D. H. (1945). *J. biol. Chem.* **161**, 21–23.
Moor-Jankowski, J. and Wiener, A. S. (1968). *Primates in Medicine* **1**, 49–67.
Moor-Jankowski, J. and Wiener, A. S. (1969). *Primates in Medicine* **3**, 64–77.
Moor-Jankowski, J., Huser, H. J., Wiener, A. S., Kalter, S. S., Pallotta, A. J. and Guthrie, C. B. (1965). *In* 'The Baboon in Medical Research' (H. Vagtborg, ed.), pp. 363–405. University of Texas Press, Austin.
Morgan, E. H. (1964). *J. Physiol.* **171**, 26–41.
Mross, G. A., Doolittle, R. F. and Roberts, B. F. (1970). *Science, N.Y.* **170**, 468–470.
Müller, E. W. (1970). *Science, N.Y.* **169**, 1001.
Nadler, C. F. (1968). *Comp. Biochem. Physiol.* **27**, 487–503.
Nadler, C. F. and Hughes, C. E. (1966). *Comp. Biochem. Physiol.* **18**, 639–651.
Nadler, C. F. and Youngman, P. M. (1969). *Can. J. Zool.* **47**, 1051–1057.
Nadler, C. F. and Youngman, P. M. (1970). *Comp. Biochem. Physiol.* **36**, 81–86.
Nadler, C. F., Hughes, C. E., Harris, K. E. and Nadler, N. W. (1967). *Comp. Biochem. Physiol.* **23**, 149–157.
Naevdal, G. (1966a). *Acta Univ. Bergen, Mat.-Naturv. Ser.* 1965, 1–20.
Naevdal, G. (1966b). *Fisk Dir. Skr. Ser. Hav. Unders.* **14**, 37–50.
Naik, S. N. and Anderson, D. E. (1970). *Biochem. Genet.* **4**, 651–654.
Naik, S. N., Bhatia, H. M., Baxi, A. J. and Naik, P. V. (1964). *J. exp. Zool.* **155**, 231–235.
Napier, J. H. and Napier, P. H. (1968). 'A Handbook of Living Primates' reprinted with corrections 1968, Academic Press, London and New York.

Narayan, K. A. (1967). *Lipids* **2**, 282–284.
Nauman, L. W. (1970). 'The Primary Structure of the Myoglobins of Six Species of the Order Pinnipedia'. Diss., Univ. Alaska.
Navarro, J., Areso, M. A. and Planas, J. (1964). *Revta esp. Fisiol.* **20**, 159–164.
Neilsen, J. T. (1969). *Hereditas* **61**, 400–412.
Nirenberg, M. (1968). *J. Am. med. Assn.* **206**, 1973–1977.
Nold, J. G., Kang, A. H. and Gross, J. (1970). *Science, N.Y.* **170**, 1096–1098.
Nute, P. E. and Buettner-Janusch, J. (1969). *Folia Primatol.* **10**, 181–194.
Nuttall, G. H. F. (1901). *J. Hyg., Camb.* **1**, 357–387.
Ohno, S. (1967). 'Sex Chromosomes and Sex-Linked Genes'. Springer-Verlag. Berlin, Heidelberg, New York.
Ohno, S. (1970). 'Evolution by Gene Duplication'. Springer-Verlag, New York, Heidelberg, Berlin.
Ohno, S., Poole, J. and Gustavsson, I. (1965). *Science, N.Y.* **150**, 1737–1738.
Osterhoff, D. R. (1966). Proc. Tenth Europ. Conf. on Blood Groups and Biochem. Polymorph. pp. 345–351.
Osterhoff, D. R. and Young, E. (1966). Proc. Tenth Europ. Conf. Blood Groups and Biochem. Polymorph. pp. 133–135.
Oudin, J. (1966). *J. Cell. Physiol.* **67** Sup. 1, 77–108.
Ovary, Z. (1966). *Ann. N.Y. Acad. Sci.* **121**, 776–786.
Parsons, R. S., Heddle, R. W. L., Flux, W. G. and Guiler, E. R. (1970). *Comp. Biochem. Physiol.* **32**, 345–351.
Pauling, L., Itano, H. A., Singer, S. J. and Wells, I. C. (1949). *Science, N.Y.* **110**, 543–548.
Pauly, L. K. (1962). *Bull. serol. Mus., New Brunsw.* **28**, 5–8.
Penhoet, E., Rajkumar, T. and Rutter, W. J. (1966). *Proc. natn. Acad. Sci. U.S.A.* **56**, 1275–1282.
Perutz, M. F. (1964). *Scient. Am.* **211**, 64–76.
Peterson, M. K. (1968). *Am. Midl. Nat.* **79**, 130–148.
Picard, J. J., Vandebroek, G., Heremans, J. F. and Defossé, G. (1966). *Biochim. Biophys. Acta* **117**, 111–114.
Podliachouk, L. (1965). Proc. Ninth Europ. Animal Blood Group Conf. pp. 229–235.
Podliachouk, L. (1970). Proc. Eleventh Europ. Conf. on Animal Blood Groups and Biochem. Polymorph. pp. 443–446.
Polonovski, M. and Jayle, M. F. (1938). *C. r. Séanc. Soc. Biol.* **129**, 457–460.
Porter, D. D. and Dixon, F. J. (1966). *Am. J. vet. Res.* **27**, 335–338.
Proceedings of the Ninth European Blood Group Conference (1965). W. Junk, Publishers, The Hague.
Proceedings of the Tenth Europ. Conf. on Animal Blood Groups and Biochem. Polymorph. (1966). Institut National de la Recherche Agronomique, Paris, France.
Proceedings of the Eleventh Europ. Conf. on Animal Blood Groups and Biochem. Polymorph. (1970). W. Junk, Publishers, The Hague.
Putnam, F. W. (1969). *Science, N.Y.* **163**, 633–644.
Radden, R. (1968). 'The Dominican Republic Hutia (*Plagiodontia aedium hylaeum*) in Captivity'. Unpubl. thesis, University of Puget Sound.
Ramakrishnan, P. and Barnabas, J. (1962). *Acta physiol. pharmac. neerl.* **11**, 328–342.
Ranney, H. M. and Gluecksohn-Waelsch, S. (1955). *Ann. hum. Genet.* **19**, 269–272.
Rapacz, J., Shackelford, R. M. and Jakóbiec, J. (1965). Proc. Ninth Europ. Animal Blood Group Conf. pp. 211–215.

Rasmusen, B. A. and Hall, J. G. (1966). Proc. Tenth Europ. Conf. on Animal Blood Groups and Biochem. Polymorph. pp. 451–457.
Rasmussen, D. I. (1961). *Genet. Res.* **2**, 449–455.
Rasmussen, D. I. (1964). *Evolution, Lancaster, Pa.* **18**, 219–229.
Rasmussen, D. I. (1968). *In* 'Biology of *Peromyscus* (Rodentia)' (J. A. King, ed.), pp. 340–372. (*Amer. Soc. Mammal. Spec. Publ.* No. 2.)
Rasmussen, D. I. and Koehn, R. K. (1966). *Genetics, Princeton* **54**, 1353–1357.
Rasmussen, D. I., Jensen, J. N. and Koehn, R. K. (1968). *Biochem. Genet.* **2**, 87–92.
Ratnoff, O. D. (1966). *Prog. Hematol.* **5**, 204–245.
Read, D. W. and Lestrel, P. E. (1970). *Science, N.Y.* **168**, 578–580.
Reichert, E. T. and Brown, A. P. (1909). 'Crystallography of the Hemoglobins'. *Publs Carnegie Instn.* No. 116.
Rice, D. W. (1967). *In* 'Recent Mammals of the World' (S. Anderson and J. K. Jones, eds.), pp. 291–324.
Richardson, B. J. and Czuppon, A. B. (1970). *Aust. J. biol. Sci.* **23**, 617–622.
Riou, G., Paoletti, G. and Truhaut, R. (1962). *Bull. Soc. Chim. biol.* **44**, 149–170.
Rivello, R. C., Cortner, A. and Schnatz, J. D. (1969). *Proc. Soc. exp. Biol. Med.* **130**, 232–235.
Robinson, A. J., Kropatkin, M. and Aggeler, P. M. (1969). *Science, N.Y.* **166**, 1420–1422.
Roig, V. (1964). *Cienc. Invest.* **20**, 270–275.
Roig, V. G. and Reig, O. A. (1969). *Comp. Biochem. Physiol.* **30**, 665–672.
Romer, A. S. (1966). 'Vertebrate Paleontology,' 3rd Ed., Univ. Chicago Press, Chicago and London.
Rosler, B. (1965). *Acta biol. med. germ.* **15**, 537–547.
Russell, M. A. and Semeonoff, R. (1967). *Genet. Res.* **10**, 135–142.
Sandberg, K. (1970). Proc. Eleventh Europ. Conf. on Animal Blood Groups and Biochem. Polymorph. pp. 447–452.
Sarich, V. (1969a). *Syst. Zool.* **18**, 286–295.
Sarich, V. (1969b). *Syst. Zool.* **18**, 416–422.
Sarich, V. (1969c). *Am. J. phys. Anthrop.* **31**, 266.
Sarich, V. M. and Wilson, A. C. (1966). *Science, N.Y.* **154**, 1563–1566.
Sarich, V. M. and Wilson, A. C. (1967a). *Proc. Nat. Acad. Sci. U.S.A.* **58**, 142–148.
Sarich, V. M. and Wilson, A. C. (1967b). *Science, N.Y.* **158**, 1200–1203.
Sarich, V. M. and Wilson, A. C. (1969). *Proc. natn. Acad. Sci. U.S.A.* **63**, 1088–1093.
Sartore, G., Stormont, C., Morris, B. G. and Grunder, A. A. (1969). *Genetics, Princeton* **61**, 823–831.
Schoeder, W. A. (1968). 'The Primary Structure of Proteins'. Harper and Row, New York, Evanston, London.
Scott, R. B. (1966). *Blut (Munich)* **12**, 340–351.
Seal, U. S. (1969). *Comp. Biochem. Physiol.* **31**, 799–811.
Seal, U. S. and Erickson, A. W. (1969). *Comp. Biochem. Physiol.* **30**, 695–713.
Seal, U. S., Swaim, W. R. and Erickson, A. W. (1967). *Comp. Biochem. Physiol.* **22**, 451–460.
Seal, U. S., Phillips, N. I. and Erickson, A. W. (1970). *Comp. Biochem. Physiol.* **32**, 33–48.
Searcy, R. L., Hayashi, S., Berk, J. E. and Stern, H. (1966). *Proc. Soc. exp. Biol. Med.* **122**, 1291–1295.
Selander, R. K. and Yang, S. Y. (1969). *Genetics, Princeton* **61**, S54.

Selander, R. K., Hunt, W. G. and Yang, S. Y. (1969). *Evolution, Lancaster, Pa.* **23**, 379–390.

Selander, R. K., Smith, M. H., Yang, S. Y., Johnson, W. and Gentry, J. B. (1971). 'Studies in Genetics' Vol VI, pp. 49–90. University of Texas Press, Austin.

Shaughnessy, P. D. (1969). *Aust. J. biol. Sci.* **22**, 1581–1584.

Shaughnessy, P. D. (1970). *Aust. J. Zool.* **18**, 331–343.

Shaw, C. R. and Barto, E. (1965). *Science, N.Y.* **148**, 1099–1100.

Shechter, Y., Landeau, J. W. and Newcomer, V. D. (1969). *J. invest. Derm.* **52**, 57–61.

Sherry, S. (1970). *Hosp. Pract.* June, 75–85.

Shifrine, M. and Stormant, C. O. (1970). *Anim. Blood Groups Biochem. Genet.* **1**, 169–170.

Shubin, P. N. (1969). *Genetika* **5**, 37–41.

Sibley, C. G. (1960). *Ibis* **102**, 215–284.

Simpson, G. G. (1945). *Bull. Am. Mus. nat. Hist.* **85**, 1–350.

Simpson, G. G. (1961). 'Principles of Animal Taxonomy'. Columbia University Press, New York.

Smith, J., Barnes, J. K., Kaneko, J. J. and Freedland, R. A. (1965). *Nature, Lond.* **205**, 298–299.

Smith, J. and Holdridge, B. A. (1967). *Comp. Biochem. Physiol.* **22**, 737–743.

Smith, M. (1960). *Science, N.Y.* **131**, 699–702.

Smith, M. J. and Beck, W. S. (1967). *Biochim. biophys. Acta* **147**, 324–333.

Smithies, O. (1955a). *Nature, Lond.* **175**, 307–308.

Smithies, O. (1955b). *Biochem. J.* **61**, 629–641.

Sober, H. A. (ed.). (1968). 'Handbook of Biochemistry Selected Data for Molecular Biology'. The Chemical Rubber Co., Cleveland, Ohio.

Spiteri, M. and Eyquem, A. (1965). Proc. Ninth Europ. Animal Blood Group Conf. pp. 205–209.

Steiner, R. F. (1965). 'The Chemical Foundations of Molecular Biology'. D. Van Nostrand Co., Inc. New York, London, Toronto.

Stormant, C. and Suzuki, Y. (1970). *Science, N.Y.* **167**, 200–202.

Studier, E. H. (1967). *J. Mammal.* **48**, 477–478.

Svehag, S-E. and Bloth, B. (1970). *Science, N.Y.* **168**, 847–849.

Swisher, S. N., Young, L. E. and Trabold, N. (1962). *Ann. N.Y. Acad. Sci.* **97**, 15–25.

Szynkiewicz, E. (1970). Proc. Eleventh Europ. Conf. on Animal Blood Groups and Biochem. Polymorph. pp. 567–570.

Taketa, F. and Morell, S. A. (1966). *Biochem. biophys. Res. Commun.* **24**, 705–713.

Taketa, F., Smits, M. R., DiBona, F. J., Lessard, J. L. (1967). *Biochemistry, N.Y.* **6**, 3809–3816.

Tamsitt, J. R. and Valdivieso, D. (1969). *Occ. Pap. R. Ont. Mus. Zool.* **14**, 1–12.

Thompson, E. O., Hosken, R. and Air, G. M. (1969). *Aust. J. biol. Sci.* **22**, 449–462.

Thompson, R. B., Hewett, H. B., Kilgore, S. S., Shepherd, A. P. and Bell, W. N. (1966). *Nature, Lond.* **210**, 1063–1064.

Thuline, H. C., Morrow, A. C., Norby, D. E. and Motulsky, H. G. (1967). *Science, N.Y.* **157**, 431–432.

Tiselius, A. and Flodin, P. (1953). *Adv. Protein Chem.* **8**, 461–486.

Trujillo, J. M., Walden, B., O'Neill, P. and Anstall, H. (1965). *Science, N.Y.* **148**, 1603–1604.

Trujillo, J. M., Walden, B., O'Neill, P. and Anstall, H. B. (1967). *Nature, Lond.* **213**, 88–90.

Tyuma, I., Enoki, Y. and Morikawa, S. (1964). *Jap. J. Physiol.* **14**, 573–586.

Vagtborg, H. (ed.). (1965). 'The Baboon in Medical Research'. University of Texas Press, Austin.

Valdivieso, D., Conde, E. and Tamsitt, J. R. (1968). *Comp. Biochem. Physiol.* **27**, 133–138.

Valdivieso, D., Tamsitt, J. R. and Pino, E. C. (1969). *Comp. Biochem. Physiol.* **30**, 117–122.

VanTets, P. and Cowan, I. McT. (1966). *Can. J. Zool.* **44**, 631–647.

Varga, M., Tolarova, M. and Tolar, M. (1965). Proc. Ninth Europ. Animal Blood Group Conf. pp. 193–196.

Vennesland, B. (1966). *In* 'Encyclopaedia Brittanica', Vol 8, pp. 620–622. Wm. Benton, Chicago, London, etc.

Vokáč, Z. (1961). *Nature, Lond.* **190**, 641–642.

Walker, E. P., Warnick, F., Lange, K. I., Vible, H. E., Hamlet, S. E., Davis, M. A. and Wright, P. F. (1964). 'Mammals of the World', 2 Vols. John Hopkins Press, Baltimore.

Wang, A-C. and Sutton, H. E. (1965): *Science, N.Y.* **149**, 435–437.

Washburn, S. L. (1963). 'Classification and Human Evolution', Aldine Publ. Co., Chicago.

Waters, J. H. (1963). *Syst. Zool.* **12**, 122–133.

Weinbren, B. (1960). *J. comp. Path. Ther.* **70**, 217–221.

Weitz, B. (1953). *Nature, Lond.* **171**, 261.

Weymyss, C. T. (1953). *Zoologica, N.Y.* **38**, 173–181.

Wiatroszak, I. (1970). Proc. Eleventh Europ. Conf. on Animal Blood Groups and Biochem. Polymorph. pp. 265–270.

Wiener, A. S. and Moor-Jankowski (1963). *Science, N.Y.* **142**, 67–69.

Wikler, M., Köhler, H., Shinoda, T. and Putnam, F. W. (1969). *Science, N.Y.* **163**, 75–78.

Wild, A. E. (1965). *Immunology* **9**, 457–466.

Williams, C. A., Jr. (1964). *In* 'Evolutionary and Genetic Biology of Primates' (J. Buettner-Janusch, ed.), Vol 2, pp. 25–74. Academic Press, New York and London.

Wilson, A. C. and Sarich, V. M. (1969). *Proc. natn. Acad. Sci. U.S.A.* **63**, 1088–1093.

Wolfe, H. R. (1939). *Zoologica, N.Y.* **24**, 309–321.

Yoshida, A., Baur, E. W. and Motulsky, A. G. (1970). *Blood* **35**, 506–513.

Zuckerkandl, E. (1965). *Scient. Am.* **212**, 110–118.

Zweig, G. and Crenshaw, J. W. (1957). *Science, N.Y.* **126**, 1065–1067.

2 Birds

C. G. SIBLEY, K. W. CORBIN,
J. E. AHLQUIST and A. FERGUSON

*Peabody Museum of Natural History,
Yale University, New Haven, Connecticut,
U.S.A.*

I. Introduction

The history of the application of biochemical and immunological techniques to problems of avian systematics parallels that for other groups of organisms. Within two years after the discovery of the precipitin reaction by Kraus (1897) the technique was applied to comparative problems by Bordet (1899), Myers (1900) and Uhlenhuth (1900). The latter prepared antisera against the egg-white proteins of the domestic fowl and the domestic pigeon. These pioneer efforts were soon followed by the extensive studies of Nuttall (1901, 1904). During the next 25 years a number of comparative serological studies involving birds were produced. In 1930 Erhardt reviewed progress to that point and concluded that the results were contradictory and lacking in convincing evidence. This pessimistic conclusion may have discouraged serological work on birds but at about this time Boyden (1926) was generating enthusiasm for systematic serology in the United States. However, the techniques were still primitive, few clear-cut results were obtained and the skeptics greatly outnumbered the disciples. The interest of ornithologists was directed primarily toward problems at the species level and not a single avian taxonomist attempted to use serological techniques during this period. Others, however, were using birds as research subjects, among them the geneticist M. R. Irwin of the University of Wisconsin. Although not primarily directed toward systematics the extensive series of papers by Irwin and his colleagues on the immunogenetics of the Columbidae have contributed to our understanding of the relationships among several genera and species of pigeons and doves.

The studies by Sievers (1939c) of egg-white proteins contain at least one significant taxonomic result and are still the most extensive serological comparisons of avian egg proteins.

The series of papers by Mainardi and his colleagues and those by Stallcup (1954, 1961) and Cotter (1957) are among the few serological studies of avian material which addressed themselves to actual taxonomic problems.

Until 1937 the immune reaction techniques were the principal methods utilized for comparative studies of proteins. The development of electrophoretic equipment (Tiselius, 1937) provided another technique and avian egg-white proteins were among the first protein systems to be examined (e.g. Landsteiner *et al.*, 1938). The Tiselius apparatus was used in several comparative studies (e.g. McCabe and Deutsch, 1952) but it was too slow for the comparison of large numbers of specimens. Although paper electrophoresis was first reported in 1937 it was not widely used until a decade later. Its greater speed permitted Sibley (1960) to examine more than 5,000 patterns produced by the egg-white proteins of nearly 700 species of birds and to report on those of 359 species of non-passerines. The equal speed and better resolution of starch gel electrophoresis have been used in several studies of avian egg-white proteins (Sibley *et al.*, 1968, 1969; Sibley, 1970; Sibley and Ahlquist, 1972) and plasma proteins (Sibley and Hendrickson, 1970).

Thus serology and electrophoresis have been the principal, although not the only, techniques employed to date in studies of avian material. Most of the studies have involved a few readily available species and have not attacked actual taxonomic problems.

In the following sections papers are considered in approximate chronological order under the techniques employed. The first period, from 1900 to 1937, was dominated by the immunological methods. In fact, except for the study by Reichert and Brown (1909) on the crystallization of hemoglobins only serological techniques were used. Electrophoresis claimed an increasing share of attention after 1937. These two techniques were merged in 1953 when Grabar and Williams developed the useful hybrid, immunoelectrophoresis. Several other techniques, including chromatography, the rate of hemolysis of erythrocytes, biochemical properties, and comparisons of peptides have been utilized but only to a limited extent. The amino acid sequences of a few avian proteins have been determined.

Nearly all of the serological studies and most of those using electrophoresis have been concerned with the relationships of the higher categories. In 1955 Smithies developed the starch gel electrophoretic technique, the resolving power of which is much better than that of the free solution (Tiselius) or filter paper systems. The improved resolution and side by side comparisons of protein patterns in starch gels made it possible to detect individual variation in homologous proteins which was expressed as differences in mobility. Subsequent studies by Smithies (1959a, 1959b) provided the basis for a new approach to the study of intraspecific variation or polymorphism in proteins.

Lush (1961, 1964a,b) was the first to examine protein polymorphism in birds but this field is expanding rapidly. With this development the biochemical techniques are now being applied to problems at all taxonomic levels.

II. Studies Using Immunological Techniques

A. SEROLOGICAL STUDIES

The discovery of the precipitin reaction (Kraus, 1897) provided a new technique which, it was soon realized, could be used to investigate systematic problems. At first it was believed that the immune reaction was absolutely specific, that is, that an antiserum would react only with the antigen that stimulated its production. It was soon discovered (Bordet, 1899) that the reaction was only relatively specific and that, in general, the degree of cross reactivity was proportional to the degree of systematic relationship between the organisms involved.

The optimism that accompanies the development of every new technique prevailed and various confident statements were recorded. Disillusionment was to set in later but for several years the apparent promise of the new method provided the inspiration that sustained Nuttall, Friedenthal, Uhlenhuth and others as they labored to collect their specimens and produce their data. In the process they refined the techniques and discovered their weaknesses and, although most or perhaps all of the early data must be viewed with caution, the real advances could have been made in no other way.

Bordet (1899) was the first to prepare an avian antiserum by injecting defibrinated chicken (*G. gallus*) blood into a rabbit. This antiserum reacted strongly with chicken blood and less strongly with pigeon blood. Bordet was thus the first to show that the precipitin reaction was not absolutely specific and from this discovery came the application of the immune reaction to problems of systematics. Myers (1900) prepared an antiserum against chicken egg-white proteins which reacted strongly with chicken egg-white and slightly with duck egg-white but gave no reaction against mammalian blood.

Uhlenhuth's (1900) study of antisera produced in rabbits against chicken and pigeon egg-white proteins showed that the anti-chicken antiserum would react strongly with chicken egg-white and would produce a weaker reaction with that of pigeon. Similarly, an anti-pigeon egg-white antiserum reacted with both pigeon and chicken egg-whites. Neither of these antisera reacted with proteins other than those of avian egg-white.

Uhlenhuth (1901) later prepared an anti-chicken egg-white antiserum and used it in tests against several species of birds. This paper was reviewed by Nuttall (1904: 204) who noted that Uhlenhuth

> 'had obtained a powerful anti-egg serum which acted on fowl blood dilutions, and produced a slight cloud in goose blood. He obtained an immediate precipitation upon adding the antiserum to 2·5% fowl blood dilution. It produced almost as powerful reactions with egg-white dilutions of goose, duck, guinea fowl, though less with the pigeon's than with its homologous (fowl) egg-white. A rabbit treated with goose egg yielded a serum which produced a precipitum in goose blood, less clouding in fowl blood, dilutions. It gave a great and immediate precipitation with goose and duck, a considerable clouding with fowl, guinea fowl, and pigeon eggs.'

Although we now know that Uhlenhuth was wrong when he concluded that it would not be possible to differentiate the egg-white proteins of different groups of birds serologically, he did demonstrate that egg-white and blood proteins would cross react.

Gengou (1902) prepared an anti-chicken (*Gallus*) egg-white antiserum and found that it would react with the egg-whites of the turkey (*Meleagris*), duck (*Anas*) and pigeon (*Columba*) as well as with that of *Gallus*. Like Uhlenhuth he concluded that it was not possible to distinguish among avian egg-whites by the precipitin test.

Uhlenhuth (1905) also prepared anti-pigeon antisera by injecting pigeon blood into chickens and anti-chicken antisera by injecting chicken blood into pigeons. In each case the antisera reacted only with the blood of the species used as the source of the antigens. Uhlenhuth was able to distinguish between chicken and pigeon blood samples. These antisera also reacted with duck blood proteins.

The dominant figure in the early history of comparative serology was George H. F. Nuttall. Following some preliminary papers (e.g. 1901, 1902) he rapidly amassed a collection of blood samples and proceeded to carry out a large number of precipitin tests.

In 1902 Nuttall published two papers which included the results of tests with avian material. In the first of these (1902a) he reported the results of tests with an anti-chicken antiserum against 250 blood samples. He summarized the tests with bird bloods as follows.

> 'Anti-fowl serum was found to produce a reaction not only with solutions of fowl blood, and that of the closely related pheasant, turkey, etc. but also with the bloods of widely divergent species, such as the parrot, various species of duck, the woodcock, sheathbill, heron, eagle, owl, condor, pigeon, a number of small Passerines, and

American rhea. A marked clouding was moreover produced in the blood of the swallow, rook, landrail, stork, swan and African ostrich. What I have termed a "marked clouding" is probably to be regarded as an indication of a more remote relationship.'

A few months later Nuttall (1902b) presented a 'progress report' on tests of more than 500 blood samples and expressed his belief that 'with care we shall perhaps be able to "measure species" by this method.' In this paper he reported on tests with an anti-ostrich antiserum.

All of Nuttall's results were brought together in his major work 'Blood Immunity and Blood Relationship' (1904) which includes extensive discussions of techniques and reviews of previous work as well as the results of some 16,000 precipitin tests upon 900 specimens of blood from more than 500 species of animals, including 219 species of birds. He also used the egg-white of four species of birds. Nuttall prepared about 30 immune sera in rabbits and used them in his tests. The antigens for four of the antisera were from avian sources, namely, chicken blood serum, ostrich blood serum, chicken egg-white and emu egg-white. A synopsis of Nuttall's results follows. Scientific names of taxa and their sequence have been brought into conformity with modern usage. Wetmore (1960) has been used as the principal reference for this purpose.

*1. Anti-Chicken (*Gallus gallus*) Blood Serum*

Seven hundred and ninety-two precipitin tests were performed. The only strong reactions were obtained with the 320 samples from 203 species of birds. Eighty-seven per cent of these 320 samples reacted to some degree with the anti-chicken antiserum. The strongest reactions were against other galliform birds closely followed by anseriform species. Other groups which gave at least one very strong reaction to the anti-chicken antiserum were the Struthioniformes, Ciconiiformes, Falconiformes, Charadriiformes, Columbiformes, Psittaciformes and Strigiformes. Strong reactions were recorded for the following groups: Podicipediformes, Pelecaniformes, Gruiformes, Musophagidae, Picidae and Passeriformes. Of the major groups available to Nuttall only the Procellariiformes (*Fulmarus*), Cuculidae, Alcedinidae, Meropidae and Bucerotidae failed to give at least one strong reaction with the anti-chicken antiserum.

These results would seem to indicate that all birds are serologically similar but even a superficial examination of Nuttall's data reveals that no conclusion can really be trusted. For example, Nuttall (1904: 282–285) recorded the results of his tests on 18 blood samples of pigeons and doves (Columbidae) representing eight species. One sample of *Columba eversmanni* gave a very strong reaction with the anti-chicken antiserum but two samples of *Columba livia* gave no reaction at all and other species of *Columba* gave intermediate degrees of reaction. Even within the Galliformes and Anseriformes the range of reactions was from weak to very strong. In the absence of consistency within closely related groups it is not possible to accept the data as valid even when they are concordant with other evidence.

2. Anti-Ostrich (Struthio camelus) *Blood Serum*

Nuttall performed 649 tests with an antiserum against ostrich blood serum prepared in a rabbit. Of these tests, 276 were with avian blood samples representing 187 species. Very strong reactions were obtained with one sample of *Anas 'boscas'* (=*platyrhynchos*) (Anatidae), *Fulica atra* (Rallidae) and *Limosa 'belgica'* (=*limosa*) (Scolopacidae). Other ratites (*Rhea, Casuarius, Dromaius*) gave only moderate reactions. Also, other samples of *Anas 'boscas'* blood gave no reaction, a weak reaction or a fairly strong reaction. These results are so contradictory that no conclusions are possible.

3. Anti-Chicken (Gallus gallus) *Egg-white*

An antiserum prepared by injecting rabbits with chicken egg-white was used in tests against 312 blood samples from 258 species of birds and the egg-whites of four species. A very strong reaction was obtained against the homologous (i.e. chicken) egg-white and fairly strong reactions occurred in the tests with the egg-whites of a pheasant, a crane and the emu. A strong reaction was obtained with the blood of a heron (*Nycticorax*) and several other avian blood samples gave moderate or weak reactions. Several reptilian blood samples also produced a precipitate when tested against the antiserum.

4. Anti-Emu (Dromaius novae-hollandiae) *Egg-white*

An antiserum prepared in a rabbit by injection of emu egg-white was used in 630 tests including 254 avian blood samples and four egg-white samples. The avian blood specimens were from 223 species. A strong homologous reaction was obtained with this antiserum and a moderately strong reaction resulted from the test with the egg-white of the crowned crane (*Balearica*). The avian blood specimens gave no strong reactions, a few gave weak or moderate reactions and many produced no reaction at all. No conclusions can be drawn from these tests.

Nuttall devised a method for measuring the amount of precipitate formed and he and T. S. P. Strangeways (Nuttall 1904: 334) reported the results of tests with anti-ostrich and anti-chicken antisera. The anti-ostrich serum was tested against eight species of birds. The homologous reaction was scored as 100% and the others were expressed as percentages as follows: *Casuarius*, 62; *Rhea*, 41; *Pelecanus*, 41; *Sarcorhamphus*, 36; *Gallus*, 17; *Anas*, 17; *Nyctea*, 17.

The anti-chicken antiserum was also scored as follows: *Gallus*, 100; *Anas*, 34; *Casuarius*, 27; *Columba*, 23; *Nyctea*, 21; *Sarcorhamphus*, 19; *Rhea*, 19; *Pelecanus*, 11.

There are some suggestive results in these tests, for example, the fact that *Casuarius* and *Rhea* gave strong reactions with *Struthio*. The *Rhea* specimen 'was putrid when received' or it might have performed even better. However, the tests are too few and, again, do not provide a basis for taxonomic speculation.

G. S. Graham-Smith (Nuttall 1904: 336–365) carried out 2,500 additional tests which included some experiments with avian material. He prepared antisera against the egg-white proteins of emu (*Dromaius*), a duck (*Anas platyrhynchos*), the domestic chicken (*Gallus*) and the crowned crane (*Balearica*). Positive reactions were obtained with avian egg-whites representing a considerable range of taxonomic diversity but the results contain many inconsistencies and the usual failure to address the experiments to defined problems rather than to a haphazard assortment of readily available samples.

Graham-Smith also carried out additional quantitative tests using Nuttall's method for measuring the precipitate. Here again there are suggestive results in several cases but also glaring inconsistencies. It is not possible to give credence to a procedure which indicates that a thrush (*Turdus merula*) is more closely related to an emu than the emu is to a crane, a rail or a pheasant or that a crane is as closely related to a finch (*Chloris*), an emu or a pheasant as it is to a rail (*Gallinula*) (Nuttall 1904: 345–347). In spite of these difficulties Graham-Smith's tables do reveal that the strongest reactions tend to occur between closely related species. The lack of consistency probably was due to the effects of poor specimens and crude techniques.

Two quotations will serve to summarize the conclusions of Nuttall and Graham-Smith about their own data. Nuttall (1904: 216) summarized the results of his precipitin studies on birds as follows.

> '1. Tests were made by means of antisera for the blood of the fowl and ostrich upon 792 and 649 bloods respectively. They demonstrated a similarity in the blood constitution of all birds, which was in sharp contrast to what had been observed with mammalian bloods when acted upon by anti-mammalian sera. Differences in the degree of reaction were observed, but did not permit of drawing any conclusions. Slight or faint reactions were observed with some reptilian bloods; only a very few mammalian bloods (0% and 0·6% respectively) showed faint reactions.
>
> 2. Tests made with antisera for egg-white of the fowl (789 tests) and emu (630 tests) gave maximum reactions with the egg-white of birds, moderate reactions with avian bloods, and distinct but slight reactions with reptilian bloods, notably those of *Crocodilia* and *Chelonia*'.

Graham-Smith (Nuttall 1904: 361) concluded that

> '(1) Powerful antisera to *Avian egg-albumins* produce very large reactions with dilutions of Birds' egg-albumins. They also give distinct positive reactions with Chelonian and Crocodilian sera, as well as with dilutions of Reptile egg-albumins. No reactions of any importance were obtained with Amphibian or Fish egg-albumins or with Lacertilian, Ophidian, Amphibian, Fish, or Crustacean sera. These tests, therefore, show a distinct relationship between the Aves and Chelonia and Crocodilia'.

These disappointing conclusions apparently disillusioned Nuttall and his colleagues for they published nothing further on comparative problems. Others, however, continued the search.

Friedenthal (1904, 1908) prepared an antiserum against ostrich (*Struthio camelus*) blood which reacted most strongly with the blood of a kiwi (*Apteryx*) and slightly less strongly with that of a cassowary (*Casuarius*). Less intensive reactions were obtained against several ducks, a stork (*Ibis*) and a pelican (*Pelecanus*). A frigate-bird (*Fregata*), a grebe (*Colymbus*) and a pigeon (*Columba*) gave moderate reactions while two hawks (*Buteo*, *Pernis*), a barn owl (*Tyto*), a parrot, two passerines and a tortoise gave no reaction.

The strong reaction between *Struthio* and *Apteryx* is of particular interest, especially since this antiserum also reacted strongly with *Casuarius*. The possibility of a relationship between *Apteryx* and the ratites is repeatedly debated and Friedenthal's data clearly bear upon this question. Because of the reasonable consistency of Friedenthal's results they can be taken seriously although they should be repeated before being fully accepted. Although Wasmann (1926) argued that Friedenthal's results were due to protein convergence rather than close relationship it is now reasonably certain that true convergence (i.e. in amino acid sequences) is unlikely to be a serious threat (Sibley, 1970). This does not eliminate the possibility of immunological cross reactivity between unrelated species.

Welsh and Chapman (1910) reported on their experiments using an antiserum against chicken (*Gallus*) egg-white proteins. Their refined technique employed known quantities of antiserum and antigen and they made rough measurements of the amount of precipitate that formed in each test. They were able to distinguish the homologous reaction from the reaction with other gallinaceous birds thus proving that avian egg-white proteins do indeed differ even among closely related species. Otherwise their results contain little of taxonomic value because only five species were involved in the tests. The egg-white of a pheasant (*Phasianus*) and that of a duck (*Anas*) gave the strongest reactions closely followed by the other galliform birds (*Coturnix*, *Perdix*). Ostrich egg-white reacted with the anti-*Gallus* antiserum but produced much less precipitate than did the other tests.

Galli-Vallerio (1911) prepared antisera against the blood and egg-white of the domestic fowl (*Gallus*). The anti-blood antiserum produced the strongest reactions with the bloods of other gallinaceous birds and lesser reactions with other non-passerines (e.g. *Columba*, *Gallinula*, *Aquila*) and passerines (e.g. *Hirundo*, *Erithacus*, *Lanius*, *Passer*). The anti-chicken egg-white antiserum reacted strongly only with gallinaceous egg-white proteins but it gave some reaction with all of the 21 species tested. Both antisera produced reactions against reptilian antigens but not with blood from 13 species of mammals, a toad or a fish.

Emmerich (1913) produced a highly active antiserum against the egg-yolk of the domestic fowl (*Gallus*) which reacted with the egg-yolk of a turtle (*Testudo*) and also with extracts of the eggs of some species of fish, namely, sturgeon, carp and redeye (*Leuciscus*) but not with those of salmon, tench or bream.

Seng (1914) also prepared an anti-chicken egg-yolk antiserum which produced strong reactions with the yolks of 15 species from widely diverse groups of birds. Although other gallinaceous birds gave strong reactions this antiserum also reacted strongly with the yolks of *Columba* and *Vanellus* and only slightly less so with the yolk of *Passer*. Seng's results seem to suggest that the proteins of avian egg-yolk are similar throughout the Class but the study should be repeated using many more species.

Graetz (1914) also prepared an anti-chicken (*Gallus*) egg-yolk antiserum but it gave equally strong reactions with the yolks of chicken, turkey, goose, duck and pigeon. This antiserum did not react with the egg-white proteins of chicken, turkey, goose or duck but it did produce a small amount of precipitate with the blood serum of chicken. It gave no reaction with the semen or muscle tissue proteins of the chicken (*Gallus*). Graetz also experimented with an anti-chicken egg-white antiserum but his results were of limited value for avian systematics since they added little to what was already known.

Glock (1914) prepared antisera against two varieties (black and Italian) of domestic fowl (*Gallus*) and attempted to differentiate between them by the precipitin reaction. This attempt failed but Glock did discover that when guinea pigs were sensitized with the serum of one of the two varieties of fowl they would develop anaphylaxis only if reinjected with the same serum and not if reinjected with the serum of the other variety. Other experiments by Glock in which he used an anti-guinea fowl (*Numida*) antiserum produced nothing of importance for avian systematics.

Reeser (1919a,b) prepared anti-chicken antisera which reacted strongly with the homologous antigens and less so with those of duck (*Anas*) and pigeon (*Columba*). He also prepared anti-pigeon and anti-duck antisera. The two anti-pigeon antisera gave contradictory results. One of them reacted more strongly with duck than with chicken, the other reacted more strongly with chicken than with duck. Neither precipitated the serum proteins of several species of mammals. Reeser's anti-duck antiserum gave approximately equal reactions with both chicken and duck sera. These experiments again illustrate the weaknesses of most of the early studies, namely, the use of a few readily available animals without regard to significant, defined problems. Even improved techniques would not compensate for these shortcomings.

Within the next few years several papers included data on immunological studies using avian material (e.g. Friedberger and Meissner, 1923; Meissner, 1923, 1926; Fröhlich, 1928) but the results were of little or no significance for systematics.

Lewis and Wells (1927) introduced some additional techniques into a comparative study of avian ovomucoids. Their fractionation procedure, although crude by modern standards, did produce an antigenic solution containing a restricted number of egg-white proteins. The ovomucoids of ostrich (*Struthio*), chicken (*Gallus*), turkey (*Meleagris*), guinea-fowl (*Numida*), goose (*Anser*), duck (*Anas*) and a turtle (genus?) were prepared. These preparations were injected into rabbits and the resulting antisera were employed in complement fixation tests and uterine strip anaphylaxis reactions. The results contained no taxonomic surprises for they showed only that the ovomucoids of goose and duck 'are perhaps identical' and that those of the chicken and guinea fowl 'are also closely related'.

Sasaki (1928) examined the degrees of relationship among domesticated ducks and certain species of wild ducks. He used the precipitin reaction and prepared 'specialized' (i.e. absorbed) antisera. An antiserum against *Anas 'domestica erecta'* (=*A. platyrhynchos*, Japanese domestic variety) reacted as strongly with wild *A. 'boschas'* (=*platyrhynchos*) and with *A. falcata* as with the homologous antigens but less strongly when reacted with the serum of the swan goose (*Anser cygnoides*). An antiserum against wild *Anas platyrhynchos* gave identical reactions with the homologous antigens and with those of the domestic variety but *A. falcata* could be distinguished although barely so. After the anti-domestic duck antiserum was absorbed against swan goose serum it no longer reacted with the goose antigens but continued to give a precipitate when tested against *Anas*.

Thus Sasaki showed clearly that *Anas* and *Anser* differed in their serum proteins. The absorption procedure did not differentiate between wild and domestic *Anas platyrhynchos* but it did indicate differences between *Anas platyrhynchos* and both *Anas crecca* and *Cairina moschata*. Sasaki made further tests involving other species of *Anas* (*falcata*, *penelope*, *formosa*) and found that, as a rule, the species of ducks available to him could rarely be differentiated using standard antisera (i.e. unabsorbed) but could be distinguished when absorbed antisera were used.

By 1929 enough data had accumulated to permit Erhardt (1929a,b) to undertake a critical review of the results. In his special review of the ornithological studies (1930) he recorded his generally pessimistic conclusions. Citing his earlier papers (1929a,b) Erhardt stated (1930: 229) that (transl.)

> 'With respect to serology birds seem to be very closely related to one another. There is no definite difference between ratites and carinates. Relationships to the reptiles have been found. However, to erect a phylogenetic tree for birds on the basis of the available serological studies would not be easy'.

Erhardt pointed out the inconsistencies and contradictions in the work of various authors and concluded (1930: 231) as follows (transl.):

> 'Since none of the authors was able to confirm his results with control experiments, since experiments using exactly the same technique turned out differently at different times and since the results frequently opposed or contradicted one another we must conclude that even a series of positive results which coincide with the system is lacking in convincing evidence . . . At least this is so in zoology . . . in summary, it becomes evident that the importance of serology in phylogeny, particularly in ornithology, must be subordinated to evidence from morphology'.

Erhardt's conclusions probably discouraged some potential work on birds which is unfortunate because the situation was not as bad as he suggested. In spite of the obvious inconsistencies it was also clear that the most closely related species gave the strongest reactions and that even closely related birds differed in their proteins. These two facts alone should have been sufficient to sustain faith in the principles while the search for better techniques continued.

Not everyone was discouraged. Boyden (1926) was just beginning what was to be more than forty years of unflagging optimism and constant searching for better techniques. In 1964 (p. 81) Boyden criticized Erhardt's reviews and reaffirmed his faith 'that better results will be attained when better methods are consistently used.'

Hektoen and Boor (1931) prepared several antisera against hemoglobins, including those of three species of birds (chicken, goose, guinea fowl) but their tests produced nothing of taxonomic interest.

Buchbinder (1934) brought a novel approach to avian systematics with his study of the occurrence of a heterophile antigen in various groups of birds. The heterophile or heterogenetic antigens (Carpenter 1956: 47–50; Boyd 1966: 195–198) are serologically related substances of uncertain composition which occur in widely separated groups of organisms. They are detected by the ability of antibodies against them to react with the tissues of other organisms which may be as distant phylogenetically as bacteria and mammals. The so-called 'Forssman antigen' is the best known of the heterophile antigens.

Buchbinder (1934) prepared antisera in rabbits against a number of species of bacteria and tested them for their capacity to agglutinate various animal erythrocytes. He found that antibodies against *Pasteurella lepisepticus* would agglutinate the red cells of the domestic pigeon (*Columba livia*) and proceeded to test 83 individual birds of 53 species representing 14 orders in Wetmore's (1960) classification. Of these, 63 individuals possessed the antigen. Buchbinder's results contain some possibly significant aspects but too many inconsistencies to be trusted. All 40 pigeons of 21 species contained the antigen as did the five species of parrots. It is tempting to consider this to be further evidence of a relationship between these groups until one notes that of six species of Galliformes two possess the antigen and four lack it. The antigen is absent in the ostrich (*Struthio*), in a tinamou (*Eudromia*), in a sand grouse (*Pterocles*) and in the two ducks (*Anas*) which were tested. Buchbinder commented on these facts as follows (p. 224, 228).

> 'Among birds of the most ancient orders i.e., the *Struthioniformes* and *Tinamiformes*, this antigen apparently does not occur. The Galliformes which are next, although some distance away in classification, include some species which do, and others which do not possess the antigen. Following this order every member of the remaining ones examined, except two, contains the antigen. These two irregular ones are the Pteroclidiformes, represented by the sand grouse, and the Anseriformes (ducks). The absence of the antigen in the latter cannot be satisfactorily explained since they have a relatively high position in the scale of evolution. These results taken *prima facie* would indicate a relatively recent appearance of this antigen or hapten in the birds with a subsequent almost universal inheritance throughout the class'.

In the reviewers' opinion it is also possible that Buchbinder's results were the product of the limited and haphazard assortment of species available to him. The impressive consistency of occurrence in the Columbiformes is offset by the variation within the Galliformes, a group which, by many other criteria, is monophyletic and consists of closely related species. It is particularly interesting to note that *Pterocles* differs from the Columbiformes and agrees with the Charadriiformes in lacking the Buchbinder antigen. This could be cited as support for the proposal by Maclean (1967) that the sand grouse are more closely related to the shorebirds than to the pigeons but it should be considered to be suggestive and not treated as an important piece of evidence.

Further evidence of the doubtful value of the occurrence of heterophile antigens as taxonomic characters is found in the data on the Forssman antigen (see Boyd 1966: 196–197). This substance is known to be present in the chicken (*Gallus*), ostrich (*Struthio*) and turkey vulture (*Cathartes*) and absent in a goose (*Anser*), an owl (genus?), a pigeon (*Columba*), the sparrow hawk (*Accipiter*), a cuckoo (*Cuculus*), a crossbill (*Loxia*) and a wagtail (*Motacilla*). Here again is a limited, haphazard assemblage of species with only two situations providing any kind of taxonomic test. The two members of the Falconiformes (vulture and hawk) differ and the two passeriforms (wagtail and crossbill) agree. Although these data are too limited to provide a basis for any conclusions they seem to suggest that, like Buchbinder's antigen, the Forssman antigen may be of little value for avian systematics. The mammalian data on Forssman's antigen reinforce this conclusion for the antigen is present in some artiodactyls, absent in others, present in some rodents, absent in others, and present in some primates but absent in others. In man it may be present in some individuals but absent in others (Boyd 1966: 196).

Beginning in 1932 and continuing to the present there has appeared a series of papers by M. R. Irwin, his colleagues and his students which bears upon the systematics of the Columbidae. This extensive work will be reviewed in Section IIB.

Sokolovskaia (1936) used the precipitin test in a study of several species of anseriform birds and certain duck hybrids. In most cases the serological data coincided with the results from hybridization, i.e. species that can hybridize also show close serological correspondence. In a muscovy duck (*Cairina moschata*) × mallard (*Anas*) cross the hybrids gave reactions intermediate between those of the parents but when antisera were prepared against the serum proteins of the hybrid some new antigens, not present in either parent, were found. Sasaki (1937) also used the precipitin test to study hybrids between guinea hen (*Numida*) and domestic fowl (*Gallus*) and (1954) between ring-necked pheasant (*Phasianus colchicus*) and domestic fowl. Spärck (1954, 1956) established the parentage of a captive hybrid duck by using immunological techniques.

Cole (1938) used crystallized ovalbumins as antigens to compare the 'pearl guinea fowl' (cf. *Numida meleagris*), amherst pheasant (*Chrysolophus amherstiae*), golden pheasant (*Chrysolophus pictus*) and domestic chicken (*Gallus*). Several serological tests were carried out which indicated that the guinea fowl and chicken are closely related to one another and that the two pheasants are closely related to one another. That the two *Chrysolophus* proved to be closely related is not surprising for two reasons. First, they are geographically complementary in the wild and differ primarily in the secondary sexual characters of the males. It would not be illogical to consider them to be conspecific. Second, captive birds of these two 'species' produce fertile hybrids and they have been hybridized so frequently that some aviculturists doubt that truly 'pure' birds of either form can be found in captivity.

The close relationship which Cole demonstrated between the ovalbumins of *Gallus* and *Numida* is of somewhat greater interest. These two species have been hybridized in captivity (Gray, 1958) but they are usually separated in different families in most classifications. The results reported by Cole provide additional evidence that *Gallus* and *Numida* should be included within the same family.

Wolfe (1939) reported on precipitin tests involving eight species of birds. Although the results contain nothing of direct interest for avian systematists Wolfe did conclude that serological relationship studies of birds seem feasible. The principal conclusion of this paper was that the most specific antisera are produced when very small quantities of antigen are used.

The egg-whites of the 16 species of northern European birds which 'by chance' were available to Turpeinen and Turpeinen (1936) were used in a comparative study. Antisera were prepared in rabbits against the egg-whites of the chicken (*Gallus*), common buzzard (*Buteo buteo*), goshawk (*Accipiter gentilis*), sparrow hawk (*Accipiter nisus*), black woodpecker (*Dryocopus martius*) and hooded crow (*Corvus cornix*). Precipitin tests were carried out against the egg-whites of these species plus those of the turkey (*Meleagris*), honey buzzard (*Pernis apivorus*), hen harrier (*Circus cyaneus*), kestrel (*Falco tinnunculus*), gray woodpecker (*Picus canus*), great spotted woodpecker (*Dendrocopos major*), lesser spotted woodpecker (*D. minor*), raven (*Corvus corax*), jay (*Garrulus glandarius*) and white wagtail (*Motacilla alba*).

In all tests it was clear that the strongest reactions occurred between the most closely related species although cross reactions were extensive when high concentrations of antigens were used.

This chance assortment of species included few possibilities for the examination of real taxonomic problems and even these were overlooked by the investigators. The kestrel (Falconidae) was tested only against *Gallus* (Phasianidae) and but one out of the seven available antisera against members of the Accipitridae. This lone test against an anti-goshawk antiserum did suggest that *Falco* is less closely related to *Accipiter* than are *Pernis*, *Buteo* and *Circus*.

Turpeinen and Turpeinen summarized their results as follows: (1) avian egg-white proteins are group specific rather than species specific; (2) avian egg-white contains some components common to all birds; (3) individual variation among rabbits results in antisera with different properties although prepared against the same antigens; (4) the egg-white and blood serum of a species contain serologically allied substances; (5) avian egg-white is a mixture of several antigens. None of these conclusions was novel and all have been examined in greater detail by other workers.

The most extensive serological study of avian egg proteins yet completed was carried out by Sievers (1939a,b,c). Two short papers (1939a,b) reported on comparisons using antisera against the egg-white of an owl (*Strix*) and certain hawks. A longer paper (1939c) contains the results of Sievers' studies using 71 antisera prepared in rabbits against the egg-white proteins of 31 species and 11 antisera prepared against the yolk proteins of six species. The yolk antisera proved to be broadly cross reacting and Sievers concluded that yolk proteins are 'organ specific' and hence of little value for systematics. He therefore gave principal attention to comparisons among the 71 anti-egg-white antisera and the egg-white proteins of 62 species of birds, including three domestic varieties of *Gallus gallus*. Sievers used the precipitin and complement fixation techniques and his stated purpose was to determine whether or not avian egg proteins could be serologically differentiated. He noted that previous studies had involved few antisera and small numbers of species for comparison.

Sievers found considerable variation in the extent of cross reactivity among his antisera. A few reacted to some degree with most of the 62 species of egg-whites available while others reacted only with their closest relatives. In general, however, the strongest reactions were among closely related species.

Although Sievers had the egg-whites of more species than any previous serologist his material was still too limited to provide data on a significant number of problems. Except for *Gallus* and *Phasianus* all of his species were native to Europe and represented only 11 orders and 23 families (Wetmore, 1960). A synopsis of his results follows.

a. Anseriformes. Thirteen antisera against five species: *Anser anser, Cairina moschata, Anas crecca, Nyroca fuligula, Somateria mollissima.*

The same six species were the only ducks and geese used as comparative antigens. The strongest reactions were those among these six species. Reactions were also obtained with some Charadriiformes, Galliformes, Falconiformes, *Podiceps* and Passeriformes. No reaction was obtained with Picidae or *Grus*. These results do not offer a basis for any conclusions about the relationships of the groups involved.

b. Falconiformes. Sixteen antisera against seven species: *Falco peregrinus, Falco columbarius, Buteo lagopus, Buteo vulpinus, Circus cyaneus, Accipiter nisus, Pernis apivorus.*

The strongest reactions were among the five species of Accipitridae ('Aquilidae' of Sievers) and between the two species of *Falco*. Reactions between *Falco* and the accipitrids were weak or lacking.

A rough measure of these reactions is obtained by calculating the percentages of the tests which resulted in a positive reaction of some degree. Of

10 tests between *Falco* antisera and *Falco* antigens 8 (80%) were positive. Similarly, 42 out of 55 (76%) tests among the species of Accipitridae were positive. However, only 28% (7/25) of the tests between *Falco* antisera and accipitrid egg-white proteins and 4.5% (1/22) of the tests between accipitrid antisera and *Falco* egg-whites were positive. That these differences may be important is indicated by the fact that tests between the Anatidae and the birds of prey gave the following results.

Falco antisera *vs* anatid egg-white, 14/30 (nearly 50%) positive.

Accipitrid antisera *vs* anatid egg-white, 21/66 (32%) positive.

These results do not mean that the birds of prey are more closely related to ducks and geese than to one another but only that the anti-*Falco* and anti-accipitrid antisera prepared by Sievers reacted as, or more, strongly with the anatids than with one another. The most conservative interpretation of these results is simply that they support the idea that the falcons and the typical hawks are not especially closely related to one another.

Sievers was aware of the significance of his results in this case and noted (p. 68) that his studies supported the separation of the Falconidae and Accipitridae (='Aquilidae'). Sievers also noted that the Falconidae and Accipitridae differ in the occurrence of the Forssman antigen.

c. Galliformes. Sixteen antisera against five species. *Lyrurus tetrix, Tetrastes bonasia, Lagopus lagopus, Phasianus colchicus, Gallus gallus.*

Sievers prepared seven antisera against *Gallus gallus* which he listed under *Gallus 'domesticus'*, 'Seidenhuhn' and 'Zwerghuhn'. The latter two are apparently varieties of bantams. It is interesting to note that no two of these seven antisera gave exactly the same distribution of positive reactions in relation to the 64 egg-whites used in comparisons.

The 16 anti-galliform antisera gave the strongest reactions with gallinaceous antigens. They also reacted with many Anseriformes and Charadriiformes and with a few Falconiformes and Passeriformes. Reactions against a woodpecker (Picidae), a cuckoo (Cuculidae), an owl (*Strix*), a crane (*Grus*), a pigeon (*Columba*) and a grebe (*Podiceps*) were also obtained. One of the anti-*Gallus* antisera was strongly cross reactive and gave reactions with nearly all groups. Another was so specific that it gave positive reactions only with the *Gallus* egg-whites.

d. Gruiformes. Two antisera against *Grus grus*. These antisera gave broad cross reactions with most of the available species except the woodpeckers and the owl.

e. Charadriiformes. Fifteen antisera against eight species: *Tringa totanus, Tringa nebularia, Numenius arquata, Larus ridibundus, Larus canus, Larus fuscus, Sterna hirundo, Sterna macrura.*

These antisera reacted most strongly with the 16 charadriiform egg-whites available and gave the next strongest reactions with the Anseriformes and the grebe (*Podiceps*). Some degree of reaction occurred with most of the other groups but no significant trends are apparent.

f. Strigiformes. Three antisera against *Strix uralensis.* This egg-white was mixed with yolk and the antiserum gave wide cross reactions from which no conclusions could be drawn.

g. Passeriformes. Six antisera against four species: *Riparia riparia, Corvus cornix, Turdus ericetorum, Turdus musicus.* The egg-whites of 19 species representing eight passerine families were used in comparisons. One antiserum of *Turdus ericetorum* gave wide cross reactions but the others reacted almost exclusively with passerine egg-whites. There seemed to be no consistent pattern to the results and no conclusions could be drawn.

Sievers felt that his data showed differences between Falconidae and Accipitridae, Tetraonidae and Phasianidae and *Larus* and *Sterna*. Of these only the first seems reasonably convincing. In spite of his meagre list of conclusions, Sievers expressed his belief that he had demonstrated that serological studies of egg proteins could be useful in avian systematics. That this conclusion was justified has been proved by subsequent studies.

As part of a study of avian ovalbumins, Landsteiner and van der Scheer (1940) prepared three antisera in rabbits against chicken (*Gallus*) ovalbumin. These antisera were tested against the ovalbumins of turkey (*Meleagris*), guinea fowl (*Numida*), duck (*Anas*) and goose (*Anser*). Although these tests produced no important taxonomic results, they provide an example of the type of variation that is encountered in serological comparisons. For each of the three antisera the homologous reaction, i.e. *Gallus* : *Gallus*, was scored as 100% and the amount of precipitate formed in the heterologous reactions with the other four species was expressed as a percentage of the amount of homologous precipitate. The results can be summarized as follows:

Antiserum	Chicken	Turkey	Guinea fowl	Duck	Goose
A	100%	35%	26%	9%	9%
B	100	67	57	42	31
C	100	61	51	30	18

It is interesting to note that although the antisera varied in the intensity of their cross reactions, the relative positions of the five species were essentially the same with each antiserum.

Landsteiner and van der Scheer used only a single antigen dilution and a single amount of antiserum in their tests and some of the discrepancies may be due to this limited examination of the potential range of reactions that would result if complete antigen titration curves were compared. Furthermore, their tests with absorbed antisera showed that their 'pure' ovalbumin preparations actually contained several antigens.

The 'photronreflectometer', a photoelectric turbidity measuring instrument, was introduced by Libby (1938) and used by DeFalco (1942) in a serological study of 10 species of birds representing six orders. DeFalco was mainly interested in technique problems and this small, haphazard assemblage of species involved no significant taxonomic problems and produced little of taxonomic interest. His principal systematic conclusion was that birds are serologically an essentially homogeneous group although (p. 211) 'chemical differences among birds may be readily demonstrated by the precipitin reaction'.

Boyden (1942) published a review of the history and techniques of systematic serology which provides an excellent summary of the first forty years. Boyden's criticisms and comments are still pertinent and should be read by anyone interested in systematic serology. Other reviews by Boyden (1926, 1934, 1943, 1953, 1958, 1963, 1964, 1965, 1966) although containing little about avian systematics, provide a complete and frequently updated picture of the activities and thinking of the Rutgers group. Certain themes are found in all of these papers. First there is the preoccupation with techniques and their improvement and second the frequent reaffirmation of faith in the principles which underlie the field of systematic serology. That these are symptoms of a widespread condition among scientists is attested by the fact that they can be found as explicit or implied aspects of every long-term research project.

The photronreflectometer was used by Martin and Leone (1952) in a study of antigenic preparations from several organ systems in four species of birds. The antigens used were serum, serum albumin, serum globulin, lens extract, gut extract and skeletal muscle extract. All of these antigen preparations were obtained from the chicken (*Gallus*) and the domestic duck (*Anas*). The gut extract only was obtained from the turkey (*Meleagris*) and all except the skeletal muscle extract were prepared from the domestic goose (*Anser*). Antisera were produced in rabbits. Whole serum and serum globulins gave similar results and both were found to be highly antigenic and specific. Serum albumin was less antigenic and also less specific. The lens extracts also produced specific antisera. The gut extracts were specific but antigenically weak and extracts of skeletal muscle were satisfactory as antigens and produced specific antisera. The gut and muscle extracts were shown to contain serum proteins, as would be expected. The authors concluded that, for practical reasons, the serum proteins are best for comparative serological studies.

The first serological study of avian material to address itself to a defined taxonomic problem was that of Stallcup (1954) who attempted to determine the relationships among the subfamilies of the Fringillidae and between these and other passerines including the estrildines and *Passer*. Stallcup used an extract of the tissues of the trunk, heart, lungs and kidneys as his antigenic material and prepared antisera in rabbits. The results of his precipitin tests suggested to Stallcup that (p. 193) '*Molothrus* and *Passer* excluded, the birds fall into two distinct groups: one includes *Piranga*, *Richmondena*, *Spiza*, *Junco*, and *Zonotrichia*; the other includes *Estrilda*, *Poephila*, *Carpodacus*, and *Spinus*'. Thus Stallcup agreed with Tordoff (1954) that the carduelines and estrildines are closely related and that *Spiza* is closer to the cardinal-bunting-tanager assemblage than to the troupials (Icteridae). Stallcup concluded that '*Passer* is not, serologically, closely related to any of the birds tested' (p. 200). *Molothrus* was found to be closest to the cardinal-bunting-tanager group but 'definitely set apart from' them (p. 200). Stallcup recommended that a family Carduelidae be recognized to include the subfamilies Estrildinae and Carduelinae and that the Fringillidae include the Richmondeninae, Thraupinae, Emberizinae, Fringillinae and Geospizinae, with *Spiza* placed in the Richmondeninae. Stallcup recognized the Icteridae as a separate family and placed the Passerinae in the Ploceidae, noting (p. 209) that 'The Estrildinae and Carduelinae are closely related subfamilies, but neither group is closely related to the Passerinae'. This conclusion means that Stallcup considers the ploceines and estrildines to be members of different families, a conclusion disputed by other facts and other authors, e.g. Sibley (1970). However, Mainardi (1958b, 1960b) supported Stallcup in his conclusion that the carduelines and estrildines are closely related.

In a subsequent study Stallcup (1961) attempted to determine the relationships among 15 families of the suborder Passeres. Stallcup used only muscle tissue as the source of antigens, otherwise his technique was essentially the same as that reported in the 1954 study. Two species of woodpeckers (*Colaptes auratus*, *Centurus carolinus*) and 18 species of passerines were collected. Antisera were prepared against these 20 species and each antiserum was tested against from eight to eleven of the 20 species. The passerines involved were: *Cyanocitta cristata* (Corvidae); *Parus carolinensis*, *P. bicolor* (Paridae); *Thryothorus ludovicianus* (Troglodytidae); *Mimus polyglottos*, *Toxostoma rufum* (Mimidae); *Turdus migratorius* (Turdidae); *Regulus calendula* (Sylviidae); *Bombycilla cedrorum* (Bombycillidae); *Lanius ludovicianus* (Laniidae); *Sturnus vulgaris* (Sturnidae); *Passer domesticus* (Ploceidae); *Vireo olivaceus* (Vireonidae); *Dendroica coronata* (Parulidae); *Agelaius phoeniceus* (Icteridae); *Piranga rubra* (Thraupidae); *Richmondena cardinalis*, *Spiza americana* (Fringillidae).

In many cases it was possible to perform reciprocal tests and these often gave unequal values which Stallcup (1961: 49) ascribed to the differential responses of the rabbits which produced the antisera. Another problem, noted by Stallcup, is that 'comparative serological studies with the photronreflectometer tend to minimize the differences between distant relatives and to exaggerate the differences between close relatives' (p. 49).

It is difficult to find significant results in the data reported by Stallcup. The two woodpeckers tended to show a low degree of serological correspondence with the passerines but in a number of tests against passerine antisera *Colaptes* gave a stronger reaction than did some of the passerines. The reaction between *Colaptes* and *Bombycilla* was stronger than between *Bombycilla* and *Turdus*, *Cyanocitta*, *Toxostoma*, *Agelaius* or *Lanius*. The reciprocal test also showed a strong reaction between *Colaptes* and *Bombycilla*.

Stallcup's data do show a convincing consistency in relation to one assemblage, the so-called 'New World nine-primaried oscines' represented by

Dendroica, *Agelaius*, *Piranga*, *Richmondena* and *Spiza*. There is a clear tendency for these genera to react most strongly with one another thus lending support to the evidence from other sources (e.g. Sibley, 1970) that they are members of a closely related cluster of 'families' that might better be treated as a single family. It is of particular interest to find that Stallcup's data suggest that *Vireo* is not closely related to *Piranga* and *Richmondena* and is not especially close to any of the species in this study. *Regulus* and *Lanius* also 'seem unlike, serologically, the other oscine species tested' (p. 58).

Stallcup's (1961) study may be criticized for the same reasons as were many earlier ones. The large array of antigens tended to produce broadly cross-reacting antisera which obscure differences and often suggest close relationships between distantly related species. Although more species were examined than in most previous studies, the number, twenty, is still smaller than required to provide significant results. Finally, although some effort was made to define problems they were still too broad.

In his study of the relationships of certain ducks, Cotter (1957) defined his problem more precisely and used single proteins as antigens. The antigens were ovalbumin, serum albumin and serum gamma globulins which were obtained from the domestic goose (*A. anser*), mallard (*Anas platyrhynchos*), domestic Pekin duck (*Anas platyrhynchos*), wood duck (*Aix sponsa*), mandarin duck (*Aix galericulata*) and Muscovy duck (*Cairina moschata*). Antisera were prepared in rabbits and precipitin tests, including the use of absorbed antisera, were carried out. Cotter concluded that *Aix* and *Cairina* are more closely related to one another than either is to *Anas* and that his data support the inclusion of *Aix* and *Cairina* in the same tribe, Cairinini.

Cotter's study demonstrates the value of attacking sharply defined problems and of using a restricted array of antigens. He also showed that the absorption technique has far greater resolving power than the cross-reacting method used alone.

Duwe (1964) prepared extracts of the pectoral muscle tissues of seven species of birds and made a single antiserum against the extract from a great blue heron (*Ardea herodias*). This antiserum was compared with the other species by the Ouchterlony double diffusion technique. The other six species, representing four orders and five families of non-passerines, were the pied-billed grebe (*Podilymbus podiceps*), white pelican (*Pelecanus erythrorhynchos*), American bittern (*Botaurus lentiginosus*), roseate spoonbill (*Ajaia ajaja*), white-faced ibis (*Plegadis (mexicana) chihi*) and mallard (*Anas platyrhynchos*). This chance assortment contains little of potential taxonomic interest and no defined problems. The anti-*Ardea* antiserum gave no reaction against *Podilymbus* or *Pelecanus*, a moderate reaction with *Anas*, and a strong reaction with *Botaurus* and *Ajaia*. The reaction between *Ardea* and *Plegadis* was less than that between *Ardea* and *Anas* but, since only one antiserum was used and no further tests were carried out, this apparent discrepancy cannot be considered significant.

In a study of six species of North American thrushes (Turdidae) Bourns

(1967) found evidence supporting Dilger's (1956) conclusion that the wood thrush (*Hylocichla mustelina*) is more closely related to the American robin (*Turdus migratorius*) than to other species of the genus *Hylocichla*. Bourns used blood serum and pectoral muscle extracts as antigens and carried out precipitin tests using the photronreflectometer. Antisera were prepared in rabbits against the American robin, wood thrush, hermit thrush (*Hylocichla guttata*), olive-backed thrush (*H. ustulata*), gray-cheeked thrush (*H. minima*) and veery (*H. fuscescens*). The serological correspondence between the wood thrush and the American robin was so high in comparison with the other species of *Hylocichla* that Bourns (1967: 99) concluded that it indicated a degree of relationship that one 'would expect between two closely related species within a genus.'

Hendrickson (pers. comm.) has compared the plasma proteins and hemoglobins of *Turdus*, *Hylocichla* and *Catharus* using acrylamide gel electrophoresis. His preliminary results (see p. 130) agree with those of Bourns in some respects and disagree in others.

It was discovered very early in the history of comparative serology that there is a relationship between the proteins of avian egg-white and those of blood. Uhlenhuth (1901) found that an antiserum against chicken egg-white would also react with the blood of chicken and, to a small degree, with that of goose (see p. 92). Several others confirmed this observation including Gengou (1902), Nuttall (1904) and Galli-Valerio (1911). Hektoen and Cole (1927, 1928) were the first to attempt to determine which of the proteins in blood and egg-white were cross reacting. They succeeded in fractionating both mixtures and prepared antisera against the fractions. Their results showed that the cross reacting proteins were found only in certain fractions which, at the time, they identified as the conalbumin of egg-white and the albumin of blood. They recognized that their conalbumin preparation also contained other proteins, including ovalbumin and ovomucoid. Kaminski and Durieux (1956) found that the embryonic blood serum of the chicken (*Gallus*) contains ovalbumin and ovomucoid and that the blood serum of both embryo and adult birds contains a protein which is immunologically identical with conalbumin. More recently Ogden *et al.* (1962), Williams (1962, 1968) and Wenn and Williams (1968) have given detailed attention to this problem and have shown that it is the transferrin fraction of serum which is similar to conalbumin. The two proteins are thought to have identical amino acid sequences and to differ only in their carbohydrate prosthetic groups. Although these facts have not yet been exploited for taxonomic purposes they suggest that both the immunological cross reactivity between them and the well-known genetic polymorphism present in both may be useful in certain types of studies. See also p. 155 of this chapter.

There are at least a dozen papers which report upon immunological studies of individual egg-white proteins but which contain little of direct systematic interest. However, because the information in these publications can be important to systematists utilizing the egg-white proteins as sources of taxonomic information, it is appropriate to cite the papers with a brief notation of their contents.

Cohn *et al.* (1949) prepared antisera in rabbits against chicken ovalbumin and conalbumin and in a horse against chicken ovalbumin. Impurities in both antigens could be detected immunologically. The ovalbumin showed the usual two electrophoretic peaks (A_1, A_2) which could not be distinguished immunologically.

Wetter and Deutsch studied the properties of chicken (*Gallus*) ovomucoid (1950) and of chicken egg-white lysozyme (1951). Although their preparations were heterogeneous, they prepared antisera which were used to assay egg-white for the two proteins and they reported on various physical properties. Wetter, Cohn and Deutsch (1952) purified chicken ovalbumin, conalbumin, lysozyme and ovomucoid and prepared antisera against each. Although these four proteins comprise approximately 92% of egg-white protein, a relatively small proportion of the total antibody in anti-egg-white serum is directed against them. Most of the response seems to be to minor components.

In a study of the cross reactions of the egg-white proteins Wetter, Cohn and Deutsch (1953) prepared antisera against the ovalbumin, conalbumin, ovomucoid and lysozyme of the chicken (*Gallus*) and tested them against the same proteins from the turkey (*Meleagris*), guinea fowl (*Numida*), 'pheasant' (presumably *Phasianus*), 'duck' (presumably *Anas*) and 'goose' (presumably *Anser*). In all tests the galliforms were, as expected, more like the chicken than were the anseriforms. Within the Galliformes, however, the relative positions of the species were not the same in all tests. The positions in terms of degrees of serological reaction were as follows.

Ovalbumin: chicken, guinea. Others not tested.

Conalbumin: chicken, turkey, guinea.

Ovomucoid: chicken, guinea, pheasant, turkey.

Lysozyme: chicken, turkey, guinea, pheasant.

These results are of little value because the tests are incomplete for ovalbumin and conalbumin. The ovomucoid and lysozyme results suggest that the guinea fowl is closer to the chicken than is the pheasant.

A series of papers by Marie Kaminski and several co-authors contain the results of immunochemical studies of avian egg-white proteins. Kaminski and Ouchterlony (1951) fractionated the egg-whites of the chicken (*Gallus*), domestic duck (*Anas*) and domestic goose (*Anser*) and carried out various serological studies, in part using the double diffusion technique of Ouchterlony. There is little of taxonomic interest in this paper but the authors found at least a dozen protein fractions in chicken egg-white, although their fractions were mostly heterogeneous.

Kaminski and Nouvel (1952) extended the earlier study to include the egg-white proteins of the white stork (*C. ciconia*), the marabou stork (*Leptoptilos crumeniferus*), and the Egyptian goose (*Alopochen aegyptiaca*) in addition to the chicken, domestic duck and domestic goose. Various fractions were prepared and antisera against the ovalbumin, conalbumin, ovoglobulin and whole egg-white of the chicken were made. Serological tests were carried out with the Ouchterlony technique and showed 'une bonne concordance avec la classification zoologique' (p. 16). The Egyptian goose and the domestic duck showed a closer relationship than did the Egyptian goose and the domestic goose. This result is in accordance with the usual classification of *Alopochen* and *Anas* in the subfamily Anatinae whereas *Anser* is a member of the Anserinae.

Kaminski and Williams (1953) reviewed various immunochemical studies and, in a section on systematics, presented further evidence that the different egg-white proteins produce antisera of differing degrees of specificity. The conalbumins produce the broadest cross reactions, ovalbumins somewhat less so and the ovoglobulins give the most specific reactions.

Kaminski and Durieux (1954) studied the immunochemical changes in the proteins of the egg-white and blood during embryonic development in the chicken (*Gallus*) and Kaminski (1957) reported on a study of the ovoglobulins. Again, the highly antigenic properties of the ovoglobulins were demonstrated. As noted above (p. 108) Kaminski and Durieux (1956) found immunological evidence of ovalbumin, ovomucoid and conalbumin in the blood of the chick embryo.

Kaminski (1960, 1962) has studied the immunochemical properties of enzymatic digests of avian ovalbumins. The results are of interest in regard to the nature of antigenic sites but contain nothing of systematic significance.

Fothergill and Perrie (1966) investigated the structural basis of the immunological similarity between the ovalbumins of the chicken (*Gallus*) and duck (*Anas*) by 'fingerprinting' the tryptic peptides of these ovalbumins. Of 23 peptides only eleven were common to both species. However, the amino acid compositions of the two ovalbumins were similar except that the value for cystine appeared to be markedly lower in duck ovalbumin.

Wiseman and Fothergill (1966) were able to demonstrate an immunological difference between two electrophoretically distinguishable chicken ovalbumin variants. The difference could be detected only by using an absorbed antiserum. When chymotryptic digests of the two genetic variant ovalbumins were 'fingerprinted' a single peptide difference was found.

B. STUDIES BY M. R. IRWIN AND ASSOCIATES ON THE IMMUNOGENETICS OF THE COLUMBIDAE

The following citations are to an important series of papers written by M. R. Irwin and his colleagues which represent more than thirty years' work on the antigenic properties of the red blood cells of several species of pigeons and doves (Columbidae). Bryan 1953; Bryan and Irwin 1961; Bryan and Miller 1953; Cumley and Cole 1942; Cumley and Irwin 1940, 1941a,b, 1942a,b, 1944, 1952; Cumley *et al.*, 1942, 1943; Gershowitz 1954; Irwin 1932a,b, 1938, 1939, 1947, 1949a,b, 1951, 1952, 1953, 1955, 1971; Irwin and Cole 1936a,b,

1937, 1940, 1945a,b; Irwin *et al.*, 1936; Irwin and Cumley 1940, 1942, 1943, 1945, 1946, 1947; Irwin and Miller 1961; Jones 1947; Miller 1953a,b, 1956, 1964; Miller and Bryan 1951, 1953; Palm 1955; Palm and Irwin, 1962; Stimpfling and Irwin 1960a,b. Many of these papers are of taxonomic interest and we shall attempt to provide an analysis of the pertinent data.

In an early paper Irwin (1932a) showed that $\frac{2}{3}$ of the antigenic specificities possessed by the pearlneck or Chinese spotted dove, *Streptopelia chinensis*, were not present on the red cells of the ring-necked dove, *S. risoria*. In contrast, only $\frac{1}{6}$ of the antigenic specificities of *risoria* red cells were not also shared with *chinensis*. From this study and others completed shortly thereafter it was clear that 'reagent' antisera specific to only one or a few of the antigenic specificities of a species could be prepared by absorbing the anti-red cell antiserum to one species (A) with the red cells of another species (B). In this way only the species-specific antibodies of species A remained in the antiserum which could then be reacted with the red cells of still other species. By making reciprocal tests using various antisera absorbed with the red cells of the different species examined it was possible to determine the extent to which the various red cell antigens and their specificities were shared among the species studied. The assumption was made that as the genetic relationship of any two species increases the more similar will be their antigens. Conversely, as the genetic relationship between any two species decreases, the proportion of antigenic specificities shared will decrease with a proportional increase in those antigens that are species-specific.

Irwin and his colleagues also showed that those red cell antigens that are species-specific, as shown by tests performed with the appropriate, absorbed antisera, segregate as unit characters in the backcross progeny of hybrid matings as expected for genetically determined phenotypes (Irwin and Cole 1936a, 1940; Irwin 1939; Irwin *et al.*, 1936). By using the techniques of genetics in conjunction with those of serology it was possible, over the years, to determine partial genotypes associated with the red cell antigenic properties of several species of pigeons and doves.

That such data should bear upon taxonomic and phylogenetic problems was obvious, and Irwin and others have dealt briefly with this aspect of their data in several papers. Cumley and Cole (1942) and Cumley and Irwin (1944) found a correlation between the geographic distribution of the species of the genus *Columba* and the red cell antigenic specificities possessed by the respective species. As in the case of the plumage characters, it was found that some antigenic characters were shared in different combinations between the Old World and New World species, some antigens were specific to either the Old or New World species, and a few antigens were common to only one or two species of both the New and Old Worlds. Thus, with one or two exceptions, it seemed that the species of either the Old World or the New World were more closely related to one another than any one of them was to a species of the other hemisphere. Furthermore, 'the evidence points definitely toward a common ancestry of these six species of the New World' (*Columba fasciata, C. leucocephala, C. picazuro, C. rufina (=cayennensis), C. flavirostris*, and *C. maculosa*), (Cumley and Irwin, 1944: 246). The evidence also indicated that the species groups of the Old and New Worlds evolved from a common ancestral stock. Within the New World species-group, *C. fasciata, C. cayennensis* and *C. leucocephala* are most like the hypothetical ancestral type whereas *C. picazuro* is somewhat different and *C. flavirostris* and *C. maculosa* are most unlike the ancestral type. Of the five Old World species examined, *C. guinea* and *C. palumbus* are the most similar to one another. *C. livia, C. oenas*, and *C. janthina* are each distinct from one another and from *guinea* and *palumbus*. The possibility of parallel evolution of characters as opposed to the lack of change in some genetic lines was recognized as an explanation for some of these results.

In contrast to the study of the genus *Columba*, studies of the red cell antigens of some species of the genus *Streptopelia* did not show a correlation between geographic distribution and antigenic phenotype (Irwin 1953, 1955). The genetic relationships within the genus *Streptopelia*, as indicated by the red cell antigen specificities, are as follows: *S. risoria* and *S. semitorquata* are more closely related to one another than either is to some other species of *Streptopelia*. *S. chinensis* and *S. senegalensis* are most similar to one another but *senegalensis* is somewhat more like other species of *Streptopelia* than is *chinensis*. *S. capicola, S. bitorquata, S. tranquebarica (=humilis)*, and *S. orientalis* are more similar to one another than any one of them is to the other four species of *Streptopelia* examined (Irwin 1953, 1955, 1971; Stimpfling and Irwin 1960a; Bryan and Irwin 1961).

The nature of these serological data, i.e. the relatively unambiguous relationship between antigenic characters and the genes controlling the synthesis of these substances, increases one's confidence in the taxonomic conclusions based upon them. Nevertheless, we believe Irwin and his colleagues have been conservative in drawing conclusions of taxonomic importance and we have some additional inferences from their data which are summarized in Fig. 1. Before presenting these inferences it is necessary to specify our assumptions concerning the serological data and to clarify our usage (as opposed to standard immunological usage) of the term 'homologous'.

Although 'very little information is available concerning the chemical bases of the specificities of the antigens of the red blood cells' (Bryan and Irwin, 1961: 333) of birds, the A, B, M, and N substances of human red blood cells are thought to be glycolipids (Watkins 1966). The data suggest that antigenicity is conferred by the carbohydrate moiety of these molecules. Some, if not all, of the antigenic specificities assigned to the pigeon species might be ascribed to the interaction of antibodies with the subunit structures of glycolipid molecules. If this is accepted, the antigenic specificities, for example those assigned to the group-1 substances (Stimpfling and Irwin, 1960a) or the C substance of *Columba guinea* (Bryan and Irwin, 1961) can be equated to variant carbohydrate chains on glycolipid molecules. However, if the C substances and group-1 substances of pigeon species are glycolipids, then the genes controlling the synthesis of the enzymes which in turn control the synthesis of those carbohydrate portions of the glycolipids that are antigenic determinants must be located within the region of a single chromosomal inversion. If located outside of an inversion it is highly probable that segregation of these genes would have been observed in some of the backcross matings carried out by Irwin and others. However, avian chromosomes are not known to contain inversions. If each specificity is under the control of a separate gene, as seems to be the case for the human blood-group antigens (Watkins 1966), and if the 21 specificities assigned to the group-1 substances are structurally comparable to the human blood-group determinants, then several genes (as few as five acting together or as many as 21 acting separately) controlling the synthesis of enzymes such as specific glycosyl transferases would be required to account for the specificities observed.

On the other hand, the antigenic specificities might be ascribed to certain amino acid residues of a single protein with differences among species being due to changes in the primary, and hence tertiary, structure of the protein. If this were true, then the serological data would provide an index to some of the amino acid substitutions that have been incorporated into two series of homologous proteins (group-1 and C) during at least part of the evolutionary history of the Columbidae.

When speaking of proteins we use the term 'homologous' in accordance with the usage generally accepted by biologists (see for example Nolan and Margoliash, 1968: 728) and thereby consider proteins of two species to be homologous if the genes controlling their synthesis were derived from a common ancestral gene. Immunologists, on the other hand, have traditionally used the term 'homologous' as though it were synonymous with 'identical'. That is, although the genes controlling the synthesis of the red cell antigens of two species may have had a common ancestor, the antigens were not considered to be homologous substances unless, in reciprocal tests, each would remove from an antiserum all of the antibodies produced in response to the other antigen.

Rather than analyze these serological data as though they pertain to either glycolipid or protein molecules, we shall make the assumption that differences result from single amino acid substitutions whether these be in a protein that (1) controls the synthesis of a glycolipid, (2) controls the modification of a glycolipid carbohydrate side chain, or (3) forms part of the structure of the red cell surface. In the first two cases the relationship between antigenic specificity and gene structure is somewhat indirect but nevertheless unambiguous, and in each of these three cases an antigenic specificity hypothetically can be related to a single nucleotide codon. It thus becomes irrelevant whether the data pertain to a single cistron or several cistrons.

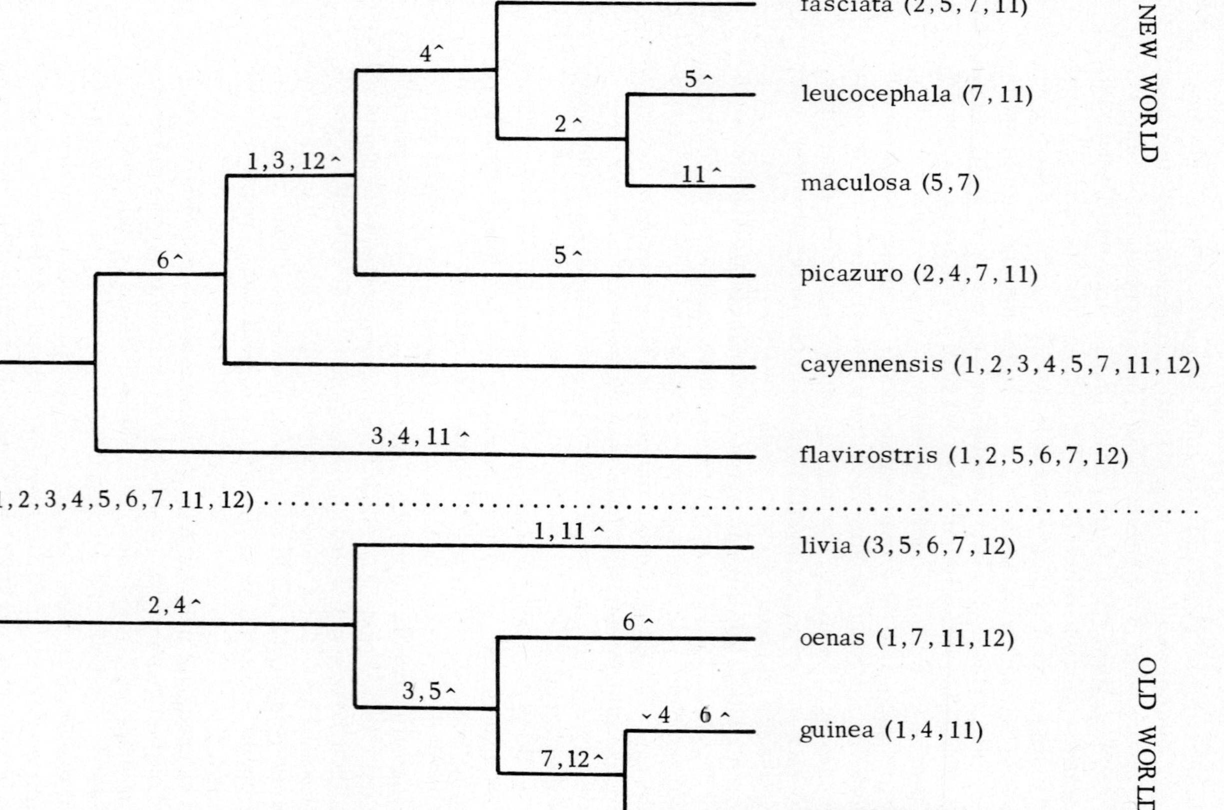

FIG. 1. An arrangement of some species of the pigeon genus *Columba* based on the distribution of the *d* antigens of *Streptopelia chinensis*. The presumed ancestral array of *d* antigens is given in parentheses at the left. The *d* antigens lost are shown at points along the branching sequence. Numbers in parentheses at the right are those *d* antigens possessed by a species. Taken from data in various papers by M. R. Irwin and colleagues.

To reconstruct the genetic relationships of the species of pigeons we assume that the most closely related species produce the most similar antigens, and that an antigenic specificity shared by all species represents the ancestral condition whereas specificities shared by a limited number of species represent more recent evolutionary variants. For example specificity *d*-1 of *Streptopelia chinensis* is shared by species of five genera (*Columba*, *Streptopelia*, *Zenaida* (=*Zenaidura*), *Phaps* and *Ocyphaps*) which means that this substance either was present in an ancestral stock prior to the divergence of these genera or has subsequently arisen independently in each lineage. The former explanation seems more likely. Most of the remaining *d* specificities are shared in various combinations by only two genera, *Columba* and *Streptopelia*, which suggests that these specificities either arose after the separation of *Columba* and *Streptopelia* lineages from the others or that these specificities were subsequently lost from the other lineages. In either case, the genus *Streptopelia* appears to be more closely related to the genus *Columba* than to any of the other genera examined.

Likewise on the basis of the distribution of antigenic specificities assigned to the C substance of *Columba guinea* (Bryan and Irwin 1961) the genus *Zenaida* (including *Zenaidura*) is more closely related to the genus *Columba* than are *Streptopelia*, *Phaps*, *Chalcophaps* and *Ocyphaps*.

The inferred genetic relationships among some of the species of the genus *Columba* are shown in Fig. 1. The lengths of the vertical lines in this figure are arbitrary and are not meant to be proportional to the time since branching occurred. The arrangement of the species shown minimizes the number of mutational events required to account for the distribution of antigens and antigen specificities. This has been accomplished in part by allowing both forward and back mutations. As represented, however, most mutations result in the loss of an antigenic determinant. Thus, species considered to be most like the ancestral type have the most antigens or antigenic specificities assigned to them. In part, this results from the reference species being a member of a genus other than that being examined. For example, the relationships of the species of the genus *Columba* are based on the distribution of the *d* antigenic substances assigned to *Streptopelia chinensis*.

The relationships of the species of *Streptopelia* also can be based on the assignment of the *d* antigens (Cumley and Irwin 1952; Irwin 1932a, 1949a, 1949b, 1953, 1955, 1971; Irwin and Cole 1940, 1945a, 1945b), the specificities assigned to the C substance of *Columba guinea* (Bryan and Irwin 1961), and the specificities assigned to the group-1 substances (Stimpfling and Irwin 1960a,b).

Data pertaining to the antigens and antigenic specificities of a pair of contrasting cellular characters specific to the muscovy duck (*Cairina moschata*) in contrast to the mallard (*Anas platyrhynchos*) were presented by Gordon (1938) and McGibbon (1944, 1945). These data alone are not of taxonomic significance but if additional anatid species are compared, this work would provide reference information.

C. SEROLOGICAL STUDIES BY D. MAINARDI AND ASSOCIATES

Between 1954 and 1963 Danilo Mainardi and several co-authors published at least 31 papers which reported upon serological studies of bird relationships. The most frequent technique employed was that also used by Irwin and his associates, namely, the injection of avian red blood cells into rabbits and the utilization of the resultant antisera to test the agglutination of red cells from the various species involved in a given problem. Absorbed antisera were often prepared. Mainardi wrote several reviews (1958a, 1961c, 1963a, 1963b) in which he described his techniques and stated the theoretical basis for the studies. Mainardi assumed that the red cell antigens are proteins and that each antigen is the product of a single gene. Mainardi based most of his assumptions upon the immunogenetic studies of Irwin and his colleagues, hence the above comments (see pp. 112–113) concerning the possibility that the red cell antigens are actually glycolipids must also apply to Mainardi's work. Like Irwin, Mainardi assumed that (1963a: 103)

> 'it seems logical to consider the erythrocyte antigens common to two species as substances derived from a common progenitor, and those (specific) antigens differing in the two species as a sign of the evolutionary progress covered. Consequently, the larger the number of common antigens, the closer to one another, in a phylogenetic sense, will the two species be'.

Mainardi developed a formula to measure the "immunological distance' between species and applied it to most of his studies. He described the procedure as follows (1963a: 103–104).

> 'An index of the relations between common antigens and specific antigens between two species may be obtained from the titers with homologous and heterologous red cells, using immune sera against the red cells of the two species. Utilizing the ratios between the titers with homologous and heterologous cells with sera against the two species under examination, a formula was obtained (Mainardi, 1957a) which gives values that increase with the diminution of immunological affinity (i.e. the higher the value, the fewer the antigens held in common). Hence, this may be used as an index to the degree of relationship between the species examined. The formula, named "immunological distance," is as follows:
>
> $$\text{I.D.} = \frac{1}{\sqrt{\frac{Oa}{Ea} \cdot \frac{Ob}{Eb}}}$$
>
> where Oa and Ob are the titers with homologous cells and Ea and Eb the titers with heterologous cells, using sera against the species a and b, respectively'.

The studies by Mainardi and co-authors fall into seven categories under which they will be reviewed.

1. Columbiformes

Four papers on pigeons and doves were published. The first of these (Cavalli-Sforza *et al.*, 1954) reported on a study of four species of *Columba* which was carried out in an attempt to determine the genetic origin of pigeons which were being used in studies of orientation and homing ability. Anti-red cell antisera against *Columba livia*, *C. guinea*, *C. palumbus* and *C. albitorques* were prepared. With absorbed antisera it was possible to distinguish among the species and to determine the number of common and specific antigens. A total of 23 antigens were detected among the four species, each species having 10. Three of the 10 were present in all species. *Livia* and *albitorques* shared seven; *livia* and *guinea* shared five; *livia* and *palumbus*, five; *guinea* and *albitorques*, five; *guinea* and *palumbus*, three, and *palumbus* and *albitorques*, five. Thus, among these four species, *livia* and *albitorques* are assumed to be the two most closely related and *guinea* and *palumbus* the two most distantly related.

A second paper on *Columba* (Cavalli-Sforza *et al.*, 1957) analysed the results of a serological study of a hybrid *livia* × *albitorques*. The hybrid contained the common and specific antigens of both species but no specific hybrid antigens were detected.

Two papers concerned with the dove genus *Streptopelia* were published by Mainardi. In the first of these (1956b) he reported that *S. risoria* and *S. decaocto* each possess many specific red cell antigens and therefore are probably not the wild and domestic forms of a single species. Instead Mainardi supported the opinion of Taibel (1951) who proposed that *S. roseogrisea* was the wild progenitor of the domestic dove. Later Mainardi (1957e) examined the relationships among *S. risoria*, *S. turtur* and *S. decaocto* and expressed them in terms of his 'immunological distance' formula. The calculated I.D.s were: *risoria-turtur* 1·4, *risoria-decaocto* 1·4, *turtur-decaocto* 8. Thus *risoria* was found to be serologically intermediate between *turtur* and *decaocto* and the latter two relatively distant from one another.

2. Passeriformes

Nine papers dealt with problems of passerine classification; eight of them concerned the Fringillidae and their relatives. Mainardi (1956a) began this series with a study of the red cell antigens of the European goldfinch (*Carduelis carduelis*), the European linnet (*Carduelis cannabina*) and the siskin (*Carduelis spinus*). Mainardi noted that these three species had been placed in different genera (*Carduelis*, *Spinus*, *Acanthis*, etc.) by various authors and this study was an attempt to resolve the question of their affinities. Using his standard anti-red cell procedure and absorbed antisera Mainardi found that *C. carduelis* and *C. cannabina* both have specific antigens not shared with the other two species but that *C. spinus* has no specific red cell antigens not shared with one of the other two species. *C. carduelis* shares more antigens with *C. cannabina* than with *C. spinus* and *C. spinus* and *C. cannabina* are especially closely related. Thus *C. cannabina* is, in a sense, between the other two.

Mainardi (1957a) used absorbed antisera in a study of the greenfinch (*Chloris chloris*), the European goldfinch (*Carduelis carduelis*) and the chaffinch (*Fringilla coelebs*). The 'immunological distances' among these species were calculated as follows: *Chloris–Carduelis* 2·8, *Chloris–Fringilla* 5·6, *Carduelis–Fringilla* 1·4. By using the data from his earlier paper (1956a) Mainardi (1957a: 156) also calculated the following I.D.s: *C. spinus–C. cannabina* 1·0, *C. cannabina–C. carduelis* 1·4, *C. spinus–C. carduelis* 8·0, *C. cannabina–Fringilla* 8·0, *C. spinus–Fringilla* 16·0. It is clear that the I.D.s are not reciprocally correct, i.e., they are not additive. In addition it is doubtful that any taxonomist would be willing to accept data that indicate that the European goldfinch is more closely related to the chaffinch than it is to the siskin. In a later paper Mainardi (1957b) again presented these same serological data plus a comparison of the paper electrophoretic patterns of the hemoglobins of the same species and several additional species. He found that *Fringilla* could be separated from the carduelines by having a different hemoglobin pattern and concluded that his data agreed 'with ethological and morphological studies'. (1957b: 180.) This is a safe conclusion but it is not supported by Mainardi's serological data as noted above.

The same three carduelines and *Fringilla* were again considered in a third paper (1957c) which presented the data from the two previous publications (1957a,b) but analyzed them in a more elaborate fashion. However, the conclusions about their relative taxonomic relationships were unchanged.

Mainardi (1957d) next prepared antisera against the red cells of the chaffinch (*Fringilla coelebs*) and also those of the brambling (*F. montifringilla*). These two species were separable only when absorbed antisera were used. Their close relationship was therefore confirmed.

Mainardi (1958b) attempted 'to erect a phylogeny of the avian family Fringillidae by a study of the immunological relationships of some species of fringillids and their allies'. (1958b: 336.) In this paper Mainardi reviewed the studies by Beecher (1953), Tordoff (1954) and Stallcup (1954) and considered their conclusions in relation to his own. Immunological and electrophoretic data for several additional species were presented and Mainardi concluded that his data supported Tordoff's work, based on skeletal characters, and Stallcup's, based on leg musculature and serology but disagreed with the conclusions that Beecher derived from his study of jaw musculature. Mainardi's conclusions were that the Carduelinae occupy a central position with relationships to the Estrildinae and Passerinae on one side and to the Fringillinae and Emberizinae on the other. A diagrammatic summary of Mainardi's conclusions follows on p. 119.

This paper was also the basis for Mainardi's later note (1960b) in which he expressed his belief that the Carduelinae should be placed in the Ploceidae.

Mainardi (1960a) questioned Vaurie's (1959) generic allocation of certain carduelines, specifically that of the European linnet which Vaurie placed in *Acanthis* while other authors had placed it in *Carduelis*. Mainardi considered

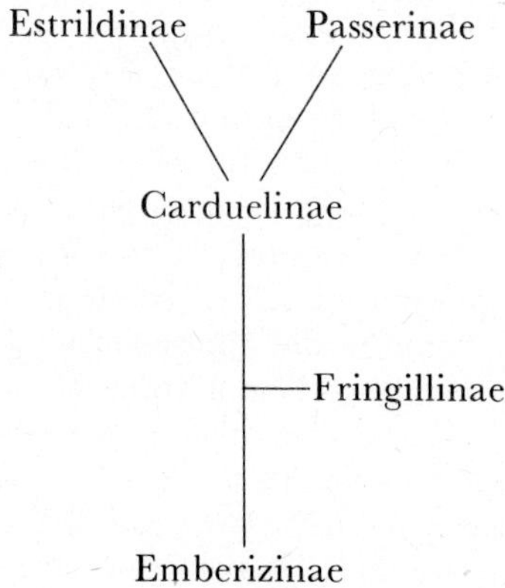

his own serological data from comparisons of the red cell antigens, the electrophoretic behavior of hemoglobins and the data on hybridization and concluded that the linnet was closely related to other species placed in *Carduelis* and that it too should be included in that genus.

From a serological study of four species representing four passerine families (Corvidae, Ploceidae, Fringillidae, Sturnidae) Mainardi (1961a) concluded that the 'immunological distances' among these groups were as follows: *Passer* (Ploceidae)–*Fringilla* 16, *Passer–Corvus* 128, *Fringilla–Corvus* 90, *Passer–Sturnus* 512, *Fringilla–Sturnus* 16. The test between *Corvus* and *Sturnus* gave no reaction and thus these two were the least related pair of genera studied. It is obvious that these results contain internal discrepancies. Mainardi suggested that his data indicated a 'phyletic series' expressed by the following diagram:

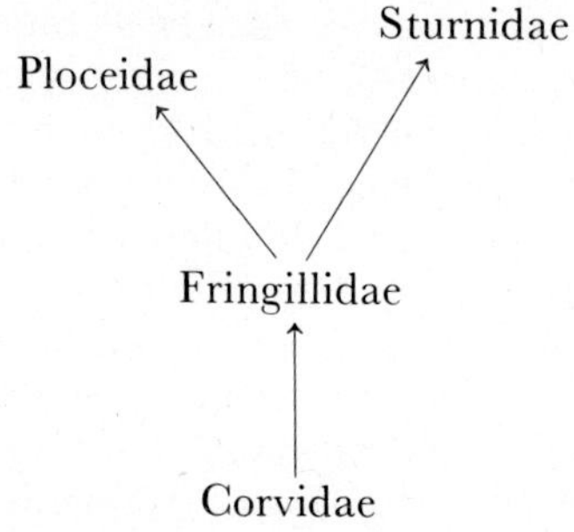

Mainardi reaffirmed his belief in the position of the Carduelinae as 'the bridge between *Fringilla* and the Ploceids' (1961a: 117).

3. Galliformes

Seven papers on gallinaceous birds were published by Mainardi and his colleagues. In the first of these Mainardi (1958c) presented the results of a serological comparison of the red cell antigens and a chromatographic comparison of the free amino acids in the pectoral muscles of *Numida meleagris*, *Gallus gallus* and *Meleagris gallopavo*. Both studies indicated a closer relationship between *Numida* and *Meleagris* than between either of these and *Gallus*. It should be noted that the chromatographic patterns produced by the free amino acids in animal muscle tissue are unlikely to contain significant taxonomic information and that if Mainardi had carried out more extensive comparisons his seemingly neat correlation would almost certainly have disappeared.

Two subsequent papers reported on additional species. Mainardi (1959b) added the Japanese quail (*Coturnix coturnix japonica*) and the ring-necked pheasant (*Phasianus colchicus*) to his study and concluded that *Numida*, *Meleagris* and *Phasianus* are closely related while *Coturnix* and *Gallus* are more remote from these and from one another. He recommended the inclusion of *Numida*, *Meleagris* and *Phasianus* within the same family.

The next paper (1959c) added the golden pheasant (*Chrysolophus pictus*) and the silver pheasant (*Gennaeus nycthemerus*) to the species being compared. These two species also showed close affinities with *Numida*, *Phasianus* and *Meleagris*. Thus the pheasants, guinea fowl and the turkey formed a fairly tight cluster with *Coturnix* somewhat distant in one direction and *Gallus* even more remote in another direction. The marginal position of *Gallus* was also supported by a study by Mainardi and Guerra (1959) in which they used the antigenic properties of bovine serum to test the agglutination of avian red cells. *Gallus*, *Meleagris*, *Numida*, *Coturnix* and *Phasianus* were compared and *Gallus* red cells were found to be readily distinguishable from those of the other four genera all of which gave essentially the same reaction.

The peafowl (*Pavo cristatus*) was next added to the list of species studied by the serological analysis of their red cell antigens (Mainardi, 1960c) and it too was found to be part of the pheasant–guinea fowl–turkey cluster. An especially close relationship was demonstrated between *Pavo* and *Numida*.

In two final papers on the Galliformes, Mainardi and Taibel (1962a,b) reviewed various sources of evidence bearing upon the relationships within the order and also presented some additional immunological data in the second paper (1962b). To the list of genera already studied they added four genera of Cracidae, *Crax*, *Mitu*, *Penelope* and *Pipile* thus giving a total of 12 genera of galliform birds which were compared on the basis of the red cell antigens. The red cells of the megapode *Alectura* were also compared but an anti-*Alectura* antiserum was not prepared. Mainardi's conclusions from his studies of the Galliformes may be summarized as follows:

i. *Numida* and *Meleagris* are true phasianids.

ii. *Meleagris* is closely related to the typical pheasants such as *Phasianus*, *Gennaeus* and *Chrysolophus*.

iii. *Pavo* is more closely related to *Numida* than to any other phasianid.

iv. *Gallus* and *Coturnix*, although well differentiated, are closer to the pheasants than to the Cracidae or Megapodiidae and should be included in the Phasianidae.

v. The four genera of Cracidae which were studied are closely related to one another and *Numida* is the phasianid which shows the greatest affinity to the cracids.

vi. The Phasianidae, Cracidae and Megapodiidae are closely related to one another and should be treated as three families having a common origin.

vii. *Numida*, *Meleagris*, *Pavo*, the typical pheasants, *Gallus* and *Coturnix* all should be included in the Phasianidae.

The studies of the Galliformes were summarized and discussed by Mainardi (1963a).

4. *Anseriformes*

Only one paper by Mainardi (1959a) concerned the ducks and geese. This study utilized a novel immunological procedure in which a mixture of the red cells of the mallard (*Anas platyrhynchos*), Muscovy duck (*Cairina moschata*) and the domestic goose (*A. anser*) was used as an antigenic solution. The resultant antiserum thus contained antibodies against the red cell antigens of all three species. Since the two most closely related species will have the greatest number of common antigens the two closest relatives should produce the highest titers of reaction. The results showed that *Anas* and *Cairina* were the most closely related, that *Anas* and *Anser* were next in degree of relationship and that *Anser* and *Cairina* were the most distantly related pair of species in this triangle. These data agreed with evidence from other sources including hybridization.

5. *Phoenicopteridae*

Mainardi's (1962b) study of the serological relationships of the flamingos lead him to conclude that the flamingos, Anseriformes and Ciconiiformes had had a common origin and that they should be treated as three separate but related orders. In a later paper (1963a) Mainardi suggested that a combination of immunological and paleontological data suggested that the Anseriformes diverged first from the common ancestral line and that the flamingos and Ciconiiformes separated from one another later. He explained his finding that the three groups share approximately equal numbers of red cell antigens as the result of different velocities of differentiation following their separation. This problem has also been studied by Sibley *et al.* (1969, see p. 113) who concluded that the total available evidence suggests that the flamingos are actually closer to the Ciconiiformes and should be included within that order.

6. *Hybrids*

Mainardi and his colleagues have published five papers concerning serological studies of avian hybrids. A study of blood groups in *Passer* (Bocchi *et al.*, 1960) led to the conclusion that the house sparrows of Sicily 'are at least in part hybrids between *Passer hispaniolensis* and *P. domesticus italiae*' (p. 63).

Mainardi (1961b, 1962a) used his usual red cell antigen technique to study a presumed wild hybrid between the European goldfinch (*C. carduelis*) and the greenfinch (*Carduelis chloris*). The bird was found to possess the specific and common antigens of both presumed parental species and it was therefore concluded that the specimen was a true F_1 hybrid between these two species.

A similar study of a presumed natural hybrid between the chaffinch (*Fringilla coelebs*) and the brambling (*F. montifringilla*) was also carried out (Lofts and Mainardi, 1963) and the hybrid origin of this specimen was also confirmed. The testes of this hybrid were normal and in full breeding condition.

Mainardi and Schreiber (1963) used the same red cell antigen procedure to confirm the origin of several suspected hybrids between the Barbary dove (*Streptopelia risoria*) and the collared dove (*S. decaocto*). Of four hybrid individuals one was apparently an F_1, the others were probably the result of backcrossing.

D. STUDIES BY A. C. WILSON AND ASSOCIATES

The serological technique of microcomplement fixation has been used by Allan C. Wilson and his colleagues. In each of their studies proteins were purified and then injected into rabbits to produce antibodies specific to a single protein.

Microcomplement fixation studies involving anti-chicken heart muscle lactic dehydrogenase (Wilson and Kaplan, 1964) showed that among orders of birds there is a decreasing similarity among muscle lactic dehydrogenases in the following sequence: Galliformes, Falconiformes, Pelecaniformes, Procellariiformes = Anseriformes, Coraciiformes, Columbiformes = Trogoniformes, Gruiformes, Apodiformes, Psittaciformes, Strigiformes = Ciconiiformes, Rheiformes, Podicipediformes, Piciformes, Cuculiformes, Charadriiformes = Passeriformes. Values for the family Musophagidae would fall within those of the Procellariiformes and Anseriformes rather than with those of the Cuculiformes.

These results are paralleled in part by studies using anti-chicken breast muscle lactic dehydrogenase, anti-chicken triosephosphate dehydrogenase, anti-chicken glucose-6-phosphate dehydrogenase and anti-chicken hemoglobin (Wilson *et al.*, 1964; Wilson and Kaplan, 1964). The antisera to these proteins were, however, tested against the appropriate protein systems of only five species; chicken (*Gallus gallus*), turkey (*Meleagris gallopavo*), duck (*Anas platyrhynchos*), ostrich (*Struthio camelus*) and pigeon (*Columba livia*).

In a study in which galliform egg-white lysozymes were compared by means of the microcomplement fixation technique (Arnheim and Wilson 1967) it was found that chicken lysozyme is more similar to the lysozymes of the bobwhite quail (*Colinus virginianus*), Sharpe's francolin (*Francolinus clappertoni*), California quail (*Lophortyx californica*), and chukar partridge (*Alectoris graeca*) than to any of eleven other species tested including six species of pheasants (*Crossoptilon auritus*, *Chrysolophus amherstiae*, *Chrysolophus pictus*,

Lophura swinhoei, *Syrmaticus reevesii*, *Phasianus colchicus*), the blue peafowl (*Pavo cristatus*), the turkey (*Meleagris gallopavo*), the ruffed grouse (*Bonasa umbellus*), the guinea-fowl (*Numida meleagris*), and the Japanese quail (*Coturnix coturnix*). Although these results were not expected from previous knowledge of these species, the results were substantiated in part by a second study (Arnheim, Prager and Wilson 1969) in which chicken lysozyme was shown to be more similar to bobwhite quail lysozyme than to ring-necked pheasant (*Phasianus colchicus*) lysozyme on the basis of amino acid compositional and tryptic peptide analyses as well as by microcomplement fixation tests.

E. STUDIES BY R. E. FEENEY AND ASSOCIATES

In addition to their investigations of the electrophoretic and biochemical properties of the egg-white proteins (see pp. 155–157) Robert E. Feeney and his colleagues have carried out immunological comparisons of avian conalbumins (=ovotransferrins) and ovomacroglobulins (=component 18).

Purified conalbumins were used as antigens in immunoelectrophoretic studies. Anti-chicken conalbumin antiserum did not form precipitin arcs in tests against Adelie penguin (*Pygoscelis adeliae*) conalbumin (Feeney *et al.*, 1965). Adelie penguin conalbumin reacted weakly with anti-cassowary conalbumin antiserum. Kiwi (*Apteryx*) conalbumin reacted strongly with anti-cassowary conalbumin antiserum. The egg-white of cassowary (*Casuarius aruensis*), rhea (*Rhea americana*), ostrich (*Struthio camelus*), and tinamou (*Eudromia elegans*) reacted with anti-cassowary conalbumin antiserum but not with anti-chicken conalbumin antiserum (Miller and Feeney 1964). The egg-white of turkey (*Meleagris*), mourning dove (*Zenaida macroura*), golden pheasant (*Chrysolophus pictus*), and green Java peafowl (*Pavo muticus*) reacted with anti-chicken conalbumin antiserum but not with anti-cassowary conalbumin antiserum. Japanese quail (*Coturnix coturnix*) egg-white reacted strongly with anti-chicken conalbumin and weakly with anti-cassowary conalbumin antiserum. Both chicken and cassowary conalbumin reacted weakly with anti-mallard (*Anas platyrhynchos*) egg-white antiserum. Using Ouchterlony plates, the egg-white of the tinamou (*Eudromia*) and the ostrich (*Struthio*) reacted weakly with anti-cassowary conalbumin antiserum (Miller and Feeney 1964). The reaction with rhea (*Rhea*) egg-white showed a spur of non-identity and there was no reaction with mallard (*Anas*) or quail (*Coturnix*) egg-white.

In another study (Feeney *et al.*, 1968) using immunoelectrophoresis, there was no cross reaction between kiwi, cassowary, emu, rhea, or ostrich conalbumins and anti-penguin conalbumin antiserum. This antiserum did cross react weakly with the egg-white of the mallard. The conalbumins of the western grebe (*Aechmophorus occidentalis*), Laysan albatross (*Diomedea immutabilis*) and pink-footed shearwater (*Puffinus creatopus*) cross reacted with anti-penguin egg-white antiserum.

Ovomacroglobulin (component 18) is a major egg-white component that is highly cross reactive in serological tests (Feeney *et al.*, 1965; Miller and

Feeney, 1966) but chicken ovomacroglobulin contains antigenic determinants not present on tinamou and duck ovomacroglobulins as determined by two dimensional immunodiffusion studies (Miller and Feeney, 1966). Chicken ovomacroglobulin is similar in amino acid composition to the homologous protein of the mallard, Adelie penguin (*Pygoscelis adeliae*), and tinamou (*Eudromia elegans*) (Miller and Feeney 1966). Immunological tests comparing the cross reactivity of all the egg-white components have been carried out, in addition to the tests comparing only the cross reactivity of conalbumins. Anti-cassowary egg-white antiserum reacted with the egg-white proteins of the following species starting with the strongest reaction and proceeding to the weakest: cassowary–emu–rhea–ostrich–kiwi–tinamou–chicken–penguin (Miller and Feeney, 1964; Feeney *et al.*, 1966). This antiserum did not cross react with either pigeon or turkey egg-white. Anti-chicken egg-white antiserum reacted well with chicken, turkey and golden pheasant egg-white proteins and weakly with rhea, emu and cassowary egg-white proteins. This cross reaction is probably due to the precipitation of ovomacroglobulin by anti-chicken ovomacroglobulin. Pigeon egg-white proteins did not cross react with anti-chicken egg-white antiserum. Anti-Adelie penguin (*Pygoscelis adeliae*) egg-white antiserum cross reacted extensively with the egg-white proteins of the pink-footed shearwater (*Puffinus creatopus*), the western grebe (*Aechmophorus occidentalis*), the Laysan albatross (*Diomedea immutabilis*) and the mallard (*Anas platyrhynchos*). Chicken egg-white protein reacted somewhat with this antiserum (Feeney *et al.*, 1966). Anti-kiwi (*Apteryx mantelli*) egg-white cross reacted to a small degree with mallard egg-white proteins. Anti-Adelie penguin serum antiserum cross reacted extensively with sera of the Emperor penguin (*Aptenodytes forsteri*) and the Humboldt penguin (*Spheniscus humboldti*).

F. AVIAN BLOOD GROUP STUDIES

Dujarric de la Riviére and Eyquem (1953) reviewed the work on the red cell antigens of chickens, pigeons and ducks. Until 1953 only two series of erythrocyte antigens were known in the chicken (*Gallus*), one consisting of a series of nine factors and another of five. Different breeds and inbred lines of chickens differed in the frequencies of these factors. Gilmour's (1962) review included data on seven separately inherited blood group systems. Other reviews of blood group systems were published by McDermid (1965) and Perramon (1965). Morton *et al.* (1965) investigated the association of the polyallelic B group locus and the polymorphism of conalbumin, G_2 globulin and G_3 globulin of the egg-white in relation to hatchability in the Light Sussex breed of chickens. Blood group genotypes of zygotes but not of dams were associated with marked differences in mortality over the entire incubation period. Blood groups in ducks are further reviewed by Podliachouk (1965).

Norris (1963), using commercially available antisera normally used for human blood typing, studied the presence of 'human ABO-like' blood groupings of 658 passerines (94 species) and 92 non-passerines (44 species).

Within given species individuals were either negative or positive with regard to their agglutination with the various antisera. Thus in avian blood group factors we have a valid, presumably genetically controlled, polymorphism.

Bush (1965) examined the ABO-like blood groups in the house sparrow (*Passer domesticus*) and suggested that in this species the blood groups are inherited differently from the ABO blood groups in man. The frequency of the blood groups in this species exhibited geographical trends. Populations from the southeast had more O-like individuals and those from the southwest, north and south Pacific coast had more AB-like individuals.

Vohs (1966) found that ring-necked pheasants (*Phasianus colchicus*) from Iowa exhibited three antigenic factors, A, B and C in their red blood cells. Antisera were prepared by iso-immunization or in rabbits. Factors A, B and C are apparently inherited in a simple Mendelian fashion and appear to segregate independently. The frequencies of the factors varied in limited samples of three widely spaced populations in Iowa. Northern and southern populations were similar but unlike the central population, factor B frequencies being 0·59, 0·49 and 0·19 respectively.

As the above studies indicate, blood groups are of importance in the investigation of the population structure of a species. In this respect they are the same as other genetically polymorphic characters such as electrophoretically determined protein variants which are discussed later (see pp. 144–153). In spite of this blood groups have received little attention in other than domestic species. This is undoubtedly due to the greater number of variant systems which can be examined by electrophoresis and also the greater technical facility of this approach. Because of the high degree of polymorphism which is present, blood groups are of limited value in determining higher category relationships.

III. Studies Using Electrophoretic Techniques

A. HIGHER CATEGORY SYSTEMATICS

1. Miscellaneous Studies

Electrophoresis has been useful in enabling the avian systematist to delimit taxonomic problems and to suggest possible solutions. It is the principal method by which many samples can be screened rapidly. With questions of relationships thus posed, the techniques of peptide analysis or amino acid sequence determination may be employed to obtain a more definitive answer. This section reviews comparative electrophoretic studies concerned with taxonomic problems above the species level.

Following the introduction of the technique of moving boundary electrophoresis by Tiselius (1937), investigators soon began to examine avian material. The first comparative study was that of Landsteiner *et al.* (1938) who examined the egg-white proteins and hemoglobins of the domestic fowl, turkey, guinea fowl, domestic duck, and domestic goose. They observed that the gallinaceous birds were alike and that the two waterfowl differed from them. Other early papers which examined egg-white proteins were those of Longsworth *et al.* (1940), Bain and Deutsch (1947), and Forsythe and Foster (1950). The first extensive study (McCabe and Deutsch, 1952) involved 37 species including a grebe, eight ducks, 16 gallinaceous birds, two rails, a tern, two doves, and seven passerines. The similarities among the patterns were in close accord with accepted classifications. The authors felt that the electrophoretic technique would be an aid to systematics.

Moore's (1945) study of serum proteins was the first to include avian material. He examined samples from the chicken and pigeon among other species of vertebrates. The survey by Deutsch and Goodloe (1945) of plasma proteins included five birds (chicken, turkey, pheasant, pigeon, and duck). Using the technique of paper electrophoresis Common *et al.* (1953) studied the sera of the domestic goose, chicken, and turkey.

Papers dealing with avian hemoglobins were published by Johnson and Dunlap (1955), Dunlap *et al.* (1956), Rodnan and Ebaugh (1956, 1957), Saha *et al.* (1957), and Dutta *et al.* (1958). The findings of these authors may be summarized as follows.

Two species, the chicken (*G. gallus*) and the little cormorant (*Phalacrocorax niger*) had patterns showing three hemoglobin components.

Those with two hemoglobin components were:

Anseriformes: mallard (*Anas platyrhynchos*);
Galliformes: turkey (*Meleagris gallopavo*), ring-necked pheasant (*Phasianus colchius*), guinea fowl (*Numida meleagris*);
Piciformes, Picidae: yellow-shafted flicker (*Colaptes auratus*);
Passeriformes, Tyrannidae: Hammond's flycatcher (*Empidonax hammondii*);
Corvidae: magpie (*Pica pica*);
Paridae: mountain chickadee (*Parus gambeli*);
Turdidae: robin (*Turdus migratorius*);
Sturnidae: starling (*Sturnus vulgaris*);
Ploceidae: house sparrow (*Passer domesticus*);
Fringillidae: evening grosbeak (*Hesperiphona vespertina*), goldfinch (*Spinus tristis*), Oregon junco (*Junco oreganus*), field sparrow (*Spizella passerina*), tree sparrow (*Spizella arborea*), song sparrow (*Melospiza melodia*), Lincoln's sparrow (*Melospiza lincolnii*), white-crowned sparrow (*Zonotrichia leucophrys*).

The two hemoglobin components of passerine species were the most distinct electrophoretically. The electrophoretic separation of the two components was less in the patterns of woodpeckers, and least in those of ducks and gallinaceous birds.

There were three species that had one hemoglobin component: a cuckoo (probably *Eudynamis scolopacea*), a parakeet (*Psittacula*), and the pigeon (*Columba livia*). However, Saha (1964) was able to separate two hemoglobin components in the pigeon by ion-exchange chromatography.

Brandt *et al.* (1952) were the first to use electrophoresis to study the proteins of avian hybrids and their parents. Paper electrophoresis was used to separate the serum proteins from hybrids of chicken × ring-necked pheasant (*G. gallus* × *Phasianus colchicus*) and ring-necked pheasant × golden pheasant (*P. colchicus* × *Chrysolophus pictus*). In most cases the patterns of the hybrids were intermediate between those of the parent species, but in a few instances the authors reported observing new components.

Wall and Schlumberger (1957) found considerable variation among the plasma protein patterns of normal budgerigars (*Melopsittacus undulatus*). Birds with tumors showed marked changes in their globulin fractions.

Conterio and Mainardi (1957) reported on a paper electrophoretic study of hemoglobins from 17 passerine species belonging to five families. They found that the two hemoglobin components of members of the Fringillidae were more widely separated electrophoretically than those of species of the other families (Turdidae, Motacillidae, Sturnidae, Ploceidae).

Using starch gel electrophoresis Beckman *et al.* (1962) examined the sera of 15 goldfinches (*C. carduelis*), 13 canaries (*Serinus canaria*) and six hybrids between these species. They observed a mobility difference in the 'main fast moving component' (= serum albumin) between the parent species. In the hybrids both proteins were present but in half the concentration found in the parent species. The same authors (1963) found a similar situation in the serum albumins of hybrids between two species of pigeons (*Columba livia* × *C. albitorques*), a ring-necked pheasant × chicken hybrid (*Phasianus colchicus* × *G. gallus*), a chicken × guinea fowl hybrid (*Gallus* × *Numida meleagris*) and three hybrids between species of guans (*Cracidae*) (*Penelope purpurascens* × *P. superciliaris*, *P. superciliaris* × *P. 'pileata'* (= *jacucaca*), and *Crax 'globicera'* (= *rubra*) × *C. fasciolata*). However, in hybrids produced by the following crosses no differences from the parents were found: spotbill duck × mallard duck (*Anas poecilorhyncha* × *A. platyrhynchos*), mallard × Muscovy duck (*A. platyrhynchos* × *Cairina moschata*) and silver pheasant × Swinhoe's pheasant (*Gennaeus* (= *Lophura*) *nycthemerus* × *G. swinhoei*).

Beckman and Nilson (1965) studied serum enzyme variations in several bird species and their hybrids. For esterases the patterns of *G. gallus* × *Numida meleagris* and *Columba livia* × *C. albitorques* hybrids were largely a summation of the patterns of the parent species. In others the hybrids differed in their esterase patterns, but interpretation was difficult because the true parents of a hybrid individual often were not known, and the esterases may be polymorphic within a species.

Indistinguishable leucine aminopeptidase patterns were obtained among parents and hybrid *Penelope superciliaris* × *P. jacucaca*, *Crax rubra* × *C. fasciolata*, and *Anas platyrhynchos* × *A. poecilorhyncha*. Hybrids of *Phasianus colchicus* × *G. gallus*, *G. gallus* × *Numida meleagris*, and *Columba livia* × *Streptopelia risoria* showed variants found in neither parent species.

Hybrids of *G. gallus* × *Numida meleagris* and *Penelope superciliaris* × *P.*

jacucaca had a single band of alkaline phosphatase which was intermediate in mobility between those of the parents. In the *Penelope purpurascens* × *P. superciliaris*, *Crax rubra* × *C. fasciolata*, and *Carduelis carduelis* × *Serinus canaria* crosses, no differences between parents and hybrids were noted. A hybrid *G. gallus* × *Phasianus colchicus* had two enzyme bands, neither of which corresponded in mobility to those of the parents.

Using paper electrophoresis Perkins (1964) studied the hemoglobins and serum proteins of seven species of gulls (*Larus glaucescens, L. argentatus, L. canus, L. occidentalis, L. californicus, L. delawarensis* and *L. philadelphia*). The hemoglobin patterns of all species were identical and showed two components. Some fractions in the serum pattern varied within a species, and the author was unable to separate the different species of gulls on this basis.

Saha and Ghosh (1965) extended their earlier studies to include 46 avian species of 13 orders. To the list of species having a single hemoglobin component were added the cattle egret (*Ardeola ibis*), three species of cuckoos (*Cuculus varius, C. micropterus, Centropus sinensis*), three pigeons (*Streptopelia orientalis, S. chinensis, Treron phoenicoptera*), two parrots (*Psittacula krameri, P. cyanocephala*) and the roller (*Coracias benghalensis*). With the exception of the egret the single hemoglobins of the other species had similar mobilities.

The other coraciiform species which they examined was a kingfisher (*Halcyon smyrnensis*) which, unlike *Coracias*, had three hemoglobins. In the mobilities of these components *Halcyon* was most like the barbets (*Megalaima lineata* and *M. asiatica*) and a woodpecker (*Dinopium benghalensis*). A rail (*Amaurornis phoenicurus*) and a coot (*Fulica atra*) also showed three hemoglobin components, as did three passerine species, an Old World oriole (*Oriolus xanthornus*), a jay (*Crypsirina* (=*Dendrocitta*) *vagabunda*) and a crow (*Corvus splendens*).

A number of species showed two hemoglobin components of about the same mobility. These included a kite (*Milvus migrans*), a duck (*Aythya ferina*), a quail (*C. coturnix*), a partridge (*Francolinus pondicarianus*) and an owl (*Otus bakkamoena*). Most passerine birds had nearly identical patterns of two components. They were two bulbuls (*Pycnonotus cafer, P. jocosus*), a cuckoo-shrike (*Coracina novaehollandiae*), two thrushes (*Copsychus saularis, Saxicola insignis*), two babblers (*Chrysomma sinensis, Turdoides somervillei*), four estrildine finches (*Estrilda amandava, Lonchura malabarica, L. punctulata, L. molucca*) and three starlings (*Sturnus pagodarum, Acridotheres tristis, Gracula religiosa*).

In the same paper Saha and Ghosh reported numerous differences in the amino acid compositions of the alpha and beta chains among duck, chicken, and pigeon hemoglobins.

Using starch gel electrophoresis Clark *et al.* (1963) and Feeney and Komatsu (1966) compared the conalbumins (=ovotransferrins) in the egg-white of several ratites and gallinaceous birds. The species varied in both the number and mobility of the conalbumins. It seems likely that electrophoretic comparisons of this protein alone are not useful for higher category systematics.

Baker and Hanson (1966) examined the hemoglobins, erythrocyte and plasma enzymes, and plasma proteins from 17 individuals of eight species of *Anser* and 31 individuals of three species of *Branta*. There were no differences among the eleven species of geese in the starch gel patterns of hemoglobin, but one of 35 tryptic peptides differed chromatographically between *Anser* and *Branta*. They also found no differences in the erythrocyte esterases or lactate dehydrogenases and saw few differences between the genera in the total plasma pattern. The writers noted that in borate gels the plasma patterns of the geese were quite similar to those of gallinaceous birds. Individual variation is discussed elsewhere in this review (pp. 144–153).

Brown and Fisher (1966) studied the serum proteins and hemoglobins of some Procellariiformes by paper electrophoresis. They found that albatrosses (*Diomedea*), shearwaters (*Puffinus*) and petrels (*Pterodroma*) were distinguishable on the basis of the serum proteins although members of the same genus had identical serum protein patterns. The hemoglobins of the albatross were different in the concentration of the two components from those of the other species.

Kitto and Wilson (1966) compared the 'supernatant' form of malate dehydrogenase (S-MDH) from over 100 species of birds. Most species had S-MDH of the same relative electrophoretic mobility as that of chicken (*G. gallus*) S-MDH. This was given a value of 100, and the mobilities of other S-MDHs were expressed as a percentage of it. Members of ten families of shorebirds (Charadriiformes) showed an S-MDH mobility of 55. The swifts (Apodidae) and hummingbirds (Trochilidae) had an enzyme of mobility 63. The authors interpreted this as additional evidence for the unity of these two groups.

Maclean (1967) reported that the electrophoretic patterns of the egg-white proteins of the sand-grouse (Pteroclidae) were more like those of shore birds of the suborder Charadrii than those of either pigeons (Columbidae) or gallinaceous birds.

Osuga and Feeney (1968) investigated the egg-white proteins of the cassowary (*C. casuarius*), emu (*Dromaius novaehollandiae*), ostrich (*Struthio camelus*), rhea (*Rhea americana*), kiwi (*Apteryx mantelli*) and tinamou (*Eudromia elegans*). Their study included electrophoresis of whole egg-white and they concluded that the ratite birds are closely related to one another but only distantly allied to the tinamous. (See additional discussion on p. 123.)

Allison and Feeney (1968) studied the sera of the emperor (*Aptenodytes forsteri*), Adelie (*Pygoscelis adeliae*) and Humboldt (*Spheniscus humboldti*) penguins. They found that the electrophoretic patterns of the penguin sera were similar to one another. The serum albumin of the penguins had a slower mobility than that of the chicken. The Adelie and emperor penguins had four or five transferrin components, the Humboldt penguin had two. The mobility of the serum transferrins of the penguins was less than that of the chicken.

Arnheim and Wilson (1967) and Arnheim *et al.* (1969) found electrophoretic differences among the egg-white lysozymes of chicken, pheasant, and bobwhite quail (*Colinus virginianus*) which were consistent with immunological differences and differences in the amino acid compositions of the tryptic peptides. Bobwhite quail lysozyme differs from that of the chicken by a minimum of two amino acid substitutions; pheasant differs from chicken by at least seven. See discussion elsewhere in this review (p. 122).

Hendrickson (in preparation) has studied the hemoglobins and plasmas of the American thrushes *Hylocichla mustelina*, *Catharus guttatus*, *C. ustulatus*, *C. minimus*, *C. fuscescens* and *Turdus migratorius*. He also has examined the plasma of *Catharus frantzii*, *Turdus grayi*, *T. pilaris*, *T. iliacus* and *T. philomelos*. The LDH of the species of *Turdus* has a slower mobility than that of *Hylocichla* or *Catharus*. *Turdus* also differs in the electrophoretic mobility of its hemoglobin from *Hylocichla* and *Catharus*. Bourns (1967) found serological evidence indicating a closer relationship between *Turdus* and *Hylocichla mustelina* than between the latter and the species of *Catharus*.

Brush (in press) compared the egg-white proteins of the red-winged blackbird (*Agelaius phoeniceus*), Brewer's blackbird (*Euphagus cyanocephalus*) and yellow-headed blackbird (*X. xanthocephalus*). *Agelaius* and *Euphagus* have a simple two allele system controlling the conalbumin locus. The morphs of both species have identical electrophoretic mobilities. *Xanthocephalus* is monomorphic for conalbumin, and the mobility is slower than that of the other two species.

2. *Studies by C. G. Sibley and Associates*

In 1956 Sibley and his collaborators began a long term study to determine avian relationships by using biochemical techniques. A number of papers have been published on electrophoretic comparisons of egg-white, hemoglobin, plasma, and eye lens proteins. Other studies involving peptide mapping and intraspecific enzyme polymorphism are discussed elsewhere in this review (pp. 152–153; 158–163).

The first paper of the series was that of Sibley and Johnsgard (1959a) who analyzed paper electrophoretic separations of over 400 serum samples from 12 species of birds. They found considerable quantitative variation with respect to the age, sex, and reproductive condition of the bird. Some qualitative differences among the species were noted, but in all the serum albumins had nearly the same relative mobility.

This was followed (Sibley and Johnsgard, 1959b) by a study of 244 samples of egg-white from 23 breeds of domestic fowl. The breeds were selected to provide maximum morphological and physiological diversity. They found that the variation in the protein patterns was no greater than would be expected from the normal sampling and measuring errors inherent in the paper electrophoretic technique.

Sibley (1960) analyzed the egg-white protein patterns of 359 species of non-passerine birds by paper electrophoresis. A portion of his evidence corroborated the standard classifications of Mayr and Amadon (1951) and Wetmore (1960), but in other instances the egg-white profiles uncovered new taxonomic problems or shed light on existing ones. The principal conclusions of this paper are summarized below.

i. Among the ratite birds the ostrich (*Struthio*), emu (*Dromaius*), and cassowary (*Casuarius*) seem to be related, but a clear decision on the probable relationships of the rheas (*Rhea*) could not be made.

ii. The patterns of the tinamous (*Tinamidae*) were unlike those of either the rheas or the gallinaceous birds.

iii. Among the Pelecaniformes the egg-white proteins suggested considerable diversity. The patterns of the water-turkey (*Anhinga*) and the cormorants (*Phalacrocorax*) were similar to one another, but differed from all other Pelecaniformes. Those of the tropic bird (*Phaethon*) and frigate bird (*Fregata*) were similar to one another and to *Pelecanus*. Three species of boobies (*Sula*) differed remarkably from one another.

iv. The patterns of the flamingos (Phoenicopteridae) are more like those of storks and herons (Ciconiiformes) than those of ducks and geese (Anseriformes).

v. The loons (Gaviiformes) may be more closely related to the shore birds (Charadriiformes) than to any other group. The shore birds, although exhibiting considerable morphological diversity, have a common egg-white pattern.

vi. The Gruiformes are heterogeneous. The limpkin (*Aramus*) appears to be more closely related to the rails (Rallidae) than to the cranes (Gruidae). There was no evidence in the egg-white patterns to suggest that the sunbittern (*Eurypyga*) or the trumpeters (*Psophia*) are closely related to the Gruidae, the Rallidae, or each other.

vii. The Galliformes and the Anatidae are both closely-knit, monophyletic groups.

viii. The patterns of the falcons (Falconidae) are unlike those of other diurnal birds of prey (Accipitridae), but the patterns of the New World vultures (Cathartidae) do resemble those of the Accipitridae.

ix. The owls (Strigiformes) and nightjars (Caprimulgidae) appear to be related.

x. The turacos (Musophagidae) appear to be related to the cuckoos (Cuculidae) through *Centropus*.

xi. The patterns of the bee-eaters (Meropidae), motmots (Momotidae) and kingfishers (Alcedinidae) are similar enough to suggest a relationship, but that of the dollar-bird (Coraciidae, *Eurystomus*) was so different as to suggest polyphylety of the Coraciiformes.

xii. The patterns of the New World members of the pigeon genus *Columba* were different from those of the Old World species, suggesting two lineages.

xiii. The colies, or mousebirds (Coliiformes) and the trogons (Trogoniformes) are each distinctive and do not seem to have close relatives.

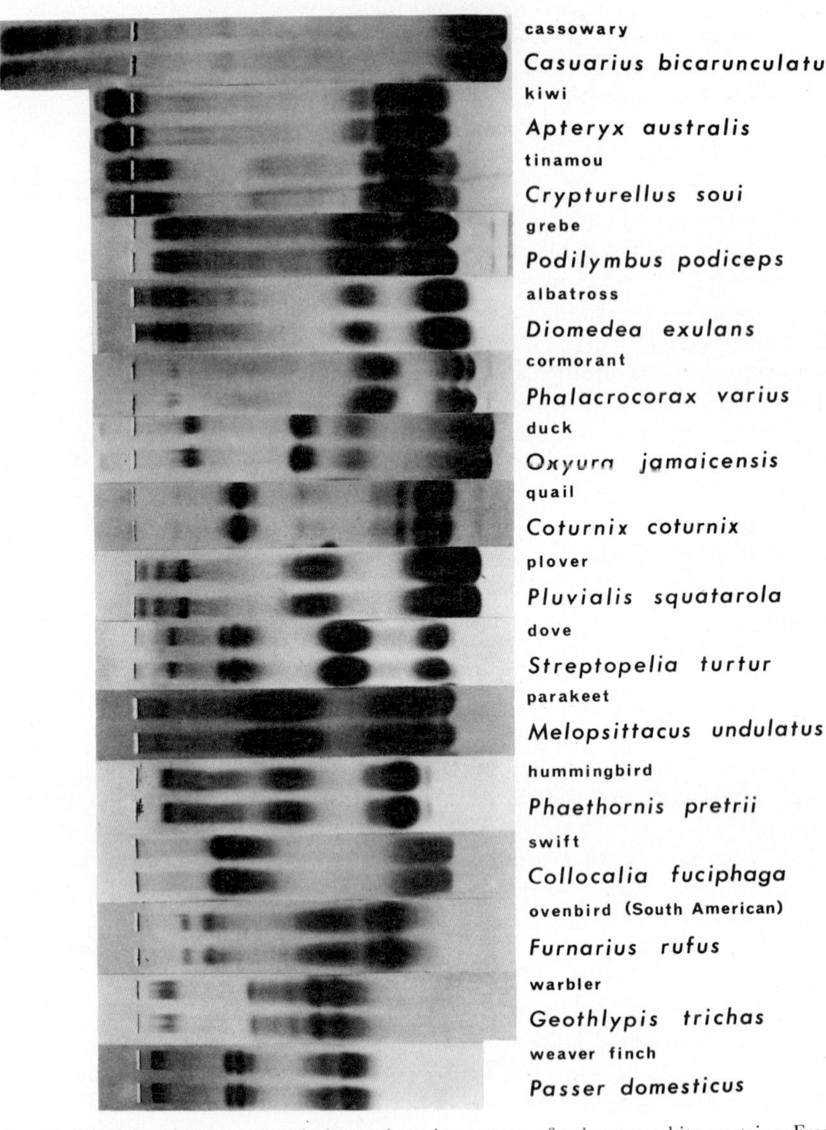

FIG. 2. Examples of the starch-gel electrophoretic patterns of avian egg-white proteins. From Sibley (1967).

xiv. The swifts (Apodidae) and the hummingbirds (Trochilidae) appear to be related to each other and both may be related to the Passeriformes.

Sibley (1962, 1965, 1967) wrote three shorter papers on the application of biochemical techniques to taxonomic problems and cited new evidence of relationships among several avian groups. The paper by Sibley and Brush (1967) on avian lenticular proteins has been discussed on pp. 143–144.

Electrophoresis of egg-white from the first egg ever found of a little-known Costa Rican bird called the wren-thrush (*Zeledonia coronata*) showed it to be a member of the New World nine-primaried oscine assemblage, and not related to the thrushes. Subsequent study of specimens and a reevaluation of previous anatomical papers indicated that *Zeledonia* is most closely allied to the wood warblers (Parulidae) (Sibley, 1968).

The acquisition of egg-white and hemoglobin from two species of seed-snipe (Thinocoridae) prompted a review of their systematic history and an electrophoretic study (Sibley *et al.*, 1968). The patterns of both egg-white and hemoglobin indicated that the seed-snipe belong in the Charadriiformes and are not close to several other groups which had been suggested as relatives. Their nearest relatives within this homogeneous order, however, were not ascertained.

Additional electrophoretic evidence from egg-white bearing on the flamingo problem (Sibley *et al.*, 1969) corroborated the earlier suggestion (Sibley, 1960) of a closer relationship of the flamingos to the Ciconiiformes than to the Anseriformes. But the electrophoretic patterns of the hemoglobins were equivocal. An analysis of the tryptic peptides of the ovalbumins and hemoglobins (see p. 162) confirmed a closer relationship to the Ciconiiformes, but suggested that all three groups may be related.

Hendrickson (1969) compared the egg-white patterns of gruiform birds by starch gel electrophoresis. The egg-white pattern of the limpkin (*Aramus*) seemed to him to be intermediate between those of cranes (Gruidae) and those of rails (Rallidae). He found that the patterns of the button quails (Turnicidae), finfoot (Heliornithidae), trumpeters (Psophiidae), and sun bitterns (Eurypygidae) were most like those of rails. The patterns of the kagu (Rhynochetidae), the seriemas (Cariamidae), and the bustards (Otididae) differed from those of rails and from one another.

Sibley and Hendrickson (1970) compared the starch gel electrophoretic patterns of the plasma proteins of about 450 species of birds. Their results showed 'a high degree of individual variation superimposed upon a common basic pattern' (p. 46). Correlations between the patterns and taxonomic groupings were not found, and the writers concluded that one-dimensional electrophoretic patterns of total plasma proteins do not provide useful information for higher category relationships. However, at lower taxonomic levels plasma patterns may be helpful. Studies of genetic polymorphisms are examples.

The egg-white studies have been expanded to include over 1,200 species from all the orders and 144 of the 165 living families recognized by Wetmore (1960). In two monographs dealing with the passerines (Sibley, 1967) and non-passerines (Sibley and Ahlquist, 1972) respectively, the literature on the relationships of the orders and families of birds has been reviewed and the egg-white evidence presented.

Sibley (1970) compared the electrophoretic patterns of nearly 650 species of passerine birds. He found that the patterns fell into four major types which he designated as 'Pattern Types' A—D. Within Pattern Types A and B he defined a series of groups and indicated the relative probability that the members of these groups are related to one another. The four categories of probability judgments which follow were based on all available evidence of relationship, 'strongly influenced by the electrophoretic patterns, but not solely dependent upon them' (p. 115).

It is highly probable that. . .

i. the New World non-oscine passerine groups are more closely related to one another than any one of them is to any Old World group. This assemblage consists of the wood-creepers (Dendrocolaptidae), ovenbirds (Furnariidae), antpittas (Conopophagidae), ant-shrikes (Formicariidae), flycatchers (Tyrannidae), cotingas (Cotingidae), manakins (Pipridae), and plantcutters (Phytotomidae);

ii. the Old World long-eared tit (*Aegithalos*) and the New World bush tit (*Psaltriparus*) are closely related to each other;

iii. the Old World weaverbirds (Ploceinae) and waxbills (Estrildinae) are more closely related to each other than either is to any other group;

iv. the cardueline finches are related to the other Fringillidae and are not especially close to the Estrildinae;

v. the wren-thrush (*Zeledonia*) is most closely allied to the wood warblers and not to the thrushes.

It is probable that. . .

i. the Cotingidae, Pipridae, and Phytotomidae are more closely related to one another than any one of them is to some other group;

ii. the New Zealand wrens (Acanthisittidae) and the pittas (Pittidae) are each more closely related to some unknown group of Old World oscines than to the New World non-oscines;

iii. the crows and jays (Corvidae) are more closely allied to the shrikes (Laniidae) than to any other group;

iv. the magpie-larks (Cracticidae) and the birds of paradise (Paradiseidae) are more closely related to one another than either is to the Corvidae;

v. the dipper (*Cinclus*) is related to the Turdidae;

vi. the Old World warblers (Sylviidae), Old World Flycatchers (Muscicapidae), and hedge sparrows (*Prunella*) are more closely related to one another than any one of them is to the Turdidae;

vii. the waxwings (*Bombycilla*) and silky flycatchers (*Phainopepla*) are more closely related to each other than either is to the palm-chat (*Dulus*);

viii. the vireos (Vireonidae) are more closely related to the New World nine-primaried oscines than to any other group;

ix. *Fringilla* (chaffinch and brambling) is more closely related to the cardueline finches than to any other group;

x. the Hawaiian honeycreepers (Drepaninini) were derived from the cardueline finches.

The most interesting taxonomic problems occur in the next group.

It is possible that . . .

i. the Malagasy Philepittidae and the broadbills (Eurylaimidae) are closely related;

ii. some genera of the Dendrocolaptidae may be transferred to the Furnariidae and vice versa;

iii. the cuckoo-shrikes (Campephagidae) are more closely related to the Corvidae than to any other group;

iv. *Parus* (tits, chickadees) and *Certhia* (brown creepers) are closely related;

v. *Parus* and *Certhia* are more closely related to *Sylvia*, *Muscicapa*, etc. than to any other group;

vi. the Wren-tit (*Chamaea*) is a sylviid;

vii. *Aegithalos* and *Psaltriparus* are sylviids;

viii. the bulbuls (Pycnonotidae) are most closely related to the drongos (Dicruridae), the starlings (Sturnidae) and/or the Old World orioles (Oriolidae);

ix. *Acanthiza* and perhaps *Sericornis* are closer to certain meliphagids (*Lichmera*, *Certhionyx*) than to the sylviids;

x. *Epthianura* is closer to *Meliphaga* than to *Sylvia*;

xi. *Rhipidura* is not a muscicapid;

xii. *Sphenostoma* is not a muscicapid;

xiii. *Rhipidura* and *Sphenostoma* are closely related;

xiv. the wagtails (Motacillidae) are most closely related to the Sylviidae and Muscicapidae;

xv. the wood swallows (*Artamus*), bulbuls (*Pycnonotus*), and drongos (*Dicrurus*) are related to *Sturnus*;

xvi. the cape sugarbird (*Promerops*) was derived from the starlings;

xvii. *Nilaus* is a laniid;

xviii. the Meliphagidae are composed of two subgroups;

xix. the diamond-birds (*Pardalotus*) and flower-peckers (*Dicaeum*) are not closely related;

xx. *Pardalotus* is more closely related to certain honeyeaters than to *Dicaeum*;

xxi. the white-eyes (*Zosterops*) are more closely related to the sylviid-muscicapid complex than to the nectariniids or dicaeids;

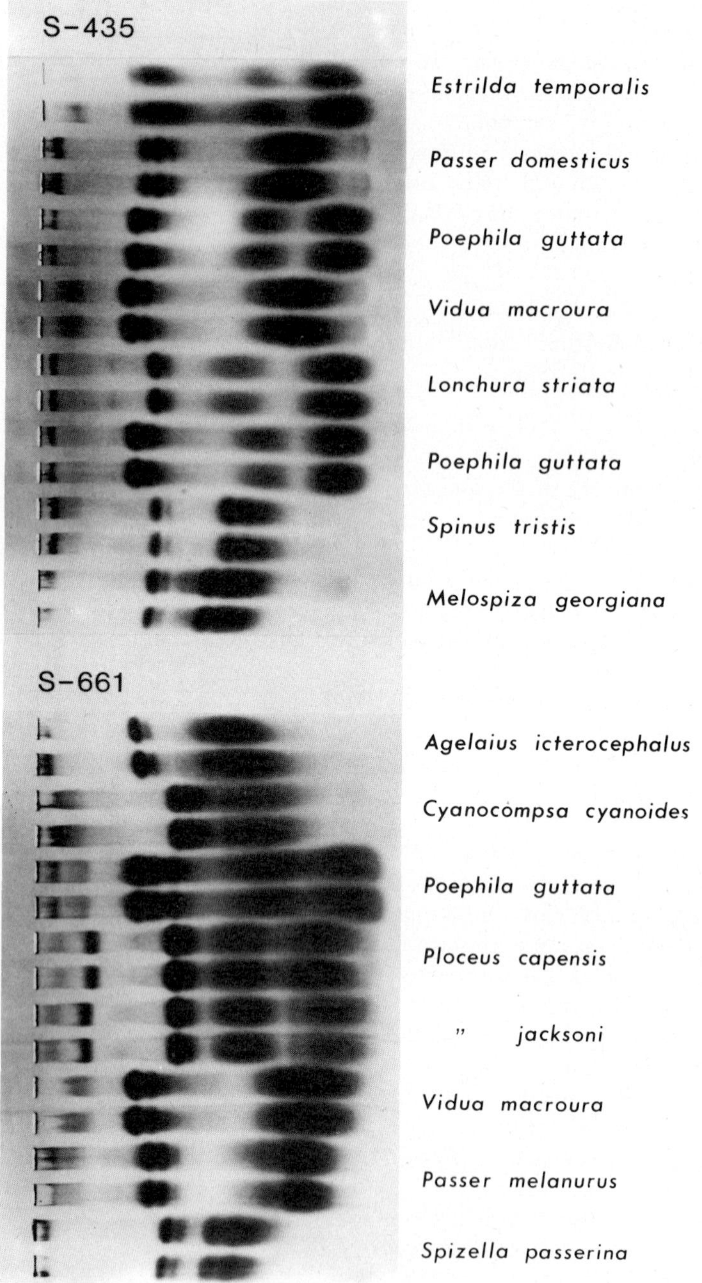

FIG. 3. Examples of two starch-gels showing that the egg-white patterns of the widow bird (*Vidua*) and the house sparrow (*Passer*) are more similar to one another than either is to those of ploceine finches (*Ploceus*) or estrildine finches (*Estrilda*, *Lonchura*, *Poephila*). The other patterns are those of members of the New World nine-primaried oscines. From Sibley (1970).

xxii. *Passer* is not ploceid;
xxiii. the widow birds (*Vidua*) are closely related to *Passer*;
xxiv. *Vidua* is not as closely related to the ploceines and estrildines as it is to *Passer*;
xxv. *Passer* and *Vidua* are more closely related to the Fringillidae than to the Ploceidae;
xxvi. *Fringilla* is as closely or more closely related to the cardinaline finches than to the carduelines.

It is improbable that . . .

i. the broadbills are closely related to the pittas;
ii. the larks are closely related to the Ploceidae;
iii. the Corvidae are closely allied to the Paridae;
iv. *Panurus* is a timaliid or that *Panurus* is a parid;
v. *Cinclus* is closely related to the wrens (Troglodytidae);
vi. the mockingbirds (Mimidae) and Troglodytidae are closely related;
vii. the Troglodytidae and Turdidae are closely related;
viii. the Motacillidae are closely related to *Panurus*;
ix. *Nilaus* is a muscicapid;
x. *Zosterops* is closely related to the dicaeids or nectariniids;

Sibley and Ahlquist (1972) have studied the egg-white proteins of over 800 species of non-passerine birds. Their conclusions corroborate and amplify those reached by Sibley (1960). The principal findings include the following.

i. The large ratites, including *Rhea*, are probably monophyletic. The relationships of the tinamous (Tinamidae) and the kiwi (*Apteryx*) to the large ratites, to each other, or to other avian groups are obscure. The egg-white patterns of the tinamous and the kiwi resemble one another and show some similarities to those of gallinaceous birds.

ii. The shearwaters and petrels (Procellariidae), albatrosses (Diomedeidae), storm petrels (Hydrobatidae), and diving petrels (Pelecanoididae) form a closely knit group to which the penguins appear to be related. There are similarities between the patterns of the Procellariiformes and those of the Charadriiformes.

iii. The Pelecaniformes are diverse and may be polyphyletic. One group appears to be composed of *Sula*, *Phalacrocorax*, and *Anhinga*; another includes the frigate birds (*Fregata*) and the tropic birds (*Phaëthon*); a third contains the pelicans (*Pelecanus*).

iv. Members of the Ciconiiformes appear to be closely related. The pattern of *Cochlearius* is most like those of the herons, especially the night herons (*Nycticorax*).

v. The screamers (Anhimidae) appear to be allied to the waterfowl.

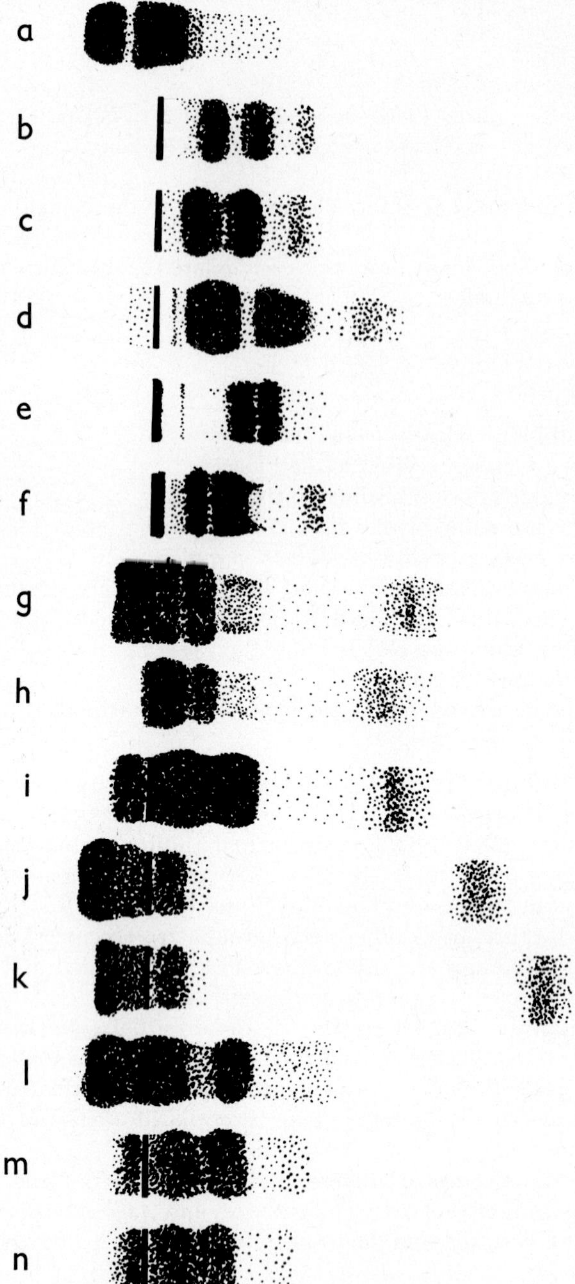

FIG. 4. Diagrammatic representations of the starch-gel electrophoretic patterns of the hemoglobins of some Ciconiiformes (*Guara* to *Ardea*, from the top), the flamingos (*Phoenicopterus* and *Phoeniconaias*), and some Anseriformes (*Cereopsis* to *Branta*). a. *Guara alba*; b. *Florida caerulea*; c. *Butorides striatus*; d. *Mycteria americana*; e. *Ardea goliath*; f. *Ardea herodias*; g. *Phoenicopterus ruber*; h. *Phoeniconaias minor*; i. *Cereopsis novaehollandiae*; j. *Anhima cornuta*; k. *Chauna chavaria*; l. *Anas acuta*; m. *Anser fabalis*; n. *Branta canadensis*. The solid vertical line represents the application point. The anode is to the right. From Sibley, Corbin and Haavie (1969).

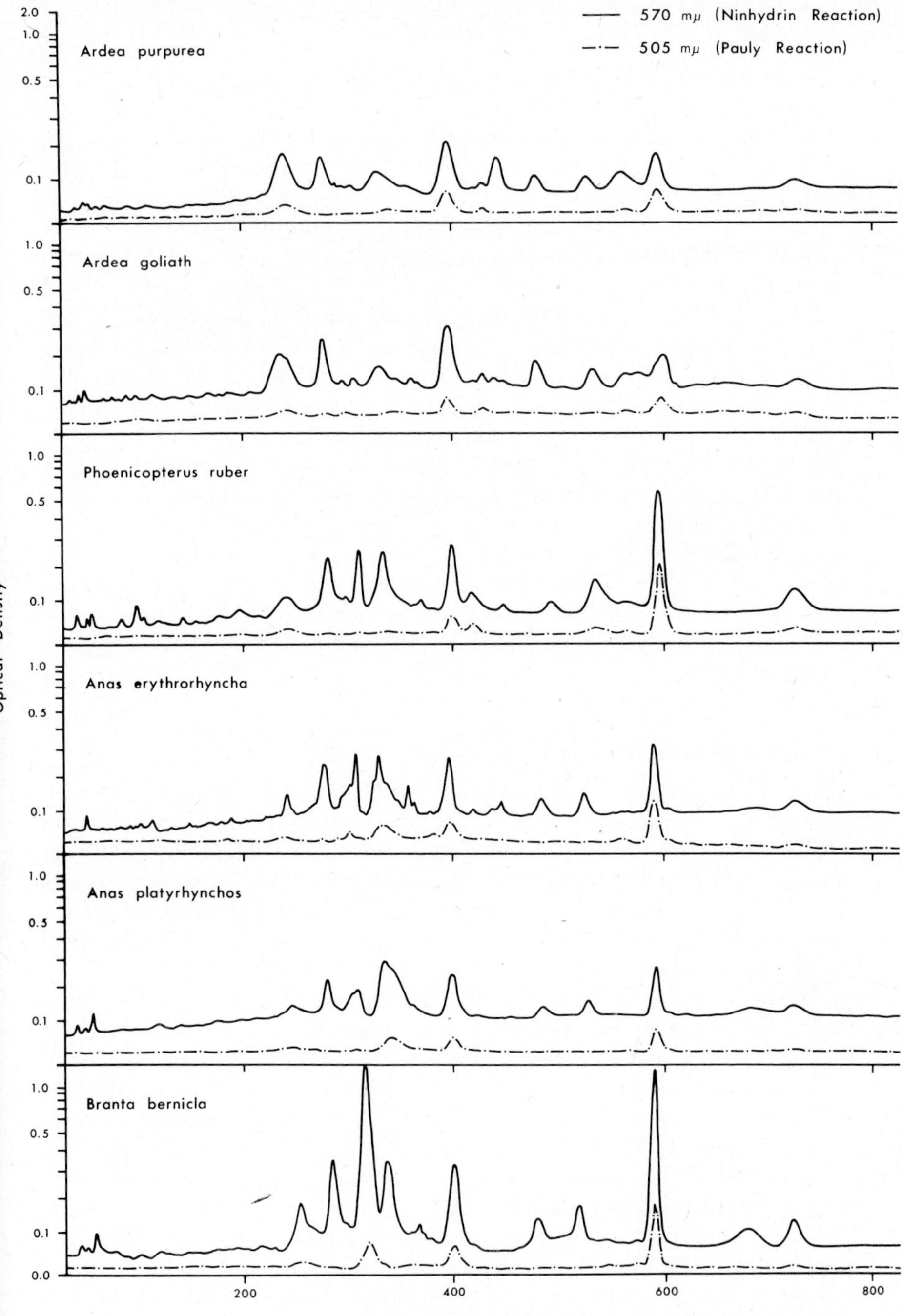

FIG. 5. Chromatograms of the tryptic peptides of the hemoglobins of (from top to bottom) two herons (Ciconiiformes), a flamingo (*Phoenicopterus*), and three Anseriformes. The ninhydrin reaction (solid line) detects all peptides; the Pauly reaction (broken line) detects only those peptides containing histidine. From Sibley, Corbin and Haavie (1969).

vi. The megapodes (Megapodiidae) are the most distinctive group of the Galliformes.

vii. The diurnal birds of prey seem to consist of two main groups. The patterns of the Accipitridae, the osprey (*Pandion*), the secretary-bird (*Sagittarius*), and the Cathartidae are similar to one another. The patterns of the Falconidae (*Falco* and *Milvago* examined) are quite different from those of the other Falconiformes and are almost identical to those of the Tytonidae and Strigidae.

viii. The Thinocoridae, sheath-bills (Chionididae), coursers and pratincoles (Glareolidae), thick-knees (Burhinidae), painted-snipe (Rostratulidae), and jacanas (Jacanidae) all have patterns resembling those of the other Charadriiformes such as plovers, sandpipers, auks, and gulls.

ix. The patterns of the sand-grouse (Pteroclidae) are more like those of the Charadriiformes than those of the Columbiformes or Galliformes.

x. The pattern of the hoazin (*Opisthocomus*) does not resemble those of the gallinaceous birds and is closest to those of the crotophagine cuckoos.

xi. The patterns of the hoopoe (Upupidae), wood hoopoes (Phoeniculidae), and hornbills (Bucerotidae) are more similar to one another than to those of any other coraciiform group. The kingfishers (Alcedinidae), todies (Todidae), bee-eaters (Meropidae) and motmots (Momotidae) share similar patterns.

3. *Studies of Eye Lens Proteins*

Marcel Rabaey and Henryk Gysels have published several papers on electrophoretic and immunoelectrophoretic comparisons of the proteins of the avian eye lens. Their work has been carried out mainly with agar gel electrophoresis. Rabaey (1959) first compared the lenticular proteins of many vertebrates, including birds, and the application of the results to avian systematics was begun by Gysels (1962, 1963) in two short papers, both of which contained electrophoretic data.

Gysels and Rabaey (1963) included immunoelectrophoretic techniques as well as specific staining techniques for myoglobins with peroxidase activity, lactate dehydrogenases, and esterases in a study of the Congo peafowl (*Afropavo congensis*). They concluded that *Afropavo* is not closely allied to the guinea fowl (Numidinae) or to many pheasant genera, but probably is distantly related to the common peafowl (*Pavo cristatus*).

Gysels (1964a) prepared an antiserum against the soluble proteins of the eye lens of the European starling (*Sturnus vulgaris*) and used immunoelectrophoresis to test it against the lens antigens of several other species. The anti-starling lens antiserum cross reacted with all species tested, including many non-passerines. Most of the groups tested gave a single precipitin line for each protein visible in the electrophoretic pattern. This type of reaction was found in the Ciconiiformes, Falconiformes, Strigiformes, Gruiformes, Psittaciformes,

Coraciiformes, Piciformes, Passeriformes, Charadriiformes, Procellariiformes, Columbiformes, Galliformes, Rheiformes, and Casuariiformes.

A second type of pattern, in which 'the precipitation lines of several antigens form one single stretched and sharp-cut line', is found in the Anseriformes, Podicipediformes, and Gaviiformes. This pattern was also found in the razorbill auk (*Alca torda*, Charadriiformes) and the boat-billed heron (*C. cochlearius*, Ciconiiformes).

A third pattern was found in the Sphenisciformes, Phalacrocoracidae, Sulidae, and in the common murre (*Uria aalge*, Charadriiformes). The European cuckoo (*Cuculus canorus*) produced a 'rather aberrant' pattern.

From these data Gysels concluded (1) that the resemblances among the electrophoretic patterns of the lens proteins of the Passeriformes are also reflected in the immunoelectrophoretic patterns, (2) that 'some presumed relationships between the birds with an Anseriform [*sic*] pattern could be confirmed (e.g., grebes-loons-auks), while also new perspectives are opened (ducks-*Cochlearius*-herons)', and (3) that 'there is no doubt that all birds with a Sphenisciform [*sic*] pattern are pretty closely related'. This group, according to Gysels, would include the penguins, cormorants, boobies, and the alcid *Uria aalge*, but not other Charadriiformes such as plovers, sandpipers, and gulls.

In a second paper (1964b) Gysels presented the same data as above plus some additional information, including the names of 13 species of non-passerines, the lens antigens of which were tested against the anti-starling lens antiserum. These were the double-wattled cassowary (*C. casuarius*), Humboldt penguin (*Spheniscus humboldti*), gannet (*Sula bassana*), little grebe (*Podiceps ruficollis*), black-throated loon (*Gavia arctica*), chicken (*G. gallus*), wood duck (*Aix sponsa*), crowned crane (*Balearica pavonina*), black-legged kittiwake (*Larus tridactylus*), razorbill auk (*Alca torda*), common murre (*Uria aalge*), European swift (*A. apus*), and a hummingbird listed by Gysels as '*Chlorestes caeruleus*'.

The results of this study are essentially as summarized above for Gysels 1964a. An addition is the statement (p. 19) that the hummingbird and swift 'are entirely different' in the electrophoretic and immunoelectrophoretic patterns of their lenses under the conditions of this study. Gysels also repeats (p. 20) his belief that 'of all living birds examined *Uria aalge* must be the most close [*sic*] related to the penguins.'

Gysels (1964c) studied a number of parrots using the same techniques. These were the blue and yellow macaw (*Ara ararauna*), peach-fronted parakeet (*Aratinga aurea*), yellow-headed parrot (*Amazona ochrocephala*), gray parrot (*Psittacus erithacus*), Senegal parrot (*Poicephalus senegalus*), masked lovebird (*Agapornis personata*), ring-necked parakeet (*Psittacula krameri*), budgerigar (*Melopsittacus undulatus*), lesser sulphur-crested cockatoo (*Cacatua sulphurea*), and red-crested cockatoo (*Cacatua moluccensis*). Gysels interpreted his data as indicating a considerable degree of heterogeneity among the members of this family but was cautious about making any formal taxonomic decisions.

Gysels (1964d) published a lengthy paper in which he compared the lens proteins of 233 species of 21 orders and 64 families as recognized by Wetmore (1960). Gysels's conclusions are summarized below.

i. The Galliformes are a group of closely related species. *Meleagris* is near *Numida*, but the Cracidae differ from the Phasianidae. The Old World quails are the most distinctive group of the order.

ii. The lens proteins were uninformative concerning possible relationships among the ratite birds.

iii. It was not possible to corroborate a relationship between the penguins and the Procellariiformes.

iv. *Sula bassana* and *Phalacrocorax* appear to be related, and both may be distantly allied to the penguins. *Pelecanus*, however, is strikingly different.

v. The Anseriformes are a group of closely related species with the tribe Dendrocygnini (tree ducks) being the most distinctive.

vi. The loons and grebes seem to be related; the grebes also may be close to the Anseriformes.

vii. *Jacana* belongs in the Charadriiformes which, with the exception of the Alcidae, seems to be a monophyletic assemblage.

viii. The fulmar (*Fulmarus glacialis*) was the only representative of the order Procellariiformes studied. Its lens electrophoretic pattern is of the charadriiform type. Glycogen is present in the lens; this is an 'advanced' character according to Gysels, therefore *Fulmarus* cannot be associated with the 'lower' non-passerine groups, as it is in most contemporary classifications. Gysels proposed (a) that *Fulmarus* is not a procellariiform and probably belongs in the Charadriiformes, (b) that the entire order Procellariiformes belongs next to the Charadriiformes, or (c) that the Procellariiformes are polyphyletic with some members closest to the penguins and others closest to the shore birds.

ix. The Columbiformes have a charadriiform lens pattern and may be placed next to that order.

x. *Cochlearius* differs from other Ciconiiformes and seems best retained as a separate family. The hammerhead (*Scopus umbretta*) is also different from the other members of the order.

xi. The falcons and accipiters differ, but Gysels was uncertain as to the taxonomic implications of the differences.

xii. The Strigiformes have a passeriform lens pattern, but no conclusions were reached regarding possible relationships either to the Falconidae or to the Coraciiformes. No Caprimulgiformes were studied.

xiii. The rails (Rallidae) and cranes (Gruidae) seem to be related, but other Gruiformes were not examined.

xiv. The parrots (Psittacidae) probably should be split up into several families, but no formal proposals were made.

xv. Only a kingfisher (*Alcedo atthis*) and a hornbill (*Bucorvus abyssinicus*) were examined from the Coraciiformes. Both had a passeriform lens pattern.

xvi. There are some similarities between the cuckoo (*Cuculus canorus*) and the turaco (*Crinifer africanus*), but the degree of relationship suggested by the patterns is uncertain.

xvii. There are differences between the swifts (Apodidae) and the hummingbirds (Trochilidae), but both groups may be related to the passerines, all of which have similar patterns.

Gysels and Rabaey (1964) repeated their assertion that *Uria aalge* is most closely related to the penguins. They also reported that the razorbill auk (*Alca torda*) and the puffin (*Fratercula arctica*) differ from each other and are unlike either *Uria* or the other Charadriiformes. In this respect it may be noted that the Alcidae comprise a family which is quite uniform in its anatomical characteristics and which has long been placed in the Charadriiformes on the basis of a number of characters. Further, the electrophoretic patterns of the egg-white proteins (Sibley and Ahlquist, 1972) of alcids are most similar to those of the Laridae and other charadriiform birds.

The 'typical song bird component' found in the lens electrophoretic patterns of 84 species of passerines (Gysels, 1964d) was the topic of a subsequent paper (Gysels, 1965) which concluded that it is an important taxonomic character. The occurrence of this component and the complex passeriform lens pattern in members of other orders was interpreted as evidence of some degree of relationship. In addition to the 20 passerine families which show the 'typical song bird component', the following orders also possess it: Ciconiiformes, Falconiformes, Gruiformes, Psittaciformes, Strigiformes, Coraciiformes, and Piciformes.

Gysels (1968) added to the material already presented on the Ciconiiformes. In this study the lenses of four species of herons, four storks, three ibises, *Scopus*, the whale-billed stork (*Balaeniceps rex*), and two flamingos were examined. Gysels found that the electrophoretic and immunoelectrophoretic patterns fell into four groups. The herons and bitterns had a passeriform pattern. The pattern of *Cochlearius* was of the anseriform type. *Scopus* resembled the storks (Ciconiidae), but Gysels was 'still in doubt, whether *Balaeniceps* should be situated nearer to the *Ardeidae* than to the *Ciconiidae*' (p. 275). There were no apparent resemblances among the lens patterns of the flamingos and those of any of the ciconiiform groups or those of the ducks. The only point he found in common between herons and flamingos is the possession of glycogen in the lens. However, considerable similarity among the ciconiiform groups as well as a clear relationship of the flamingos to this order has been demonstrated by a study of the tryptic peptides of ovalbumin and hemoglobin (Sibley *et al.*, 1969; also see p. 162 of the present review).

An independent study of lenticular proteins, that of Sibley and Brush (1967), produced results contrary to those of Gysels and Rabaey. Although Sibley and Brush examined more than 1,400 specimens representing over 400 species and 22 of the Wetmore orders, they were unable to find taxonomically meaningful similarities within groups or consistent differences between groups. Comparisons of absolutely fresh lenses showed similarities among most birds.

Upon storage denaturation was rapid and resulted in an unusual amount of random heterogeneity. Sibley and Brush pointed out, however, that since they had used a different technique, starch gel electrophoresis, they could not claim to have disproved the results of Gysels and Rabaey.

When one examines the table of mobilities in Gysels 1964d, it is apparent that the major aspects of the lens pattern are basically similar among all birds. The taxonomic decisions seem often to have been made on the basis of small, and frequently not statistically significant, mobility differences and the presence or absence of minor protein fractions. Thus the interesting proposals made by Gysels merit investigation by additional biochemical means.

B. INTRASPECIFIC VARIATION IN PROTEINS

1. Introduction

Electrophoretic studies of homologous proteins of birds are of importance not only in determining interspecific relationships but also in revealing genetically controlled variants among populations of the same species. Such genetic variants provide markers for the investigation of the relationships of populations, races and naturally hybridizing species. Comparison of gene frequencies for these variant proteins reveals the extent of gene flow and hence the degree of reproductive isolation among the populations comprising a species. Differing selection pressures may result in unique morphs of a protein having arisen in populations which are geographically isolated but more often results in differences in the frequency of specific genes. The amount of divergence brought about by selective advantage depends on the function of a particular gene. Thus the degree of genetic differentiation suggested by the gene frequencies may not be a simple reflection of the relationship between populations. This means that individuals do not necessarily belong to the same population even if no differences in gene frequency are found. Where formerly isolated species have been brought into secondary contact by habitat changes the effects of introgression may be revealed by the pattern of gene frequencies throughout the range of the species concerned.

As well as direct comparisons of gene frequencies in samples from different localities, application of methods of population genetical analysis can give much information. The most widely used test in this respect is the Hardy-Weinberg equilibrium. This says that if two alleles, A and B, occur in gametes with frequencies p and q then the frequencies of AA, AB and BB individuals which arise through random fertilization will be p^2, $2pq$ and q^2. Agreement with the expected occurs in populations of individuals that mate at random, are isolated from other populations and are not under the influence of selection, mutation or genetic drift. This then can be used to test for panmixia within a population of any size.

A deficit of heterozygotes is taken to mean that random mating does not take place and that the individuals are from two or more isolated or semi-isolated populations. The coefficient of inbreeding (f) can be calculated from the formula $(He-Ho)/He$, where He is the expected number of heterozygotes and Ho is the observed number. If individuals are from a panmictic population, 'f' is zero, and if no interbreeding takes place 'f' is equal to 1·0. Failure to demonstrate that a genetic system is in Hardy-Weinberg equilibrium may also mean that the genetic hypothesis proposed for the control of the system may not be valid or that certain ecological factors may be changing the balance. If selection favors the heterozygotes they will be present in greater numbers than expected. Such hybrid vigor or heterosis is well known in animal breeding.

Manwell and Baker (1969) have discussed several examples where heterozygote deficiency, which could not be explained on the basis of inbreeding of distinct populations, was found. European quail (*Coturnix coturnix*) carry individual variation at 58% of their protein loci (Baker and Manwell, 1967) and have such inbreeding depression that full-sib-mated families usually die out after only three generations (Sitmann *et al.*, 1966). Clearly this species carried a heavy 'genetic load'. Manwell and Baker (1969) postulate that 'occasional negative heterosis at a few loci would equalize the "genetic load" by decreasing the variance of fitness within the population'. A detailed discussion of negative heterosis and its possible role in race and species formation is given in that paper.

The occurrence of protein variants in proportions that cannot be maintained by recurrent mutation is referred to as genetic or balanced polymorphism. Studies attempting to detect intraspecific variation for the differentiation of populations should take into account variability which is not polymorphism but which is produced under the influence of physiological or environmental conditions. Several studies on avian proteins have shown qualitative variation with age and sex (Amin, 1961; Bush, 1967; Ferguson, 1969). Removal of phosphate or carbohydrate attached to the protein molecule can alter the electrophoretic mobility, for example, removal of sialic acid from transferrin (Williams, 1962) or dephosphorylation of ovalbumin (Baker *et al.*, 1966). Bacterial enzymes can bring about these changes and hence care is required when dealing with stored samples. Metal binding by proteins can increase or decrease mobility, for example, iron binding by transferrin and conalbumin (Baker, 1967).

In view of these artifact polymorphisms it is desirable to have breeding data to support claims for genetic polymorphism. In practice, when dealing with natural populations of birds, this type of information is almost impossible to obtain but other criteria can be applied, as follows:

i. the protein concerned has been shown to exhibit polymorphism in a species where breeding experiments were possible;

ii. a genetic hypothesis can be proposed to account for the variation;

iii. the variation is constant under different electrophoretic conditions that is, buffer type, pH and ionic strength;

iv. the observed variation cannot be correlated with age or sex of the individuals;

v. variation in transferrin and conalbumin must be maintained on saturation of the protein with iron.

In most cases physiological and environmental influences result in quantitative not qualitative changes in the proteins (Sibley and Johnsgard, 1959a; Kartashev *et al.*, 1966). Quantitative and environmentally determined qualitative variation may be of value in the separation of populations. Choogoonov *et al.* (1965) compared the serum proteins of lowland and mountain great spotted woodpeckers (*Dendrocopos leucotus*) and observed quantitative differences in the albumin. In the mountain population the albumin made up 22.1% of the total protein whereas in the lowlands it represented 17·2%. In the mountain population of the same species one of the hemoglobin fractions was increased almost 1·5 times.

At the intraspecific level protein variants are probably due to structural differences in the molecule involving no more than the subsitution of a single amino acid residue. If this substitution results in a change in the net charge or in the configuration of the protein molecule the variant will be detectable by its altered electrophoretic mobility. Increased sensitivity of this methodology has been achieved by work dealing with specific enzymes rather than general protein patterns. Using standard histochemical reactions many enzymes are easily localized on an electrophoretic plate. Some enzymes have been found to be polymorphic in many of the species which have been examined, for example, esterase and glucose-6-phosphate dehydrogenase. This greater degree of variation in enzymes as opposed to general proteins is due to their narrower functional limits. Variants adapted to particular environmental conditions have been selected for in different populations.

The most frequently occurring type of polymorphism is the one locus-two allele type producing three different electrophoretic phenotypes. In the genotype homozygous for each allele the proteins are shown on the gel as single fractions which differ in their migration rate. The heterozygote has both fractions present, each in half the amount of the homozygotic condition. An artificial mixture of the two homozygotes gives an electrophoretic pattern which is identical to that of the natural heterozygote. In some cases minor bands accompany the main fraction or the homozygote may consist of several electrophoretic fractions. Polymorphism controlled by three or more alleles is also known. The more alleles at a locus the greater the chance will be of finding heterogeneity of the frequencies between populations.

The majority of intraspecific studies to date have been carried out on domestic or feral species and have established little more than the existence and extent of polymorphism. Although *per se* they are of little interest to the avian taxonomist they have provided certain basic information especially in view of the difficulty of carrying out breeding experiments on most species. These studies are thus an essential prerequisite to work on natural populations.

It is proposed to review briefly these basic studies and then to deal in detail with the few studies which have so far been carried out on wild species. Unfortunately few authors mention the number of individuals and proteins examined when polymorphism was not found. It must be realized that this 'negative information' is important and a necessary part of the data of any study.

2. *Variation in Egg-white Proteins*

The first avian system in which intraspecific variation was shown was in the egg-white proteins of the chicken (*Gallus gallus*). This was first reported by Lush (1961, 1967) who found variation in the electrophoretic mobilities of four egg-white proteins. Similar variation was also noted by Baker and Manwell (1962) and Feeney *et al.* (1963). Each of the variant proteins is under the control of codominant alleles at a separate autosomal locus (Lush, 1964a,b).

Chicken ovalbumin has at least three electrophoretically distinguishable phenotypes (Lush, 1961, 1964b; Baker and Manwell, 1962). In the homozygous condition chicken ovalbumin comprises several electrophoretic fractions due partly to differing amounts of phosphate in the molecule. The degree of phosphorylation is greater in the faster migrating morph than in the slower one, resulting in the corresponding electrophoretic fractions differing in amount as well as in mobility (Lush, 1964b). Baker (1968) has described additional variation in chicken ovalbumin but without further information the possibility that this may involve changes in the phosphate or carbohydrate moieties cannot be excluded. A type of ovalbumin polymorphism similar to that in the chicken has been reported in the egg-white of the golden pheasant (*Chrysolophus pictus*), in Lady Amherst's pheasant (*C. amherstiae*) (Baker, 1965) and in the European quail (*Coturnix coturnix*) (Haley, 1965; Baker and Manwell, 1967). In the case of the European quail ovalbumin phosphorylation or dephosphorylation may be involved. Further enzymes which are polymorphic in the European quail are lysozyme (two alleles), postalbumin (two alleles), X-protein (three alleles) and Y-protein (three alleles) (Baker and Manwell, 1967). The observed frequencies of the X-protein differed significantly from that expected on the basis of the Hardy-Weinberg equilibrium and there was a striking deficit of heterozygotes.

Both egg-white globulins 2 and 3 are polymorphic (Lush, 1961). Baker (1964) found a third allele of protein 2 in a flock of jungle fowl (*Gallus gallus*). This may be the same as the jungle fowl variant noted by Croizier (1966). Protein 2 of Lush (1961, 1964a, 1964b) corresponds to the G_3 globulin and protein 3 to the G_2 globulin of Baker and Manwell (1962) and Baker (1964, 1968). Baker (1968) has reported further variants in some chickens at both of these loci but no genetic information is given.

Baker (1968) has noted that out of 100 chicken eggs two had a second lysozyme (G_1 globulin) fraction migrating more slowly than the usual lysozyme fraction. These two phenotypes parallel two of the three European quail lysozyme phenotypes which have been shown to be a genetic polymorphism (Baker and Manwell, 1967).

The extent of polymorphism in the egg-white and egg-yolk proteins of some domestic geese and ducks has been examined by Stratil and Valenta (1966). In the egg-white from geese (*Anser anser*) two variant regions were found. Region 1 comprises six phenotypes explicable on the basis of one genetic locus with three codominant alleles. Differences in the gene frequencies for these morphs were found in the Italian goose, Bohemian white goose and hybrids between these two domestic varieties of *A. anser*. The egg-white of Peking and Indian ducks (= domestic varieties of *Anas platyrhynchos*) revealed three polymorphic regions which varied in the usual way.

3. *Variation in Plasma Proteins*

Polymorphism of the plasma albumin has been noted in the chicken (McIndoe, 1962; Stratil, 1968). The variation is under the control of three codominant alleles. Each variant of the plasma albumin is accompanied by a minor, faster fraction which varies in mobility in step with the main fraction. Baker *et al.* (1966) have noted variation in the prealbumins of the ring-necked pheasant (*Phasianus colchicus*) which may represent a genetic polymorphism.

Márkus *et al.* (1965) examined serum samples from four breeds of domestic geese (*A. anser*) (Rheinland, Landes, Hungarian and Dutch Landrace) using starch gel electrophoresis. They observed genetic polymorphism of a group of serum protein fractions showing peroxidase activity. Breeding experiments confirmed that the variation is under the control of two codominant alleles. Gene frequencies were found to differ significantly among the breeds studied.

4. *Transferrin and Conalbumin Variants*

Ogden *et al.* (1962) have shown that in the chicken the transferrin of the serum and the conalbumin of the egg-white are controlled by the same genetic locus and that genetic variation in the electrophoretic mobility of the transferrin is accompanied by an identical variation in the conalbumin. Williams (1962) has shown that chicken serum transferrin and egg-white conalbumin are glycoproteins which have the same amino acid sequence, but differ in their carbohydrate moieties causing a difference in their electrophoretic mobilities. The electrophoretic heterogeneity of the homozygotic transferrin in the chicken is due to different amounts of sialic acid attached to the protein molecule. Removal of this sialic acid with the enzyme neuraminidase results in the transferrin appearing as a single fraction with the same mobility as the conalbumin. The identity of the serum transferrin and egg-white conalbumin of the common pigeon or rock dove (*Columba livia*) has been demonstrated by immunological and electrophoretic methods. In this species treatment of the transferrin with neuraminidase results in an overall decrease in mobility to that of the conalbumin but does not remove the electrophoretic heterogeneity or the quantitative relationship of the fractions to each other (Ferguson, 1969).

Transferrin and conalbumin polymorphism has been found in some 60% of the bird species which have been examined in sufficient numbers. Longsworth *et al.* (1940) using free boundary electrophoresis noted heterogeneity of chicken egg-white conalbumin. That this is a genetic polymorphism was noted by Lush (1961) for conalbumin and by Ogden *et al.* (1962) for transferrin. Further variants at this locus have been described by Stratil (1966), Croizier (1966) and Baker (1967). Polymorphism of transferrin and conalbumin has been found in the common pigeon (*Columba livia*) and speckled pigeon (*Columba guinea*) (Mueller *et al.*, 1962); red-collared dove (*Streptopelia 'humilis'* = *S. tranquebarica*) (Desborough and Irwin, 1966); domestic goose (*Anser anser*), muscovy duck (*Cairina moschata*), laughing gull (*Larus atricilla*) (Stratil and Valenta, 1966); in several species of geese of the genus *Branta* (Baker and Hanson, 1966); ring-necked pheasant (*Phasianus colchicus*) (Baker and Manwell, 1967); chukar partridge (*Alectoris graeca*) and red-winged blackbird (Baker, 1967); magpie (*Pica pica*), hooded crow (*Corvus cornix*), black-headed gull (*Larus ridibundus*), common scoter (*Melanitta nigra*), wood pigeon (*Columba palumbus*) and barbary dove (*Streptopelia risoria*) (Ferguson, 1969).

In the egg-white of the eider duck (*Somateria mollissima*) Milne and Robertson (1965) found a protein variant which, judging from its position in their published diagram, is conalbumin. In their search for egg-white protein polymorphisms they also examined eggs from 196 herring gulls (*Larus argentatus*), 100 lesser black-backed gulls (*Larus fuscus graellsii*), 52 gannets (*Sula bassana*), 30 starlings (*Sturnus vulgaris*) and 19 great tits (*Parus major*) but they failed to find any variation. Baker (1967) mentions the effect of the addition of iron to conalbumin in a number of species where no polymorphism was noted. No record of the numbers examined for these species is given and they cannot be taken as lacking this polymorphism.

5. *Variation in Enzymes*

In a number of vertebrates, extracts of mature testes contain a lactate dehydrogenase fraction 'X' in addition to the usual forms (M and H) which are found in other tissues (Zinkham *et al.*, 1963). Zinkham and his colleagues suggested that fraction 'X' is a tetramer of polypeptides whose synthesis is governed by a third locus. They found 'X' polymorphism in several populations of wild and domesticated pigeons (*Columba livia*) (Zinkham *et al.*, 1964). The three phenotypes which they observed can be explained on the basis of two alleles at an autosomal locus which controls the synthesis of polypeptide chains of a type different from A and B and which they call C. The distribution of the three morphs in six populations agreed with that expected according to the Hardy-Weinberg equilibrium for a single pair of alleles. Their observations suggest that there is considerable variation in the frequency of the C′ mutant allele in different breeds of pigeons. The frequency of C′ varied from 0·13 in wild birds to 0·76 in the white carneau domestic variety and 0·52 in the silver king variety. Polymorphism has also been noted at the LDH B locus (Zinkham *et al.*, 1966). In this case no significant difference between the breeds in the frequency of the alleles was noted.

Kaminski (1964), in a study of several species of birds, noted that variation exists in a serum esterase of European quail (*C. coturnix*). Beckman and Nilson (1966) examined the electrophoretic patterns of esterase, alkaline phosphatase and leucine aminopeptidase in the serum of 18 species of birds and their hybrids. The hybrids often, but not always, showed enzyme patterns representing a combination of the parental patterns. Bush (1967) has noted esterase variation in the house sparrow (*Passer domesticus*) but like the other esterase studies mentioned no breeding data are given and hence these may or may not be true polymorphisms. The ontogeny and characterization of muscovy duck (*Cairina moschata*) esterases has been investigated by Holmes and Masters (1968).

In the European quail (*Coturnix coturnix*) Manwell and Baker (1969) have shown genetic polymorphisms of two esterases, one glucose-6-phosphate dehydrogenase and one 6-phosphogluconate dehydrogenase. Cooper *et al.* (1969a, 1969b) have shown polymorphisms in the glucose-6-phosphate dehydrogenases of the laughing turtle dove (*Streptopelia senegalensis*) and the barbary dove (*S. risoria*) and in the 6-phosphogluconate dehydrogenases of *S. risoria* and *Columba livia*.

Boehm and Irwin (1970) investigated the plasma esterases of a number of species of Columbidae. Particular emphasis was placed on the polymorphic esterase isozymes in two interfertile species, laughing turtle dove (*S. senegalensis*) and barbary dove (*S. risoria*), hybrids of these two and backcross hybrids. Nine phenotypes were observed in the plasma esterase patterns of *S. senegalensis* but no evidence of a heterozygotic pattern was found suggesting that nine alleles control the nine patterns. The isozyme patterns could be divided into slow and fast regions, both being present in *S. senegalensis* but only in the fast region in *S. risoria*. Species hybrids and backcross offspring from matings of species hybrids to *S. risoria* showed overlapping of the bands in the fast region.

Esterase polymorphism was also noted in cape ring dove (*S. capicola*) with five patterns among 15 birds, spotted turtle dove (*S. chinensis*) six patterns in 31 individuals and half-collared ring dove (*S. semitorquata*) with three patterns among eight individuals. All of the 19 species of Columbidae which they examined could be distinguished from each other on the basis of their esterase patterns.

Kimura (1969a, 1969b) has shown genetically controlled variations at two esterase loci of the chicken. Ferguson (1969) found intraspecific variations in the esterases, phosphatases and leucine aminopeptidases of *Columba livia*, *C. palumbus*, *S. decaocto* and *S. risoria*.

Lush (1966) has found individual variation in the amount of catalase activity in the egg-white of the Japanese quail (*Coturnix coturnix japonica*). The activities of this enzyme fell into three groups: low, medium and high. Lush tentatively suggested that these types are under the genetic control of two alleles at an autosomal locus. Homozygotes for one allele, Oct^L, have a low ovocatalase activity while homozygotes for the other allele, Oct^H, have a high ovocatalase activity. Heterozygotes have a medium ovocatalase activity. No mating data were available but the number of phenotypes in two populations did not differ significantly from the numbers to be expected in a Hardy-Weinberg equilibrium. Two populations differed slightly in gene frequencies.

Electrophoretic patterns of single serum samples of chicken contain either a slow or a fast moving alkaline phosphatase fraction or no bands at all (Law and Munro, 1965; Wilcox, 1966). Fast and slow bands have been shown to be controlled by a dominant and a recessive allele respectively. Law (1967) has shown that two electrophoretic variants of leucine aminopeptidase are directly associated with the two alkaline phosphatase types. Treatment of the plasma with neuraminidase converted the faster migrating band of both enzymes to slow moving forms, but the slow moving forms were unaffected by the treatment. The two forms of both of these enzymes are probably due to the presence or absence of a gene controlling the attachment of sialic acid to the enzyme molecules. This would explain why the fast fraction appears to be dominant to the slow one.

6. *Population Studies*

Milne and Robertson (1965), using starch gel electrophoresis examined the egg-white of eider ducks (*Somateria mollissima*) from five populations in Iceland, Holland and Scotland. They found an egg-white protein, possibly conalbumin, with variation explicable on the basis of three alleles. The gene frequencies showed slight differences among the populations. Of particular interest is the variation within the colony at Forvie in Aberdeenshire. This colony is composed of two groups, one of which remains at Forvie all year while the other group overwinters farther south. Individual birds were shown to be consistent in their behavior. By the time the migrant birds return to the breeding area both they and the sedentary birds are already paired. Field observations suggested that migrant and sedentary birds occupy separate segments of the breeding ground with some overlapping in the intermediate area. Milne and Robertson found that the gene frequencies were significantly different for the two groups, being respectively 0·14 and 0·27 for one of the alleles. In the region of overlap the frequency was estimated at 0·21. This gene frequency evidence shows clearly that there is at least partial reproductive isolation between the two groups.

Brush (1968) found that in the red-winged blackbird (*Agelaius phoeniceus*) the conalbumin gene frequencies varied among different populations on a transcontinental basis. The greatest heterogeneity of gene frequencies was found to occur in the only sedentary population in an otherwise migratory species. This sedentary group belongs to the subspecies *A. p. nevadensis* whereas the others belong to the subspecies *A. p. caurinus* and *A. p. phoeniceus*. Thus, within the species *A. phoeniceus*, subspecific groups differ in their genetic variability.

Conalbumin polymorphism was also noted in Brewer's blackbird (*Euphagus cyanocephalus*) (Brush, 1970). In this case the locus is di-allelic. The yellow-headed blackbird (*X. xanthocephalus*) was found to be monomorphic with regard to conalbumin.

Baker (1968) has reviewed the intraspecific variation of the egg-white proteins in the genus *Gallus*. Twenty six of the 37 breeds of chickens which have been examined show protein polymorphisms at one or more of their egg-white protein loci. No variants which were unique to a particular breed were found although the G_3^b allele, with a few exceptions, appears to be associated with Asiatic breeds and breeds known to have Asiatic stock in their composition.

Vohs and Carr (1969) have reported on the frequency of four alleles in 869 wild ring-necked pheasants (*Phasianus colchicus*) captured in Iowa. The most common allele, Tf^c, varied in frequency from 0·82 to 0·93 in different populations and different years. No significant difference in gene frequencies was found among populations from four localities in the State. They also presented breeding evidence to support the hypothesis that the polymorphism is controlled by four codominant alleles at one autosomal locus.

In wood pigeons (*Columba palumbus*) collected from several areas in Ireland, Ferguson (1969) found significant differences in the transferrin and conalbumin gene frequencies in two populations about 100 miles apart, being 0·47 in the west and 0·72 in the east for a particular allele. Significant differences in frequency of the alleles were also noted between breeding and wintering birds, the latter being of Continental or Scandinavian origin. The frequencies of the three phenotypes observed in the breeding birds were in close agreement with those expected on the basis of the Hardy-Weinberg equilibrium. In the winter sample, including both resident and immigrant birds, there was a significant deficit of heterozygotes suggesting non-panmictic populations.

The collared dove (*Streptopelia decaocto*) was found to be monomorphic with regard to transferrin but the closely related barbary dove (*S. risoria*) had the usual type of polymorphism (Ferguson, 1969). The collared dove has shown a remarkable extension of range in the past thirty years. Originally an Asiatic species, it bred for the first time in England in the early 'fifties and in Ireland in 1960. In such a rapid spread new populations were probably set up by a small number of individuals of one genotype. 'Founder' effects can thus complicate population comparisons by accentuating a particular genotype.

Corbin *et al.* (in preparation) examined 354 specimens of the metallic starling (*Aplonis metallica*) collected from 15 localities in the Territory of Papua and New Guinea. Both heart muscle lactate dehydrogenase-H and plasma esterase-1 were polymorphic. In both cases an hypothesis of three codominant alleles controlling the synthesis of each of the enzymes accounts for the observed polymorphisms. The frequencies of two of the LDH-H alleles varied clinally among populations on islands of the Bismarck Archipelago. The most common LDH allele varied in frequency from 0·45 in southern Papua to 0·96 in a population on Manus Island in the Admiralty group. The Manus population also lacked homozygotes of the second most common allele and lacked the third allele. This third allele was found only in the eastern part of the main island at Lae and Karema where it occurred at a low frequency. No definite cline was noted in the esterase frequencies and again the rarest one was restricted to a few populations.

Four populations of the singing starling (*Aplonis cantoroides*) were found to be essentially monomorphic with regard to LDH (Corbin *et al.*, in preparation). Esterase variation, probably controlled by two alleles, was found. One of these alleles was rare in all but a limited area. Thirteen other loci in both *A. metallica* and *A. cantoroides* were found to be monomorphic.

The above examples illustrate the types of polymorphism to be found in avian proteins. Are these examples exceptions or are they typical of all proteins? To date there are few estimates of the number of loci at which individuals of a species are heterozygous. The European quail (*Coturnix coturnix*) has 14 polyallelic loci out of a total of 24 (58%) (Baker and Manwell, 1967) and the ring-necked pheasant (*Phasianus colchicus*) has 19 polyallelic loci out of a total of 44 (43%) (Baker and Manwell, 1967). This compares with 39% for *Drosophila pseudoobscura* (Lewontin and Hubby, 1966) and 30% for *Mus musculus* (Selander *et al.*, 1969). From a taxonomist's point of view this means that there is a high probability of finding a suitable marker for any particular intraspecific problem providing a large number of proteins is examined.

A considerable amount of speculation about the functional significance of polymorphism has arisen recently. A discussion of mutation rates, gene flow, and selective advantage versus selective neutrality is not relevant in this review but can be found elsewhere. See for example Shaw (1965), Lewontin and Hubby (1966) and Harris (1969).

IV. Miscellaneous Techniques

A. CRYSTALLIZATION OF HEMOGLOBINS

Hemoglobin solutions when dried or frozen will form hemoglobin crystals which vary in size and shape depending upon the species sampled. The variation among hemoglobin crystals of different species was recognized and studied by Reichert and Brown (1909) who examined the hemoglobin crystals of a number of vertebrates including 10 species of birds. Avian hemoglobins examined were those of the ostrich (*Struthio camelus*), cassowary (*Casuarius 'galeatus' = casuarius*), domestic goose (*Anser anser*), trumpeter swan (*Olor buccinator = Cygnus buccinator*), whistling swan (*Olor columbianus = Cygnus colum-*

bianus), chicken (*Gallus gallus*), bob-white quail (*Colinus virgianianus*), guinea fowl (*Numida meleagris*), rock dove (*Columba livia*), and American crow (*Corvus americanus* = *Corvus brachyrhynchos*).

It is impossible to compare Reichert's and Brown's data on the hemoglobin crystals of the ostrich and cassowary since methemoglobin crystals were prepared from the ostrich and oxyhemoglobin crystals from the cassowary. The oxyhemoglobin crystals of *Anser anser*, *Cygnus buccinator*, *Cygnus columbianus*, *Gallus gallus*, *Colinus virginianus* and *Columba livia* are essentially identical. The oxyhemoglobin crystals of *Casuarius* and *Corvus* are each unique and the methemoglobin crystals of *Columba* differ from those of *Struthio*.

B. LYSIS OF RED BLOOD CELLS

The properties of the erythrocyte membrane and its permeability to penetrating non-electrolytes were studied by Jacobs (1931) and by Jacobs *et al.* (1950). An aliquot of the erythrocytes of the species being examined was suspended either in water or in a 0·3*M* solution of one of four non-electrolytes (glycerol, ethylene glycol, urea, and thiourea). The time required for each solution to cause 75% lysis of the erythrocytes was recorded and the log of these times compared. For all except one of the avian species tested, which included the pigeon (presumably *Columba*), herring gull (*Larus argentatus*), tern (possibly *Sterna*), pheasant (presumably *Phasianus colchicus*), turkey (*Meleagris*), chicken (*Gallus*), guinea fowl (*Numida*), sparrow (presumably *Passer*), robin (presumably *Turdus migratorius*), and starling (*Sturnus*), these values were essentially identical. The exceptional erythrocytes were those of the chicken. Excepting chicken erythrocytes, avian red blood cells are unique in comparison with those of other vertebrate classes as regards hemolysis time in solutions of glycerol, ethylene glycol, urea or thiourea.

Johnston and Hochman (1953) used the same technique to study the differences among three species of the family Fringillidae, the house finch (*Carpodacus mexicanus*), fox sparrow (*Passerella iliaca*) and song sparrow (*Melospiza melodia*). They concluded that the cell membrane properties of *Melospiza* and *Passerella* are more similar to one another than either is to the red cell membrane properties of *Carpodacus*. This conclusion is in agreement with current opinion regarding the genetic relationships of these three genera.

C. STUDIES OF FEATHER PROTEINS

The chemical structure of feathers is only generally known, but some of the principal components are proteins. The possibility that the primary structures of the protein components might vary among avian species was indicated by the study of Wilson and Lewis (1927) in which it was shown that the cystine content of the keratins of the barbs and rachi of turkey, goose, and duck feathers differed. Due to the error inherent in the experimental procedures used, one could not be certain that these differences were significant.

However, Schroeder *et al.* (1955) and Harrap and Woods (1964a,b, 1967) have since isolated some feather proteins and showed that their amino acid compositions differ among species. Schroeder *et al.* (1955) showed that the proteins of turkey feather barbs differ in amino acid composition from those of goose feather barbs and goose down. Harrap and Woods (1964a,b, 1967) were able to show that the *S*-carboxymethyl (SCM) feather proteins of chicken (*Gallus gallus*), domestic turkey (*Meleagris gallopavo*), mallard duck (*Anas platyrhynchos*), domestic goose (*Anser anser*), emu (*Dromaius novae-hollandiae*) and silver gull (*Larus novae-hollandiae*) varied both in electrophoretic mobility and in amino acid composition. Since several proteins were included in each compositional analysis this aspect of these studies is less relevant taxonomically than are the electrophoretic analyses. Using both moving boundary and acrylamide gel electrophoresis, it was shown that the patterns of the SCM feather proteins of chicken and turkey were most similar to one another. Likewise the patterns of the duck and goose SCM proteins were most similar. The electrophoretic patterns of each of the other two species were unique. The SCM proteins of the rachis, calamus, barb or medulla portions of feathers may not be homologous. However, concordant results were obtained when the SCM proteins of each of these feather parts was examined individually.

D. BIOCHEMICAL PROPERTIES AND COMPOSITIONS OF EGG-WHITE PROTEINS

The biochemical properties of the avian egg-white proteins have been studied by Robert E. Feeney and his associates. Such characteristics as isoelectric points were measured by their electrophoretic mobilities or the proteins were isolated, purified and characterized individually.

The comparative aspects of the electrophoretic analyses and serological studies are dealt with elsewhere in this review (see p. 123). In this section the comparative and taxonomic aspects of the compositional and biochemical properties of the egg-white proteins are considered.

Ovalbumin is the most abundant protein in avian egg-white. Since ovalbumin is identified by its isoelectric point and by its molecular weight it is not as easy to identify as are certain enzymes such as lysozyme or other proteins with a known function such as ovomucoid which inhibits the activity of trypsin or chymotrypsin. Consequently ovalbumin was not examined in detail by Feeney and his associates. Much is known about conalbumin and ovomucoid, however.

Of 34 species of birds studied the percentage of conalbumin in the egg-white was found to be roughly correlated with genetic relationships. For example, the percentage of conalbumin is more or less uniform for all of the galliform species sampled. However, the percentage of conalbumin is sometimes similar in two unrelated groups. For example, 10% of total protein in the egg-white of the cassowary (*Casuarius aruensis*), the emu (*Dromaius novae-hollandiae*) and the kiwi (*Apteryx mantelli*) is conalbumin while 11% of turkey (*Meleagris gallopavo*) and 12% of chicken (*Gallus*) egg-white protein is conalbumin (Feeney and Allison 1969; Osuga and Feeney 1968; Feeney *et al.* 1965; MacDonnell *et al.* 1954; Feeney *et al.* 1960a; Clark *et al.* 1963).

Windle *et al.* (1963) found the electron paramagnetic resonance spectra of the iron and copper complexes of chicken, turkey and Japanese quail (*Coturnix*) conalbumins to be identical to one another. The iron binding and thermal denaturation properties of the conalbumins of the ostrich (*Struthio*), rhea (*Rhea*), cassowary (*Casuarius*), emu (*Dromaius*), tinamou (*Eudromia*), chicken (*Gallus*), turkey (*Meleagris*), ring-necked pheasant (*Phasianus colchicus*), lady Amherst pheasant (*Chrysolophus amherstiae*), golden pheasant (*Chrysolophus pictus*), blue-eared pheasant (*Crossoptilon auritum*), California quail (*Lophortyx californica*), Japanese quail (*Coturnix coturnix*), and chukar partridge (*Alectoris graeca*) were essentially identical (Clark *et al.*, 1963). The chromatographic properties of the conalbumins could be grouped as being either galliform or ratite in character. There were no differences in the visible absorption spectra of the iron-conalbumin complexes of these 14 species.

The ovomucoids of different avian species have been compared primarily on the basis of their amino acid compositions and on their ability to inhibit the activity of trypsin and chymotrypsin. The amino acid compositions of nine species were determined (Feeney and Allison 1969; Osuga and Feeney 1968; Haynes *et al.*, 1967; Stevens and Feeney 1963; Rhodes *et al.*, 1960). The following comparisons can be drawn from these studies. Of the species sampled, the composition of chicken ovomucoid is most similar to that of the turkey and decreases in similarity to other species in the following order: penguin–emu–tinamou–rhea–cassowary–ostrich–duck. The order of similarity of turkey ovomucoid to that of other species decreases as follows: chicken–emu–penguin–tinamou–ostrich–cassowary–rhea–duck. For the emu the order is: cassowary–rhea–turkey–penguin–ostrich–chicken–tinamou–duck. For the cassowary the order is: emu–rhea–penguin–ostrich–tinamou–turkey–duck–chicken. For the rhea the order is: penguin–cassowary–emu–ostrich–tinamou–turkey–chicken–duck. For the ostrich the order is: penguin–rhea–cassowary–tinamou–emu–turkey–duck–chicken. For the tinamou the order is: cassowary–ostrich–duck–penguin–chicken–turkey–emu–rhea. For the mallard the order is: tinamou–ostrich–cassowary–rhea–penguin–chicken–emu–turkey. The order for the penguin is: ostrich–rhea–cassowary–emu–chicken–tinamou–turkey–duck. The order for the chicken is: turkey–penguin–emu–tinamou–rhea–cassowary–ostrich–duck. The order for the turkey is: chicken–emu–penguin–tinamou–ostrich–cassowary–rhea–duck.

There is no apparent correlation between species relationships and the ability of ovomucoids to inhibit trypsin or chymotrypsin (Osuga and Feeney, 1968; Rhodes *et al.*, 1960; Feinstein and Feeney, 1966; Feinstein *et al.*, 1966a,b; Osuga and Feeney, 1967). No taxonomic conclusions can be based on the percentage of ovomucoid in egg-white (Feeney and Allison, 1969; Feeney *et al.*, 1965; MacDonnell *et al.*, 1954). Chicken, turkey, and pheasant ovomucoids could not be differentiated with respect to their properties after exposure to 9 *M* urea, trichloroacetic acid, iodination, acetylation, or carbamylation (Stevens and Feeney 1963).

Few comparative data are available for the other egg-white proteins. The percentage of lysozyme in egg-white varies considerably but is generally within an order of magnitude when families within an order are compared whereas the difference is sometimes even greater between orders (Feeney and Allison, 1969; MacDonnell *et al.*, 1954; Feeney *et al.*, 1965; Feeney *et al.*, 1960). Comparative data on lysozyme amino acid sequences are presented elsewhere (see p. 163).

The few data available on the amino acid compositions of ovomucins pertain to the relationships of the ratites, but the ovomucins of the species studied (chicken, cassowary, ostrich, rhea, tinamou) are not essentially different (Feeney and Allison, 1969; Feeney and Osuga, 1968).

The following papers which relate to the above work of Robert Feeney and his colleagues, contain information on the techniques of protein purification (Rhodes *et al.*, 1958, 1959; Azari and Feeney, 1958, 1961; Greene and Feeney, 1968).

Peakall (1963) hydrolyzed purified ovalbumins from several species of birds and separated the amino acids by ascending paper chromatography. The densitometric traces for three different samples of the American crow (*Corvus brachyrhynchos*) were virtually identical. As expected, different taxa had dissimilar patterns. Three species of accipitrids (*Accipiter gentilis, Buteo jamaicensis,* and *Aquila chrysaetos*) were compared to the cathartid vultures *Coragyps atratus* and *Cathartes aura*. The data suggested that the cathartids are not closely related to other falconiforms.

E. SUBSTRATE AND INHIBITOR SPECIFICITIES OF ENZYMES

Bamann *et al.* (1967) studied the hydrolysis of racemic mandelic acid-ethyl ester by esterases purified from the livers of several species of ducks (*Anatidae*). Since esterases, like most other enzymes, are stereo-specific only one isomer of a racemic substrate is hydrolysed and the change in optical rotation which results is a measure of the degree and rate of hydrolysis. It is assumed that esterases of similar structure will have similar stereochemical specificities and activities. Hence closely related species should have similar optical selectivity or 'esterase type'. Bamann *et al.* (1967) judged the 'esterase type' of species to be the same if variation of substrate concentration, addition of the hydrolysis product or addition of an optically active foreign substance such as ethanol or strychnine resulted in a consistent optical rotation curve. Individuals of the same species were found to have constant activity.

The taxonomic findings of this study can be summarized as follows:

i. species of the genus *Branta* were alike in their stereochemical specificity but were significantly different from *Anser* species in the optical selectivity following the addition of ethanol;

ii. the following groups of species were similar in their 'esterase types' and therefore are assumed to be closely related:

(a) Mallard (*Anas platyrhychos*) and teal (*A. crecca*),

(b) Swans (Cygninae) and geese (Anserinae),

(c) Pintail (*A. acuta*), garganey (*A. querquedula*) and shoveler (*A. clypeata*). These three species differed from wigeon (*A. penelope*),

(d) Shelduck (*Tadorna tadorna*) and Andean goose (*Chloëphaga melanoptera*),

(e) Muscovy duck (*Cairina moschata*) and mandarin duck (*Aix galericulata*),

(f) Common eider (*Somateria mollissima*), common scoter (*Melanitta nigra*) and velvet scoter (*Melanitta fusca*),

(g) Goldeneyes (*Bucephala*), mergansers (*Mergus*), smew (*Mergellus*) and long-tailed duck (*Clangula*).

It is of particular interest to note that these results agree in many respects with data from other sources or with commonly accepted arrangements. For example Cotter (1957) also found *Aix* and *Cairina* to be closely related (see p. 107) and most of the other groups are not surprising. However, the alliance of *Tadorna* and *Chloëphaga* is unusual for these genera are customarily placed in different subfamilies.

Ferguson (1969) examined the substrate and inhibitor specificities of individual electrophoretic fractions with esterase activity in four species of Columbidae. Location of non-specific esterase activity after starch gel electrophoresis using 1-naphthyl acetate as substrate revealed 18 esterase fractions in the feral domestic pigeon (*Columba livia*), 15 in wood pigeon (*Columba palumbus*), 14 in barbary dove (*Streptopelia risoria*) and 13 in collared dove (*S. decaocto*). Tests were made using the following substrates: 2-naphthyl acetate, 1-naphthyl butyrate, 1-naphthyl propionate, 1-naphthyl laurate, 1-naphthyl stearate, naphthol AS acetate, and 2-carbonaphthoxy choline iodide. Inhibitory substances as follows were used at the concentrations indicated: physostigmine sulphate $10^{-4} M$, D.F.P. (di-iso-propyl fluorophosphate) $10^{-3} M$, E600 (diethyl-*p*-nitrophenyl phosphate) 10^{-4} to $10^{-5} M$ and *p*CMB (*p*-chloro-mercuri-benzoic acid) 10^{-2} to 10^{-3}. Esterase fractions in the different species were judged to be identical if their activity and inhibition was consistent to each of the substrates and inhibitors.

On these criteria 10 fractions (61%) were identical between *Columba livia* and *C. palumbus*; between *Streptopelia risoria* and *S. decaocto* 89%; between *C. livia* and *S. risoria* 31%; between *C. livia* and *S. decaocto* 32%; between *C. palumbus* and *S. risoria* 28% and between *C. palumbus* and *S. decaocto* 25%.

F. COMPARISONS OF PEPTIDES

Protein molecules may be chemically or enzymatically cleaved into smaller subunits known as peptides. The resulting peptide mixture, in turn, can be characterized in a variety of ways including one-dimensional electrophoresis, two-dimensional chromatography and electrophoresis, and ion-exchange column chromatography. Sibley (1964) used the latter technique to explore the possibility of using peptide chromatograms in comparative taxonomic studies. He showed that replicate chemical cleavages of chicken ovomucoid using 0·03 *N* HCL gave reproducible chromatograms when column elution conditions were standardized. This was also true for peptic digests of chicken lysozyme, ovomucoid, and ovalbumin. Peptic digests of the ovalbumins of mallard (*Anas platyrhynchos*), American coot (*Fulica americana*), domestic turkey (*Meleagris gallopavo*), ring-necked pheasant (*Phasianus colchicus*), ostrich (*Struthio camelus*), herring gull (*Larus argentatus*) and ring-billed gull (*L. delawarensis*) were also compared. The first portion of the chromatograms of these species had peptide peaks in similar if not identical positions. On the other hand, many differences in the chromatograms were present in the remaining major portion of the chromatograms. Thus, peptide homologies were more readily assigned to the peptides eluted during the early part of each separation.

Nevertheless, in the comparison of the latter portion of the peptic peptide chromatograms of chicken, turkey and ring-necked pheasant ovalbumins there are several peaks that can be tentatively identified as being homologous.

No homologies between the last portion of the chromatograms of the ovalbumin peptic peptides of the remaining species (ostrich, mallard, etc.) and the chromatograms of the galliforms can be detected. At least partial amino acid compositional analyses of each peptide would have to be carried out before additional homologies could be ascertained. The peptide chromatograms of the two species of gulls are also more similar to one another than either of them is to one of the other species examined. However, the ninhydrin reaction with the peptides in the terminal portions of the gull chromatograms was weak and the complexity of the patterns is consequently very much reduced. This makes it difficult to compare these chromatograms with those of the other species.

The ovalbumin tryptic peptides of 18 species of pigeons, the chicken and the mallard duck were compared using two-dimensional thin-layer chromatography and electrophoresis (Corbin 1965, 1967, 1968). Again, using standardized techniques it was possible to obtain reproducible patterns of replicate digests of the same ovalbumin preparation (Corbin 1965, 1967), but variation among four genetic strains of the domestic pigeon or rock dove (*Columba livia*) was detected. The variation among individuals of the white-crowned pigeon (*C. leucocephala*) was no greater than that among separate digests of the same preparation.

Interspecific comparisons of the tryptic peptide maps indicated that the primary structure of the ovalbumin of each species was unique. Analysis of these data indicated that the Old World pigeons of the genus *Columba* are indeed related to the New World species of the genus. *C. cayennensis* is perhaps most like the ancestral stock from which the species of *Columba* evolved. Among the species of the New World *C. leucocephala*, *maculosa*, and *cayennensis* are most closely related, and *picazuro*, *fasciata*, and *flavirostris* are species which either

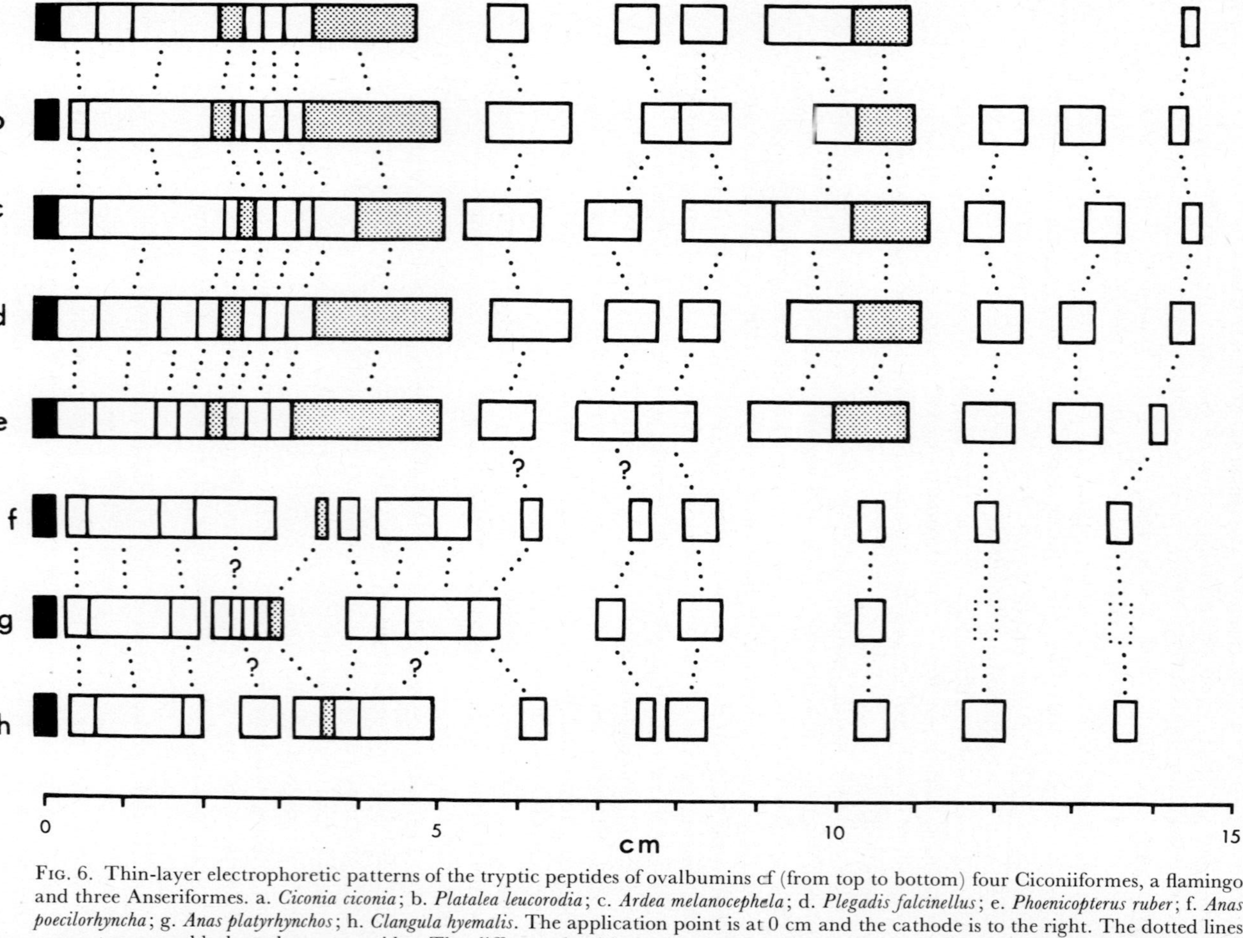

FIG. 6. Thin-layer electrophoretic patterns of the tryptic peptides of ovalbumins of (from top to bottom) four Ciconiiformes, a flamingo and three Anseriformes. a. *Ciconia ciconia*; b. *Platalea leucorodia*; c. *Ardea melanocephala*; d. *Plegadis falcinellus*; e. *Phoenicopterus ruber*; f. *Anas poecilorhyncha*; g. *Anas platyrhynchos*; h. *Clangula hyemalis*. The application point is at 0 cm and the cathode is to the right. The dotted lines connect presumably homologous peptides. The different densities approximate differences in the intensity of the ninhydrin staining reaction in the original analysis. From Sibley, Corbin and Haavie (1969).

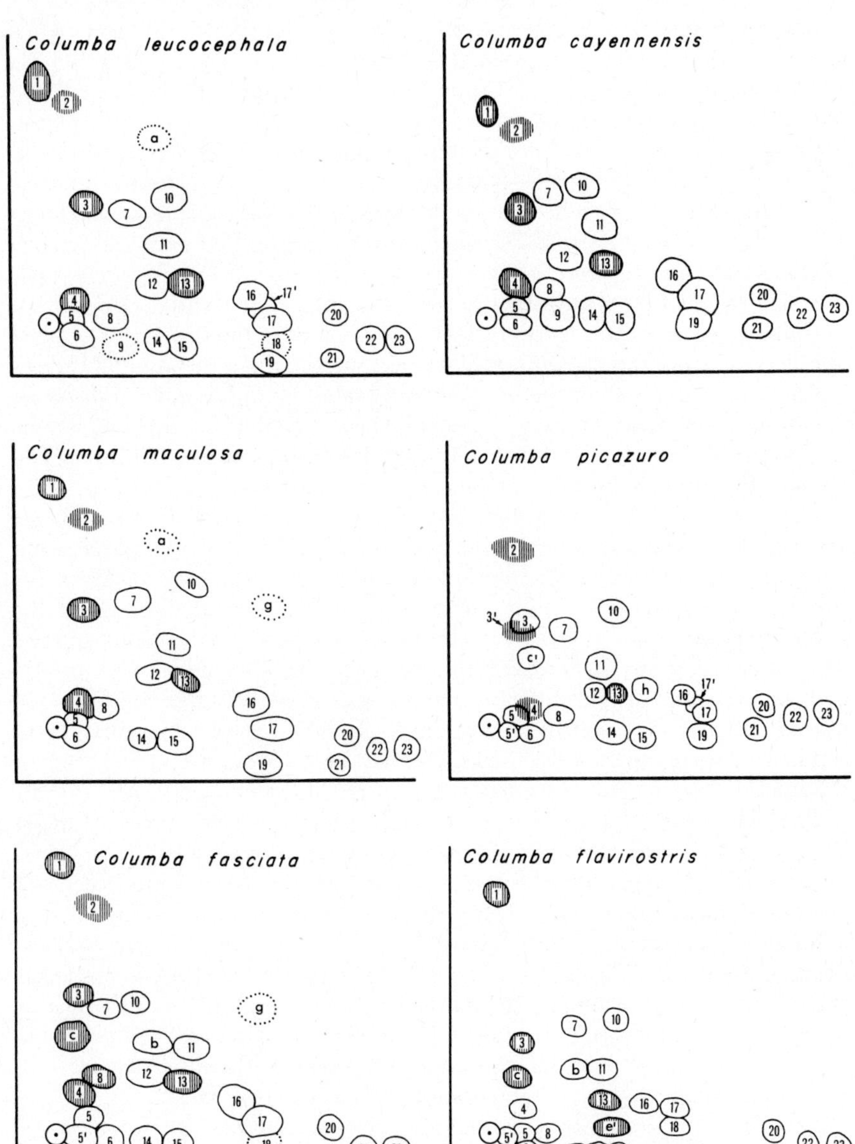

FIG. 7. Tryptic peptide maps of ovalbumins of representative species of the pigeon genus *Columba*. Peptides were separated electrophoretically in the horizontal direction and chromatographically in the vertical dimension. Peptides circled by a solid line reacted with ninhydrin. Peptides circled by a dotted line showed very little color after reaction with ninhydrin. Shaded peptides reacted with a reagent specific for the amino acid tryptophan. From Corbin (1968).

diverged from the ancestral type early during the evolutionary history of the genus or evolved much more rapidly. Of the Old World species studied *C. guinea* is most similar to *C. oenas*. These two species are in turn more similar to *C. palumbus* than to *C. livia*. It is perhaps a moot question whether the Old World species were derived from or gave rise to the New World species, although the peptide data favor the latter interpretation. Of the peptide maps of the other genera of pigeons examined, that of the genus *Leptotila* was most similar to those of *Columba*. The peptide map of *Streptopelia* was next most similar. The peptide map of *Zenaida macroura* was most like those of *Leptotila rufaxilla* and *Columbigallina talpacoti*. The peptide map of *Ducula goliath* is not particularly like that of any other species examined. The tryptic peptide maps of *Gallus gallus* and *Anas platyrhynchos* ovalbumin were too different from one another and from the pigeon species to be compared in a meaningful way.

The tryptic peptides of the hemoglobins and ovalbumins of several species of herons, ducks and a flamingo were analyzed by Sibley, Corbin and Haavie (1969) to obtain data which bear on the relationships of the flamingos. The tryptic peptides of the hemoglobins of nine species (*Ardea purpurea*, *Ardea goliath*, *Phoenicopterus ruber*, *Anas platyrhynchos*, *Anas erythrorhyncha*, *Branta bernicla*, *Gallus gallus*, *Larus argentatus* and *Gavia stellata*) were compared using ion-exchange column chromatography. The peptide chromatograms of the two species of *Ardea* differ from one another by one or two peptides and the chromatograms of the anatids differ from one another by only a few peptides. The chromatograms of *Phoenicopterus* and *Ardea* sp. have at least 17 tryptic peptides in common and differ by a minimum of six peptides, while *Phoenicopterus* differs from *Anas* by a minimum of eight peptides, and has 14 in common. The patterns of the ducks, flamingo and herons are unlike those of the chicken, herring gull (*Larus argentatus*) or red-throated loon (*Gavia stellata*).

The ovalbumin tryptic peptides of eight species (*Ciconia ciconia, Platalea leucorodia, Ardea melanocephala, Plegadis falcinellus, Phoenicopterus ruber, Anas poecilorhyncha, Anas platyrhynchos* and *Clangula hyemalis*) were compared using one-dimensional thin-layer electrophoresis. The chromatogram of the flamingo (*Phoenicopterus*) is much more like those of the ciconiiforms (*Ciconia, Platalea, Ardea* and *Plegadis*) than like the chromatograms of the ducks (*Anas* and *Clangula*). For example the patterns suggest that the ovalbumins of *Phoenicopterus* and *Plegadis* contain 17 homologous if not identical tryptic peptides or groups of peptides whereas *Phoenicopterus* ovalbumin contains only 3–5 tryptic peptides that may be homologous to those of *Anas*.

These data suggested that flamingos, ducks and herons are more closely related to one another than any one of them is to some other order and that the Phoenicopteridae are more closely related to the Ciconiiformes than to the Anseriformes.

As discussed above (see p. 110) Fothergill and Perrie (1966) compared the ovalbumin tryptic peptides of the chicken (*Gallus*) and the duck (*Anas*) using two-dimensional chromatography and electrophoresis. With the same technique, Wiseman and Fothergill (1966) were able to distinguish two variants of chicken ovalbumin that differed by a single chymotryptic peptide, and presumably by a single amino acid.

G. AMINO ACID SEQUENCES

The lysozymes and cytochromes C of a few species of birds have been isolated, purified and sequenced. The amino acid sequence analyses of the cytochromes C were performed by Margoliash and his associates (Chan and Margoliash 1966; Chan *et al.*, 1963, 1969a,b,c,d; Margoliash *et al.*, 1963). In these studies the sequences of chicken, turkey, king penguin, (*Aptenodytes patagonicus*), mallard and pigeon (*Columba*) cytochromes C were determined. The amino acid sequence of chicken and turkey are identical and differ from the cytochrome C of the king penguin by two amino acid substitutions, from that of the mallard duck (*Anas*) by three substitutions, and from that of the pigeon by four substitutions. Penguin cytochrome C differs from that of the mallard by three substitutions and from pigeon cytochrome C by four substitutions. Mallard duck cytochrome C differs from pigeon cytochrome C by three amino acid substitutions. Each of these substitutions can be accounted for by single nucleotide base-pair changes. From these data it appears that of the four orders sampled the Columbiformes are most closely related to the Anseriformes, the Sphenisciformes most closely related to the Galliformes, the Galliformes most similar to the Sphenisciformes and the Anseriformes equally distantly related to the other three orders.

Lysozyme amino acid sequences or compositions of four species have been determined. These species are chicken (Jollès and Jollès, 1961; Jollès *et al.*, 1963; Canfield, 1963), turkey (LaRue and Speck, 1970), mallard duck (*Anas platyrhynchos*) (Jollès *et al.*, 1967; Herman and Jollès, 1970; Imanishi *et al.*, 1966; Niemann *et al.*, 1968), and domestic goose (*Anser anser*) (Canfield and McMurry, 1967; Canfield and Kammerman, unpublished data; Dianoux and Jollès, 1967; Kammerman and Canfield, 1969). Chicken and turkey lysozymes differ by seven amino acid substitutions and a minimum of seven nucleotide base-pair substitutions would be required to convert one sequence into the other. Amino-terminal and carboxy-terminal sequences of the duck lysozyme III indicate that the C-terminal sequence of duck lysozyme III differs at residue 122 from the sequences of chicken and turkey lysozymes (Jollès *et al.*, 1967). Two nucleotide base pair substitutions are required to account for the variation at position 122. Compositional analyses of duck lysozyme III indicate that other differences are present.

The complete sequence of duck lysozyme II differs from the lysozyme of chicken by 19 amino acid substitutions (minimum mutation distance = 24) and from the lysozyme of turkey by 19 amino acid substitutions (minimum mutation distance also equals 24 but different nucleotide base-pair substitutions are required) (Herman and Jollés, 1970).

Compositional analyses and preliminary sequence analyses of goose lysozyme indicate that it is very different from the lysozymes of chicken, turkey and duck (Canfield and McMurry, 1967; Dianoux and Jollès, 1967; Kammerman and Canfield, 1969; Canfield and Kammerman, pers. comm.). This apparently anomalous situation may be explained by recent data of Arnheim and Steller (pers. comm.). They found evidence for two antigenically dissimilar lysozyme molecules in the egg-white of the black swan (*Cygnus atratus*). One of these is antigenically similar to but not identical with the lysozyme of the chicken and the other is antigenically similar to the lysozyme of the goose. Thus there appear to be two lysozyme-like molecules which differ both in their antigenic properties and in their primary structures.

Addenda

Arnheim and Steller (1970) found two lysozymes in the egg-white of the black swan (*Cygnus atratus*). The evidence indicated that the two lysozymes may be products of non-allelic genes and the authors suggested caution in the interpretation of data on the rate of protein evolution based upon proteins which may be the products of multiple gene loci.

Baker *et al.* (1970) have reviewed the occurrence of protein polymorphisms in chicken egg-white and blood serum and have suggested a standard nomenclature for the various morphs. This was made possible by the exchange of samples among laboratories and by working out the correspondence of variants given alternative names when discovered by different workers.

Polymorphism of ovoglobulin G_2, ovoglobulin G_3 and ovotransferrin was noted in egg whites of indigenous Indian domestic chickens by Baker *et al.* (1971). All three known variants of ovotransferrin and three (J, A and B) of the G_3 ovoglobulin variants were found in these samples. The occurrence of the same polymorphisms in these indigenous birds as in western commercial breeds suggests that these polymorphisms are selectively maintained. *Gallus sonneratii* is the wild species in the area in which the samples were obtained. *Gallus gallus* and *G. sonneratii* can be distinguished on the basis of electrophoretic mobility of their G_2 ovoglobulins. None of the Indian eggs examined had proteins characteristic of any species other than *G. gallus*, thus supporting the latter as the main ancestor of the domestic fowl.

Csuka and Petrovsky (1970) found that the manifestation of polymorphic plasma esterase fractions in chickens (*Gallus*) is influenced by physiological processes. A change from active esterase phenotypes to a type without esterase activity was found in hens which produce significantly more eggs.

Fothergill and Fothergill (1970) compared the amino acid compositions and tryptic peptide maps of the ovalbumins from four galliforms (chicken; turkey; golden pheasant, *Chrysolophus pictus*; and ptarmigan, *Lagopus lagopus*) and five anseriforms (Aylesbury duck, *Anas platyrhynchos*; fulvous whistling duck, *Dendrocygna bicolor*; magpie goose, *Anseranas semipalmata*; Chinese goose, *Anser cygnoides*; and grey lag goose, *Anser anser*). Within the galliforms the ovalbumin of the ptarmigan was most similar to that of the chicken, followed by those of the golden pheasant and the turkey. On the basis of the tryptic peptide data the magpie goose, Chinese goose and grey lag goose form a subgroup among the anseriforms. The amino acid composition data, however, suggest that the fulvous whistling duck ovalbumin is more similar to those of the two *Anser* species than is the ovalbumin of the magpie goose.

Fothergill and Fothergill (1971) cross reacted the purified ovalbumins with anti-duck and anti-chicken ovalbumin antisera by the Ouchterlony double diffusion method and determined the amount of cross reactivity by quantitative antigen-antibody precipitin assays. In general their results agreed with those of the previous study.

In the chicken (*Gallus gallus*) Fujio (1971) has shown that heterozygotes of B blood group alleles have a greater hatchability than homozygotes.

Gysels (1970) published a study on the relationships of the tinamous based on zone electrophoresis and immunoelectrophoresis of lenticular proteins. The species involved included a rhea (*Rhea*), cassowary (*Casuarius*), tinamou (*Crypturellus*), chicken (*Gallus*), pheasant (*Chrysolophus pictus*) and loon (*Gavia*). He concluded that the tinamou is more closely related to the rhea than to the Galliformes, but that close ties exist between *Casuarius* and the galliforms.

Kaminski (1970) reported the presence of five lipoesterases and two non-lipoesterases in avian sera (five species of galliforms, two anatids and two columbids). The lipoesterases were mainly function-specific while the non-lipoesterases were species-specific.

Using starch gel electrophoresis, Karig and Wilson (1971) examined the supernatant malate dehydrogenase of 35 pitohuis (*Pitohui kirhocephalus*) from Papua and New Guinea. All but two individuals showed single bands of supernatant malate dehydrogenase activity with identical mobilities. Two individuals showed three electrophoretic fractions with this enzyme activity and which were present in the concentration ratio of 1:2:1. Since malate dehydrogenase is a dimer these two individuals seem to be heterozygotes for two alternative alleles at the supernatant malate dehydrogenase locus. The authors report that they failed to find supernatant malate dehydrogenase variation in 123 *Aplonis metallica*, 43 *Aplonis cantoroides* and 11 *Rhipidura leucophrys* from Papua and New Guinea or in 100 *Gallus gallus*, 50 *Columba livia* and 12 *Coturnix c. japonica*.

Kimura (1969c) studied genetic polymorphism in the plasma esterases of the chicken (*Gallus*). Three phenotypes, probably controlled by a pair of co-dominant autosomal alleles, were found.

In another study Kimura (1969d) examined genetic variants of the eserine resistant esterases in *Gallus*.

Using starch-gel electrophoresis Kimura and Makita (1968) compared the egg-white proteins of 13 species representing five families of non-passerines. They found specific variations in the mobilities of several proteins but the results are of little taxonomic interest.

A more detailed discussion of negative heterosis and its possible role in race and species formation is given by Manwell and Baker (1970). They review many examples of polymorphism in avian proteins and suggest that highly polymorphic species are 'pre-adapted' to a wider range of habitats and environmental conditions than are monomorphic ones. In many domestic species and undoubtedly some wild ones introgression is a major source of this increased variability.

Paulov (1971) has noted that paper, cellulose acetate and acrylamide gel zymograms of duck (*Anas platyrhynchos*) and turkey (*Meleagris gallopavo*) sera are species-specific. He failed to find any difference between white Peking and black Rouen ducks and between white and bronze turkeys. In turkey serum 16 protein components were resolved and 18 in duck serum.

Prager and Wilson (1971) used micro-complement fixation to study the relationship between cross-reactivity and differences in the amino acid sequences of the egg-white lysozymes of six galliforms and the domestic duck (*Anas*). Although this study was not primarily taxonomic the authors found that the lysozyme of the chachalaca (*Ortalis vetula*) 'is actually less related to the lysozymes of gallinaceous birds than are the lysozymes of ducks . . . Thus the lysozyme results do not suggest inclusion of the chachalaca in the order Galliformes' (p. 5,987). The authors also concluded that lysozyme is unlikely to be a useful molecule for the development of phylogenies.

Schram (1970) compared the serum proteins of the silver pheasant (*Lophura nycthemera*) to those of the chicken and ring-necked pheasant (*Phasianus colchicus*). His results indicated that pheasant serum has a higher albumin content than that of chicken and that the pheasant albumin has a higher content of aromatic amino acids.

In the ruff (*Philomachus pugnax*), a species in which many plumage morphs are found, Segre *et al.* (1970) have noted a high degree of heterozygosity of the serum proteins. In a captive flock of 22 birds they found polymorphism in five out of nine enzyme systems examined by starch gel electrophoresis. There was no obvious correlation, however, between a particular plumage morph and the type of protein polymorphism present.

Sibley and Frelin (1972) studied the egg-white proteins of the large ratites (*Struthio, Rhea, Casuarius, Dromaius*), the kiwis (*Apteryx*) and the tinamous (Tinamidae) using isoelectric focusing in acrylamide gel and the electrophoretic patterns of the tryptic peptides of ovalbumins. They concluded that the large ratites form a monophyletic group that should be treated as a single order, but that *Apteryx* and the tinamous should not be included within it. The tinamous may be related to the Galliformes.

Smith and Back (1970) determined the amino acid compositions of chicken, turkey, and duck ovalbumins. Difference indices between species calculated from their data are as follows: chicken to turkey, 4·6; chicken–duck, 6·8; turkey–duck, 8·0.

Vyshinsky and Muraviev (1970) studied genetic polymorphisms in the serum transferrins of various breeds of domestic chickens (*Gallus*) using disc electrophoresis. Five transferrin phenotypes were found, apparently controlled by a triallelic system at a single locus. One of the six theoretically possible types was not found. The triallelic hypothesis was tested by mating parents of known transferrin phenotypes.

Weintraub and Schlamowitz (1970) compared the ovalbumins of chicken (*Gallus*), turkey (*Meleagris*) and duck (*Anas platyrhynchos*). All three have similar molecular weights (*ca* 45,000) and phosphorus content. Chicken and turkey ovalbumins are electrophoretically similar and the patterns reflect the phosphorus content of the proteins. Turkey ovalbumin has the most prosthetic carbohydrate, with large amounts of galactose. The amino acid compositions of the ovalbumins of chicken and turkey are more similar to one another than either is to the duck. Similar results were obtained in a subsequent immunochemical study (Weintraub and Schlamowitz, 1971).

Woods (1971) observed at least ten components of the solubilized proteins of feather rhachis following chromatography on DEAE cellulose. Patterns of elution of these proteins varied among chicken, turkey, duck, goose, and sulphur-crested cockatoo (*Cacatua galerita*) and suggested possible taxonomic usefulness.

References and Bibliography

Allison, R. G. and Feeney, R. E. (1968). *Archs Biochem. Biophys.* **124**, 548–555.

Amin, A. (1961). *Nature, Lond.* **191**, 1208.

Arnheim, N. and Steller, R. (1970). *Archs Biochem. Biophys.* **141**, 656–661.

Arnheim, N., Jr. and Wilson, A. C. (1967). *J. biol. Chem.* **242**, 3951–3956.

Arnheim, N., Prager, E. M. and Wilson, A. C. (1969). *J. biol. Chem.* **244**, 2085–2094.

Azari, P. R. and Feeney, R. E. (1958). *J. biol. Chem.* **232**, 293–302.

Azari, P. R. and Feeney, R. E. (1961). *Archs Biochem. Biophys.* **92**, 44–52.

Bain, J. A. and Deutsch, H. F. (1947). *J. biol. Chem.* **171**, 531–541.

Baker, C. M. A. (1964). *Comp. Biochem. Physiol.* **12**, 389–403.

Baker, C. M. A. (1965). *Comp. Biochem. Physiol.* **16**, 93–101.

Baker, C. M. A. (1967). *Comp. Biochem. Physiol.* **20**, 949–973.

Baker, C. M. A. (1968). *Genetics* **58**, 211–226.

Baker, C. M. A. and Hanson, H. C. (1966). *Comp. Biochem. Physiol.* **17**, 997–1006.

Baker, C. M. A. and Manwell, C. (1962). *Br. Poult. Sci.* **3**, 161–174.

Baker, C. M. A. and Manwell, C. (1967). *Comp. Biochem. Physiol.* **23**, 21–42.

Baker, C. M. A., Manwell, C., Labisky, R. F. and Harper, J. A. (1966). *Comp. Biochem. Physiol.* **17**, 467–499.

Baker, C. M. A., Croizier, G., Stratil, A. and Manwell, C. (1970). *Adv. Genet.* **15**, 147–174.

Baker, C. M. A., Manwell, C., Jayaprakash, N. and Francis, N. (1971). *Comp. Biochem. Physiol.* **40B**, 147–153.

Bamann, E., Haggag, A. and Lang, R. (1967). *Hoppe Seyler's Z. physiol. Chem.* **348**, 983–989.

Beckman, L. and Nilson, L. R. (1965). *Hereditas* **53**, 221–230.

Beckman, L., Conterio, F. and Mainardi, D. (1962). *Nature, Lond.* **196**, 92–93.

Beckman, L., Conterio, F. and Mainardi, D. (1963). *Bull. serol. Mus., New Brunsw.* **29**, 5–8.

Beecher, W. J. (1953). *Auk* **70**, 270–333.

Bocchi, G. D., Mainardi, D. and Orlando, C. (1960). *Rc. Ist. lomb. Sci. Lett.* B**94**, 63–74.

Boehm, L. G. and Irwin, M. R. (1970). *Comp. Biochem. Physiol.* **32**, 377–386.

Bordet, J. (1899). *Annls Inst. Pasteur, Paris* **13**, 225–250.

Bourns, T. K. R. (1967). *Can. J. Zool.* **45**, 97–99.

Boyd, W. C. (1966). 'Fundamentals of Immunology'. Fourth edition. Interscience Publ., New York.

Boyden, A. (1926). *Biol. Bull. mar. biol. Lab., Woods Hole* **50**, 73–107.

Boyden, A. (1934). *Amer. Nat.* **68**, 516–536.

Boyden, A. (1942). *Physiol. Zool.* **15**, 109–145.

Boyden, A. (1943). *Am. Nat.* **77**, 234–255.

Boyden, A. (1953). *Syst. Zool.* **2**, 19–30.

Boyden, A. (1958). *In* 'Serological and biochemical comparisons of proteins' (W. H. Cole, ed.), pp. 3–24. Rutgers University Press, New Brunswick, New Jersey.

Boyden, A. (1963). *Syst. Zool.* **12**, 1–7.

Boyden, A. (1964). *In* 'Taxonomic Biochemistry and Serology' (C. A. Leone, ed.), pp. 75–99. The Ronald Press Co., New York.

Boyden, A. (1965). *Bull. serol. Mus., New Brunsw.* **33**, 5–8.

Boyden, A. (1966). *Bull. serol. Mus., New Brunsw.* **34**, 1–4.

Brandt, L. W., Smith, H. D., Andrews, A. C. and Clegg, R. E. (1952). *Archs Biochem. Biophys.* **36**, 11–17.

Brown, L. E. and Fisher, H. I. (1966). *Auk* **83**, 111–116.

Brush, A. H. (1968). *Comp. Biochem. Physiol.* **25**, 159–168.

Brush, A. H. (1970). *Occ. Papers Zool. U. Conn.* (In press.)

Bryan, C. R. (1953). Ph. D. thesis, Univ. of Wisconsin.

Bryan, C. R. and Irwin, M. R. (1961). *Genetics, Princeton* **46**, 323–337.

Bryan, C. R. and Miller, W. J. (1953). *Proc. natn. Acad. Sci. U.S.A.* **39**, 412–416.

Buchbinder, L. (1934). *J. Immun.* **26**, 215–231.

Bush, F. M. (1965). *Year Book Amer. Phil. Soc.* 302–304.

Bush, F. M. (1967). *Comp. Biochem. Physiol.* **22**, 273–287.

Canfield, R. E. (1963). *J. biol. Chem.* **238**, 2698–2707.

Canfield, R. E. and McMurry, S. (1967). *Biochem. biophys. Res. Commun.* **26**, 38–42.

Carpenter, P. L. (1956). 'Immunology and Serology'. W. B. Saunders, Philadelphia.

Cavalli-Sforza, L., Mainardi, D. and Schreiber, B. (1954). *Boll. Zool.* **21**, 253–260.

Cavalli-Sforza, L., Mainardi D. and Schreiber, B. (1957). *Pubbl. Staz. zool. Napoli* **29**, 323–328.

Chan, S. K. and Margoliash, E. (1966). *J. biol. Chem.* **241**, 507–515.

Chan, S. K., Needleman, S. B., Stewart, J. W., Walasek, O. F. and Margoliash, E. (1963). *Fedn Proc. Fedn. Am. Socs exp. Biol.* **22**, 658.

Chan, S. K., Tulloss, I. and Margoliash, E. (1969a). *In* 'Atlas of Protein Sequence and Structure 1969' (cited by M. O. Dayhoff). Silver Spring, Maryland.

Chan, S. K., Tulloss, I. and Margoliash, E. (1969b). *In* 'Atlas of Protein Sequence and Structure 1969' (cited by R. O. Dayhoff). Silver Spring, Maryland.

Chan, S. K., Tulloss, I. and Margoliash, E. (1969c). *In* 'Atlas of Protein Sequence and Structure 1969' (cited by M. O. Dayhoff). Silver Spring, Maryland.

Chan, S. K., Tulloss, I. and Margoliash, E. (1969d). *In* 'Atlas of Protein Sequence and Structure 1969' (cited by M. O. Dayhoff). Silver Spring, Maryland.

Choogoonov, N. D., Kooshniruk, V. A. and Kooshniruk, I. F. (1965). *Ornithology News.* 408–409.

Clark, J. R., Osuga, D. T. and Feeney, R. E. (1963). *J. biol. Chem.* **238**, 3621–3631.
Cohn, M., Wetter, L. R. and Deutsch, H. F. (1949). *J. Immun.* **61**, 283–296.
Cole, A. G. (1938). *Archs Path.* **26**, 96–101.
Common, R. H., McKinley, W. P. and Maw, W. A. (1953). *Science, N.Y.* **118**, 86–89.
Conterio, F. and Mainardi, D. (1957). *Ricerca scient. (Convegno Genet.)*, **29** *Suppl.*, 71–79.
Cooper, D. W., Irwin, M. R. and Stone, W. H. (1969a). *Genetics, Princeton* **62**, 597–606.
Cooper, D. W., Irwin, M. R. and Stone, W. H. (1969b). *Genetics, Princeton* **62**, 607–617.
Corbin, K. W. (1965). Ph.D. thesis. Cornell Univ.
Corbin, K. W. (1967). *Evolution, Lancaster, Pa.* **21**, 355–368.
Corbin, K. W. (1968). *Condor* **70**, 1–13.
Cotter, W. B. (1957). *Wilson Bull.* **69**, 291–300.
Croizier, C. (1966). *Annls biol. anim. Biochim. Biophys.* **6**, 379–388.
Csuka, J. and Petrovsky, E. (1970). *In* 'Proc. XI European Conference on Animal Blood Groups and Biochemical Polymorphism,' pp. 433–437. W. Junk, Publisher, The Hague.
Cumley, R. W. and Cole, L. J. (1942). *Am. Nat.* **76**, 570–581.
Cumley, R. W. and Irwin, M. R. (1940). *Proc. Soc. exp. Biol. Med.* **44**, 353–355.
Cumley, R. W. and Irwin, M. R. (1941a). *J. Hered.* **32**, 178–182.
Cumley, R. W. and Irwin, M. R. (1941b). *J. Hered.* **32**, 429–434.
Cumley, R. W. and Irwin, M. R. (1942a). *Genetics* **27**, 177–192.
Cumley, R. W. and Irwin, M. R. (1942b). *J. Hered.* **33**, 357–365.
Cumley, R. W. and Irwin, M. R. (1944). *Am. Nat.* **78**, 238–256.
Cumley, R. W. and Irwin, M. R. (1952). *Genetics, Princeton* **37**, 396–412.
Cumley, R. W., Irwin, M. R. and Cole, L. J. (1942). *Genetics, Princeton* **27**, 139.
Cumley, R. W., Irwin, M. R. and Cole, L. J. (1943). *J. Immun.* **47**, 35–51.
DeFalco, R. J. (1942). *Biol. Bull. mar. biol. Lab., Woods Hole* **83**, 205–218.
Desborough, S. and Irwin, M. R. (1966). *Physiol. Zool.* **39**, 66–69.
Deutsch, H. F. and Goodloe, M. B. (1945). *J. biol. Chem.* **161**, 1–20.
Dianoux, A.-C. and Jollès, P. (1967). *Biochim. biophys. Acta* **133**, 472–479.
Dilger, W. C. (1956). *Syst. Zool.* **5**, 174–182.
Dujarric de la Rivière, R. and Eyquem, A. (1953). *Éditions médicales Flammarion, Paris,* 132–149.
Dunlap, J. S., Johnson, V. L. and Farner, D. S. (1956). *Experientia* **12**, 352–356.
Dutta, R., Ghosh, J. and Guha, B. C. (1958). *Nature, Lond.* **181**, 1204–1205.
Duwe, A. E. (1964). *In* 'Taxonomic Biochemistry and Serology' (C. A. Leone, ed.), pp. 501–504. The Ronald Press Co., New York.
Emmerich, E. (1913). *Z. ImmunForsch. exp. Ther.* **17**, 299–304.
Erhardt, A. (1929a). *Z. ImmunForsch. exp. Ther.* **60**, 156–166.
Erhardt, A. (1929b). *Ergebn. Fortschr. Zool.* **7**, 279–321.
Erhardt, A. (1930). *J. Orn., Lpz.* **78**, 214–234.
Feeney, R. E. and Allison, R. G. (1969). 'Evolutionary Biochemistry of Proteins: Homologous and Analogous Proteins from Avian Egg Whites, Blood Sera, Milk, and other Substances.' Wiley-Interscience. New York.
Feeney, R. E. and Komatsu, S. K. (1966). The transferrins. *In* 'Structure and Bonding' (C. K. Jorgensen, J. B. Neidlands, R. S. Nyholm, D. Reinen and R. J. P. Williams, eds.), Vol. 1, pp. 149–206. Springer-Verlag, Berlin.
Feeney, R. E., Allison, R. G., Osuga, D. T., Bigler, J. C. and Miller, H. T. (1968). *In* 'Antarctic Bird Studies' (O. L. Austin, ed.), Antarctic Research Studies Vol 12, pp. 151–165. Amer. Geophys. Union, Washington, D.C.
Feeney, R. E., Anderson, J. S., Azari, P. R., Bennett, N. and Rhodes, M. B. (1960a). *J. biol. Chem.* **235**, 2307–2311.

Feeney, R. E., Abplanalp, H., Clary, J. J., Edwards, D. L. and Clark, J. R. (1963). *J. biol. Chem.* **238**, 1732–1736.
Feeney, R. E., Osuga, D. T., Lind, S. B. and Miller, H. T. (1965). *Fedn. Proc. Fedn. Am. Socs exp. Biol.* **24**, 419.
Feeney, R. E., Osuga, D. T., Lind, S. B. and Miller, H. T. (1966). *Comp. Biochem. Physiol.* **18**, 121–130.
Feinstein, G. and Feeney, R. E. (1966). *J. biol. Chem.* **241**, 5183–5189.
Feinstein, G., Osuga, D. T. and Feeney, R. E. (1966a). *Biochem. biophys. Res. Commun.* **24**, 495–499.
Feinstein, G., Osuga, D. T. and Feeney, R. E. (1966b). *Fedn. Proc. Fedn. Am. Socs exp. Biol.* **25**, 589.
Ferguson, A. (1969). Ph.D. thesis. The Queen's University of Belfast.
Forsythe, R. G. and Foster, J. H. (1950). *J. biol. Chem.* **184**, 377–383.
Fothergill, L. A. and Fothergill, J. E. (1970). *Eur. J. Biochem.* **17**, 529–532.
Fothergill, L. A. and Fothergill, J. E. (1971). *Comp. Biochem. Physiol.* **40**A, 445–451.
Fothergill, J. E. and Perrie, W. T. (1966). *Biochem. J.* **99**, 58P.
Friedenthal, H. (1904). *Berl. Klin. Wschr.* Nr. **12**, 339.
Friedenthal, H. (1908). *Arbeit. Gebeit. exper. Physiol.* **1**, 340.
Friedberger, E. and Meissner, G. (1923). *Z. ImmunForsch. exp. Ther.* **36**, 233–271.
Fröhlich, H. (1928). *Z. ImmunForsch. exp. Ther.* **55**, 236–257.
Fujio, Y. (1971). *Jap. J. Genet.* **46**, 181–190.
Galli-Valerio, B. (1911). *Z. ImmunForsch. exp. Ther.* **9**, 313–320.
Gengou (1902). *Annls Inst. Pasteur, Paris* **16**, 734–755.
Gershowitz, H. (1954). Ph.D. thesis. Calif. Inst. Tech.
Gilmour, D. G. (1962). *In* 'Blood Groups in Infrahuman Species'. *Ann. New York Acad. Sci.* **97**, 166–172.
Glock, H. (1914). *Biol. Zbl.* **34**, 385.
Gordon, C. D. (1938). Ph.D. Thesis. Univ. of Wisconsin, Madison.
Grabar, P. and Williams, C. A. Jr. (1953). *Biochim. Biophys. Acta* **10**, 193–194.
Graetz, F. (1914). *Z. ImmunForsch. exp. Ther.* **21**, 150.
Gray, A. P. (1958). 'Bird Hybrids'. Commonwealth Agricultural Bureaux, England.
Greene, F. C. and Feeney, R. E. (1968). *Biochemistry, N. Y.* **7**, 1366–1370.
Gysels, H. (1962). *Gerfaut* **52**, 577–581.
Gysels, H. (1963). *Experientia* **19**, 107–111.
Gysels, H. (1964a). *Experientia* **20**, 145–146.
Gysels, H. (1964b). *Gerfaut* **54**, 16–28.
Gysels, H. (1964c). *Bull. Soc. r. Zool. Anvers* **33**, 29–41.
Gysels, H. (1964d). *Natuurwet. Tijdschr. Ned. Indie* **46**, 43–178.
Gysels, H. (1965). *J. Orn., Lpz.* **106**, 208–217.
Gysels, H. (1968). *Ardea* **56**, 267–280.
Gysels, H. (1970). *Acta Zool. Pathol. Antverp* **50**, 3–13.
Gysels, H. and Rabaey, M. (1963). *Bull. Soc. r. Zool. Anvers* **26**, 72–85.
Gysels, H. and Rabaey, M. (1964). *Ibis* **106**, 536–539.
Haley, L. E. (1965). *Genetics* **51**, 983–986.
Harrap, B. S. and Woods, E. F. (1964a). *Biochem. J.* **92**, 8–18.
Harrap, B. S. and Woods, E. F. (1964b). *Biochem. J.* **92**, 19–26.
Harrap, B. S. and Woods, E. F. (1967). *Comp. Biochem. Physiol.* **20**, 449–460.
Harris, H. (1969). *Brit. med. Bull.* **25**, 5–13.
Haynes, R., Osuga, D. T. and Feeney, R. E. (1967). *Biochemistry, N.Y.* **6**, 541–547.
Hektoen, L. and Boor, A. K. (1931). *J. infect. Dis.* **49**, 29–36.

Hektoen, L. and Cole, A. G. (1927). *J. infect. Dis.* **40**, 647–655.
Hektoen, L. and Cole, A. G. (1928). *J. infect. Dis.* **42**, 1–24.
Hendrickson, H. T. (1969). *Ibis* **111**, 80–91.
Herman, J. and Jollès, J. (1970). *Biochim. Biophys. Acta* **200**, 178–179.
Holmes, R. S. and Masters, C. J. (1968). *Biochim. Biophys. Acta* **159**, 81–93.
Imanishi, M., Shinka, S., Miyagawa, N. and Amano, T. (1966). *Biken's J.* **9**, 107–114.
Irwin, M. R. (1932a). *Int. Conf. Genet. 6th.* **2**, 103–104.
Irwin, M. R. (1932b). *Proc. Soc. exp. Biol. Med.* **29**, 850–851.
Irwin, M. R. (1938). *J. Genet.* **35**, 351–373.
Irwin, M. R. (1939). *Genetics, Princeton* **24**, 709–721.
Irwin, M. R. (1947). *In* 'Advances in Genetics' Vol. 1, pp. 131–159.
Irwin, M. R. (1949a). *Q. Rev. Biol.* **24**, 109–123.
Irwin, M. R. (1949b). *Genetics, Princeton* **34**, 586–606.
Irwin, M. R. (1951). *In* 'Genetics in the 20th Century', pp. 173–319. Macmillan, New York.
Irwin, M. R. (1952). *In* 'Heterosis' (J. W. Gowen, ed.). Iowa State College Press, Ames.
Irwin, M. R. (1953). *Evolution, Lancaster, Pa.* **7**, 31–50.
Irwin, M. R. (1955). *Evolution, Lancaster, Pa.* **9**, 261–279.
Irwin, M. R. (1971). *Genetics, Princeton* **68**, 509–526.
Irwin, M. R. and Cole, L. J. (1936a). *J. exp. Zool.* **73**, 85–108.
Irwin, M. R. and Cole, L. J. (1936b). *J. exp. Zool.* **73**, 309–318.
Irwin, M. R. and Cole, L. J. (1937). *J. Immun.* **33**, 355–373.
Irwin, M. R. and Cole, L. J. (1940). *Genetics, Princeton* **25**, 326–336.
Irwin, M. R. and Cole, L. J. (1945a). *Genetics, Princeton* **30**, 439–447.
Irwin, M. R. and Cole, L. J. (1945b). *Genetics, Princeton* **30**, 487–495.
Irwin, M. R. and Cumley, R. W. (1940). *Am. Nat.* **74**, 222–231.
Irwin, M. R. and Cumley, R. W. (1942). *Genetics, Princeton* **27**, 228–237.
Irwin, M. R. and Cumley, R. W. (1943). *Genetics, Princeton* **28**, 9–28.
Irwin, M. R. and Cumley, R. W. (1945). *Genetics, Princeton* **30**, 363–375.
Irwin, M. R. and Cumley, R. W. (1946). *Genetics, Princeton* **31**, 220.
Irwin, M. R. and Cumley, R. W. (1947). *Genetics, Princeton* **32**, 178–184.
Irwin, M. R. and Miller, W. J. (1961). *Evolution, Lancaster, Pa.* **15**, 30–43.
Irwin, M. R., Cole, L. J. and Gordon, C. D. (1936). *J. exp. zool.* **73**, 285–308.
Jacobs, M. H. (1931). *Proc. Am. phil. Soc.* **70**, 363–370.
Jacobs, M. H., Glassman, H. N. and Parpart, A. K. (1950). *J. exp. Zool.* **113**, 277–300.
Johnston, R. F. and Hochman, B. (1953). *Condor* **55**, 154–155.
Johnson, V. L. and Dunlap, J. S. (1955). *Science, N.Y.* **122**, 1186.
Jollès, J. and Jollès, P. (1961). *C. r. hebd. Séanc. Acad. Sci., Paris* **253**, 2773–2775.
Jollès, J., Jauregui-Adell, J., Bernier, I. and Jollès, P. (1963). *Biochim. Biophys. Acta* **78**, 668–689.
Jollès, J., Herman, J., Niemann, B. and Jollès, P. (1967). *Eur. J. Biochem.* **1**, 344–346.
Jones, L. M. (1947). M.S. thesis, Univ. Wisconsin.
Kaminski, M. (1957). *Annls Inst. Pasteur, Paris* **92**, 802–; **93**, 102.
Kaminski, M. (1960). *Annls Inst. Pasteur, Paris* **98**, 51–69.
Kaminski, M. (1962). *Immunology* **5**, 322–332.
Kaminski, M. (1964). *Experientia* **20**, 286–287.
Kaminski, M. (1970). *In* 'Proc. XI Eur. Conf. on Animal Blood Groups and Biochemical Polymorphism,' pp. 429–431. W. Junk, Publishers, The Hague.
Kaminski, M. and Durieux, J. (1954). *Bull. Soc. Chim. Biol.* **36**, 1037–1051.

Kaminski, M. and Durieux, J. (1956). *Exp. Cell Res.* **10**, 590–618.
Kaminski, M. and Nouvel, J. (1952). *Bull. Soc. Chim. Biol.* **34**, 11–19.
Kaminski, M. and Ouchterlony, O. (1951). *Bull. Soc. Chim. Biol.* **33**, 758–770.
Kaminski, M. and Williams, C. (1953). *J. Méd. Bordeaux* **130**, 291–299.
Kammerman, S. and Canfield, R. E. (1969). *Fedn. Proc. Fedn. Am. Socs exp. Biol.* **28**, 343.
Karig, L. M. and Wilson, A. C. (1971). *Biochem. Genet.* **5**, 211–221.
Kartashev, N. N., Ghelfon, I. A. and Gromakova, S. P. (1966). *Zool. Zh.* **45**, 1843–1851.
Kimura, M. (1969a). *Jap. J. Genet.* **45**, 107–108.
Kimura, M. (1969b). *Nippon Kakin Gakkaishi* **6**, 68–72.
Kimura, M. (1969c). *Jap. Poultry Sci.* **6**, 68–72.
Kimura, M. (1969d). *Jap. J. Genet.* **44**, 107–108.
Kimura, M. and Makita, N. (1968). *Res. Bull. Fac. Agric. Gifu-ken prefect. Univ.* **24**, 258–270.
Kitto, G. B. and Wilson, A. C. (1966). *Science, N.Y.* **153**, 1408–1410.
Kraus, R. (1897). *Wien. klin. Wochenschr.* **10**, 736.
Landsteiner, K., Longsworth, L. G. and van der Scheer, J. (1938). *Science, N.Y.* **88**, 83–85.
Landsteiner, K. and van der Scheer, J. (1940). *J. exp. Med.* **71**, 445–454.
La Rue, J. N. and Speck, J. C. Jr. (1970). *J. biol. Chem.* (In press.)
Law, G. R. J. (1967). *Science, N.Y.* **156**, 1106–1107.
Law, G. R. J. and Munro, S. S. (1965). *Science, N.Y.* **149**, 1518.
Lewis, J. H. and Wells, H. G. (1927). *J. infec. Dis.* **40**, 316–325.
Lewontin, R. C. and Hubby, J. L. (1966). *Genetics, Princeton* **54**, 595–609.
Libby, R. L. (1938). *J. Immun.* **34**, 71–73.
Lofts, B. and Mainardi, D. (1963). *Riv. Ital. Orn., Ser. II* **33**, 1–5.
Longsworth, L. G., Cannan, R. K. and MacInnes, D. A. (1940). *J. Am. chem. Soc.* **62**, 2580–2590.
Lush, I. E. (1961). *Nature Lond.* **189**, 981–984.
Lush, I. E. (1964a). *Genet. Res.* **5**, 39–49.
Lush, I. E. (1964b). *Genet. Res.* **5**, 257–268.
Lush, I. E. (1966). *Life Sci.* **5**, 1537–1542.
Lush, I. E. (1967). *In* 'Frontiers of Biology' (A. Neuberger and E. L. Tatum, eds.), Vol. 3. John Wiley, New York.
MacDonnell, L. R., Ducay, E. D., Sugihara, T. F. and Feeney, R. E. (1954). *Biochim. Biophys. Acta* **13**, 140–141.
McCabe, R. A. and Deutsch, H. F. (1952). *Auk* **69**, 1–8.
McDermid, E. M. (1965). *In* 'Blood Groups of Animals' (J. Matousek, ed.). W. Junk, Publishers, The Hague.
McGibbon, W. H. (1944). *Genetics, Princeton* **29**, 407–419.
McGibbon, W. H. (1945). *Genetics, Princeton* **30**, 252–265.
McIndoe, W. M. (1962). *Nature, Lond.* **195**, 353–354.
Maclean, G. L. (1967). *J. Orn., Lpz.* **108**, 203–217.
Mainardi, D. (1956a). *Re. Ist. Lomb. Sci. Lett. B* **90**, 122–130.
Mainardi, D. (1956b). *Re. Ist. Lomb. Sci. Lett. B* **90**, 448–452.
Mainardi, D. (1957a). *Archo. zool. Ital.* **42**, 151–159.
Mainardi, D. (1957b). *Re. Ist. lomb. Sci. Lett. B.* **92**, 180–186.
Mainardi, D. (1957c). *Re. Ist. lomb. Sci. Lett. B.* **91**, 565–569.
Mainardi, D. (1957d). *Re. Ist. lomb. Sci. Lett. B.* **91**, 570–573.
Mainardi, D. (1957e). *Boll. Soc. ital. Biol. sper.* **33**, 829–831.
Mainardi, D. (1958a). *Riv. Ital. Orn.* **28**, 114–124.

Mainardi, D. (1958b). *Re Ist. lomb. Sci. Lett.* B **92**, 336–356.
Mainardi, D. (1958c). *Nature, Lond.* **182**, 1388–1389.
Mainardi, D. (1959a). *Re. Ist. lomb. Sci. Lett.* B **93**, 91–96.
Mainardi, D. (1959b). *Nature, Lond.* **184**, 913–914.
Mainardi, D. (1959c). *Boll. Zool.* **26**, 207–211.
Mainardi, D. (1960a). *Convegno Genet* **5**, 3–8.
Mainardi, D. (1960b). *Br. Birds* **53**, 238–239.
Mainardi, D. (1960c). *Misc. Reports Yamashina's Inst. Orn. Zool.* **2**, 64–66.
Mainardi, D. (1961a). *Rc. Ist. lomb. Sci. Lett.* B **95**, 117–122.
Mainardi, D. (1961b). *Riv. Ital. Orn., II S.* **31**, 175–178.
Mainardi, D. (1961c). *Archo zool. Ital.* **46**, 273–290.
Mainardi, D. (1962a). *Nature, Lond.* **194**, 111.
Mainardi, D. (1962b). *Ibis* **104**, 426–428.
Mainardi, D. (1963a). *Int. orn. Congr.* **13**, 103–114.
Mainardi, D. (1963b). *Int. Congr. Zool.* **16**, Vol 2, p. 178. Washington, D.C.
Mainardi, D. and Guerra, F. (1959). *Boll. Soc. ital. Biol. sper.* **35**, 1805–1806.
Mainardi, D. and Schreiber, B. (1963). *Monitore zool. ital.* **71**, 408–415.
Mainardi, D. and Taibel, A. M. (1962a). *Re. Ist. lomb. Sci. Lett.* B**96**, 118–130.
Mainardi, D. and Taibel, A. M. (1962b). *Re. Ist. lomb. Sci. Lett.* B**96**, 131–140.
Manwell, C. and Baker, C. M. A. (1969). *Comp. Biochem. Physiol.* **28**, 1007–1028.
Manwell, C. and Baker, C. M. A. (1970). 'Molecular Biology and the Origin of Species.' Sidgwick and Jackson, London.
Margoliash, E., Needleman, S. B. and Stewart, J. W. (1963). *Acta chem. scand.* **17**, 5250–5256.
Márkus, S., Fésüs, L. and Kovács, G. (1965). *Acta Veterinaria (Magyar Tudomanyos Akad.)* **15**, 461–463.
Martin, E. P. and Leone, C. A. (1952). *Trans. Kansas Acad. Sci.* **55**, 439–444.
Mayr, E. and Amadon, D. (1951). *Am. Mus. Novit.* No. 1496, 42 pp.
Meissner, W. (1923). *Z. ImmunForsch. exp. Ther.* **36**, 281.
Meissner, W. (1926). *Zentbl. Bakt. ParasitKde (Abt 1)* **100**, 258.
Miller, H. T. and Feeney, R. E. (1964). *Arch. Biochem. Biophys.* **108**, 117–124.
Miller, H. T. and Feeney, R. E. (1966). *Biochemistry, N.Y.* **5**, 952–958.
Miller, W. J. (1953a). Ph.D. thesis, Univ. of Wisconsin.
Miller, W. J. (1953b). *Physiol. Zool.* **26**, 124–131.
Miller, W. J. (1956). *Genetics, Princeton* **41**, 700–714.
Miller, W. J. (1964). *Science, N.Y.* **143**, 1179–1180.
Miller, W. J. and Bryan, C. R. (1951). *Genetics, Princeton* **36**, 566–567.
Miller, W. J. and Bryan, C. R. (1953). *Proc. natn. Acad. Sci.* **39**, 407–412.
Milne, H. and Robertson, F. W. (1965). *Nature, Lond.* **205**, 367–369.
Moore, D. H. (1945). *J. biol. Chem.* **161**, 21–32.
Morton, J. R., Gilmour, D. G., McDermid, E. M. and Ogden, A. L. (1965). *Genetics, Princeton* **51**, 97–107.
Mueller, J. O., Smithies, O. and Irwin, M. R. (1962). *Genetics, Princeton* **47**, 1385–1392.
Myers, W. (1900). *Lancet* **2**, 98–100.
Niemann, B., Hermann, J. and Jollès, J. (1968). *Bull. Soc. Chim. biol.* **50**, 923–924.
Nolan, C. and Margoliash, E. (1968). *A. Rev. Biochem.* **37**, 727–790.
Norris, R. A. (1963). *Bull. tall Timb. Res. Stn No. 4*, 1–71.
Nuttall, G. H. F. (1901). *Proc. R. Soc.* **69**, 150–153.
Nuttall, G. H. F. (1902a). *Trans. Camb. phil. Soc.* **11**, 334–336.
Nuttall, G. H. F. (1902b). *Br. med. J.* **3**, 825–827.

Nuttall, G. H. F. (1904). 'Blood Immunity and Blood Relationship'. Cambridge University Press.
Ogden, A. L., Morton, J. R., Gilmour, D. G. and McDermid, E. M. (1962). *Nature, Lond.* **195**, 1026–1028.
Osuga, D. T. and Feeney, R. E. (1967). *Arch. Biochem. Biophys.* **118**, 340–346.
Osuga, D. T. and Feeney, R. E. (1968). *Arch. Biochem. Biophys.* **124**, 560–574.
Palm, J. E. (1955). Ph.D. thesis, University of Wisconsin.
Palm, J. E. and Irwin, M. R. (1962). *Genetics, Princeton* **47**, 1409–1426.
Paulov, S. (1971). *Comp. Biochem. Physiol.* **40**B, 313–315.
Peakall, D. B. (1963). *In* 'Proceedings XIII International Ornithological Congress' Vol 1, pp. 135–140. Amer. Orn. Union, Lawrence, Kansas.
Perkins, R. (1964). *Murrelet* **45**, 26–28.
Perramon, A. (1965). *In* 'Blood Groups of Animals' (J. Matousek, ed.). pp. 179–186. W. Junk, Publishers, The Hague.
Podliachouk, L. (1965). *In* 'Blood Groups of Animals' (J. Matousek, ed.). pp. 187–191. W. Junk, Publishers, The Hague.
Prager, E. M. and Wilson, A. C. (1971). *J. biol. Chem.* **246**, 5978–5989.
Rabaey, M. (1959). Aggr. thesis Rijksuniv. Gent.
Reeser, H. E. (1919a). *Meded. Rijksseruminricht.* **2**, (2) 54.
Reeser, H. E. (1919b). *Meded. Rijksseruminricht.* **2**, (2) 83.
Reichert, E. T. and Brown, A. P. (1909). *Publs Carnegie Instn. No. 116*, 161–171.
Rhodes, M. B., Azari, P. R. and Feeney, R. E. (1958). *J. biol. Chem.* **230**, 399–408.
Rhodes, M. B., Bennett, N. and Feeney, R. E. (1959). *J. biol. Chem.* **234**, 2054–2060.
Rhodes, M. B., Bennett, N. and Feeney, R. E. (1960). *J. biol. Chem.* **235**, 1686–1693.
Rodnan, G. P. and Ebaugh, F. G. Jr. (1956). *Fedn. Proc. Fedn. Am. Socs exp. Biol.* **15**, 155–156.
Rodnan, G. P. and Ebaugh, F. G. Jr. (1957). *Proc. Soc. exp. Biol. Med.* **95**, 397–401.
Saha, A. (1964). *Biochim. Biophys. Acta* **93**, 537–584.
Saha, A., Dutta, R. and Ghosh, J. (1957). *Science, N.Y.* **125**, 447–448.
Saha, A. and Ghosh, J. (1965). *Comp. Biochem. Physiol.* **15**, 217–235.
Sasaki, K. (1928). *J. Dep. Agric., Kyushu imp. Univ.* **2**, 117–132.
Sasaki, K. (1937). *Z. Zücht.* B**38**, 361–365.
Sasaki, K. (1954). *Tenth Worlds' Poultry Congress, Edinburgh.* 91–93.
Schram, Alfred C. (1970). *Comp. Biochem. Physiol.* **36**, 481–492.
Schroeder, W. A., Kay, L. M., Lewis, B. and Munger, N. (1955). *J. Am. chem. Soc.* **77**, 3901–3908.
Segre, A., Richmond, R. C. and Wiley, R. H. (1970). *Comp. Biochem. Physiol.* **36**, 589–595.
Selander, R. K., Hunt, W. G. and Yang, S. Y. (1969). *Evolution, Lancaster, Pa.* **23**, 379–390.
Seng, H. (1914). *Z. ImmunForsch. exp. Ther.* **20**, 355–366.
Shaw, C. R. (1965). *Science, N.Y.* **149**, 936–942.
Sibley, C. G. (1960). *Ibis* **102**, 215–284.
Sibley, C. G. (1962). *Syst. Zool.* **11**, 108–118.
Sibley, C. G. (1964). *In* 'Taxonomic Biochemistry and Serology' (C. A. Leone, ed.). pp. 435–450. The Ronald Press Co., New York.
Sibley, C. G. (1965). *Oiseau Revue fr. Orn.* **35**, 112–124.
Sibley, C. G. (1967). *Discovery New Haven, Conn.* **3**, 5–30.
Sibley, C. G. (1968). *Postilla* **125**, 12 pp.
Sibley, C. G. (1970). *Bull. Peabody Mus. nat. Hist. No.* **32**. 131 pp.

Sibley, C. G. and Ahlquist, J. E. (1972). *Bull Peabody Mus. nat. Hist.* No. 39. 276 pp.
Sibley, C. G. and Brush, A. H. (1967). *Auk* **84**, 203–219.
Sibley, C. G. and Frelin, C. (1972). *Ibis* **114**, 377–387.
Sibley, C. G. and Hendrickson, H. T. (1970). *Condor* **72**, 43–49.
Sibley, C. G. and Johnsgard, P. A. (1959a). *Condor* **61**, 85–95.
Sibley, C. G. and Johnsgard, P. A. (1959b). *Am. Nat.* **93**, 107–115.
Sibley, C. G., Corbin, K. W. and Ahlquist, J. E. (1968). *Bonn. zool. Beitr.* **19**, 235–248.
Sibley, C. G., Corbin, K. W. and Haavie, J. H. (1969). *Condor* **71**, 155–179.
Sievers, O. (1939a). *Ornis fenn.* **16**, 13–21.
Sievers, O. (1939b). *Ornis fenn.* **16**, 45–51.
Sievers, O. (1939c). *Acta path. microbiol. scand.* **16**, 44–98.
Sittmann, K., Abplanalp, H. and Fraser, R. A. (1966). *Genetics, Princeton.* **54**, 371–379.
Smith, M. B. and Back, Joan F. (1970). *Aust. J. biol. Sci.* **23**, 1221–1227.
Smithies, O. (1955). *Biochem. J.* **61**, 629–641.
Smithies, O. (1959a). *Biochem. J.* **71**, 585–587.
Smithies, O. (1959b). *In* 'Advances in Protein Chemistry', Vol 14, pp. 65–113.
Sokolovskaia, I. I. (1936). *Iz Akad. Nauk SSSR ser. biol.* **2–3**, 465–489.
Spärck, J. V. (1954). *Vidensk. Meddr. dansk naturh. Foren.* **116**, 399–410.
Spärck, J. V. (1956). *In* 'Proceedings 14th International Congress of Zoology' pp. 335–336.
Stallcup, W. B. (1954). *Univ. Kans. Publs Mus. nat. Hist.* **8**, 157–211.
Stallcup, W. B. (1961). *J. Grad. Res. Center (Sth. Methodist Univ. stud.)* **29**, 43–65.
Stevens, F. C. and Feeney, R. E. (1963). *Biochemistry, N.Y.* **2**, 1346–1352.
Stimpfling, J. H. and Irwin, M. R. (1960a). *Evolution, Lancaster, Pa.* **14**, 417–426.
Stimpfling, J. H. and Irwin, M. R. (1960b). *Genetics, Princeton* **45**, 233–242.
Stratil, A. (1966). *In* 'Polymorphismes Biochmiques des Animaux' Proc. Tenth Eur. Conf. on Animaux. pp. 241–243. Institut Nat. Recherche Agron., Paris.
Stratil, A. (1968). *Comp. Biochem. Physiol.* **24**, 113–121.
Stratil, A. and Valenta, M. (1966). *Folia biol., Praha* **12**, 307–309.
Taibel, A. M. (1951). *Rend. Congr. dell' U.Z.I., Boll. Zool.* **18**, 375.
Tiselius, A. (1937). *Trans. Faraday Soc.* **33**, 524–531.
Tordoff, H. B. (1954). *Misc. Publs Mus. Zool., Univ. Mich.* **81**, 77 pp.
Turpeinen, O. and Turpeinen, K. (1936). *Acta Soc. Med. fenn. 'Duodecim', Ser. A* **19**, 1–24.
Uhlenhuth, P. (1900). *Dt. med. Wschr.* **26**, 734–735.
Uhlenhuth, P. (1901). *Dt. med. Wschr.* **27**, 260–261.
Uhlenhuth, P. (1905). *Dt. med. Wschr.* **31**, 1673–1676.
Vaurie, C. (1959). 'The Birds of the Palearctic Fauna, A Systematic Reference: Order Passeriformes.' Witherby, London.
Vohs, P. A. (1966). *J. Wildl. Mgmt* **30**, 745–753.
Vohs, P. A. and Carr, L. R. (1969). *Condor* **71**, 413–417.
Vyshinsky, F. S. and Muraviev, V. I. (1970). *In* 'Proc. XI Eur. Conf. on Animal Blood Groups and Biochemical Polymorphism,' pp. 425–428. W. Junk, Publishers, The Hague.
Wall, R. L. and Schlumberger, H. G. (1957). *Science, N.Y.* **125**, 993–994.
Wasmann, E. (1926). *Naturwissenschaften* **14**, 504.
Watkins, W. M. (1966). *Science, N.Y.* **152**, 172–181.
Weintraub, M. S. and Schlamowitz, M. (1970). *Comp. Biochem. Physiol.* **37**, 49–58.
Weintraub, M. S. and Schlamowitz, M. (1971). *Comp. Biochem. Physiol.* **38** B, 513–522.
Welsh, D. A. and Chapman, H. G. (1910). *J. Hyg., Camb.* **10**, 177–184.

Wenn, R. V. and Williams, J. (1968). *Biochem. J.* **108**, 69–74.
Wetmore, A. (1960). *Smithson. misc. Collns.* **139**, 37 pp.
Wetter, L. R. and Deutsch, H. F. (1950). *Archs Biochem.* **28**, 399–404.
Wetter, L. R. and Deutsch, H. F. (1951). *J. biol. Chem.* **192**, 237–242.
Wetter, L. R., Cohn, M. and Deutsch, H. F. (1952). *J. Immun.* **69**, 109–115.
Wetter, L. R., Cohn, M. and Deutsch, H. F. (1953). *J. Immun.* **70**, 507–513.
Wilcox, F. H. (1966). *Genetics, Princeton* **53**, 797–805.
Williams, J. (1962). *Biochem. J.* **83**, 355–364.
Williams, J. (1968). *Biochem. J.* **108**, 57–67.
Wilson, A. C. and Kaplan, N. O. (1964). *In* 'Taxonomic Biochemistry and Serology' (C. A. Leone, ed.) pp. 321–346. The Ronald Press Co., New York.
Wilson, A. C., Kaplan, N. O., Levine, L., Pesce, A. Reichlin, M. and Allison, W. S. (1964). *Fedn Proc. Fedn. Am. Socs exp. Biol.* **23**, 1258–1266.
Wilson, R. H. and Lewis, H. B. (1927). *J. biol. Chem.* **73**, 543–553.
Windle, J. J., Wiersma, A. K., Clark, J. R. and Feeney, R. E. (1963). *Biochemistry, N.Y.* **2**, 1341–1345.
Wiseman, R. L. and Fothergill, J. E. (1966). *Biochem. J.*, **99**, 58 P.
Wolfe, H. R. (1939). *Biol. Bull. mar. biol. Lab., Woods Hole* **76**, 108–120.
Woods, E. F. (1971). *Comp. Biochem. Physiol.* **39** A, 325–331.
Zinkham, W. H., Blanco, A. and Kupchyk, L. (1963). *Science, N.Y.* **142**, 1303–1304.
Zinkham, W. H., Blanco, A. and Kupchyk, L. (1964). *Science, N.Y.* **144**, 1353–1354.
Zinkham, W. H., Kupchyk, L., Blanco, A. and Isensee, H. (1966). *J. exp. Zool.* **162**, 45–56.

3 Biochemical and Immunological Evidence of Relationships in Amphibia and Reptilia

H. C. DESSAUER

Louisiana State University Medical Center, New Orleans, Louisiana, U.S.A.

I. Introduction

Herpetologists are applying chemical and immunological methods with increasing frequency to taxonomic problems. This molecular approach has made possible an extensive comparison of organisms at levels that closely reflect gene activity (Leone, 1964; Bryson and Vogel, 1965; Dayhoff and Eck, 1967/1968; Dessauer, 1968; Yunis, 1969).

The need for these additional criteria for evaluating systematic relationships is particularly critical for herpetology, since in this field traditional methods have often proved inadequate for the solution of many taxonomic problems. The paleontological approach, so useful in establishing evolutionary events within other groups, is often of little use with amphibians and reptiles because of a lack of critical fossils (Dowling, 1959; Parsons and Williams, 1963; Hecht, 1969). Lacking a meaningful fossil record, the herpetologist must estimate relationships and postulate evolutionary patterns from evidence on living forms. Unfortunately, morphological features of many reptiles and amphibians are reduced and simplified. Crocodilians have been conservative in skeletal structure and in external morphology during their long existence (Romer, 1966). Conversely, limbless forms have arisen many times within the

Squamata, and difficulties arise in assessing their affinities, especially those of snakes which have undergone extreme modifications and specializations (Underwood, 1967). Blair (1962) has commented on how the paucity of morphological features has affected the taxonomy of frogs, 'The classification of anuran amphibians from the level of suborder to species population from purely morphological criteria may be best described as chaotic'.

II. Relationships of Higher Categories

A. INTRODUCTION

With current biochemical and immunological methods, the herpetologist can obtain evidence useful in: (1) characterizing taxa, (2) estimating the extent of their divergence, (3) determining their sequence of origin, and (4) postulating the time when their phyletic lines diverged in the past.

Structures of homologous proteins reflect the relationships of different taxa and indicate the extent of their divergence. Once two taxa are reproductively isolated, their genomes evolve independently, and their proteins become progressively more unlike. However, proteins seem to undergo structural change at different rates. Those which have changed slowly with time, e.g. lens proteins, are most useful in taxonomic studies concerned with phyletic lines of ancient divergence; those which have changed more rapidly with time, e.g. transferrins, are most useful in taxonomic studies concerned with recent divergences (Dessauer, 1968). Measures of structural differences between homologous proteins serve as useful evolutionary clocks, if some fossils are available for calibrating the clock (Salthe and Kaplan, 1966; Wilson and Sarich, 1969).

Characteristics of proteins are of great variety but of different taxonomic value. Physicochemical properties such as electrophoretic mobilities, inactivation temperatures and substrate affinities are often similar in proteins of related taxa and serve to group them (Wilson, *et al.*, 1964; Kaplan, 1965; Salthe, 1965; Dessauer, 1966a). These characteristics, however, afford little indication of the magnitude of the structural differences between proteins. Comparison of amino acids sequences are necessary for the precise estimation of the divergence of taxa (Margoliash and Smith, 1965; Dayhoff and Eck, 1967/1968). Unfortunately, since formidable problems are involved in obtaining such data, little is currently available on proteins of the Amphibia and the Reptilia. More easily acquired 'peptide fingerprints' substitute to a limited extent. These are stained patterns of fragments of partial digests of proteins spread over sheets of filter paper by means of electrophoresis and chromatography. The number of peptide differences between fingerprints of a pair of homologous proteins from two organisms is a crude estimate of the magnitude of sequence differences between their primary structures (Sutton, 1969).

Quantitative immunological data afford the best available evidence of the divergence of proteins in the Amphibia and Reptilia. In such studies the magnitude of the cross reaction of a protein with an antiserum to another protein is determined. Tests performed with mixed antigens and mixed antibodies are useful in aligning organisms in series which reflect their taxonomic affinities to a test organism (Boyden, 1963). Microcomplement fixation titrations, performed with mixed antigens but with antibodies against a purified antigen, yield 'indices of dissimilarity' (Sarich and Wilson, 1966), the magnitude of which seems to correlate with the number of sequence differences between the cross reacting and reference proteins (Arnheim *et al.*, 1969). The immunological distance, defined as $100 \times \log_{10}$ of the index of dissimilarity of two proteins, seems to increase linearly with the length of the period of divergence of the species from which the proteins originated. The microcomplement fixation technique is particularly useful since it requires a minimal expenditure of reagents and is capable of detecting the slightest affinities between antibody and antigen (Salthe and Kaplan, 1966; Sarich and Wilson, 1966; Wilson and Sarich, 1969; George and Dessauer, 1970).

The evolution of metabolic systems not only involved protein alterations but also gains and losses of enzymes and other proteins. New proteins may have been acquired at different times in the past. The immunochemical absorption technique furnishes a means for differentiating primitive from more recently evolved antigens, data useful in estimating the sequence of origin of different taxa (Manski *et al.*, 1964, 1967). The apportionment of enzymes involved in the synthesis of steroid toxins (Michl and Kaiser, 1962/1963; Shoppee, 1964) and bile salts (Haslewood, 1967, 1968) varies among amphibians and reptiles, as do the enzymes required for the synthesis of urea (Cohen and Brown, 1960) and aromatic amines (Cei *et al.*, 1967) and for the degradation of uric acid (Florkin, 1949). Even qualitative analyses of the patterns of tissue products which these enzymes make possible are useful in characterizing taxa. Such patterns are easily obtained using simple chromatographic and electrophoretic procedures (Cei and Erspamer, 1966; Haslewood, 1968; Wittliff, 1968).

B. AMPHIBIAN AND REPTILIAN LINES OF DESCENT

Biochemical and immunological studies support the hypothesis that tetrapods evolved from some ancient fresh water, jawed fish. An origin from a fresh water ancestor is supported by evidence on osmoregulation (Smith, 1951), visual systems (Wald, 1960) and pteridine pigments (Bagnara and Obika, 1965). Immunochemical absorption analyses of lens proteins, among the most conservative structures known, show that lenses of land vertebrates contain antigens found in four major groups of fish: the Agnatha, Choanichthyes, Chondrichthyes and Actinopterygii. Lenses of the Choanichthyes, however, lack a number of antigens present in the Chondrichthyes, Actinopterygii and some land vertebrates, suggesting that vertebrates have descended from a jawed fish of later origin than the line leading to the lungfish (Fig. 1, Manski *et al.*, 1967; see Hecht, 1969). Although the lack of these lens proteins excludes lungfish from the direct line of tetrapod ancestry, lungfish resemble amphibians in respect to their bile steroid pathways (Haslewood, 1967, 1968) and strong tendencies toward polyploidy (Britten and Kohne, 1968; Britten and Davidson, 1969).

The Amphibia and Reptilia represent the earliest living members of the tetrapod line. The order of evolution of lens proteins shows that amphibians appeared before reptiles and other tetrapods. All antigens that are common to both fish and tetrapods occur also in lenses of amphibians. However, many of

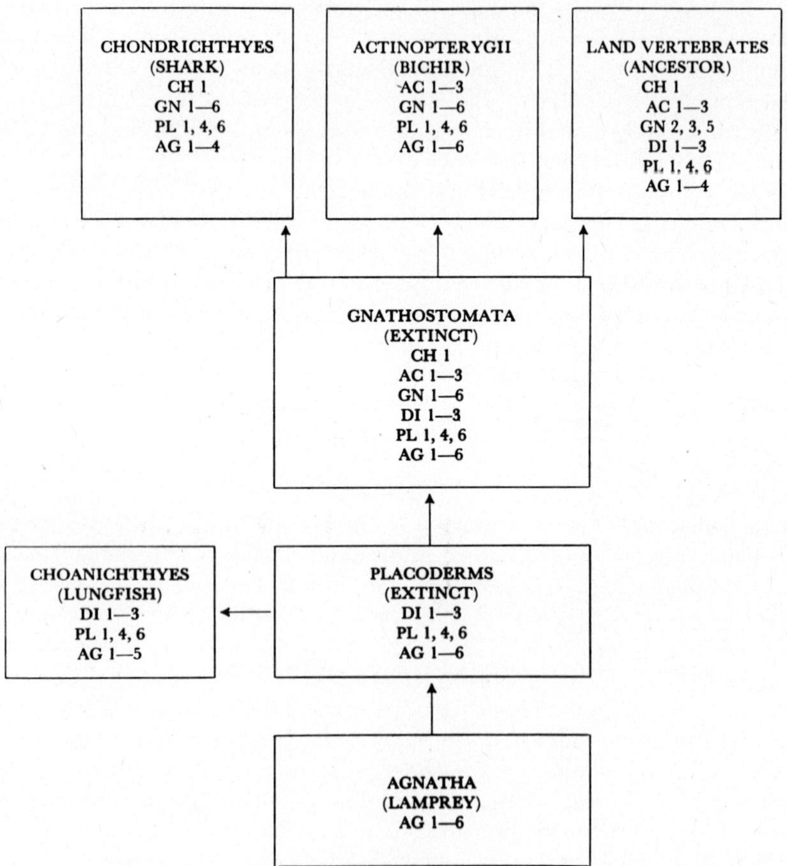

FIG. 1. The phylogeny of land vertebrates based on the distribution of primitive lens antigens. Abbreviations indicate primitive lens antigens shared by: AG-agnatha with jawed fish and land vertebrates; Pl-all classes of jawed fish with land vertebrates; GN-chondrichthyes and actinopterygii with land vertebrates; DI-choanichthyes with land vertebrates; CH-chondrichthyes with land vertebrates; AC-actinopterygii with land vertebrates. Reproduced with permission of Manski *et al.* (1967).

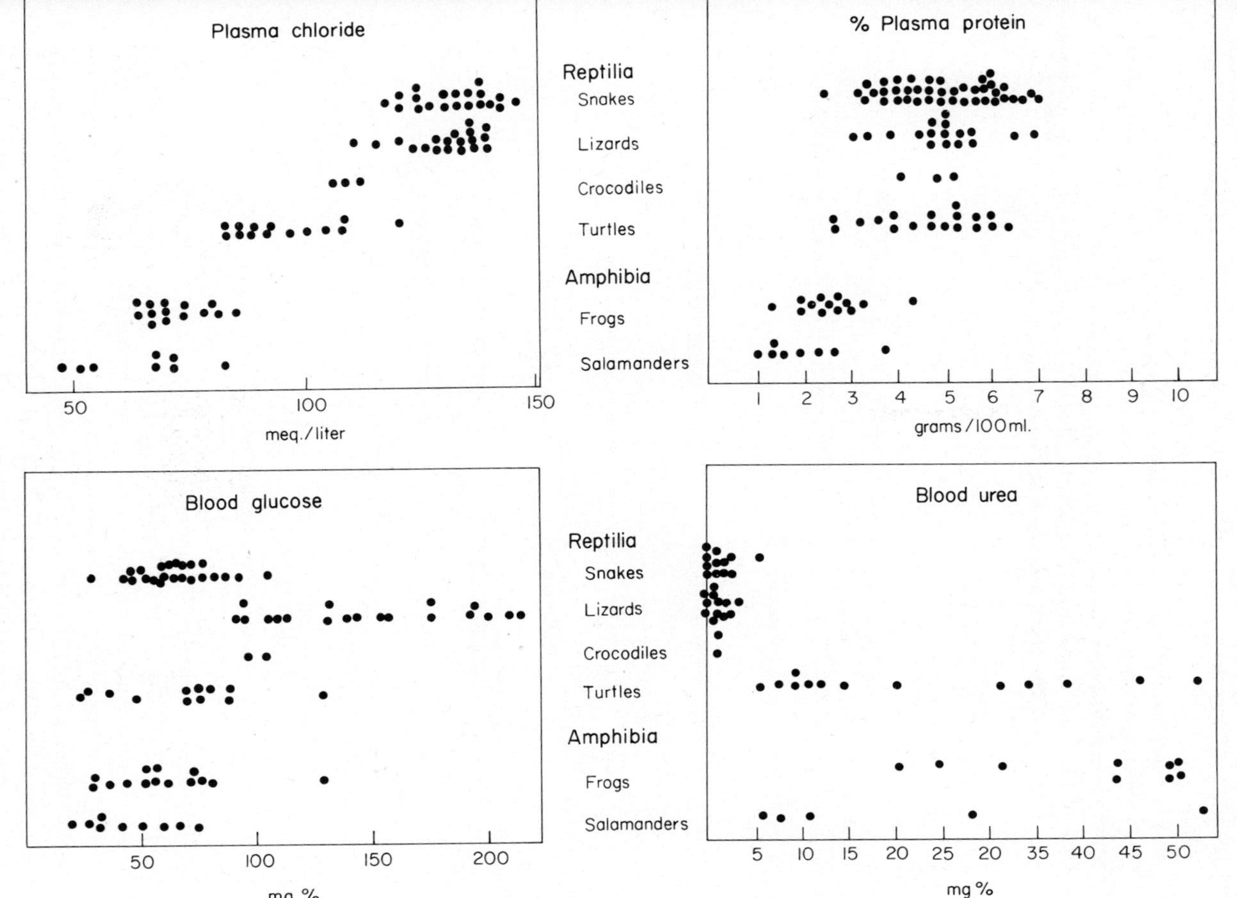

FIG. 2. Levels of various constituents found in blood of Amphibia and Reptilia. Each point represents a value found in a different species. Data were taken from various compilations (Dessauer *et al.*, 1961; Dessauer, 1961, 1970.

these components are absent from lenses of reptiles, birds and mammals. All fish antigens present in lenses of birds and mammals occur in lenses of reptiles (Manski *et al.*, 1964; Manski and Halbert, 1964).

Other immunological evidence is in keeping with the very ancient divergence of amphibian and reptilian lines of descent. Graham-Smith (1904) could detect no cross reactivity between antiserum against the plasma proteins of *Rana temporaria* and plasma proteins of 31 species of reptiles, representing all major orders. Similarly, he detected no cross reactivity between antisera

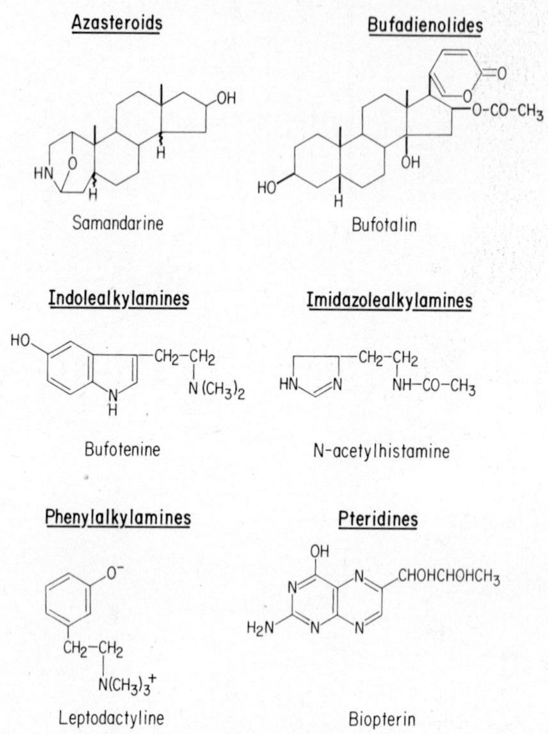

FIG. 3. Classes of complex derivatives found in amphibian skin.

against plasma proteins of turtles, crocodiles, snakes or lizards and plasma proteins of salamanders (4 families) and anurans (3 families). Using the highly sensitive microcomplement fixation technique, Salthe and Kaplan (1966) obtained indices of dissimilarity for muscle lactic dehydrogenases which suggested that phyletic lines leading to Amphibia and Reptilia diverged over 300 million years in the past (Figs 5 and 6).

Metabolic systems also reflect the ancient divergence of the Amphibia and Reptilia. The composition of their tissues is so different that the source of a tissue specimen can be identified by gross chemical analysis (Fig. 2). With

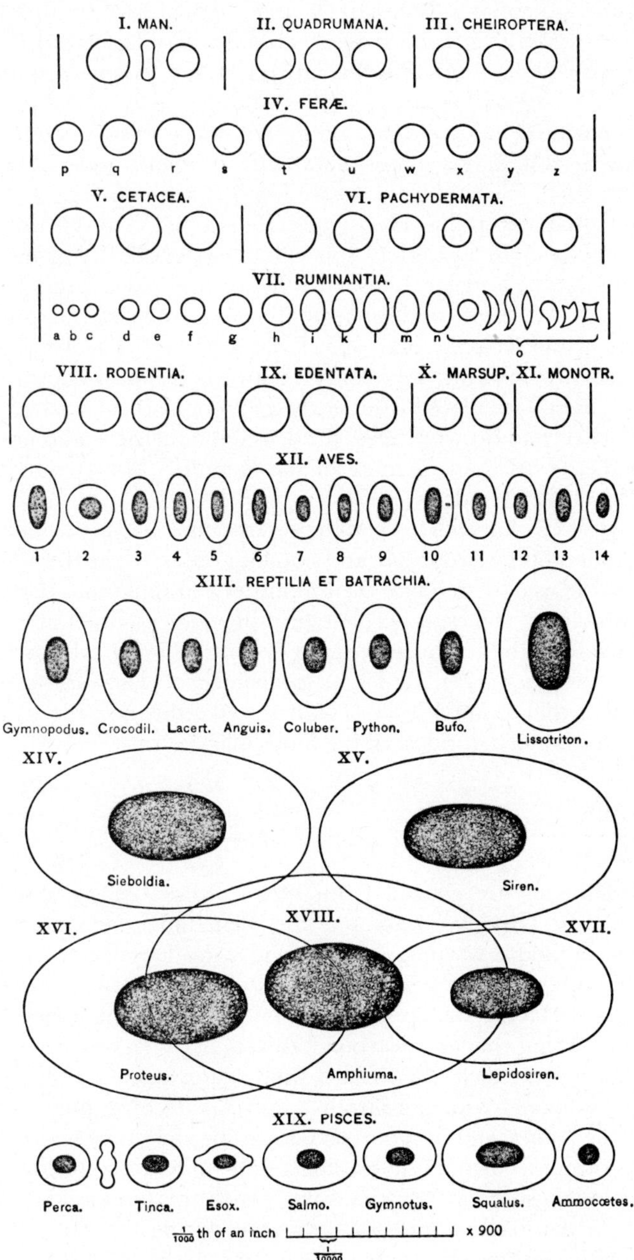

Fig. 4. Relative sizes of erythrocytes and erythrocyte nuclei. (From Reichert and Brown, 1909.)

some notable exceptions (Schmidt-Nielsen, 1962/1963) the body fluids of amphibians are more dilute than those of reptiles, having lower salt (Dessauer, 1961; Dessauer *et al.*, 1961; Lockwood, 1961) and plasma protein contents (Brown, G. W., 1964; Dessauer, 1970). Although urea is the principal end product of purine (Florkin, 1949) and protein metabolism in amphibians and turtles, it is, at best, only a minor metabolite in other reptiles (Fig. 2; Cohen and Brown, 1960). Sterol alcohols are common in the bile of Amphibia but are absent in that of Reptilia (Haslewood, 1967, 1968). The skin of amphibians is unique insofar as it has many specialized metabolic roles, including the exchange of fluid and electrolytes with the environment, a respiratory function (Brown, G. W., 1964) and the secretion of a number of very unique toxins (Fig. 3; Michl and Kaiser, 1962/1963; Cei *et al.*, 1967). The Amphibia have a marked tendency towards polyploidy (Brawerman and Shapiro, 1962). As might have been anticipated from early measurements of nuclear diameters (Fig. 4, Reichert and Brown, 1909), the deoxyribonucleic acid is usually much higher in their cells than in those of the Reptilia (Mirsky and Ris, 1951; Gerzeli *et al.*, 1956; Vendrely, 1958).

Living species with which herpetologists are concerned are usually classified into three orders of the Amphibia and into four orders of the Reptilia (Romer, 1966). The origins of these taxa, their affinities and radiations pose numerous unsolved problems. 'There is no biological phenomenon to mark the generic or familiar status; the higher categories are in this sense arbitrary; they are creations of the classifier made for his convenience' (Dobzhansky, 1968). In the discussions that follow I have usually selected the most recent taxonomic schemes as frameworks for describing molecular data.

C. AMPHIBIA

1. Anura, Apoda and Urodela

Living amphibians are classified into three orders: Anura (Salientia) the frogs and toads; Urodela (Caudata) the salamanders and newts; and the Apoda, the caecilians. Although there is widespread agreement on this classification, the origin of the orders and their interrelationships are in dispute (Romer, 1967; Hecht, 1969). Many believe that Amphibia rose as a single phyletic line during the later Carboniferous or early Devonian and diverged into modern orders much later. Other investigators contend that amphibian origin was diphyletic with the Urodela and Apoda diverging from the line leading to the Anura and other tetrapods during the Carboniferous, and the Anura separating from the line leading to the amniotes a geologic era or so later. Paleontologists have not been able to settle this question because a long gap occurs in the fossil record of amphibians during the period connecting stem and modern forms (Parsons and Williams, 1963; Hecht, 1969).

Immunological evidence gives strong support for the monophyletic view, although no data are available on the Apoda. An antiserum against muscle type lactic dehydrogenase (M-LDH) of the urodele, *Amphiuma tridactylum* and one against M-LDH of the anuran, *Rana pipiens*, were used in microcomplement fixation titrations against M-LDHs of numerous amphibians and reptiles. Indices of dissimilarity (ID) from a reference anuran established with the *Rana* antiserum and the IDs from a reference urodele established with *Amphiuma* antiserum are shown in the semilog plots of halves of Figs 5 and 6. The lines relating ID to time connect the average IDs of groups of organisms (open circles on Figs 5 and 6) whose time of divergence from the reference taxon is known. Taxa whose times of divergence are in question are positioned on the basis of the average of their IDs from the M-LDH of the reference species. These positions suggest that the Anura and Urodela diverged in the Permian or early Triassic, long after their separation from the phyletic line which led to reptiles and birds (Salthe and Kaplan, 1966).

Qualitative immunological and biochemical evidence distinguish the Anura and Urodela. Graham-Smith (1904) could detect no cross reactivity between an antiserum to plasma proteins of *Rana* and plasma antigens of a number of the Urodela. With more modern immunological methods weak cross reactions can be demonstrated (Boyden and Noble, 1933). Urodela, except the Cryptobranchidae, can be distinguished from adult Anura by the very slow electrophoretic migrations of their albumin-like proteins (Dessauer and Fox, 1956, 1964; Hebard, 1964). Electrophoretic patterns of plasma proteins of anuran tadpoles and adult salamanders are similar in having very short total migrations. The 'fast' albumins of the adult Anura appear at metamorphosis (Herner and Frieden, 1960). The amount of DNA/cell is usually higher in the Urodela than in the Anura (Fig. 4; Goin *et al.*, 1968; Goin and Goin, 1968). It has been reported that the DNA of the Urodela has a higher guanylic plus cytidylic acid content than the DNA of other vertebrates (Wolstenholme and Dawid, 1968).

2. *Relationships within the Anura*

The set of families chosen by Kluge and Farris (1969) and the dendrogram which they developed of Anuran phylogeny will be used as a basis for discussing the contribution of molecular studies to Anuran systematics. Immunological evidence offers some idea of the affinities of the various families and their evolution (Fig. 5). Microcomplement fixation titrations using antibodies against the muscle type lactic dehydrogenases of *Rana pipiens* (Fig. 5) support a Jurassic divergence of Ranidae from the other families. Only the Rhacophoridae show a close relationship to Ranidae, with the two families diverging in the Cretaceous (Salthe and Kaplan, 1966). Precipitin titrations, using plasma proteins as the source of immunological reagents (Fig. 5), also showed the Ranidae as widely divergent from the Hylidae, Leptodactylidae and Bufonidae. The Hylidae, Leptodactylidae and Bufonidae showed moderate affinities to each other, the Bufonidae being most

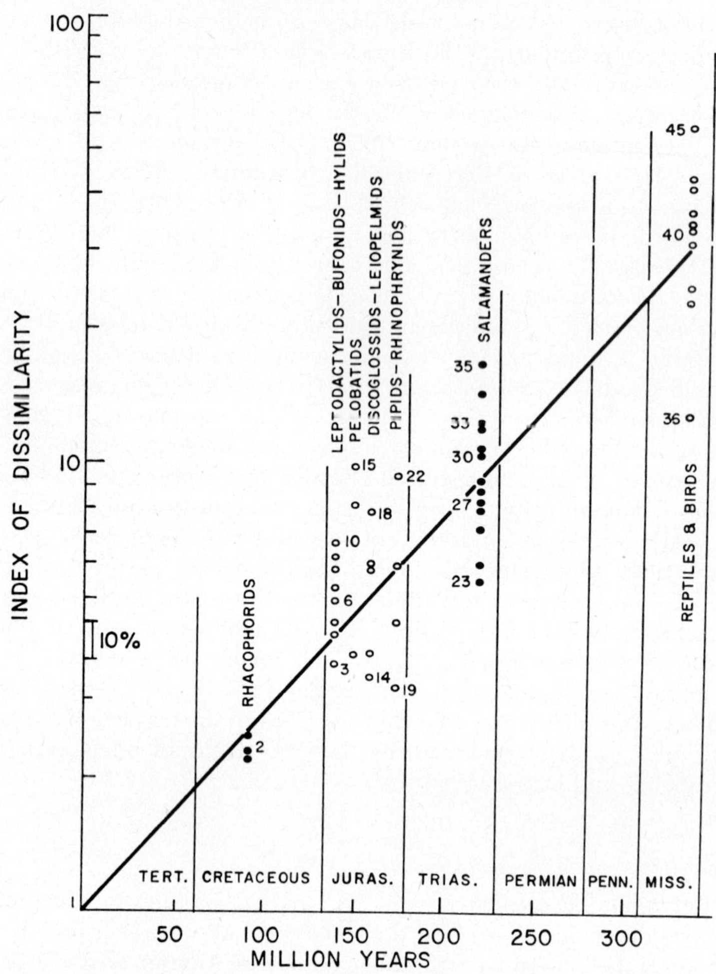

FIG. 5. Immunological titrations of proteins of the Anura. Above: Microcomplement fixation experiments using antiserum against the muscle-LDH of *Rana pipiens* in titrations against muscle-LDHs of various vertebrates. Each dot represents evidence on a different species. The line relating ID to time connects the average IDs of groups of organisms (open circles) whose time of divergence from the reference taxon *Rana pipiens* is known. Reproduced with permission of Salthe and Kaplan (1966). Opposite: Precipitin titrations using antiserum against plasma proteins of *Rana palmipes* (top) or *Bufo spinulosus* in reaction with plasma proteins from various other anurans. Reproduced with permission of Cei (1963).

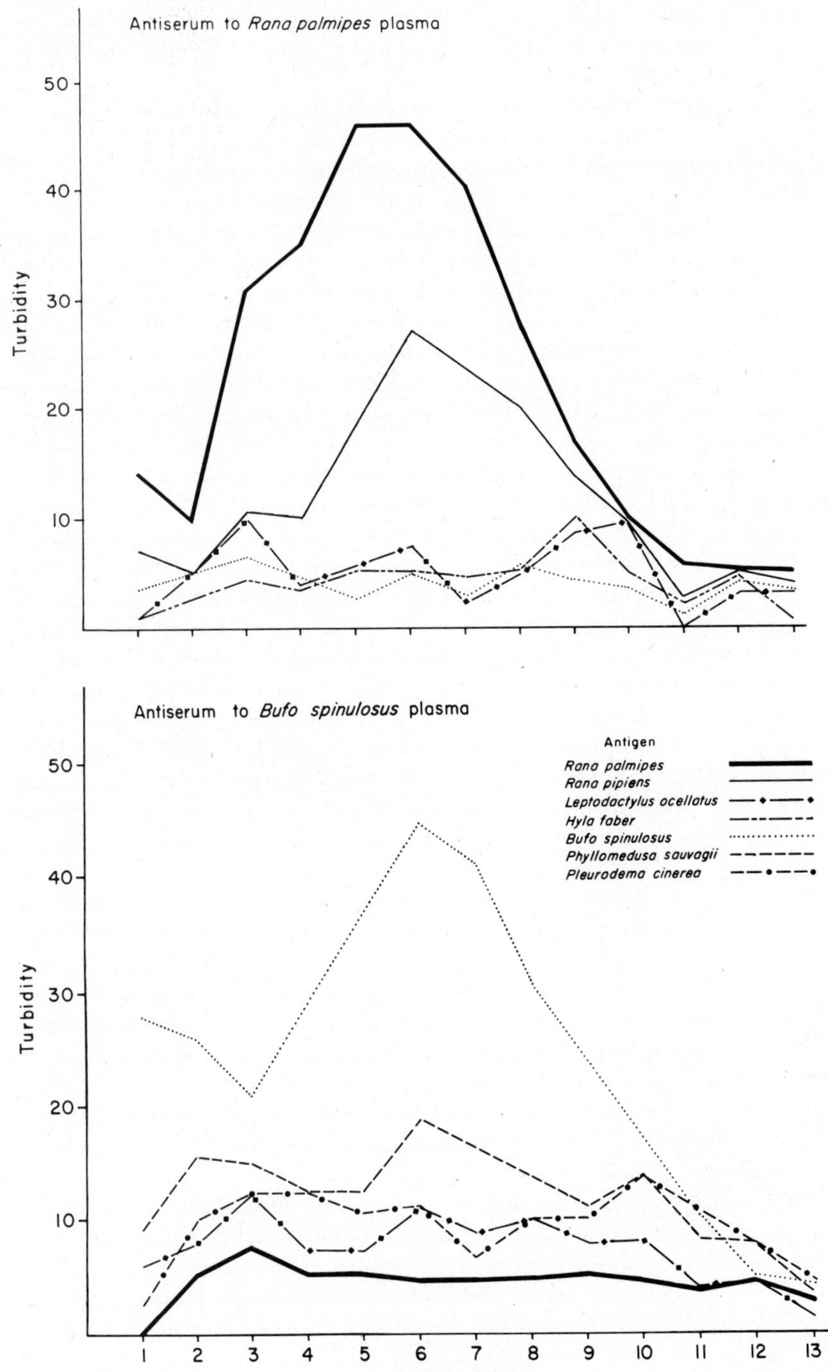
Antiserum to Rana palmipes plasma
Turbidity
50
40
30
20
10
Antiserum to Bufo spinulosus plasma
Antigen
Rana palmipes
Rana pipiens
Leptodactylus ocellatus
Hyla faber
Bufo spinulosus
Phyllomedusa sauvagii
Pleurodema cinerea
1 2 3 4 5 6 7 8 9 10 11 12 13
Antigen Dilution

closely allied to the Hylidae (Cei, 1963). *Rana* and *Bufo* plasma proteins and antibodies against these proteins show little tendency to cross-react in immunodiffusion tests (Lewis, 1964).

Components of the skin (Fig. 3) are particularly useful in characterizing anuran families (Michl and Kaiser, 1962/1963; Cei and Erspamer, 1966). Patterns of these constituents and other varieties of molecular evidence are helpful in distinguishing and clarifying relationships within certain families. Skin secretions of the Bufonidae lack biologically active polypeptides and imidazole amines but contain phenylalkyl amines and large concentrations of a great variety of indole amines. Among the latter, bufotenine is usually present in large amounts (Cei and Erspamer, 1966; Cei *et al.*, 1968). Numerous complex steroids are present in the skin secretions (Michl and Kaiser, 1962/1963; Shoppee, 1964) and the characteristic steroid of the bile is the sterol alcohol, 5-beta-bufol (Haslewood, 1967, 1968). Large quantities of a number of pteridines including bufochrome are present in the skin. Based upon the similarities in skin pteridines, Bagnara and Obika (1965) believe that the Bufonidae and the Pelobatidae are more closely related than present taxonomy indicates. Differences in patterns of distribution of these various metabolites are further useful in distinguishing certain genera (Cei *et al.*, 1968).

The Leptodactylidae show remarkable differences in the quantity and variety of biologically active polypeptides and amines present in skin secretions (Erspamer *et al.*, 1964a,b). Certain subfamilies lack both types of constituents; in the subfamilies, Leptodactylinae and the Ceratophrydinae some generic groups lack the substances whereas others produce them. The skin of *Physalaemus* contains physalaemin, a unique decapeptide which lowers blood pressure and stimulates smooth muscle. Some species of *Leptodactylus* synthesize large amounts of leptodactyline (Fig. 4) and a great variety of imidazole alkylamines. Skin of *Odontophrynus* contains large amounts of 5-hydroxytryptamine (Cei and Erspamer, 1966; Cei *et al.*, 1967). Precipitin titrations using antisera to plasma proteins of various species of the Leptodactylidae in both homologous and heterologous reactions show that *Ceratophrys*, *Lepidobatrachus* and *Odontophrynus*, sometimes placed in a separate family (Ceratophryidae) are divergent from *Leptodactylus*, *Pleurodema* and *Calyptocephalella*. *Ceratophys* and *Lepidobatrachus* are close serologically, but both are relatively widely divergent from *Odontophrynus*. *Calyptocephalella* is distant from all genera of the Leptodactylidae tested (Cei, 1965, 1969).

The skin of the Atelopoidae contains a unique potent toxin of unknown structure. The toxin is quite different from those found in related amphibians (Fuhrman *et al.*, 1969).

Data on skin components and the proteins of the Hylidae are available. The bright yellow, orange and red pigmentation of hylid skin is largely due to high concentrations of a variety of pteridines (Fig. 3; Bagnara and Obika, 1965). The amines and polypeptides of the skin have a curious generic distribution. With the exception of *Phylomedusa*, both are lacking in species of western hemisphere genera. *Phylomedusa* not only resembles Australian hylids in possessing polypeptides with bradykinin-like activities, but in addition, has a very active physalaemin-like, polypeptide. Both its specialized skin components (Cei and Erspamer, 1966, in press) and the immunological affinities of its plasma proteins (Cei, 1963) show *Phylomedusa* to be an independent branch of the Hylidae. Plasma albumins of species of *Hyla* from Europe (Chen, 1967), New Guinea and the western hemisphere (Dessauer and Fox, 1964 and unpublished) have faster electrophoretic mobilities than albumins of most frogs of other families.

Extracts of the skin of the Ranidae only contain low concentrations of a few simple indole alkylamines; phenylalkylamines, and imidazoleamines are absent. They contain polypeptides with potent bradykinin-like activities (Cei and Erspamer, 1966), and pteridines, especially ranochrome-3, are present in high concentrations (Bagnara and Obika, 1965). Plasma of *Rana* from North America is unique among vertebrates in its high amylase activity (Dessauer and Fox, 1964). The types of steroids found in the bile, as well as their great variety suggest that the Ranidae is a very heterogeneous family, more advanced biochemically than the Bufonidae (Haslewood, 1967, 1969).

3. *Relationships within the Urodela*

Noble (1931) proposed a scheme of the phylogeny of the Urodela, which Romer (1966) now divides into eight families. Immunological evidence suggests that members of these families arose from a common evolutionary line. Serological affinities indicate that the Amphiumidae, Proteidae and Sirenidae are relatively closely related and about equally distant from the Cryptobranchidae (Fig. 6; Boyden and Noble, 1933). Indices of dissimilarities of muscle lactate dehydrogenases (Fig. 6) suggested that the Urodela underwent its major radiation during the late Permian and the Triassic. Amphiumidae appear to be more closely related to the Plethodontidae than to any other family. The amphiumid-plethodontid line probably diverged from other salamanders in the Triassic and into amphiumids and plethodontids in the Cretaceous (Salthe and Kaplan, 1966).

Tetrodotoxin, one of the most poisonous nonproteinaceous substances known (Brodie, 1968), occurs in the Salamandridae but is absent in the Cryptobranchidae, Proteidae, Sirenidae, Ambystomidae, Amphiumidae and the Plethodontidae. Within the Salamandridae it has been found in *Taricha*, *Notophthalamus*, *Cynops*, *Triturus* and *Dimictylus*. Its absence in *Salamandra* adds an additional character to the series used by Wake and Ozeti (1969) which suggested that *Salamandra* has branched away from other genera of the Salamandridae (Mosher *et al.*, 1964; Wakeley *et al.*, 1965/1966). The skin toxin of *Salamandra* is unique in that it contains the complex azasteroids, samandarin and samandariden (Michl and Kaiser, 1962/1963; Fig. 3; Shoppee, 1964).

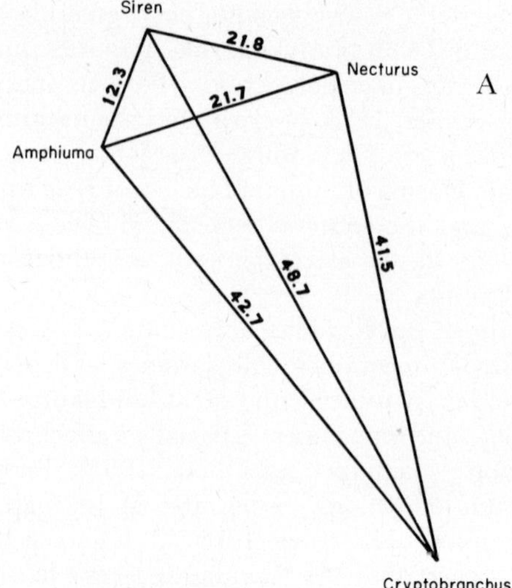

Fig. 6. Immunological titrations of proteins of Urodela. A (above): Immunological correspondence of plasma proteins of members of the Urodela based on reciprocal titrations using antisera to the plasma proteins of each form. Reproduced with permission of Boyden and Noble (1933). B (opposite): Microcomplement fixation experiments using antiserum against the muscle-LDH of *Amphiuma means* in titrations against muscle-LDHs of various vertebrates. Each dot represents evidence on a different species. The line relating ID to time connects the average IDs of groups of organisms (open circles) whose time of divergence from the reference taxon *Amphiuma means* is known. Reproduced with the permission of Salthe and Kaplan (1966).

D. REPTILIA

1. Chelonia, Squamata, Rhynchocephalia and Crocodilia

Data on both the primary structure and immunological affinities of their proteins emphasize the ancient divergence of the four reptilian orders. Cytochrome-C of the turtle *Chelydra serpentina* (Chan *et al.*, 1966) and of the snake *Crotalus adamanteus* (Bahl and Smith, 1965) seem to differ by 21 amino acid residues. To acquire the specific differences would require a minimum of twenty-eight mutations, which, based on the average mutation rate of cytochrome-C (Margoliash and Smith, 1965), is greater than expected for cytochrome-C of animals whose evolutionary lines probably separated during the Carboniferous. Comparisons of fingerprints of the tryptic peptides of hemoglobins of Crocodilia, Chelonia and Squamata indicate that the sequences of their amino acids are very different. For example (Fig. 7), the only spot common to the fingerprint of the arginine peptides of *Alligator* and those of the squamatan reptiles (*Iguana*, *Constrictor*, *Crotalus*, *Pituophis* and *Natrix*) is free arginine (spot-a). The fingerprints of the digests of the turtle, alligator and squamatan hemoglobins do not have a single peptide in common (Sutton, 1969).

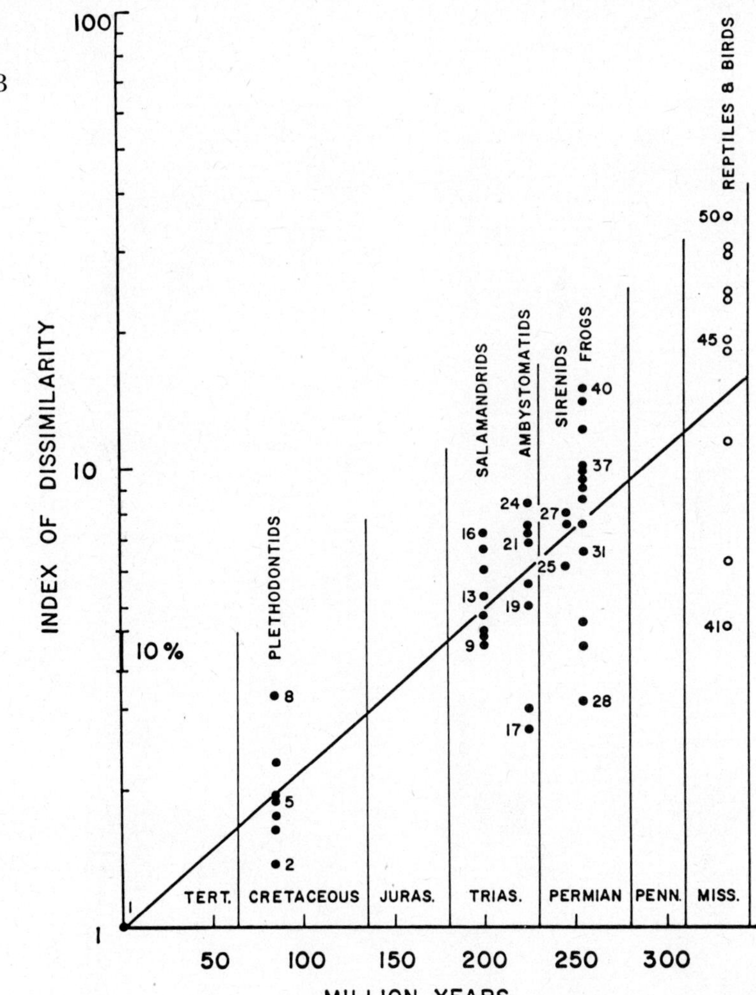

Immunological data on the proteins also attest to the remoteness of reptilian orders. Gorman *et al.* found that titrations of heart lactic dehydrogenases of Crocodolia, Rhynchocephala and Aves with an antiserum to heart lactic dehydrogenase of *Iguana iguana* (Squamata) yielded immunological distances of the high magnitude expected of forms divergent since at least the Permian. Other immunological studies using plasma proteins have led to similar conclusions (Graham-Smith, 1904; Cohen, 1955; Lewis, 1964).

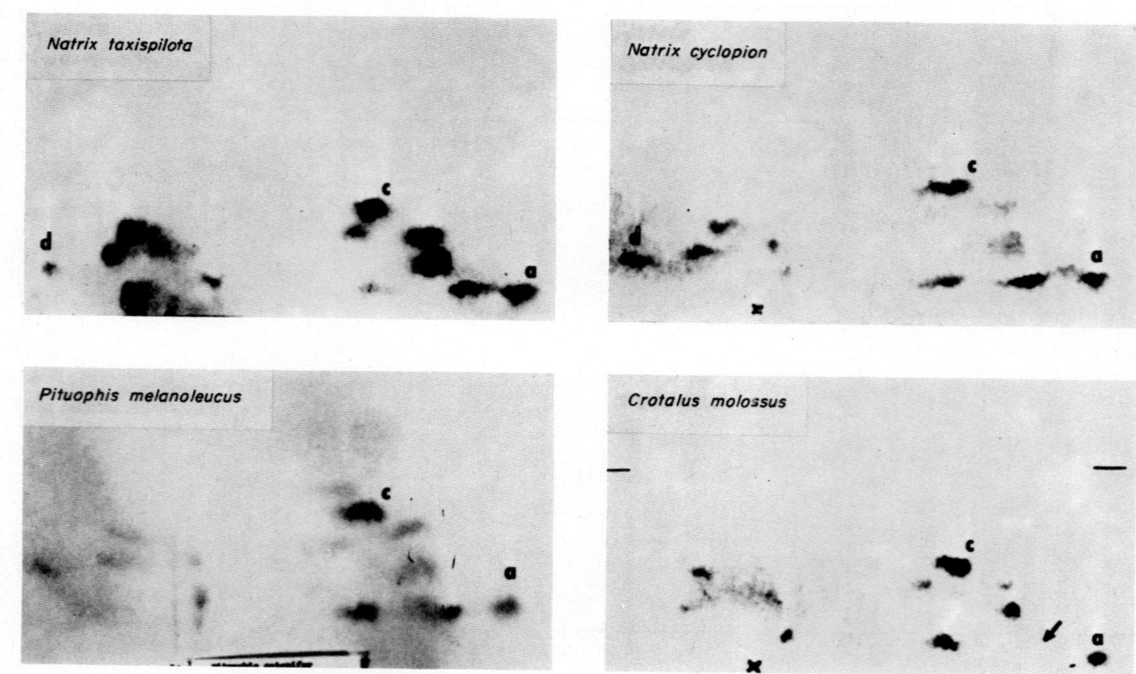
Natrix taxispilota
d
c
a
Natrix cyclopion
c
d
a
x
Pituophis melanoleucus
c
a
Crotalus molossus
c
a
x

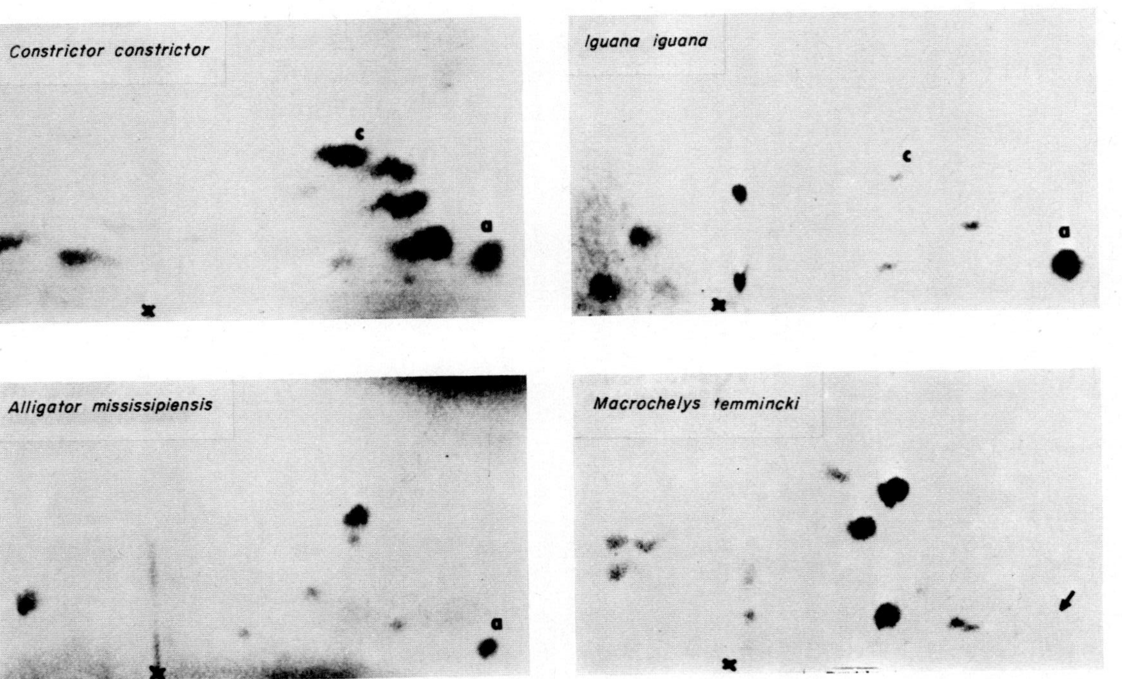

Fig. 7. Fingerprints of arginine peptides of hemoglobins of reptiles of different degrees of divergence. Letters designate peptides with identical properties; a = arginine; x = site of sample application. Electrophoresis was carried out in horizontal and chromatography in the vertical dimension. Reproduced with permission of Sutton (1969).

The three common orders of living reptiles are characterized by the electrophoretic behavior and temperature stability of their blood and tissue proteins. For example, albumins and transferrins of the Chelonia migrate at slower rates in alkaline buffers than do those of the Crocodilia and Squamata. Red cell hemolysates from turtles resolve into complex patterns of multiple hemoglobins. Hemolysates from snakes, lizards, and crocodiles resolve into simple patterns of one or two hemoglobin fractions. Hemoglobins of the Squamata have uniquely slow mobilities. Electrophoretic patterns of plasma proteins of lizards and snakes usually lack fractions with the mobility of mammalian gamma-globulin (Dessauer and Fox, 1956, 1964). Heart-LDHs of turtles denature at lower temperatures than those of Crocodilia and Squamata (Wilson *et al.*, 1964).

The orders can even be distinguished on the basis of their low molecular weight constituents. The latter reflect the many gains and losses in metabolic capabilities which have occurred in species of the different evolutionary lines during their long periods of separation. The chief constituent of crocodilian bile, a C-27 trihydroxy cholestanoic acid, may represent the precursor metabolite from which arose the unique C-27 tetrahydroxy cholestanoic acid of turtles and the C-24 bile acids of lizards and snakes (Haslewood, 1968). The high concentration of urea in the blood of turtles reflects the presence of active urea-cycle enzymes. The absence or low concentration of urea in blood of the Crocodilia and Squamata results from either the loss or suppression of urea-cycle enzymes (Fig. 2; Cohen and Brown, 1960). Other metabolic specializations are probably responsible for the uniquely high concentration of high energy phosphates in red cells of the Squamata, the presence of phytic acid and phytase in erythrocytes of turtles (Rapoport *et al.*, 1941, 1942), and the high levels of 3-methylhistidine in cells of the Crocodilia (Herbert *et al.*, 1966). Crocodilians also are unique among tetrapods in their acid base mechanisms, secreting an alkaline urine which contains a high concentration of ammonium bicarbonate (Coulson and Hernandez, 1964). Turtles are far more tolerant of anoxia than are other reptiles (Belkin, 1963; Dessauer, 1970).

2. *Relationships within the Chelonia*

Paleontological evidence suggests that turtles of the Pleurodira and Cryptodira separated during the Jurassic (Romer, 1966). Plasma proteins of members of the two suborders reflect this early divergence, showing very low immunological correspondence (Table I). For example, antiserum against plasma proteins of *Chelus fimbriatus* of the Pleurodira reacts weakly with plasma proteins of species of *Chelydra*, *Kinosternon*, *Terrapene* and *Caretta* of the Cryptodira (Frair, 1964). Globins of species of the Chelidae, a family of the Pleurodira, have unique electrophoretic properties (Dozy *et al.*, 1964; Sullivan and Riggs, 1967b).

The Cryptodira radiated into three divergent groups: (1) marine, (2) softshell, and (3) fresh water and land turtles (Table I). Living genera of marine turtles *Chelonia*, *Caretta*, *Dermochelys*, *Eretmochelys* and *Lepidochelys* are classified in two families, the Cheloniidae and Dermochelyidae. The proteins of species of these five genera cross react strongly in immunological tests but show only weak affinities to proteins of species of the Pleurodira and even other Cryptodira. Antiserum to plasma proteins of Trionychidae, the softshell turtles, also cross reacts only weakly with serum of species of other families. Within the Testudinidae (Emydidae) immunological cross reactions span a wide range, suggesting that this family of fresh water and land turtles is a very heterogeneous group. For example (Table I), antiserum against proteins of *Pseudemys*

Table I
Taxonomic Affinities of Turtles Based on Immunological Titrations Using Plasma Proteins (Frair, 1964)

	Antisera[1]								
Test Antigens	Css	Css	So	Pse	Tcc	Ccc	Tf	Ps	Cf
Suborder Cryptodira									
Family Chelydridae									
Chelydra s. serpentina	100[2]	100	43	6	36		8	18	5
Macroclemys temmincki				10					
Family Kinosternidae									
Kinosternon bauri palmarum			96						
Kinosternon f. flavescens	61							20	
Kinosternon scorpioides	62								
Kinosternon sonoriense	67			0					
Kinosternon subrubrum steindachneri							3		11
Kinosternon s. subrubrum		27			25				
Sternotherus odoratus		29	100		18	17			
Family Dermatemydidae									
Dermatemys mawi	68	34			29		0		
Family Emydidae									
Chrysemys p. picta				55					
Clemmys insculpta	60						3		
Cuora amboinensis	65								
Geoemyda trijuga				10					
Kachuga t. tecta			90	7					
Malayemys subtrijuga	35			12					
Pseudemys scripta elegans		41		100	63			22	
Terrapene c. carolina		41	59	45	100				12
Family Testudinidae									
Gopherus polyphemus berlandieri		32	36	2	37		2		

cont. p. 196.

Table I—*contd.*
Taxonomic Affinities of Turtles Based on Immunological Titrations Using Plasma Proteins (Frair, 1964)

	Antisera[1]								
Test Antigens	Css	Css	So	Pse	Tcc	Ccc	Tf	Ps	Cf
Family Cheloniidae									
Caretta c. caretta						100	6	18	9
Chelonia m. mydas						90			
Eretmochelys i. imbricata						80			
Family Dermochelyidae									
Dermochelys coriacea						42	4		
Family Trionychidae									
Trionyx ferox		17	14			10	100	0	
Trionyx spiniferus emoryi							76		
Suborder Pleurodira									
Family Pelomedusidae									
Pelusios subniger						17	2	100	33
Podocnemis unifilis								9	
Family Chelidae									
Batrachemys nasuta								42	57
Chelodina longicollis							0	13	23
Chelus fimbriatus	14							23	100

[1] Antisera to plasma proteins of: Css = *Chelydra s. serpentina*; So = *Sternotherus odoratus*; Pse = *Pseudemys scripta elegans*; Tcc = *Terrapene c. carolina*; Ccc = *Caretta c. caretta*; Tf = *Trionyx ferox*; Ps = *Pelusios subniger*; Cf = *Chelus fimbriatus*.

[2] Results obtained in the heterologous titrations are recorded as percentages of the magnitude of the homologous titration which is expressed as 100%.

scripta reacts strongly with plasma proteins of *Chrysemys* and *Terrapene*, but its reaction with those of *Kachuga* is barely perceptible (Frair, 1962, 1964, 1969). Species of fresh water and land turtles form two groups based on the electrophoretic behavior of their plasma albumin. The albumins of the Testudinidae (Emydidae) and Chelydridae migrate slowly in an electric field but those of the Kinosternidae and Dermatemydidae have relatively fast migration rates (Crenshaw, 1962; Dessauer and Fox, 1964; Frair, 1964).

3. *Relationships within the Squamata*

The Squamata includes two suborders, Sauria (lizards) and Serpentes (snakes). Romer (1966) classifies living lizards, a highly diverse group, into five infraorders but groups living snakes, a more uniform taxon, in only a single comparable category. Molecular evidence shows that snakes and lizards are far more closely related to each other than to other reptiles. Serologists since Graham-Smith (1904) have obtained weak immunological cross reactions in tests involving material from lizards and snakes. Physicochemical properties of blood proteins do not distinguish snakes from lizards. Electrophoretic mobilities of the albumins, transferrins, hemoglobins (Dessauer *et al.*, 1957, 1962a; Dessauer and Fox, 1964) and heart-LDHs (Wilson *et al.*, 1964) of a wide variety of snakes and lizards form overlapping series. The heart type LDHs of species of both suborders also have high inactivation temperatures (Wilson *et al.*, 1964). Some subordinal nonprotein differences have been noted, however. Blood of lizards commonly contains higher concentrations of glucose than blood of snakes (Fig. 2). Snake plasma contains a higher concentration of sialic acid than plasma of other reptiles, including iguanid lizards (Seal, 1964).

Although most herpetologists agree that snakes are derived from lizards, they have various opinions as to which line of lizards was their point of origin (Bellairs and Underwood, 1951). Most recently the Varanidae has been suggested as the lizard family closest to the direct line of snake evolution (McDowell and Bogert, 1954; Romer, 1966). Biochemical and immunological data do not conclusively support this contention. Fingerprints of hemoglobin tryptic peptides of snakes (Fig. 7, *Natrix*, *Pituophis*, *Crotalus* and *Constrictor*) show many similarities to those of lizards of the Iguanidae (Fig. 7, *Iguana*), Agamidae and Gekkonidae but show fewer similarities to fingerprints of the hemoglobins of the Varanidae and Anguidae (Sutton, 1969). Antiserum to the heart-LDH of the iguanid lizard *Iguana iguana* cross reacts more strongly with the heart-LDHs of snakes (Natricidae, Viperidae, Acrochordidae, Homalopsidae) than with the heart-LDHs of lizards of the Scincidae and Teiidae (Gorman *et al.*, 1971b). Data on bile steroid pathways, which have been advanced in support of a varanid origin, are inconclusive. In bile acid synthesis *Varanus* (and *Heloderma*) resembles snakes in utilizing the 'five-beta pathway', but it differs from snakes in producing C-27 rather than C-24 final products. Conversely, although members of the Iguanidae, Gekkonidae, Amphisbaenidae, Chameleontidae, Agamidae and Teiidae synthesize C-24 bile acids, they seem to utilize, primarily, the 'five-alpha pathway' (Haslewood, 1967, 1968). Without knowing whether the alpha or beta stereospecific reductions at position five of the sterol is more critical than the oxidations leading to the decrease in side chain length from C-27 to C-24, it is impossible to assess the taxonomic significance of these data. In summary, the molecular evidence suggests that lizards of the infraorders IGUANIA and GEKKODA should also be given careful consideration in studies designed to delineate the closest living relative of snakes.

a. Sauria. Surprisingly few molecular studies have been concerned with the affinities of the many families of lizards and of genera within these families. Microcomplement fixation titrations, using antisera against the heart-LDH and against the plasma albumin of *Iguana iguana* (Gorman *et al.*, 1971b), have indicated that the Iguanidae have a close affinity to the Agamidae, moderate affinities to the Helodermatidae, Amphisbaenidae, and the Anniellidae, and diverge most widely from the Scincidae and Teiidae. Using albumin immunological distances as an evolutionary clock (as in Figs 5 and 6) suggests that the Iguanidae diverged from other families of lizards in the early Cretaceous. The very wide range of immunological differences between the albumins of *I. iguana* and other species of the Iguanidae suggests, however, that the order underwent a number of radiations shortly after its origin. Fingerprints of tryptic peptides of the hemoglobins of different species of the Iguanidae also show a wide range of patterns, certain of which differ in more peptides than patterns of snakes of different families. Fingerprints of tryptic peptides of the Iguanidae, Agamidae and Gekkonidae resemble each other but show few similarities to fingerprints of hemoglobins of the Varanidae and Anguidae (Sutton, 1969).

Lizards of the infraorder GEKKOTA seem to have unique retinal chromoproteins. Their rhodopsins contain retinine, but absorb light maximally between 518 and 530 mμ rather than near 500 mμ like 'classic' rhodopsin. Slight differences in the position of the absorption maximum of rhodopsins (Crescitelli, 1958) may be useful in grouping species within the GEKKOTA.

b. Serpentes. Probably because of the medical importance of poisonous forms, snakes have been studied more extensively than have lizards by the biochemist and immunologist. I will describe the molecular data in the framework of the classification of Underwood (1967).

Molecular evidence of systematic relationships is compatible with his grouping of living snakes into three infraorders: Scolecophidia, the blind snakes; Henophidia, the primitive snakes; and, Cenophidia, the advanced snakes. Precipitin titrations indicate that a family of the Scolecophidia (Typhlopidae) is distant from families of both the Henophidia (Boidae) and Caenophidia (Elapidae, Viperidae, Natricidae and Colubridae). The Henophidia show some affinities to the Scolecophidia, but are more closely allied to the Caenophidia (Pearson, 1966). In an immunodiffusion experiment, antiserum to transferrin of *Thamnophis* (Caenophidia) did not yield precipitin lines with transferrins of *Typhlops* and *Leptotyphlops* (Scolecophidia) or with transferrins of *Cylindrophis* and *Xenopeltis* (Henophidia). In quantitative microcomplement fixation experiments, using antisera against the transferrins of three species of the Caenophidia, high indices of dissimilarity were obtained in titrations against transferrin of *Acrochordus* of the Henophidia (Table II; George and Dessauer, 1970).

Evidence of relationships within infraorders comes from a variety of studies. Within the Henophidia immunological evidence shows the close affinities of the Pythoninae and Boinae (Graham-Smith, 1904; Cohen, 1955; Pearson, 1966). The Aniliidae (*Cylindrophis*) and the Boidae (*Corallus*, *Eunectes*, *Python*, *Eryx* and *Boa*) have a unique alpha-hydroxylase capable of catalyzing the hydroxylation of deoxycholic acid to pythocholic acid. Intestinal microorganisms synthesize the substrate, deoxycholic acid, from cholic acid, the principal bile acid of the animal's metabolic system. Pythocholic acid is not present in the bile of *Corallus enhydris* or *Acrochordus javanicus*, a member of another family of the Henophidia, the Acrochordidae. Its absence may indicate that the proper micro-organisms are not present in the intestinal tract (Haslewood, 1967, 1968).

The immunological and biochemical approach may help unravel the interrelationship of the advanced snakes (Caenophidia). Taxonomists generally agree on the placement of most genera of elapid, hydrophid and viperid snakes, but few agree on the classification of rear-fanged and nonvenomous species, the so called colubrid snakes. In an experiment designed to test the naturalness of various classifications of colubrid snakes (Table II), the transferrins of a number of species were compared using the quantitative micro-complement fixation method (George and Dessauer, 1970). From the immunological distances obtained, natricine and colubrine genera seem to have diverged as far from each other as has either group from the Viperidae, which

Table II
Indices of Dissimilarities of Transferrins and Recent Classifications of Advanced Snakes

Genus	Classification			Immunological Titrations[1]	
	Romer (1956)	Dowling (1967)	Underwood (1967)	$ID_{Col.}$	$ID_{Tham.}$
Coluber (racer)	COLUBROIDEA Colubridae Colubrinae	COLUBROIDEA Colubridae Colubrinae	COLUBROIDEA[2] Colubridae Colubrinae	1·0	9·5
Pituophis (bull snake)	COLUBROIDEA Colubridae Colubrinae	COLUBROIDEA Colubridae Colubrinae	COLUBROIDEA Colubridae Colubrinae	3·1	7·1
Heterodon (hognose snake)	COLUBROIDEA Colubridae Colubrinae	COLUBROIDEA Colubridae Colubrinae	COLUBROIDEA Dipsadidae Xenodontinae	15	9·2
Thamnophis (garter snake)	COLUBROIDEA Colubridae Colubrinae	COLUBROIDEA Colubridae Natricinae	COLUBROIDEA Natricidae	16	1·0
Homalopsis (rearfanged water snake)	COLUBROIDEA Colubridae Homalopsinae	COLUBROIDEA Colubridae Natricinae	COLUBROIDEA Homalopsidae	>16	>14
Acrochordus (wart snake)	COLUBROIDEA Colubridae Acrochordinae	COLUBROIDEA Colubridae Natricinae	BOOIDEA[2] Acrochordidae	15	>14
Crotalus horridus (canebrake rattlesnake)	COLUBROIDEA Viperidae	COLUBROIDEA Viperidae Crotalinae	COLUBROIDEA Viperidae Crotalinae	>16	9·5

[1] $ID_{Col.}$ = index of dissimilarity obtained in titrations using antiserum against transferrin of *Coluber*. $ID_{Tham.}$ = those obtained with *Thamnophis* antitransferrin.
[2] Underwood uses different superfamily names for primitive and advanced snakes.

virtually all taxonomists recognize as a distinct family. Xenodontine snakes (*Heterodon* and *Farancia*), as well as *Homalopsis* and *Acrochordus* (sometimes grouped with the natricine snakes) were remote from both colubrine and natricine genera.

The classification of Underwood (1967) appears to have great predictive value. The transferrin study as well as the earlier serological work of Graham-Smith (1904), Cohen (1955) and Pearson (1966) all indicate that his families, Viperidae, Colubridae, Elapidae, Natricidae and Dipsadidae, have diverged enough to be recognized as taxons of at least equivalent levels of distinction. Such studies have also documented the relatively close interfamily relationships of the Elapidae, Viperidae, Natricidae, and Colubridae.

The relatively small differences in their transferrins and hemoglobins support the view that these groups of advanced snakes separated as different evolutionary lines well within the Cenozoic Period. The fossil record is insufficient to calibrate properly the rate of change of transferrin in snakes. However, based on the rate obtained for primates (Wilson and Sarich, 1969), the magnitude of the immunological distances between transferrins of species of the Natricidae, Colubridae and Viperidae (Table II) indicate that these families diverged well within the last 80 million years. The great similarity in the fingerprint patterns of tryptic peptides of hemoglobins of species of Natricidae (*Natrix*), Colubridae (*Pituophis*) and Viperidae (*Crotalus*) also indicate a relatively modern radiation (Fig. 7).

Considerable molecular data are available on inter and intrafamilial relationships within the two families of venomous snakes (Elapidae and Viperidae). Typically, the venoms of elapids exhibit marked esterase activities whereas those of the viperids usually contain very active proteases and peptidases and lack choline esterases (Michl and Molzer, 1964/1965; Tu and Toom, 1967; Tu *et al.*, 1965/1966, 1966/1967; Delpierre, 1968; Mebs, 1968). Further, although proteins of the elapid venoms move primarily toward the cathode during electrophoresis in alkaline buffers, those of the viperid venoms migrate primarily toward the anode (Fischer and Neumann, 1954; Neelin, 1963; Jimenez-Porras, 1963; Bertke *et al.*, 1966/1967; Fisher and Kabara, 1967). There is considerable overlap, however, in the enzyme composition of viperid and elapid venoms so that 'the day is long past when damage due to an injection of venom was assumed to be primarily necrotic in the case of viperid venom and primarily neurotoxic in the case of elapid venom' (Gans and Elliott, 1968).

Within the Elapidae the similarity of both the plasma and venom proteins confirm the close affinity of the sea snakes (Hydrophinae) and cobras (Elapinae). However, although antiserum against venoms of the sea-snake *Enhydrina* gives some protection against venoms of many elapids of both the Hydrophinae and the Elapinae, antivenins from the Elapinae fail to completely neutralize the toxicity of *Enhydrina* venom (Minton, 1967/1968). Based on the antigenic affinities of their venom proteins, mole vipers (Atractaspidinae) probably should be classified in the Elapidae rather than in the Viperidae (Minton, 1968/1969).

The Viperidae include the true vipers (Viperinae) and the pit vipers (Crotalinae). Although both plasma and venom proteins reflect the close relationship of snakes of these subfamilies, they differ in pathways of bile steroid synthesis. The bile acid, bitocholic acid, is synthesized by most Viperinae but not by the Crotalinae (Haslewood, 1968). Additional molecular data indicate that within the Crotalinae, *Agkistrodon* is a more widely divergent genus than *Crotalus*, and is less closely related to *Sistrurus* than is the latter (Cohen, 1955), and on Formosa is closely related to *Trimeresurus* (Kuwajima, 1953). Fingerprints of tryptic peptides of hemoglobins of both *Crotalus* and *Agkistrodon* lack an arginine peptide (Fig. 7, *Crotalus*, arrow) commonly found on fingerprints from species of Natricidae and Colubridae (Sutton, 1969).

Both physicochemical and immunological data indicate that the Natricidae is a natural group of genera. Their plasma resembles that of the Colubridae, insofar as it contains active leucine amino peptidases. It is characterized however by the presence of an electrophoretically unique esterase which has been found in plasmas of African, Asian, European and North American species. Although one of their hemoglobin polypeptides is also unique in electrophoretic mobility (Dessauer, 1967), fingerprints of these hemoglobins (Fig. 7) do not exhibit any 'family-specific' peptide. Fingerprints for species of five genera were found to be very similar, with no two patterns showing more than seven differences (Sutton, 1969). Similarly, the transferrins of *Natrix*, *Regina*, *Amphiesma*, *Seminatrix*, *Storeria*, *Tropidoclonion* and *Thamnophis* had low indices of dissimilarity when tested with an antiserum to the transferrin of *Thamnophis couchi*. Species of *Natrix* from Europe and *Xenocrophis* and *Amphiesma* from India showed the greatest immunological distances, suggesting a wide divergence of these forms from the North American genus, *Thamnophis* (George and Dessauer, 1970; Mao and Dessauer, 1971).

Immunological data on the total plasma proteins suggests that the Colubridae contains at least two groups of related genera. One, including *Elaphe*, *Lampropeltis*, *Pituophis* and *Rhinocheilus*, is made up of remarkably closely allied species. The second, more variable group includes *Coluber*, *Masticophis*, and *Opheodrys* (Pearson, 1966). A specific study of the transferrins of species of the first group also indicates that they are remarkably similar to each other but differ somewhat from *Pituophis* of the second group (George and Dessauer, 1970). Leucine amino peptidase activity of the plasma of species of Colubridae is very high (Dessauer, 1967).

4. *Relationships within the Crocodilia*

Immunological comparisons of plasma proteins of members of the two families of Crocodilia, the Crocodilidae and Alligatoridae, are in keeping with their presumed early Tertiary divergence (Romer, 1966). Antiserum against either total plasma proteins (Graham-Smith, 1904; Lewis, 1964) or against purified plasma albumin (Gorman *et al.*, 1971b) of *Alligator* cross-reacted moderately with antigens from *Caiman*, another genus of the Alligatoridae. However, the two antisera reacted much more weakly with plasma proteins of *Crocodylus* of the Crocodylidae.

III. Species Groups and Speciation

A. INTRODUCTION

Numerous fundamental questions concerning the structure of species and processes of speciation seem capable of being answered by studies concentrated at the molecular level. Collectively, these should lead to more precise definitions of species and to clarification of their relationships to closely allied forms.

The study of electrophoretic patterns of enzymes and other proteins is a promising approach toward these problems (Dessauer, 1966a; Hubby and Lewontin, 1966). High-resolution electrophoretic techniques make it possible to detect slight differences in proteins. One can compare large numbers of proteins of numerous individuals with small expenditure of tissue and time. Electrophoretic patterns developed with general protein stains are useful, but those treated to identify specific proteins are more meaningful. With them the investigator can compare proteins showing signs of common ancestry instead of nonspecific fractions of diverse origin. Combined with histochemical and radiochemical methods for localizing specific proteins (Dessauer, 1966b; Yunis, 1969), electrophoresis is capable of uncovering a large portion of the allelic variation at a specific locus, making it possible to calculate the frequency of specific alleles in different populations. Peptide fingerprinting (Sutton, 1969), immunodiffusion (Schenberg, 1962/1963), immunochemical absorption (Abramoff, *et al.*, 1964) and cell agglutination (Frair, 1963), afford additional sensitive means of detecting both similarities and differences in polymer structure. Even metabolic intermediates are useful genetic indicators, as their presence depends upon the enzymes present in a tissue (Cei *et al.*, 1967).

B. BIOCHEMICAL GENETICS AND MOLECULAR PHENOTYPES

There are few specific data on mechanisms which control the protein structure and metabolic capabilities of species of amphibians and reptiles. Experimental crosses of amphibians are readily carried out, but great care is required to raise tadpoles to maturity. Even maintaining nonbreeding reptiles in the laboratory requires special effort. Because of such difficulties most of the biochemical evidence on inheritance is indirect and arises from the examination of protein phenotypes of hybrids and of individuals of populations polymorphic relative to a particular protein. Most commonly such studies require the interpretation of electrophoretic patterns of proteins of individual organisms.

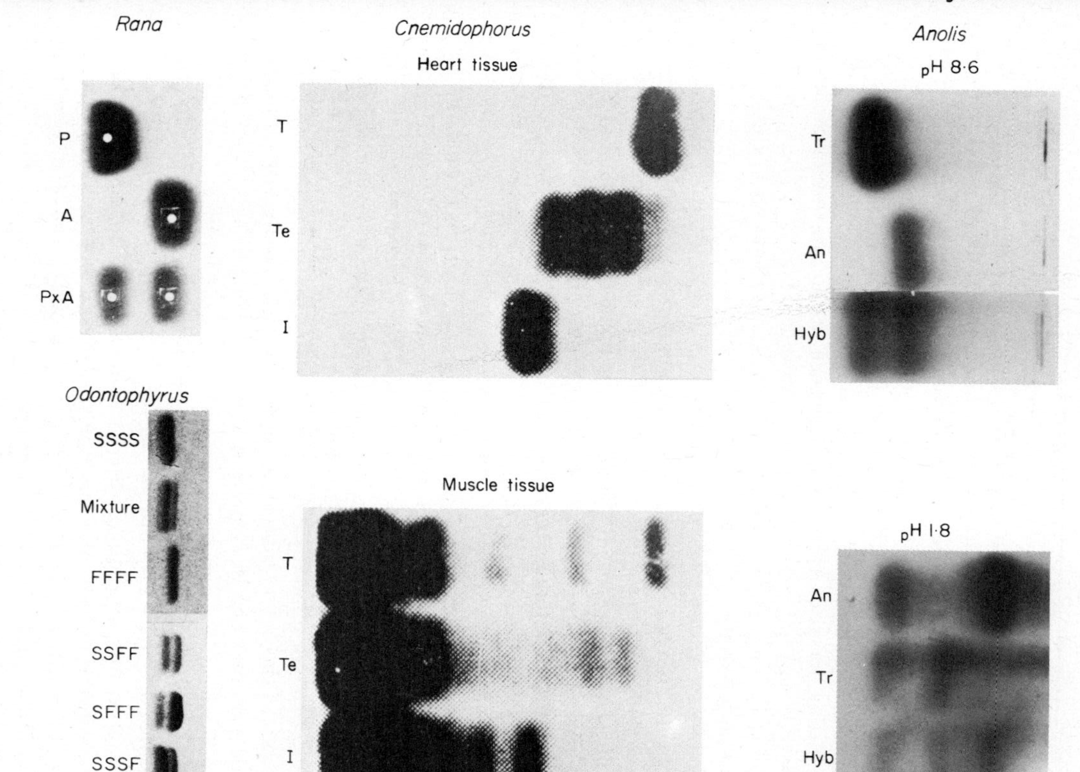

FIG. 8. Inheritance and electrophoretic phenotypes of proteins. Abbreviations are as follows: P = *Rana pipiens*, A = *Rana areolata*, P × A = F_1 hybrid of a cross between *R. pipiens* and *R. areolata*; S = one dose of an allele yielding a slow albumin; F = one dose of an allele yielding a fast albumin; T = *Cnemidophorus tigris*, Te = *Cnemidophorus tesselatus*; I = *Cnemidophorus inornatus*, Tr = *Anolis trinitatus*, An = *Anolis aeneus*. Photographs of the *Odontophrynus* gels are reproduced with permission of Beçak *et al.* (1968); those of the *Cnemidophorus* gels with permission of Neaves and Gerald (1968).

The electrophoretic patterns of many proteins coded for by only two allelic genes are relatively simple to analyze. They exhibit either one or two bands of activity. Apparently, if the genes are present in the homozygous state, the one band pattern results. In the heterozygous state the two band pattern appears, since the genes code for different polypeptides. Therefore, in populations in which two variant genes occur at a specific locus, individuals with three different protein phenotypes are found, and the distribution of these phenotypes may approximate Hardy-Weinberg ratios. The inheritance of leucine amino peptidase (Pough and Dessauer, unpublished) amino acid dehydrogenase (Jimenez-Porras, 1964/1965a,b), glucokinase (Dessauer and Nevo, 1969), albumin (Beçak *et al.*, 1968) and transferrin (Dessauer *et al.*, 1962a,b; Fox *et al.*, 1961; Guttman, 1967, 1969, 1972; Dessauer and Nevo, 1969) are usually controlled in this manner. Albumin inheritance in the progeny of the cross *Rana pipiens* × *R. areolata* offers a typical example (Fig. 8). The parental species have electrophoretically distinct albumins and the hybrid exhibits the albumins of both parents.

Inheritance patterns of other proteins may be very complex. Polypeptide products of the same or of different loci may aggregate as series of isozymes, physicochemically distinct proteins with qualitatively identical activities. Lactic, malic, alpha glycerol phosphate, isocitric, alcohol (Salthe and Kitto, 1966), and glyceraldehyde-3-phosphate (Lebherz and Rutter, 1967) dehydrogenases, adenosine deaminases (Fig. 13; Ma and Fisher, 1968; Neaves, 1969) creatine kinases (Eppenberger *et al.*, 1967) and many esterases (Fig. 11; Dessauer and Nevo, 1969; Shontz, 1968) form multibanded electrophoretic patterns. Although single alleles determine single transferrins in most species, they appear to determine multiple transferrins in New World toads of the *cognatus*-group and in *Bufo alvarius* (Guttman, 1972). Isozyme patterns often vary from tissue to tissue (Wright and Moyer, 1966; Eppenberger *et al.*, 1967; Chen, 1968; Ma and Fisher, 1968). The number of isozymes as well as their variety are dependent upon the number of gene loci that are active in a tissue, the tendencies of the polypeptide gene products to aggregate, the number of polypeptide subunits per isozyme and the resolving power of the electrophoretic method.

The complexity of such interactions and the numerous factors involved are well illustrated by studies on the lactic dehydrogenases (Fig. 8). In many species of reptiles and amphibians the LDH phenotypes are similar to those commonly observed in birds and mammals. LDH-polypeptides are produced by at least two different loci, a heart type (H-LDH) and a muscle type (M-LDH). Polypeptide products of either or of both loci commonly polymerize randomly in aggregates of four subunits. The electrophoretic pattern of LDH in tissues in which only one locus is active is a single band if the locus is homozygous (Fig. 8, heart-tissue T + I) and shows five bands if the locus is heterozygous. If two subunits are present in equal concentrations, the staining intensity of the five isozymes will approximate binomial proportions,

1:4:6:4:1 (Fig. 8, heart tissue, Te; Neaves and Gerald, 1968). When both the H-LDH and M-LDH loci are homozygous and active, the tissue electrophoretic pattern exhibits five isozymes (Fig. 8, muscle tissue, T) from the random aggregation of the two types of polypeptides. If both loci are active and one is heterozygous, three peptides are involved in the polymerization. The phenotype will consist of a maximum of 15 bands, if all isozymes are resolved during electrophoresis; when both loci are active and heterozygous, the four varieties of polypeptides polymerize into as many as 30 isozymes (Adams and Finnegan, 1965; Markert, 1965; Wright and Moyer, 1966). Interesting exceptions to these generalizations occur in amphibians. In *Acris crepitans* of certain populations the H-LDH has the same mobility as the M-LDH; isozymes may be present but, if so, are undetectable electrophoretically (Salthe and Nevo, 1969). Many of the H-LDH variants of *Rana pipiens* will polymerize with each other but not with M-LDH polypeptides (Salthe *et al.*, 1965; Salthe, 1969).

Inheritance of hemoglobins is especially complex. Blood of individual amphibians and reptiles commonly contains one but may contain two or more hemoglobins distinguishable by differences in surface charge, molecular weight and chemical properties. Hybrids of some crosses exhibit two hemoglobins, one from each parental species (Fig. 8), but other hybrids possess a hemoglobin with a mobility similar to that of one parent or different from that of both parents (Guttman, 1972). Electrophoretic patterns of turtle red cell hemolysates are especially complex (Dessauer *et al.*, 1957; Sullivan and Riggs, 1967b and 1967d). Molecules of 'double' hemoglobins are common (Svedberg and Hedenius, 1934), apparently formed by the linkage of two 'single' hemoglobins by a disulfide bridge (Sullivan and Riggs, 1967a). If hemolysates are subjected to electrophoresis in a buffer which contains a reducing agent such as dithioerythritol, the disulfide bond apparently breaks simplifying and sharpening the hemoglobin pattern.

The functioning hemoglobin molecule is a heteropolymer of two or more different gene products (Dozy *et al.*, 1964; Gillespie and Crenshaw, 1966; Manwell, 1966; Sullivan and Riggs, 1967b,d; Dessauer, 1967; Dessauer and Nevo, 1969; Gorman *et al.*, 1971a). When electrophoresed in formate buffer of low pH, each hemoglobin commonly resolves into two polypeptides (Fig. 8). Additional bands that are sometimes present may be hydrolytic fragments formed when acid sensitive bonds of the two originally intact polypeptides are exposed to the low pH of the buffer (Fig. 10).

Polyploidy, common in amphibians and in some reptiles, has a marked influence upon protein phenotypes. In polyploid salamanders (Uzzell and Goldblatt, 1967) and in frogs heterozygous for albumin genes, each allele leads to a protein variant but the relative concentration of each variant in the tissue depends upon the dose of the corresponding gene. For example, in a population of tetraploid frogs, *Odontophrynus americanus*, in which two albumin alleles are common, individuals of five different albumin phenotypes were found (Fig. 8). The electrophoretic pattern of plasma proteins of homozygous individuals exhibit either a single albumin of relatively slow mobility, or a single albumin of relatively fast mobility. Patterns of plasma proteins of heterozygous individuals exhibit both the fast and the slow albumin but in different ratios: 3 fast to 1 slow, 2 fast to 2 slow or 1 fast to 3 slow (Beçak *et al.*, 1968; Beçak, 1969).

Parthenogenetic teiid lizards offer additional evidence of the effect of ploidy upon protein phenotypes. *Cnemidophorus exsangius* is the only animal for which three variant alleles are known (triploid heterozygosity). The electrophoretic pattern of adenosine deaminase of this triploid lizard shows 3 anodal bands, each apparently the expression of a different allele (Fig. 13, Ex; Neaves, 1969). *Cnemidophorous tesselatus* may be either diploid or triploid. Both forms are heterozygous at the heart-LDH locus for the same two alleles, so that electrophoretic patterns of homogenates of tissue in which only genes for heart type of LDH are expressed, show the same five band electrophoretic pattern (Fig. 13). The concentration of the five isozymes, however, varies in the two forms. For diploids, the proportion of the different isozymes approximates the binomial expansion; but for triploids their proportions reflect the double expression of the allele yielding the variant of slow mobility (Neaves and Gerald, 1969).

C. DEVELOPMENTAL AND PHYSIOLOGICAL CHANGES IN MOLECULAR PHENOTYPES

The herpetologists using molecular data in making taxonomic decisions must be aware that the genetic system is dynamic and that shifts in the activities of genes affect tissue composition throughout the life cycle of the organism (Manwell and Baker, 1966; Dessauer, 1968; Britten and Davidson, 1969). It is well documented that marked changes in the distribution of different proteins accompany development in certain species of Amphibia. For example, the heart-LDH of *Rana* persists from the egg to the adult in at least some tissues. Isozymes containing muscle-LDH subunits appear later, at the neurula stage, and persist in varied combinations with heart-LDH from then until hatching (Adams and Finnegan, 1965; Wright and Moyer, 1966; Kunz and Hearn, 1967; Chen, 1968). Only the maternal gene of the heart-LDH locus is active until the stage of the first heart beat (Wright and Moyer, 1966). Enzymes involved in pteridine synthesis become active concomitant with xanthophore differentiation (Bagnara and Obika, 1965).

Many dramatic metabolic changes take place at metamorphosis. These include the appearance of an albumin-like protein in the blood, the attainment of functional activity by urea-cycle enzymes, shifts in enzymes of pteridine synthesis in skin (Bagnara and Obika, 1965), a shift of eye pigments from porphyropsin to rhodopsin (Wald, 1960; Frieden, 1961; Herner and Frieden, 1960; Bennett and Frieden, 1962; Chen, 1968), and the change from a fetal to an adult type hemoglobin (Hahn, 1962; Manwell, 1966). Although there is far less evidence on the Reptilia, the indications are strong that similar developmental changes occur in their tissues and secretions (Kaplan, 1960; Minton, 1967).

The process of differentiation endows each tissue with a characteristic complement of enzymes and with specialized metabolic capabilities. In making decisions as to which tissue to analyze in studies of a particular protein, it is important to know those tissues in which activity is high and free from interfering compounds. For example (Fig. 8), the heart of many forms contains primarily heart type-LDH and skeletal muscle primarily muscle type-LDH, but the liver and many other organs contain complex mixtures of LDH isozymes in varied ratios. Some evidence on the tissue distribution of enzymes in amphibians and reptiles can be found in studies of G. W. Brown (1964), Chen (1968), Eppenberger *et al.* (1967), Lebherz and Rutter (1967), Holmes *et al.* (1968), Ma and Fisher (1968), Neaves and Gerald (1968).

The protein complement and metabolic potential of many tissues probably remain fairly constant in the adult organism, but some adjustments in activity do accompany different physiological states. The rate of synthesis of antibodies (Hildemann, 1962; Evans, 1963; Evans *et al.*, 1965), venoms (Kochva and Gans, 1967) and probably many other proteins is strongly temperature dependent. As part of an adaptive response to salinity changes, liver enzymes of the urea-cycle shift in distribution in anurans (McBean and Goldstein, 1967; Harpur, 1968). Seasonal metabolic cycles probably result from such adjustments (DiMaggio and Dessauer, 1963; Barwick and Bryant, 1966). During the reproductive season especially, a number of striking changes occur in the composition of the blood and liver as well as the reproductive organs. The complex yolk precursor, plasma vitellin appears in plasma of estrous females (Dessauer and Fox, 1959; Simkiss, 1967; Dessauer, 1970), and may complicate analysis of plasma proteins by electrophoresis or immunology. The liver, which synthesizes this protein, undergoes a series of marked structural changes and a decrease in glutamic dehydrogenase activity (Rosenquist, 1969).

D. SPECIES STRUCTURE

1. Introduction

The genome of the species exhibits a high degree of uniformity at the biochemical level. With our ability to detect variation, it is easy to de-emphasize the similarities between the proteins and metabolic systems of members of the same species. In most individuals, even those from populations spread over a wide range, many proteins are identical in structure (invariant proteins). Other proteins are polymorphic (variant), with certain variants frequently limited to populations from a specific geographic area. Most of the genes leading to variant forms of a protein are probably isoalleles. The amino acid sequences of their products, series of homologous proteins, are of almost identical structure. Generally, individual proteins of a series seem to differ

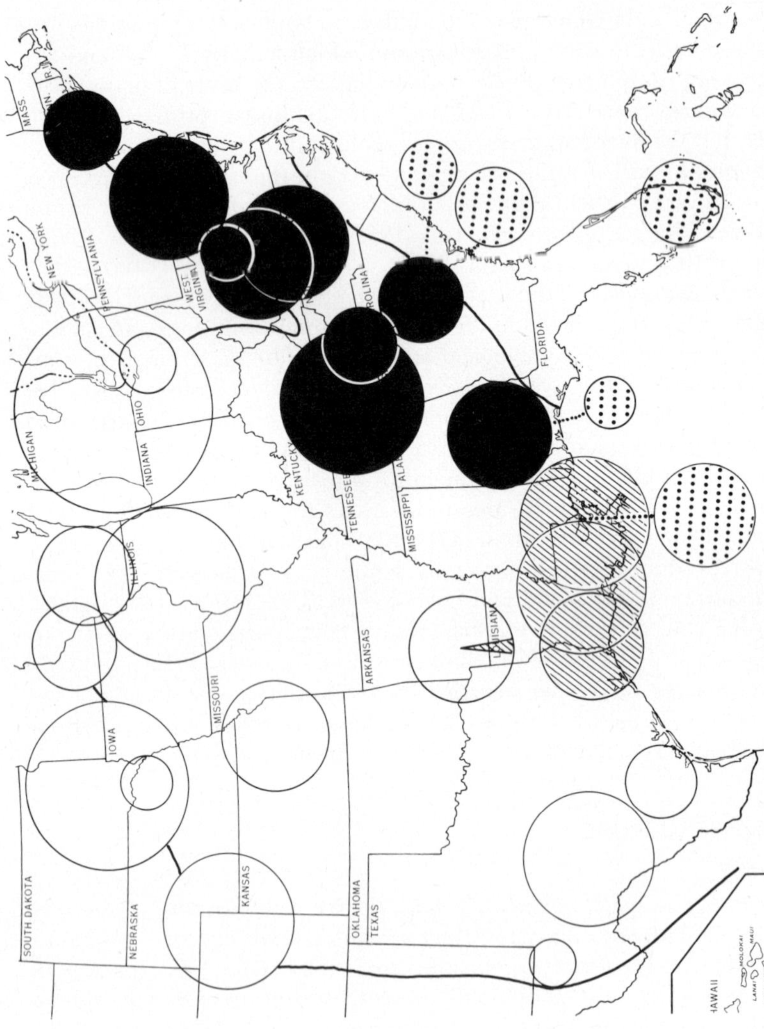

Fig. 9. Geographic distribution of globin polypeptide variants among populations of *Acris crepitans* and *A. gryllus*. Pie graphs are centered over collecting sites, except for those for certain populations of *A. gryllus*. White represents the frequency of polypeptides 1, 2, 3, 4 and 5; black indicates presence of 1s, 2s and 3s variants; lined graphs show the distribution of variant 5s (see Fig. 10); dots = *A. gryllus* patterns. Lines outline range of *A. crepitans*. (From Dessauer and Nevo, 1969).

from each other by only a small number of amino acid residues. Perhaps the uniform, more invariant proteins 'co-adapt' the species for high average fitness in a slowly, but inevitably, changing environment (Ehrlich and Raven, 1969), whereas the more variant proteins, controlled by genes more sensitive to environmental feed back, provide a 'fine adjustment' to the immediate environment.

Variant proteins and other evidence of diversity at the biochemical level have been found in populations of all species of amphibians and reptiles carefully examined. Protein polymorphism in esterases (Fig. 12; Dessauer *et al.*, 1962b; Coulson and Hernandez, 1964, p. 130; Shontz, 1968; Dessauer and Nevo, 1969) and transferrins (Fig. 17; Dessauer *et al.*, 1962a, 1962b; Gorman and Dessauer, 1965, 1966; Guttman, 1967, 1969; Dessauer and Nevo, 1969; Gorman *et al.*, 1971a) is the rule in most species. Other proteins known to occur in variant forms within the same species include hemoglobin (Fig. 10; Gillespie and Crenshaw, 1966; Manwell and Schlesinger, 1966; Newcomer, 1967; Sullivan and Riggs, 1967b; Shontz, 1968; Dessauer and Nevo, 1969), plasma albumin (Fig. 8; Zweig and Crenshaw, 1957; Dessauer and Fox, 1958; Beçak *et al.*, 1968), heart type lactic dehydrogenase (Wright and Moyer, 1966; Salthe, 1969; Salthe and Nevo, 1969), muscle type lactic dehydrogenase (Shontz, 1968), venom proteases and amino acid dehydrogenase (Fig. 17; Jimenez-Porras, 1963, 1964/1965a,b), crotoxin (Gonçalves and Vieira, 1950), leucine amino peptidase (Dessauer, 1966a; Pough and Dessauer, unpublished), glucokinase, malic dehydrogenase and alpha-glycerol-phosphate dehydrogenases (Dessauer and Nevo, 1969).

Studies of amphibians and reptiles at the species level (see the following sections), collectively emphasize the complexity of the biological species, showing that it is poorly characterized by the static '*type specimen*' concept. Since the activity of the genome of an organism varies with its stage of development and physiology, the molecular systematist must consider age and condition of each experimental animal. Further, as individuals of a population exhibit heritable differences in their protein complements, he must also recognize the population nature of the species (Ehrlich and Raven, 1969).

2. *Anura*

a. Acris. Population and species structure of the North American cricket frogs, *Acris crepitans* and *A. gryllus*, have been examined in terms of 21 different proteins and their variants (Dessauer and Nevo, 1969; Salthe and Nevo, 1969). *Acris crepitans* is distributed over central and southeastern United States from the Canadian border to northern Mexico; *A. gryllus* is restricted to Florida and the southeastern lowlands (Fig. 9).

Despite their extremely similar morphology, these two species of hylid frogs are easily distinguished on the basis of their blood proteins. They have different albumins, transferrins and hemoglobin polypeptides. These protein differences and the lack of frogs with protein patterns expected of hybrids in areas where the ranges of the two species overlap confirm the reproductive isolation of these sibling species.

Protein distribution emphasizes the essential genetic unity of the widespread species, *A. crepitans*. Patterns of eight different proteins, including plasma albumin, one of four hemoglobin polypeptides and liver lactic dehydrogenase, glutamic dehydrogenase and the glutamic-oxaloacetic transaminases, did not vary in animals from populations across the geographic range of the species. The polytypic nature of the species, with its divergence into Plains, Delta and Appalachian groups was suggested by the geographic distribution of specific variants of hemoglobin polypeptides (Figs 9 and 10),

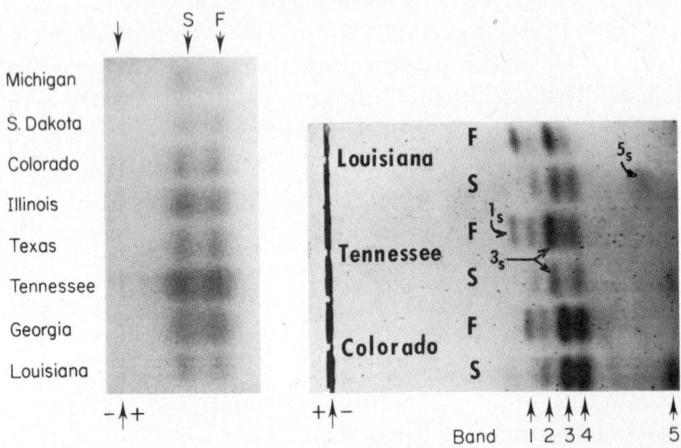

Fig. 10. Hemoglobins and hemoglobin substructures of *Acris crepitans*. Left: Hemoglobin electrophoresis of hemolysates of individual frogs in an alkaline buffer. F = hemoglobin of rapid mobility; S = hemoglobin of slow mobility. Right: Electrophoretic patterns of hemoglobin polypeptides separated in formate buffer, pH 1·8. Gel segments containing either the slow or fast hemoglobin were cut from the gel shown above and inserted into sample slots of a formate gel. Although the intact hemoglobins were electrophoretically identical in animals from many geographic areas, their polypeptide substructure varied (Dessauer and Nevo, 1969).

transferrins, esterases and heart LDHs. 'Within population' polymorphism was exhibited by transferrins, heart-LDH, malic dehydrogenases and three complexes of liver esterases. Protein variability was greatest in populations from the center of the range in southcentral and southeastern United States, suggesting that the genomes of individuals from that area were more heterozygous than frogs from more peripheral populations (Fig. 11).

Biochemical evidence supports the hypothesis that the two species of *Acris* diverged during a Pleistocene disjunction of proto-*Acris* populations in Florida and Mexican refuges. With recession of the glaciers, *A. crepitans* presumably moved from its Mexican refuge northward and northeastward to become adapted to both grassland and deciduous forest biomes and to attain contact and overlap with *A. gryllus* in southeastern United States (Dessauer and Nevo, 1969).

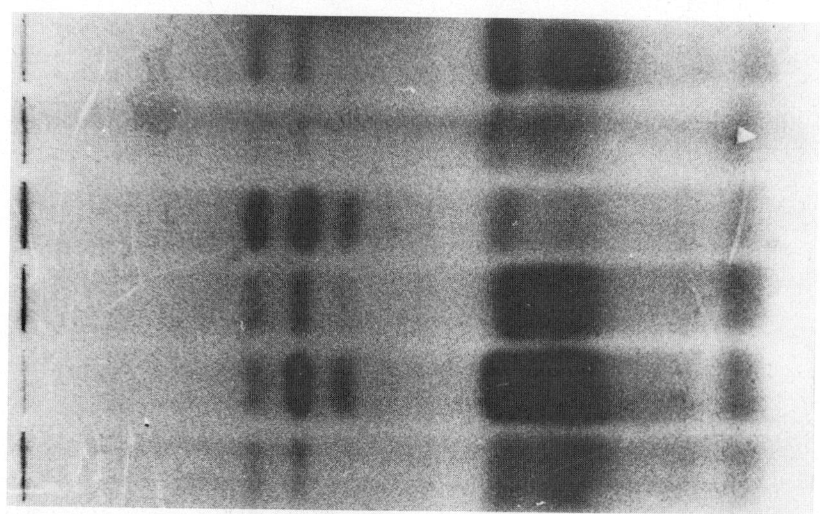

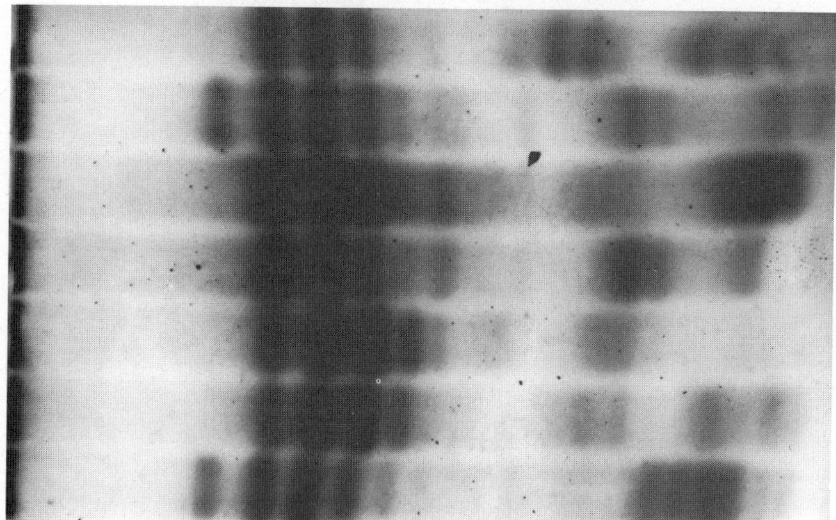

FIG. 11. Liver esterases of *Acris crepitans* from a central and a peripheral population. Top: Esterase patterns of liver homogenates of six frogs from Ann Arbor, Michigan, near the periphery of the range. Bottom: Patterns of liver homogenates of individuals from New Orleans, Louisiana, near the center of the range. Stains were developed using alpha naphthol acetate as the substrate for the esterases.

b. Rana. Electrophoretic and immunological studies of tissue proteins are clarifying our understanding of the *Rana pipiens* complex. Wells (1964) has presented a key to five named subspecies based upon electrophoretic patterns of their plasma proteins. Gillespie and Crenshaw (1966) have discovered hemoglobin variants in populations from the northern area of the range. Salthe (1969) has carried out an extensive study of lactic dehydrogenases of frogs from numerous populations in the United States, Canada, Mexico and

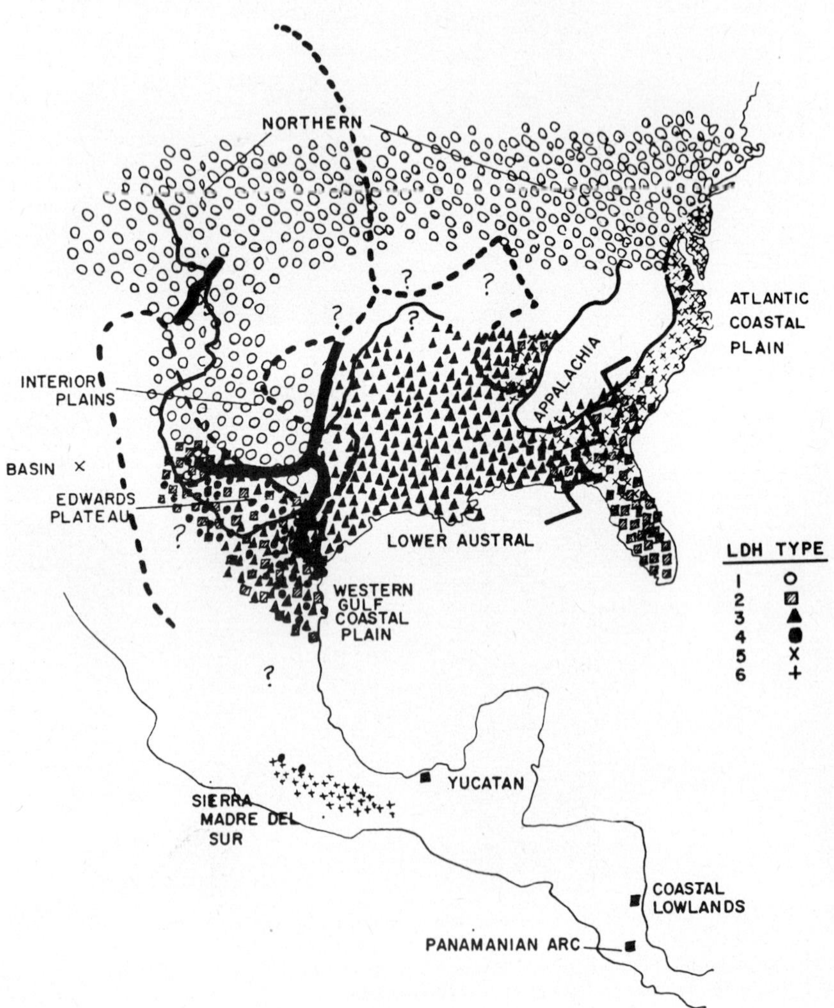

FIG. 12. Geographic distribution of heart-LDH variants of *Rana pipiens*. Dotted lines outline subspecies ranges (see text); the heavier solid lines delimit populations with different call note types. Reproduced with permission of Salthe (1969).

Central America. Muscle type LDHs did not vary in the many population samples; however 12 variant heart type LDHs were discovered. Eleven variants were electrophoretically distinct, one of these was found to include molecules of equal mobility but of different immunological and hybridization properties. In immunological titrations using antiserum against the heart-LDH of *R. pipiens* (from Wisconsin), immunological distances of the heart-LDHs of *R. catesbeiana, R. virgatipes, R. aurora* and *R. sylvatica* were no greater than IDs obtained for certain heart-LDH variants of *R. pipiens*.

The distribution of the seven most common alternative heart-LDHs (Salthe, 1969) coincided with some subspecific ranges and could be correlated with mating call distribution and data derived from laboratory hybridization studies (Fig. 12). Four population groups were defined by heart-LDH complexes: (Group-1) H-LDH-1 characterized populations extending across the northern tier of the range, dipping southward in the region of the interior plains; (Group-2) H-LDH-5 and the Florida 'immunological-variant' of H-LDH-2 were distributed in populations from the Atlantic Coastal Plains; (Group-3) H-LDH-3 was concentrated in populations of the Lower Austral region of south central United States; (Group-4) H-LDH-4 and the Texas 'immunological-variant' of H-LDH-2 occurred in high frequency in populations of the Western Gulf Coastal Plain and the Edwards Plateau. Hemoglobin data (Gillespie and Crenshaw, 1966), though less thoroughly examined, are concordant with the LDH evidence. Frogs of a Wisconsin and a Maryland population of Group-1 were characterized by hemoglobin variant Hb-Is, whereas populations of Group-2 from the coastal plain in Maryland were characterized by the variant Hb-If.

The protein evidence is compatible with the suggestion that *Rana pipiens* is two species, a northern complex of populations and a southern complex. The northern form (Group-1 populations) would include *R. p. pipiens* and *R. p. brachycephala*. The southern species would include *R. p. berlandieri* (Group-4 populations) and *R. p. sphenocephala* is further differentiated into two biochemically distinct subspecies: one composed of Group-3 populations of southcentral United States and the other, of Group-2 populations of the Atlantic Coastal Plain (Salthe, 1969).

Electrophoresis of plasma and tissue proteins have been used to distinguish many species of *Rana* from Europe and from North America (Lanza and Antonini, 1955; Paulov and Kmeťová, 1964; Hebard, 1964; Manwell, 1966; Chen, 1967; Salthe, 1969). The process could be very useful in identifying tadpoles, particularly those of newly discovered species of frogs. Although many changes in the qualitative distribution of proteins accompanies development (see Section III C), the pattern of proteins of the tadpole is probably 'species specific' (Chen, 1968). A number of proteins, e.g. LDH isozymes (Salthe *et al.*, 1965) and soluble muscle proteins (Manwell, 1966), have identical electrophoretic properties in both the tadpole and adult stages. For example, the tadpoles of *R. catesbeiana*, probably, can be easily identified and distinguished from tadpoles of *R. clamitans* on the basis of LDH patterns alone (Manwell, 1966).

c. Bufo. This world wide genus occurs on all major land masses inhabited by amphibians, except Madagascar, New Zealand and Australia. Its diversification is great, especially in Africa and in the Americas (Blair, 1963). Cei *et al.* (1968) have presented a phylogenetic arrangement of species groups of American toads, that differs from that of Blair (1963) in that it incorporates biochemical and immunological data.

The genus *Bufo* apparently originated in South America and has evolved along two major phyletic lines: a Northern-line of cold adapted toads with narrow skulls and a Southern-line of warm adapted toads with broad skulls (Blair, 1972). Cross reactions are relatively weak in immunological titrations which involve antisera to proteins of toads of one line and antigens from toads of the other line (Cei and Cohen, in press). Toads of the Northern-line, especially those of the *boreas*-group, show affinities to the *calamita*-group of Europe in components of the skin (Cei *et al.*, 1968) and in blood proteins (Guttman, 1972). Immunological correspondence of plasma proteins and patterns of skin amines (Cei *et al.*, 1968) show that the *spinulosus*-group of the Andes is of the Northern-line. Toads of the Southern-line and the *regularis*-group of Africa show closest affinities to the *melanostictus*-like stock of southern Asia. The skin of these toads characteristically contains moderate to high concentrations of 5-hydroxytryptamine.

Metabolites of the skin and proteins of the blood are useful in assigning toads to species-groups. The skin of members of each group possess a fairly unique spectrum of indole alkylamines and other metabolites (Hunsaker *et al.*, 1961; Wittliff, 1962, 1964; Porter, 1964; Cei and Erspamer, in press; Cei *et al.*, 1968; Low, 1968). Immune cross reactions are high between proteins of toads of the same species-group (Cei and Cohen, in press). Their plasma protein patterns (Cei and Cohen, 1963; Brown, L. E. 1964; Hebard, 1964) and the electrophoretic mobilities of their transferrins and hemoglobins are usually very similar (Fox *et al.*, 1961; Marchlewski-Koj, 1963; Guttman, 1967; Brown and Guttman, 1970; Guttman, 1972). For example, species included in the *americanus*-group show a great number of biochemical similarities. Hunsaker and colleagues (1961) and Wittliff (1962, 1964) found that secretions of the paratoid glands of species of the group have many common components. Using paper electrophoresis, Brown, L. E. (1964) could not distinguish the plasma protein patterns of *B. americanus* from those of *B. woodhousei.* Guttman (1969) using gel-electrophoresis found thirteen electrophoretically distinct transferrins and ten distinct hemoglobins in six species of the group. Many of these electrophoretic variants were common to a number of members of the species group. Although proteins of identical electrophoretic mobility are not necessarily of identical structure (see Fig. 10; Dessauer and Fox, 1964) the possession of such proteins, when of common occurrence, is suggestive of a close relationship of the species involved.

Protein polymorphism is very common in many species of *Bufo* (Cei, 1959; Guttman, 1972). In a recent survey, Guttman (1972) found alternative forms of hemoglobin in populations of 19 of 37 species and variant transferrins in 30 of 46 species. Polymorphism was especially extensive in the *americanus*-group of North America (Guttman, 1969) and in the *regularis*-group of Africa (Guttman, 1967). Within the *americanus*-group, a population sample of 16 *B. microscaphus* from Virgin, Utah, possessed 6 molecular types of transferrins distributed in 8 phenotypes. In a population of *B. regularis* from El Mohalla el Kubra in Egypt 11 hemoglobins were distributed in 8 phenotypes in a population sample of 13 specimens.

Interspecific hybridization probably contributes to this polymorphism. Natural hybridization is common among toads and species exhibit all degrees of genetic compatibility (Blair, 1963). If the mating yields fertile offspring, the hybridization can result in gene exchange between species. For example, *Bufo regularis* and *B. rangeri* hybridize in an area of sympatry near Port St. John's, South Africa. In that area transferrin and hemoglobin variants, typical of one species are common in the other (Guttman, 1967). Minor transfers of genetic material, however, may occur even between members of divergent phyletic lines. For example, *B. valliceps* of the Southern-line and *B. fowleri* of the Northern-line hybridize in areas of sympatry along the Gulf Coast of the United States. Only male hybrids of the cross of *B. fowleri* (♀) and *B. valliceps* (♂) survive to maturity. The testes of these hybrids are atrophic and viable sperm can be found only by careful examination (Volpe, 1960 and personal communication). Although such evidence suggests that the hybrids have a low reproductive potential, two rare backcross individuals were detected on the basis of their blood proteins (Fox *et al.*, 1961; Dessauer and Fox, 1964) and paratoid gland components (Witliff, 1964).

Guttman (1972) suggests that much of the interpopulation protein polymorphism may be only apparent, reflecting deficiencies in taxonomy. Morphologically distinct species can generally be clearly distinguished by biochemical or immunological means (Bertini and Cei, 1959, 1962; Fox *et al.*, 1961; Bertini and Gustavo, 1962; Wittliff, 1962, 1964; Michl and Kaiser, 1962/1963; Porter, 1964; Cei and Erspamer, 1966; Erspamer *et al.*, 1967; Chen, 1967; Low, 1968; Guttman, 1972). Patterns of skin metabolites (Cei and Erspamer, 1966; Cei *et al.*, 1968) and electrophoretically distinct transferrins distinguish *Bufo spinulosus* of Chili, Argentina and Peru, suggesting that these forms may be distinct species (Guttman, 1972). Biochemical data also support species status for the Peruvian toad, *Bufo trifolium* which is often considered a subspecies of *B. spinulosus* (Cei *et al.*, 1968). Similarly protein electrophoretic patterns (Guttman, 1967) and skin constituents (Low, 1968) suggest that the *B. regularis* complex of Africa may include two species.

Abramoff *et al.* (1964) have utilized intraspecific differences in plasma proteins to estimate paths taken by *B. americanus* in populating islands of Lake Michigan. Plasma proteins of population samples from five islands and five mainland sites on adjacent shorelines were compared. The majority of the proteins were similar in all populations, but at least three components varied geographically and were useful in estimating the affinities of the different populations. The mainland populations on opposite sides of Lake Michigan showed definite signs of divergence. Toads from the Wisconsin shoreline had antigens absent in toads of the eastern mainland in Michigan. Toads of High Island, showed more affinity to the upper peninsula population of the western mainland whereas those of the other islands were more similar to Michigan populations. The counterclockwise swirl of surface currents around the islands and the proximity of the islands to eastern mainland are probably the major factors responsible for the affinities observed. High Island is closest to, and thus more apt to be colonized by toads from the upper peninsula. Furthermore, the counterclockwise swirl of surface currents around the islands tend to favor dispersal from Garden and Beaver Islands toward High Island, whereas material from High Island would more likely be conveyed toward open water, bypassing the other islands. Animals of mainland populations exhibited a greater variety of proteins than island populations.

d. Leptodactylus, *Odontophrynus* and *Lepidobatrachus*. Species of *Leptodactylus* exhibit highly varied complexes of skin biogenic amines and polypeptides. The distribution of specific components in different species of *Leptodactylus* corresponds closely with Lutz's (1931) arrangement of the genus. Species of his Cavicola-section completely lack biologically active amines and polypeptides. Species of his Patchypus-section fall into two groups: the skin of members of one group (e.g. *L. pentadactylus* and *L. laticeps*) contains an extraordinary variety of indole and imidazole amines, low concentrations of phenylalkylamines and high concentrations of caerulin-like polypeptides; skin of members of the other group (*L. ocellarus* and *L. bolivianus*) contains large amounts of leptodactyline (Fig. 3). Within the Patchypus-section the distribution of pattern types parallels morphological divergence. Skin of members of the Platymantis-section contains high concentrations of both leptodactyline and a number of indolealkylamines (Cei *et al.*, 1967).

Electrophoretic patterns of plasma proteins (Cei and Bertini, 1960, 1961) and patterns of skin amines (Cei *et al.*, 1967) were useful in distinguishing the sibling species *L. chaquensis* and *L. ocellatus* in areas of sympatry. Plasma protein patterns also served to distinguish the two ceratophid species *Odontophrynus cultripes* and *O. americus*. A population of the tetraploid species, *O. americus*, was polymorphic for albumin (Fig. 8). Components of fast and of slow mobility were distributed in five phenotypes of one or two bands of varying concentrations. DNA analysis served to determine the ploidy of species of this genus in which polyploidy is common (Beçak *et al.*, 1968; Beçak, 1969). Plasma proteins of *Lepidobatrachus asper*, *L. laevis* and *L. ilanensis* were distinct by both electrophoretic and immunological criteria. Quantitative immunological evidence indicated that these three species are about equally divergent from each other (Cei, 1969).

3. Caudata

Biochemical evidence supports the contention that interspecific hybridization is implicated in the origin of parthenogenic species (Macgregor and Uzzell, 1964; Uzzell and Goldblatt, 1967). Two members of the *Ambystoma jeffersonium* complex of salamanders, *A. platineum* and *A. tremblayi*, are parthenogenic and triploid. The remaining two members of the complex, *A. jeffersonium* and *A. laterale*, breed sexually, are diploid and have electrophoretically different albumins. Both triploid species exhibit the *A. jeffersonium* and *A. laterale* albumin variants but in different concentrations. In *A. platineum* the *jeffersonium*-variant predominates over the *laterale*-variant, whereas in *A. tremblayi* the *laterale*-variant predominates.

This evidence, combined with data on morphology, geographic distribution and mating behavior, suggests that both salamanders arose as hybrids of the diploid species. The concentration of the albumin bands probably reflects gene dosage. Both albumins are present in equal concentrations in plasma of laboratory hybrids between the diploid species. The variable density of the two albumin bands in the natural hybrids indicate that *A. platineum* has a double dose of *jeffersonium*-genes and a single dose of *laterale*-genes; whereas, *A. trembalayi* probably has a double dose of *laterale*-genes and a single dose of *jeffersonium*-genes (Uzzell and Goldblatt, 1967).

Study of the electrophoretic properties of tissue proteins offers a much needed additional means of characterizing salamanders (Guttman, 1965). Patterns of malic, lactic and alcohol dehydrogenase distinguish three western species, *Taricha torosa*, *T. rivularis* and *T. granulosa* (Salthe and Kitto, 1966). Plasma protein patterns offer additional criteria for distinguishing these species as well as for identifying their hybrids (Coates, 1967). Chen (1967) used protein patterns to characterize *Triturus alpestis*, *T. cristatus*, *T. vulgaris* and *T. helvitius* captured near Zurich, Switzerland. Females of *T. vulgaris* and *T. helvitius* are difficult to distinguish morphologically. Kiortsis and Kiortsis (1960 and personal communication) used plasma protein patterns to demonstrate the genetic affinities of the natural hybrid *Triturus blasii* to its parental species. Shontz (1968), however, was only partially successful in identifying four species of *Desmognathus* of North America using hemoglobin, lactic dehydrogenase and esterase phenotypes.

Intraspecific variants of proteins have been detected in a number of urodeles. Three of the ten classes of bands, which characterize electrophoretic patterns of plasma proteins of *Taricha*, vary within single species. Distinct qualitative differences were found between subspecies, between populations from different geographic areas of the same subspecies and, in some cases, between individuals of the same population (Coates, 1967). A population of ringed salamanders, *Ambystoma annulata*, from Fayetteville, Arkansas includes members with three electrophoretically different hemoglobin phenotypes of one or two bands. Small population samples of this species from several sites in Oklahoma and Missouri exhibited only the slow variant (Newcomer, 1967). Two muscle-LDH variants were found in populations of *Desmognathus fuscus* in southeastern United States. The esterase phenotypes of the liver homogenates of individuals of species of *Desmognathus* from that area were highly variable (Shontz, 1968).

4. Chelonia

Electrophoretic patterns of plasma proteins distinguish species of slider turtles, *Pseudemys* (Zweig and Crenshaw, 1957) and indicate the occurrence of the hybridization of certain species. Crenshaw (1965) discovered a hybrid swarm of turtles in an area of North Carolina where the ranges of *P. floridana* and *P. rubriventris* overlap. Variations in albumins of specimens in this population indicated that gene exchange was taking place between these species, which are reproductively isolated over most of their ranges. Intraspecific variation in blood proteins have been demonstrated for other genera (Crenshaw, 1962; Frair, 1964; Masat and Musacchia, 1965).

Observations of Frair (1963) suggest that blood typing methods could be useful in studies of processes of speciation involving turtles. He was able to distinguish individuals and to characterize populations by differences in the response of red cells to a variety of agglutinins. Intraspecific variation was great in some cases. For example, each of 24 tortoises, *Testudo hermanii*, could be identified by this technique. Geographic differences of a genetic nature were evident. For example, serum from individuals of a northern New Jersey population of painted turtles, *Chrysemys picta*, agglutinated cells of turtles from the more southern population. If such incompatibilities lead to developmental problems, they could be an important cause of reproductive isolation.

Hemoglobin polymorphism also exists within the painted turtle, *Chrysemys picta*. A D-variant differed from an S-variant both in electrophoretic mobility and in oxygen affinity. The D-variant predominated in populations of the southern subspecies, *C. p. dorsalis*, and the S-variant predominated in populations from the northcentral United States of the subspecies, *C. p. belli* and *C. p. marginata* (Manwell and Schlesinger, 1966). In addition, variation in electrophoretic patterns of plasma proteins has been observed in individual *C. picta* (Masat and Musacchia, 1965).

5. Sauria

a. Cnemidophorus. Parthenogenetic, all female species are of common occurrence in the North American genus of teiid lizard. Electrophoretic patterns of tissue proteins serve to distinguish species and support the theory that the parthenogenetic forms originated as hybrids of bisexual species. Further, the presence of particular protein variants in tissues of each parthenogenome suggests the identity of the bisexual forms that may have been

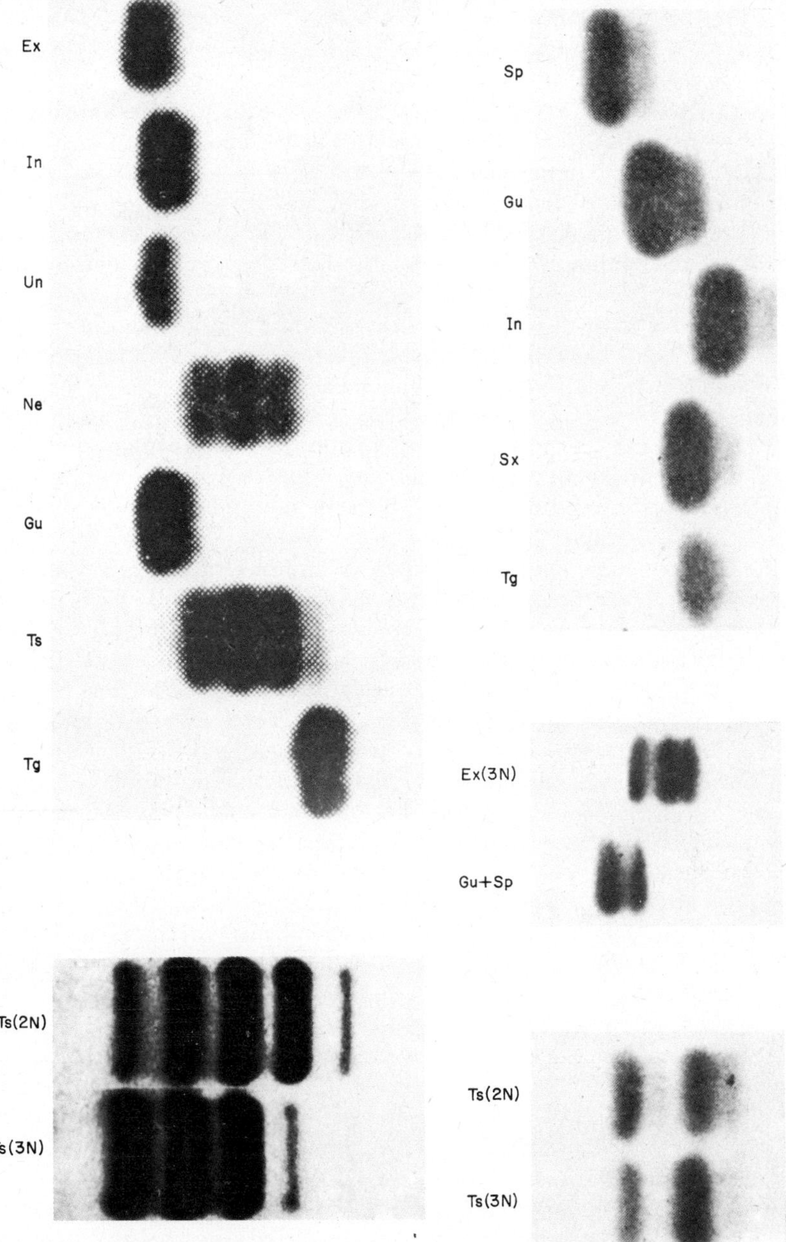

FIG. 13. Tissue enzymes and the origin of parthenogenetic lizards (*Cnemidophorus*). Electrophoretic phenotypes of heart type LDHs are shown on the left and those of the adenosine deaminases appear on the right; anode is to the right in each case. Abbreviations are as follows: Ex = *C. exsanguis*, In = *C. inornatus*, Un = *C. uniparens*, Ne = *C. neomexicanum*, Gu = *C. gularis*, Ts = *C. tesselatus*, Tg = *C. tigris*, Sp = *C. septemvittatus*, Sx = *C. sexlineatus*. Reproduced from Neaves and Gerald (1968, 1969) and from Neaves (1969) with permission.

involved in the hybridizations (Neaves and Gerald, 1968, 1969; Neaves, 1969).

The five bisexual species whose tissues were examined were easily distinguished on the basis of their heart-LDHs (Fig. 13, left), adenosine deaminases (ADA) (Fig. 13, right) and phosphogluconate dehydrogenases (PGD). Four different ADA phenotypes were found; only those of *C. tigris* (Tg) and *C. sexlineatus* (Sx) were of identical mobility. The unique heart-LDH of *C. tigris* and the unique PDG of *C. sexlineatus*, however, served to identify these species.

The tissues of parthenogenetic species contained combinations of many of the same variants. The evidence obtained on *C. tesselatus* (Ts) illustrates how such findings enlarge our understanding of parthenogenetic species (Fig. 13). Animals classified as *C. tesselatus* may be either diploid (2N) or triploid (3N). ADA patterns of tissues from both diploid and triploid individuals have two major anodal bands, one with the mobility of the variant found in bisexual *C. septemvittatus* (Sp) and the other with the mobility of the variant found in both *C. tigris* (Tg) and *C. sexlineatus* (Sx). The enzyme activity of the two variants is about equal in diploid individuals but the *C. tigris*-like variant is of higher activity in the triploid form. Heart-LDH patterns exhibit the five bands typical of a heterozygous genotype, but the proportion of the different isozymes reflects the number of the different alleles. Both diploid and triploid *C. tesselatus* appear to have one *C. tigris*-like heart-LDH allele. The patterns show that it represents one half of total heart-LDH activity in the diploid and one third of total heart-LDH activity in the triploid form (Fig. 13, bottom left). The PGD pattern of the diploid is the same as that of all bisexual species tested except *C. sexlineatus* (Sx); the PGD pattern of the triploid, however, exhibits both known variants, the common type and the unique *C. sexlineatus* variant. Collectively, phenotypes of the three enzymes suggest that the diploid *C. tesselatus* arose as a hybrid of *C. tigris* and *C. septemvittatus* and that the triploid arose from combinations of haploid complements from each of these species and *C. sexlineatus*. Evidence on the same enzymes also suggests that parthenogenetic *C. neomexicanus* (Ne) is a hybrid of *C. tigris* and *C. inornatus* (In) and that the triploid parthenogenome *C. exsanguis* (Ex) arose from the combination of haploid contributions from *C. gularis* (Gu), *C. inornatus* and some other species.

Biochemical studies of *Cnemidophorus tigris* from an area in southeastern Arizona, where two previously isolated forms, *C. t. marmoratus* and *C. t. gracilis*, had renewed contact (Dessauer *et al.*, 1962b), substantiated a decision of Zweifel (1962) to maintain them as subspecies rather than give them specific status. Twelve of sixteen protein fractions of the electrophoretic patterns of their plasma proteins as well as the patterns of their hemoglobins were identical in animals of the both 'pure' populations and specimens from the area of introgression. Four pairs of plasma proteins were variable in different individuals. One protein member of each pair was either restricted to, or of higher

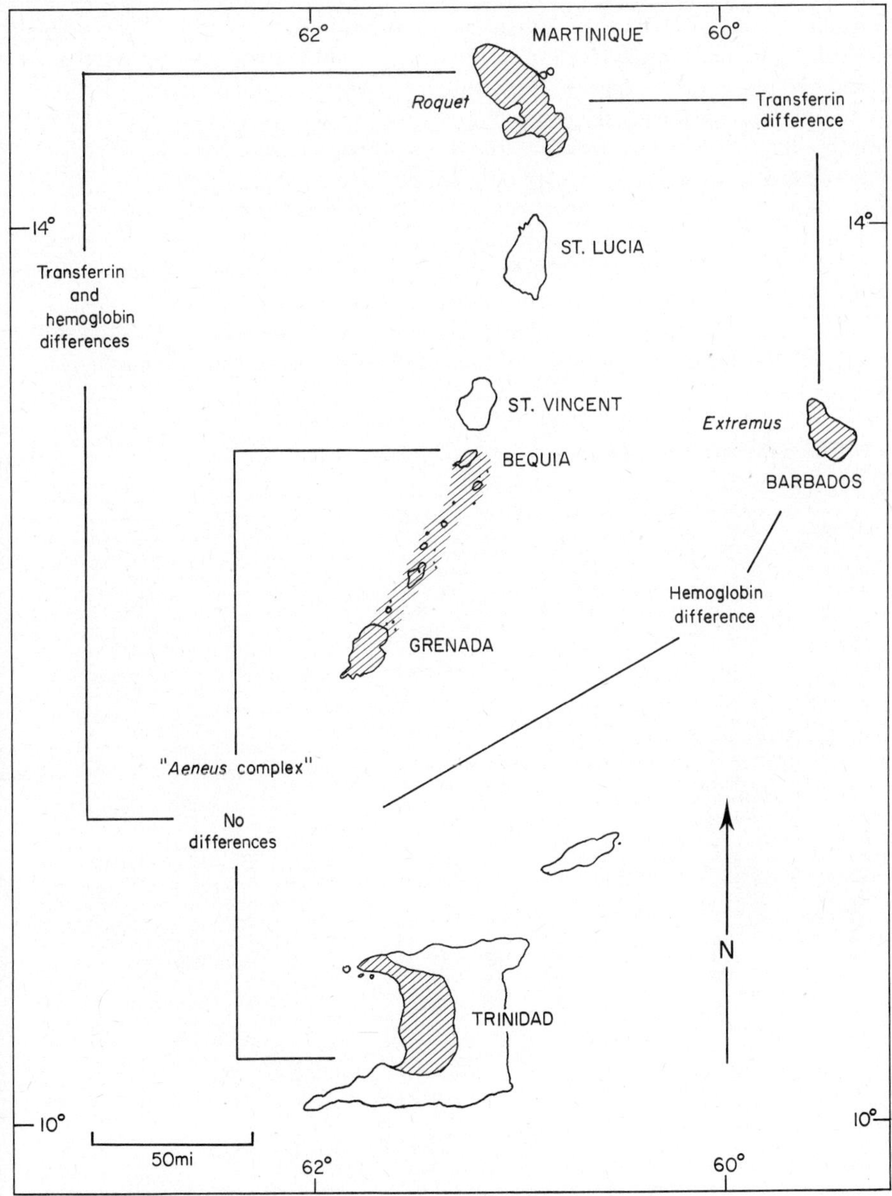

FIG. 14. Hemoglobin and transferrin distribution among Carribean island populations of *Anolis roquet* and *Anolis aeneus*. (From Gorman and Dessauer, 1965.)

frequency, in one subspecies than in the other. For example, different transferrins were found in each subspecies, but one esterase variant was more frequent in one subspecies than in the other. Hybrid indices were calculated for each lizard on the basis of its complement of proteins, and average hybrid indices depicted the distribution of *gracilis*-type and *marmoratus*-type proteins in population samples from the area of introgression. Biochemical identifications correlated well with morphological identifications of animals away from the area of contact but were less close for animals in the zone of contact. Although the transition between the 'pure' subspecies was a relatively narrow 'suture-zone' (Remington, 1968) protein distribution revealed that it was broader than expected from morphological evidence.

b. Anolis. Members of this genus occur in enormous numbers in a great variety of forms throughout the islands of the Carribean Sea and on adjacent

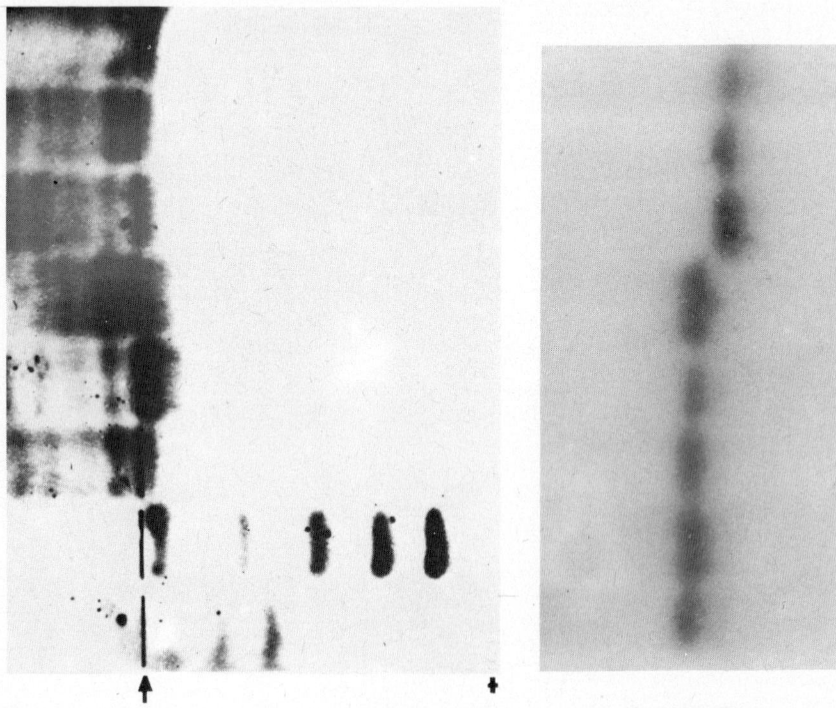

FIG. 15. Grouping taxa with electrophoretic data. Left: Lactic dehydrogenase patterns of red cells of different species of *Anolis*. The six topmost patterns characterize six species of the *roquet*-group; the bottom patterns of species of other species groups (from Gorman and Dessauer, 1966). Right: Transferrins of subspecies of west coast garter snakes. The three top patterns characterize subspecies of terrestrial ecology (*Thamnophis elegans*); the five lower patterns characterize subspecies of aquatic ecology (*T. couchi*). Anode is to the right on both patterns (Dessauer *et al.*, 1962a).

mainland areas. Many forms have differentiated into distinct species; others represent populations in intermediate stages of species formation, posing formidable taxonomic problems. 'It is more evident in this genus than in others that we cannot become prisoners of our conventional museum techniques' (Williams, E. E. quoted by Gorman and Atkins, 1969; see also Williams, 1969).

Supporting their suggestion with a variety of evidence, including such unconventional criteria as the electrophoretic behavior of tissue proteins, Gorman and Atkins (1969) postulate that *Anolis* from South America colonized the islands of the Carribean in two major movements. One to the Greater Antilles, gave rise to species groups of those islands and to the *bimaculatus*-group of the northern Lesser Antilles. The second movement was to the southern Lesser Antilles and gave rise to the *roquet*-group. These authors have also proposed a phylogeny of the species groups which evolved from the first movement, suggesting a course of inter-island migrations to account for differentiation of members of the *roquet*-group.

The pattern of distribution of proteins among anoles of the *roquet*-group illustrates how electrophoretic evidence can help unravel the systematics of closely related, insular organisms (Fig. 14). The lactic dehydrogenases of species of this group have uniquely slower mobilities in alkaline buffers (Fig. 15) than do those of *Anolis* of other species groups (Gorman and Dessauer, 1966; Gorman and Atkins, 1969; McLaughlin and Conde del Pino, 1968). Hemoglobins are of intermediate variability, they usually differ from species to species but rarely within the same species. Transferrins are most variable; populations of close relatives, occupying different island banks have different transferrins. For example, *Anolis richardi griseus* from St. Vincent has a different transferrin than *A. r. richardi* from Bequia which is only 11·2 km to the south. *Anolis aeneus* of three islands and mainland South America were classified into three named forms until blood protein patterns proved their identity. *A. aeneus* is endemic to islands of the Grenada Bank. Apparently, the populations on Trinidad and the mainland have been introduced (Gorman and Dessauer, 1965, 1966).

In port cities of Trinidad, *Anolis aeneus* is sympatric with another introduced species of the *roquet*-group, *A. trinitatus*, setting up a natural test of the reproductive isolation of the two forms. In areas where populations of the species come into contact, they hybridize. Electrophoresis of blood proteins served to describe the structure of the hybridizing population. Species specific albumins, hemoglobins (Fig. 8) and transferrins characterized the parental species. Blood of the hybrids contained two hemoglobins and two albumins, which had mobilities identical to the 'species specific' variants characteristic of the parental forms. The distribution of protein variants throughout the hybridizing populations proved that backcrosses were rare, if they occurred at all. Molecular data correlated beautifully with evidence on color pattern, scalation, karyotypes, behavioral and ecological evidence and with experimental studies on reproductive fitness (Gorman and Dessauer, 1966; Gorman *et al.*, 1971a). Electrophoretic patterns of plasma proteins have also served to distinguish certain species of *Anolis* of Puerto Rico (Maldonado and Ortiz, 1966).

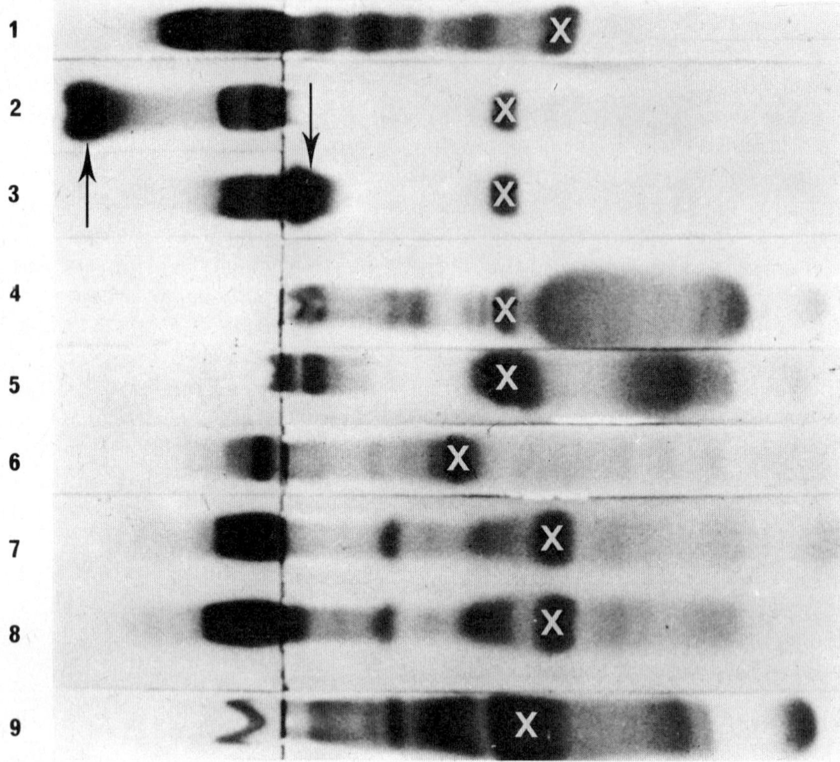

Fig. 16. Species differences in starch-gel electrophoretic patterns of venom proteins of viperid snakes of the genus *Bothrops*.1. *B. atrox*, 2. *B. nummifer* (Pacific), 3. *B. nummifer* (Atlantic), 4. *B. picadoi*, 5. *B. nasuta*, 6. *B. godmani*, 7. *B. schlegeli*, 8. *B. schlegeli*, 9. *B. laterals*. X = position of L-amino acid dehydrogenases; arrows point to protease variants of *B. nummifer*; anode is to the right. Reproduced with permission of Jimenez-Porras (1963).

6. *Serpentes*

Electrophoretic and immunological properties of blood and venom proteins have furnished evidence on the affinities of snakes of different populations and on the relationships of species within certain genera. Snake venoms, which are complex mixtures of a variety of proteins, have been used in comparative studies since the discovery of the species specificity of immunological reactions (Lamb, 1902). Many comparative studies at the protein level demonstrate the importance of geographic factors in the speciation of snakes.

Venom proteins, in particular, are so variable that venomologists emphasize the importance of preparing regional types of antivenins (Gonçalves and Vieira, 1950; Gonçalves, 1956; Schenberg, 1959; Jimenez-Porras, 1963, 1964/1965a,b).

a. Bothrops. Biochemical and immunological studies on venoms of *Bothrops*, a crotalid genus found throughout South and Central America, illustrate the marked variability of venom proteins. Species of *Bothrops* are usually distinguishable by electrophoretic patterns of these proteins (Fig. 16; Jimenez-Porras, 1963; Drujan *et al.*, 1963); even species of markedly similar morphology, such as *B. nummifer* and *B. picadoi*, are easily differentiated (Jimenez-Porras, 1963, 1967).

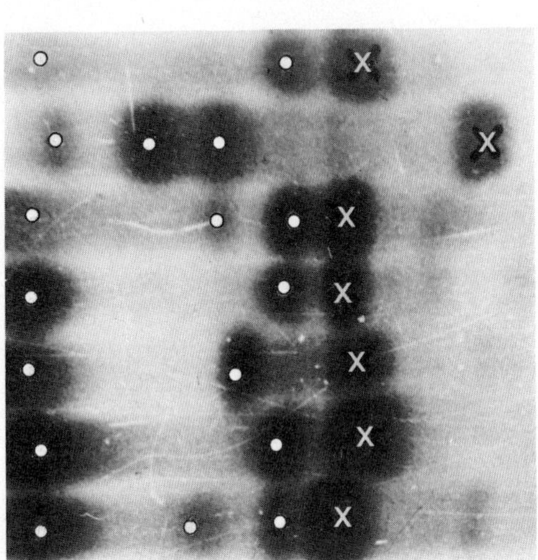

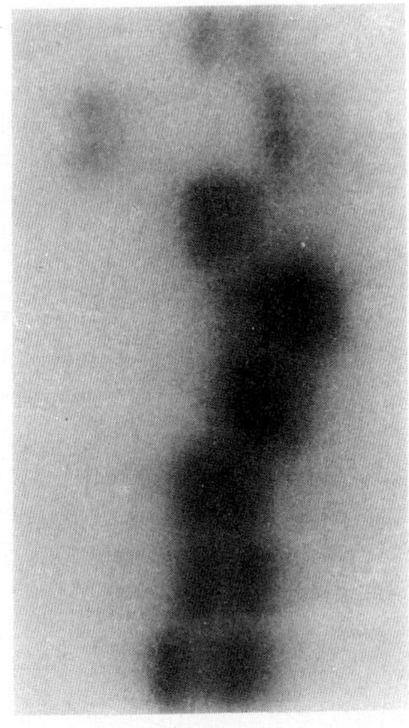

FIG. 17. Individual variation in electrophoretic properties of proteins of snakes. Left: L-amino acid dehydrogenases (X) and proteases (o) in venom samples from eight fer-de-lances (*Bothrops atrox*) captured in Golfito, Costa Rica, on the Pacific side of the central mountain range. Reproduced with permission from Jimenez-Porras (1964/1965a). Right: Transferrins of plasma of eight racers (*Coluber constrictor*). Snakes whose plasma gave the lower six patterns were captured in a region of small area in northern Indiana, USA. (From Dessauer *et al.*, 1962a.)

Topography probably has had a strong influence on the speciation of *Bothrops* in the Central American nation of Costa Rica. High mountains separate the tropical rain forest of its Caribbean coast from the tropical rain forests and tropical savannahs of the Pacific coast. *Bothrops nummifer*, the jumping viper, is found on both sides of the central range. Electrophoretic patterns of its venom proteins are very similar for individuals from all populations. However, *Bothrops nummifer* from the Caribbean side of the central range (Fig. 16) have an electrophoretically different protease than those from the

Pacific side. This protease apparently is the factor most responsible for the deadly effects of the venom of this snake (Jimenez-Porras, 1964/1965b, 1967).

The fer-de-lance, *B. atrox*, has a wide distribution in South and Central America. The electrophoretic variations in venom proteins were so extensive that individuals from Costa Rica could be characterized biochemically (Fig. 17). Variations in the frequency of the different proteases (dots, Fig. 17) and L-amino acid dehydrogenases (Xs, Fig. 17), however, distinguished Atlantic and Pacific coast populations. Four proteases were found in venoms of most snakes of the Atlantic coastal region but only two were common in venoms of snakes of the Pacific Coast. Two variant L-amino acid dehydrogenases occurred in three phenotypes of one or two bands in venoms of different individuals. The estimated gene frequency of the fast variant was high (94%) in Atlantic populations, whereas the slow variant was more frequent in Pacific populations (83%) (Jimenez-Porras, 1964/1965a).

Six venom antigens varied in populations of *Bothrops neuwiedi* from southeastern Brazil (Schenberg, 1962/1963). Twenty combinations of antigens were identified on immunodiffusion patterns of venom proteins of different individuals. The population distribution of the 20 types varied geographically. The protein composition of the venom of an individual remained constant from 'milking to milking'; feeding, time in captivity and season did not alter the immunodiffusion pattern.

b. Crotalus and other viperid genera. Qualitative differences in plasma and venom proteins also distinguish species of other viperid genera (Deutsch and McShan, 1949; Fine *et al.*, 1954; Cohen, 1955; Uriel *et al.*, 1957; Strauss and Grunbaum, 1961; Bertke *et al.*, 1966/1967; Voris, 1967; Johnson, 1968/1969) and in addition furnish further examples of intraspecific differentiation (Gonçalves, 1956; Drujan *et al.*, 1963; Leviton *et al.*, 1964). Crotamine is a strongly basic, toxic polypeptide the presence of which is readily detected in venom by electrophoresis or toxicity tests. Gonçalves (1956) has designated a biochemical subspecies of the South American rattlesnake, *C. durissus crotaminicus* (*C. terrificus crotaminicus*) on the basis of its presence. Certain regions seem to have only one type of snake, whereas other regions have both. However, the distribution of the crotamine positive and negative types shows no geographic continuity over long distances but appears to vary from population to population in an almost mosaic pattern (Schenberg, 1959).

c. Naja and other elapids. Electrophoretic and immunological evidence suggests that the common cobra, *Naja naja*, of southeast Asia comprises a number of divergent forms. Although the venom proteins of *N. n. atra* of Taiwan exhibit many similarities to those of other species of *Naja*, they show a number of important differences. At least one toxic protein of the Taiwan subspecies is not completely neutralized by antivenins produced with venoms of cobras from Thailand, Iran, Pakistan and Africa. Similarly, antivenins produced with the venom of the Taiwan cobra fail to neutralize all toxic proteins of venoms of cobras from Thailand and Iran. The major toxic component of the Taiwan cobra venom also differs electrophoretically from that of cobras from India and West Pakistan (Minton, 1967/1968; Tu and Ganthavorn, 1967/1968). Species of *Naja* and other elapids have been distinguished by immunological (Kuwajima, 1953; Kellaway and Williams, 1931; Christensen and Anderson, 1967), electrophoretic (Jouannet, 1968; Fisher and Kabara, 1967) and thin layer chromatographic (Fisher and Kabara, 1967; Kabara and Miceli, 1967/1968) properties of their venom proteins and polypeptides.

d. Natrix. This highly successful genus, with a world wide distribution, included an unwieldly and confusing assemblage of 86 species before Malnate (1960) began its reorganization. Immunological evidence on the transferrins (Mao and Dessauer, unpublished) shows that the genus, even as presently constituted, includes widely divergent species. Quantitative titrations suggest that species from North America, Taiwan and Europe are widely divergent.

Comparative studies on proteins serve to distinguish species from North America and to prove their close relationship to each other. Transferrins of the different species (*N. sipedon*, *N. rhombifera*, *N. valida*, *N. erythrogaster* and *N. taxispilota*) have indices of dissimilarity of less than 1·4 when titrated with antiserum against *N. cyclopion* transferrin (Mao and Dessauer, 1971). Fingerprints of hemoglobins of any two species differ by fewer than five peptides, differences distinct enough to identify the species source of the hemoglobin (Sutton, 1969). Patterns of the histidine peptides of hemoglobins clearly illustrate such species differences (Fig. 18). Species are also readily distinguished by electrophoretic patterns of their plasma proteins. Individual and population differences have also been noted in transferrins and in patterns of 'fast migrating proteins' of *N. sipedon* (Dessauer and Fox, 1958, 1964; Dessauer *et al.*, 1962a; Seniów, 1963; Kmeťová and Paulov, 1966).

e. Thamnophis. Garter snakes occur in a wide variety of forms throughout North and Central America. This genus, a near relative of North American *Natrix*, is a complex of very closely related species. The range of immunological distances between the transferrins of different species of *Thamnophis* is small, indicating that species formation in the genus has been relatively recent (George and Dessauer, 1970). Many of the species are distinguishable by the electrophoretic patterns of their plasma proteins (Dessauer and Fox, 1958, 1964; Dessauer *et al.*, 1962a). Certain groups within the genus appear to be rapidly evolving. Studies at the molecular level have clarified the systematics of one such complex, the garter snakes of the Pacific coast of the United States.

Polytypic complexes of closely related forms such as the Pacific Coast garter snakes have challenged the astuteness of systematists for years. Morphological variation among these snakes is very great and few conservative characters are available to properly group the many races. In 1951 Fox revised the

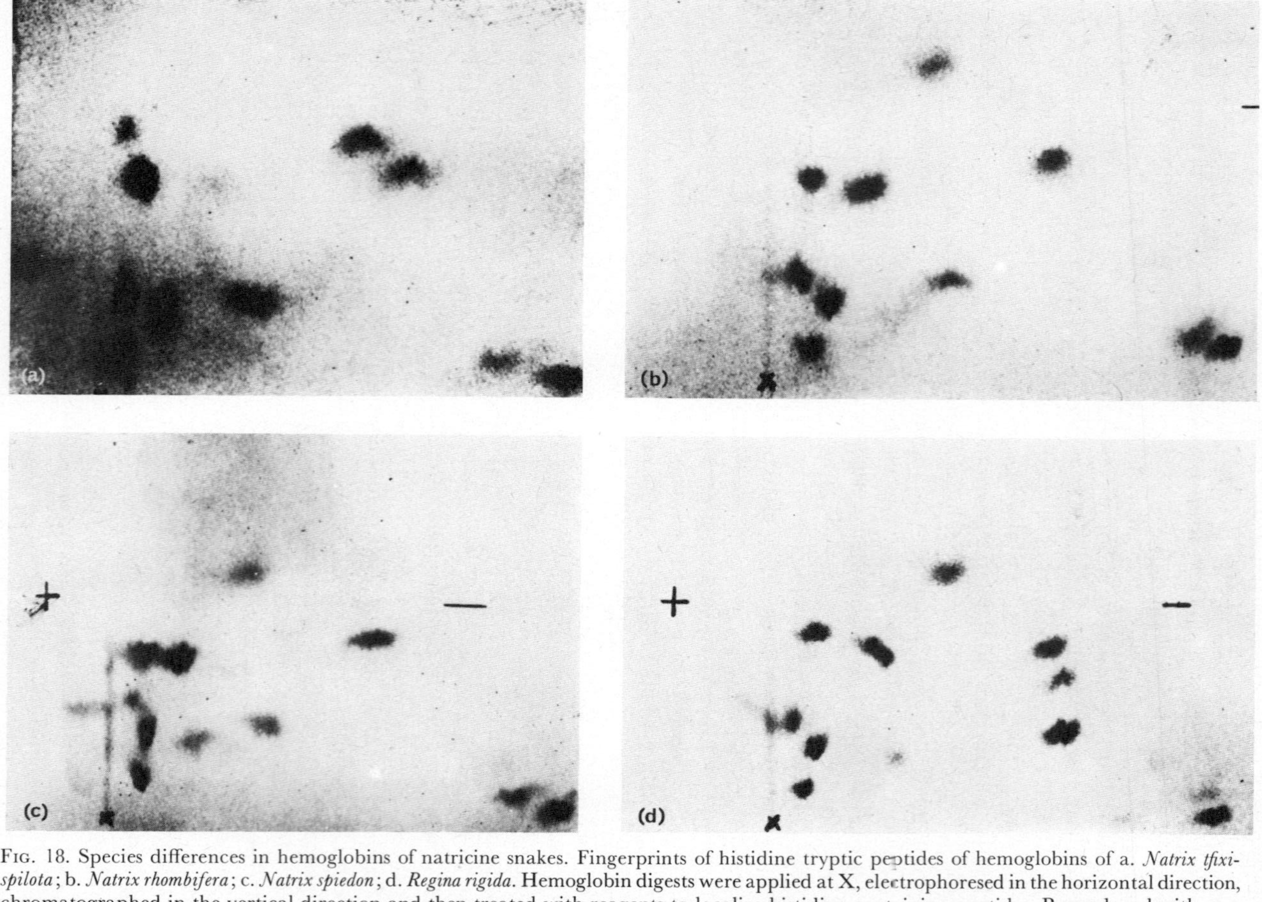

FIG. 18. Species differences in hemoglobins of natricine snakes. Fingerprints of histidine tryptic peptides of hemoglobins of a. *Natrix tfixispilota*; b. *Natrix rhombifera*; c. *Natrix spiedon*; d. *Regina rigida*. Hemoglobin digests were applied at X, electrophoresed in the horizontal direction, chromatographed in the vertical direction and then treated with reagents to localize histidine containing peptides. Reproduced with permission of Sutton (1969).

group by placing the 14 distinct forms into two ecological groups, one terrestrial in habits and the other aquatic. Races of the two groups, sympatric over most of California and Oregon, usually behave as two species. On the basis of what appeared to be morphological evidence of intergradation between the races in northern California, forms of the two ecological groups were placed in a single species, *Thamnophis elegans*.

Specimens of the two ecological groups are distinguished readily by the mobility of their transferrins (Fig. 15; Dessauer *et al.*, 1962a; Fox and Dessauer, 1965). Three variant transferrins occur in the montane, aquatic subspecies *couchi*, but the most common of these variants has the mobility of the transferrin found in the other aquatic subspecies. Specimens from the area in northern California where terrestrial and aquatic forms were presumed to intergrade have a single transferrin, that of the ecological group which they resemble. If free interbreeding had occurred between the aquatic and terrestrial forms in that area, some specimens of the population would have had two transferrins, one characteristic of the aquatic parent and the other of the terrestrial parent. Only 7 animals of 411 sampled across the entire range had atypical transferrin patterns; 2 of 175 specimens of the aquatic group possessed the transferrin variant typical of the terrestrial group; 5 of 236 specimens of the terrestrial group possessed a transferrin characteristic of the aquatic group. Snakes with these atypical transferrin patterns were collected from widespread localities outside of the presumed area of intergradation. Even if the 7 atypical snakes arose from hybrid matings, the low frequency and spotty distribution of the hybrids give little justification for grouping these snakes in a single species.

The many forms of the two ecologic groups are obviously related. Thus, immunoelectrophoretic studies of their plasma proteins reveal that they have many proteins in common (Fox and Dessauer, 1965). Antiserum to transferrin of an aquatic subspecies *gigas* cross reacted strongly with transferrins of snakes of many subspecies of both ecologic groups (George and Dessauer, 1970). Molecular data thus confirm the close genetic affinities of the many forms and prove that at least two different species *T. elegans* and *T. couchi* should be recognized.

f. Lampropeltis, Coluber and other colubrid snakes. Numerous studies have demonstrated the species specificity of electrophoretic patterns of colubrid blood proteins (see, Latifi *et al.*, 1965); Dessauer and Fox, 1964). Other studies have uncovered geographically limited protein variants in different populations of species of wide range. For example, both for *Lampropeltis getulus* and *Coluber constrictor*, western populations are readily distinguished from eastern populations by differences in the electrophoretic patterns of their plasma proteins (Dessauer and Fox, 1958). *Lampropeltis getulus californiae, L. g. yumenensis* and *L. g. splendida* from California and Arizona have different albumins, leucine naphthylamidases and haptoglobins than *L. g. holbrooki, L. g. getulus, L. g. brooksi* from Louisiana, Georgia and Florida. The striped and the banded forms of *L. g. californiae* have identical plasma protein electrophoretic patterns. Intraspecific differences occur in transferrin and leucine naphthylamidases. Two variants of the latter were distributed in three phenotypes of one or two bands in a population sample of *L. g. getulus* captured in northwest Georgia (Pough and Dessauer, unpublished). Intraspecific variation in transferrins is common in many colubrid snakes. The extent of variability may be very great, e.g. five transferrins were distributed in five phenotypes of one or two bands in a population sample of six snakes, *Coluber constrictor*, collected in the sand dune area south of Lake Michigan (Fig. 17; Dessauer *et al.*, 1962a).

IV. Biochemical Adaptations

Presumably structural differences between homologous proteins evolve from the interaction of mutation and natural selection. Present theory suggests that, on the average, each protein accumulates amino acid sequence changes at a constant rate. The mutation factor is judged equally strong for each protein, but the selectivity factor is thought to vary from protein to protein. For each protein, the selectivity of a mutant variant must depend upon its overall structure, the tissue and animal in which it occurs and the environment in which the animal lives. If a sequence change resulting from a mutation does not drastically alter three-dimensional configuration, selection may not act strongly against the mutant protein. Some conservative mutations may be treated neutrally by natural selection. From what is known of the properties of different amino acids, it is not difficult to understand why certain substitutions do not significantly alter the properties of the protein (Granick, 1965; Zuckerkandl and Pauling, 1965; Dixon, 1966).

The cytochromes-C and transferrins represent extreme examples. Cytochromes-C, as units of mitochondrial structure, apparently have very rigid sequence requirements. They must conserve structures concerned with the electron exchange function as well as structures affecting specific binding to other polymers of the mitochondrial architecture. For transferrins, on the other hand, structural requirements probably are much more flexible. They must retain the integrity of structures involved in Fe-binding, but probably have few structural limitations related to interactions with other polymers. Apparently, in most species functional transferrins are single polypeptides that float free in blood plasma.

Protein structures of heterotherms may be expected to reflect environmental selective pressures more directly than do proteins of homotherms. Proteins of reptiles and amphibians experience greater extremes in oxygen tension, pH, salt content and temperature than do those of mammals and birds (Dessauer, 1970). For example, turtle blood during extended diving periods is characterized by a virtual absence of oxygen and a pH as low as 6·5 whereas under other conditions the blood pH approaches 8 (Robin *et al.*,

1964). The blood pH of alligators may rise from 7 to near 8 simply following a meal (Coulson *et al.*, 1950). The body temperature of desert lizards may vary over 30°C during a single day. What structural characteristics of dehydrogenases, transferrins, globins and albumins allow these polymers to function effectively as oxidation-reduction catalysts, in iron and oxygen transport and in conserving blood volume under such variable conditions?

The evolution of plasma albumin appears to reflect the metabolic activity of the animal and the availability of water in its environment. The 'volume expanding' function of albumin relates to its hydrophilic nature, its relatively high net charge and to its concentration and relatively low molecular weight. Plasma of tadpoles, which spend most of their lives in water, has a very low protein level and lacks a protein comparable to albumin. Plasma of frogs, which are capable of moving about in drier areas, does contain an albumin (Herner and Frieden, 1960). Anurans which live in dry regions seem to have higher concentrations of plasma albumin than those from more moist areas (Bertini and Cei, 1960; Cei *et al.*, 1961). Albumin of reptiles with the highest metabolic rates, with high preferred body temperatures and especially those living in desiccating environments (e.g. desert lizards) usually has a high net charge and is present in high concentration in the blood. Reptiles with lower metabolic rates, low preferred body temperatures and especially those which live in fresh water have low blood levels of albumins and the latter have a relatively low net charge (Khalil and Abdel-Messeih, 1963; Masat and Dessauer, 1968). The albumin of many fresh water turtles has an extremely low charge and is present in low concentrations in the blood (Cohen and Stickler, 1958). The high volume of extravascular-extracellular fluid observed in such turtles (Smith, 1932) may be correlated with driving physiology (Robin *et al.*, 1964; Dessauer, 1970).

The structure of hemoglobin reflects adaptations to diving, to shifts in the availability of oxygen and to the activity temperature of the organism. For example, soon after aquatic turtles dive, their blood pH drops to below 7 and they utilize all oxygen bound to hemoglobins (Robin *et al.*, 1964). The marked Bohr effect of their hemoglobins as compared to that of terrestrial species, facilitates this release of oxygen (Sullivan and Riggs, 1967c). Hemoglobin variants, which readily oxidize to methemoglobin, are lethal in the homozygous condition in mammals but apparently are normal structures in many turtles (Riggs *et al.*, 1964; Sullivan and Riggs, 1967a). Oxygen affinities of red cells of the lizards *Uma*, *Sceloporous*, *Dipsosaurus* and *Gerrhonotus*, though very different when tested at the same temperature, are remarkably similar at the activity temperature of each species (Pough, 1969).

Lactic dehydrogenase structure reflects the range of temperatures likely to be encountered by the animal and the oxygen requirements of different tissues. Muscle-LDH apparently has more rigid structural requirements than heart-LDH. Muscle-LDH activity is high in tissues and in organisms geared to requirements of bursts of energy, whereas heart-LDH activity predominates in tissues and organisms which require steady supplies of energy (Salthe, 1965). The thermostability of LDH as well as other proteins probably has been an important factor in the evolution of amphibians and reptiles (Ushakov, 1964). Heart-LDHs of squamatan reptiles, which are subjected often to body temperatures above 40°C, have high temperature stabilities (Wilson *et al.*, 1964). The thermostability of myosin ATPase correlates generally with the thermal relations of lizards (Licht, 1964; Dawson, 1965).

The occurrence of certain hydrolytic enzymes and their tissue distribution relate to food sources and to the presence of parasites. For example, the antibiotic and cytolytic activities of protein and peptides of the skin secretions control skin microorganisms of the amphibians, *Bombina* and *Triturus* (Bachmayer *et al.*, 1967). The battery of digestive enzymes of insectivorous lizards includes a chitinase (Jeuniaux, 1961). The venom of vipers contains hydrolytic enzymes which initiate the digestion of their victim as well as the toxins which kill them (Porges, 1953).

The adaptive significance of most structural differences in proteins, however, is not known. Transferrins and esterases furnish good examples. One of the two transferrins found in turtles does seem to be adapted to binding iron at the relatively low pHs sometimes reached during diving, but this is the only known correlation of transferrin structure with physiology (Barber and Sheeler, 1963). Why are transferrins of low net charge found in turtles, whereas in some lizards these proteins are as highly charged as plasma albumin? Is there any adaptive significance to the pronounced intraspecific polymorphism of transferrins so common in species of toads and snakes (Fig. 17)? The pronounced diversity of the esterases in *Acris* (Fig. 11) and in many other vertebrates raises questions of considerable biological importance (Augustinsson, 1959; Holmes *et al.*, 1968). Does such great variability represent balanced polymorphism, selection for the heterozygote, or a slow rate of elimination of mutants from the populations? Although the molecular approach is enlarging our understanding of evolutionary biology, such questions show that it is posing many new problems.

Addendum

Polypeptides homologous to the alpha and beta globins of mammals may contribute to the subchain structure of hemoglobins of amphibians and reptiles. The two globins isolated from the hemoglobin of *Triturus cristatus* have the same C-terminal dipeptides as those of human hemoglobin (Sorcini *et al.*, 1970). The alpha chain of the hemoglobin of the snake *Vipera aspis* was also identified and partially sequenced (Duguet *et al.*, 1971). Supernatant and mitochondrial types of malate dehydrogenase of turtles apparently are homologous with the corresponding enzymes of birds and mammals (Karig and Wilson, 1971).

Comparative immunological studies are extending knowledge of affinities of higher taxa and generic groups. Microcomplement fixation (MC'F) titrations involving antisera against albumins of frogs of the Ranidae, Hylidae and Pipidae support an ancient divergence for evolutionary lines leading to modern species of these families (Section II C, 2) and indicate that frogs of very similar morphology within these families are often widely divergent genetically (Wallace *et al.*, 1971). Quantitative precipitin reactions involving plasma proteins of species within the Leptodactylidae indicate a low affinity of *Calyptocephalella* to other genera of Telmatobini stock, a close relationship for *Telmatobufo* and *Telmatobium* (Cei, 1970a), and support specific status for three named forms of *Leptodactylus* which differ only slightly in morphology (Cei, 1970b).

Precipitin data suggest that turtles of the Chelydridae are genetically closer to the Testudinidae (Emydidae) than to the Kinosternidae, contrary to widely held views (Frair, 1972). MC'F titrations involving antisera against albumin of the lizard *Anolis carolinensis* demonstrate the wide divergence of species classified in genus *Anolis*, suggesting that a number of species-series were independent lines as early as the late Eocene (Shochat *et al.*, 1972). Immunoelectrophoretic evidence supports the views that snakes of the Boidae and Caenophidia are widely divergent and that natricines should be placed into a distinct family (Minton and Salanitro, 1972; Section IID, 3b). MC'F titrations involving transferrins suggest that modern genera of Asiatic natricine snakes originated during the Miocene, that one of these, *Natrix*, had spread to Europe, Africa and North America by the end of the Miocene, and that *Natrix* underwent extensive speciation in North America during the Pleistocene (Mao and Dessauer, 1971).

Electrophoretic phenotypes of proteins are being used increasingly to group closely related forms and to obtain estimates of the genetic structure of individual species. Using a nuclear transplant technique Aimare and Chalumeau-Le Foulgoc (1969) presented strong evidence that the nucleus controls the two phenotypes of transferrin observed in populations of the salamander *Pleurodeles waltii* (Fine *et al.*, 1967). Patterns of LDH served to assemble North American *Desmognathus* into three species groups. Multiple alleles were detected at the heart LDH and the muscle LDH loci in a number of the species. The specific muscle LDH variants found in individual *D. quadramaculatus* correlated with their site of capture and morphology (Hinterstein, 1971). A number of different albumin phenotypes were detected among salamanders of populations of *Ensatina eschscholtzi* of North America (Petrakis and Brown, 1970) and *Salamandra salamandra* of Europe (Gasser and Cahet, 1967). The distribution of the variants showed geographic continuity in both species. Albumin phenotypes also furnished additional evidence concerning the genetic structure of the *Rana pipiens* species complex (see Section III D, 2b). Frogs of the interior plains and of southcentral United States have different albumin phenotypes. Although individuals with hybrid albumin patterns are rare in Western Texas where they are sympatric, evidence suggests that some gene flow does occur between the two forms (Platz, 1972). Isoalleles were detected at 19 of 22 loci in populations of *Bufo viridis* sampled in Israel alone; the number of heterozygous loci/individual (HL/I) averaged 11·2% in these populations. Frequencies of two isocitric dehydrogenase and two transferrin alleles followed a north-south cline and appeared to vary with the aridity of the geographic area (Chuang *et al.*, 1972). Estimates of genic variability of many species of *Anolis* lizards ranged between 5 and 10% HL/I, however, some anoles from small islands showed much less variability (Webster, 1972). Populations of *Uta stansburiana* exhibited between 6 and 8% HL/I (Selander *et al.*, 1970); and exposure to high level radiation caused no detectable increase in estimates of variability (McKinney and Turner, 1971). Seven species of *Agama* and two *Uromastix* were distinguished and grouped on the basis of phenotypes of four proteins (Gorman and Shochat, 1972a). Four isoalleles of heart LDH were found in a population of *Agama stellio* collected in a small area of the Golan Heights of Israel. Gorman and Shochat (1972b) suggested that the large number of alleles present in this sample might represent a natural example of the phenomenon 'polymorphism breeds polymorphism' (see Ohno *et al.*, 1969).

Although high resolution electrophoretic techniques are opening up many new areas of study concerning speciation processes, investigators must realize that many factors influence the protein phenotypes which they interpret (Section III B). Polyploidy, for example, may (Schwantes *et al.*, 1969) or may not influence the patterns. Apparently it does not complicate those of a number of proteins of *Amphiuma means*, the species having the largest amount of DNA/cell of any vertebrate (Comings and Berger, 1969). A serious caution to those using electrophoretic data in studies of population genetics is raised by Hinterstein's (1971) finding that only one allele at a heterozygous locus of heart LDH may be active in adult salamanders.

Acknowledgements

I have become indebted to a large number of friends and colleagues while putting together this manuscript. My wife Frances, Patricia Hicks, Nicholas Nicosia and Caral Tate gave of their time and skills in producing the figures and in typing the manuscript. José Cei, George Gorman, Sheldon Guttman, Vassili Kiortsis, Shou-hsian Mao, Donald Sutton and Allan Wilson allowed me to utilize material from unpublished manuscripts; W. Beçak, Alan Boyden, Jésus Jimenez-Porras, W. Manski, William Neaves and Stanley Salthe gave permission to reproduce certain illustrative material from their publications. Marjorie Fox spent many hours in the thankless task of reading an early copy of the manuscript with a thoughtful and critical eye. The National Science Foundation (Grant # GB-7294X) supported this study.

References and Bibliography

Abramoff, P., Darnell, R. M. and Balsano, J. S. (1964). *In* 'Taxonomic Biochemistry and Serology' (C. A. Leone, ed.), pp. 515–525. The Ronald Press Co., New York.
Adams, E. and Finnegan, C. V. (1965). *J. exp. Zool.* **158**, 241–252.
Aimare, M. C. and Chalumeau-Le Foulgoc, M. T. (1969). *C. r. hebd. Seanc. Acad. Sci., Paris*, Ser. D. **268**, 368–370.
Arnheim, N., Prager, E. M. and Wilson, A. C. (1969). *J. biol. Chem.* **244**, 2085–2094.
Augustinsson, K. (1959). *Acta Chem. Scand.* **13**, 1081–1096.
Bachmayer, H., Michl, H. and Roos, B. (1967). *In* 'Animal Toxins' (F. E. Russell and P. R. Saunders, eds.), pp. 395–399. Pergamon Press, New York and London.
Bagnara, J. T. and Obika, M. (1965). *Comp. Biochem. Physiol.* **15**, 33–49.
Bahl, O. P. and Smith, E. L. (1965). *J. biol. Chem.* **240**, 3585–3593.
Barber, A. A. and Sheeler, P. (1963). *Comp. Biochem. Physiol.* **8**, 115–122.
Barwick, R. E. and Bryant, C. (1966). *Physiol. Zool.* **39**, 1–20.
Beçak, W. (1969). *Genetics* **61** (Suppl.), 183–190.
Beçak, W., Schwantes, A. R. and Schwantes, M. L. B. (1968). *J. exp. Zool.* **168**, 473–475.
Belkin, D. A. (1963). *Science* **139**, 492–493.
Bellairs, A. d'A and Underwood, G. (1951). *Biol. Rev.* **26**, 193–237.
Bennett, T. P. and Frieden, E. (1962). *In* 'Comparative Biochemistry' (M. Florkin and H. S. Mason, eds.), Vol 4, pp. 483–556. Academic Press, New York.
Bertini, F. and Cei, J. M. (1959). *Actas Trab. Congr. Sud-am. Zool.* **4**, 161–172.
Bertini, F. and Cei, J. M. (1960). *Revta. Soc. argent. Biol.* **36**, 355–362.
Bertini, F. and Cei, J. M. (1962). *Herpetologica* **17**, 231–238.
Bertini, F. and Gustavo, R. (1962). *Copeia* **1962**, 181–185.
Bertke, E. M., Watt, D. D. and Tu, T. (1966/1967). *Toxicon* **4**, 73–76.
Blair, W. F. (1962). *Syst. Zool.* **11**, 72–84.
Blair, W. F. (1963). *Evolution, Lancaster, Pa.* **17**, 1–16.
Blair, W. F. (1972). 'Evolution in the Genus *Bufo*'. University of Texas Press, Austin.
Boydon, A. (1963). *Syst. Zool.* **12**, 1–7.
Boyden, A. and Noble, G. K. (1933). *Am. Mus. Novit.* **606**, 1–24.
Brawerman, G. and Shapiro, H. S. (1962). *In* 'Comparative Biochemistry' (M. Florkin and H. S. Mason, eds.), Vol 4, pp. 107–183. Academic Press, New York.
Britten, R. J. and Davidson, E. H. (1969). *Science, N.Y.* **165**, 349–357.
Britten, R. J. and Kohne, D. E. (1968). *Science, N.Y.* **161**, 529–540.
Brodie, E. D., Jr. (1968). *Copeia* **1968**, 307–313.
Brown, G. W., Jr. (1964). *In* 'Physiology of the Amphibia' (J. A. Moore, ed.), pp. 1–98. Academic Press, New York.
Brown, L. E. (1964). *Syst. Zool.* **13**, 92–95.
Brown, L. E. and Guttman, S. I. (1970). *Am. Midl. Nat.* **83**, 160–166.
Bryson, V. and Vogel, H. J. (1965). 'Evolving Genes and Proteins'. Academic Press, New York.
Cei, J. M. (1959). *Evolution, Lancaster, Pa.* **13**, 532–536.
Cei, J. M. (1963). *Bull. serol. Mus., New Brunsw.* **30**, 4–6.
Cei, J. M. (1965). *Herpetologica* **20**, 217–224.
Cei, J. M. (1969). *Physis, B. Aires.* **28**, 273–277.
Cei, J. M. (1970a). *Acta zool. lilloana* **27**, 181–192.
Cei, J. M. (1970b). *Acta zool. lilloana* **27**, 299–306.
Cei, J. M. and Bertini, F. (1960). *Actas Trab. Congr. Sud-am. Zool* **4**, 189–194.

Cei, J. M. and Bertini, F. (1961). *Copeia* **1961**, 336–340.
Cei, J. M. and Cohen, R. (1963). *Bull. serol. Mus., New Brunsw.* **30**, 6–8.
Cei, J. M. and Cohen, R. (in Press). *Actas Trab. Congr. Sud-am. Zool.*
Cei, J. M. and Erspamer, V. (in Press). *Actas Trab. Congr. Sud-am. Zool.*
Cei, J. M. and Erspamer, V. (1966). *Copeia* **1966**, 74–78.
Cei, J. M., Bertini, F. and Gallopin, G. C. (1961). *Revta. Soc. argent. Biol.* **37**, 215–225.
Cei, J. M., Erspamer, V. and Roseghini, M. (1967). *Syst. Zool.* **16**, 328–342.
Cei, J. M., Erspamer, V. and Roseghini, M. (1968). *Syst. Zool.* **17**, 232–245.
Chan, S. K., Tulloss, I. and Margoliash, E. (1966). *Biochemistry, N.Y.* **5**, 2586–2597.
Chen, P. S. (1967). *Experientia* **23**, 483–485.
Chen, P. S. (1968). *J. exp. Zool.* **168**, 337–350.
Christensen, P. A. and Anderson C. G. (1967). *In* 'Animal Toxins' (F. E. Russell and P. R. Saunders, eds.), pp. 223–234. Pergamon Press, New York and London.
Chuang, K. C., Dessauer, H. C. and Nevo, E. (1972). *Abst. 52nd Meeting Am. Soc. Icthy. Herp.* 38–39.
Coates, M. (1967). *Evolution, Lancaster, Pa.* **21**, 130–140.
Cohen, E. (1955). *Biol. Bull. mar. biol. Lab., Woods Hole* **109**, 394–403.
Cohen, E. and Stickler, G. B. (1958). *Science, N.Y.* **127**, 1392.
Cohen, P. P. and Brown, G. W., Jr. (1960). *In* 'Comparative Biochemistry' (M. Florkin and H. S. Mason, eds.), Vol 2, pp. 161–244. Academic Press, New York.
Comings, D. E. and Berger, R. O. (1969). *Biochem. Genet.* **2**, 319–333.
Coulson, R. A. and Hernandez, T. (1964). 'Biochemistry of the Alligator'. Louisiana State Univ. Press, Baton Rouge, Louisiana.
Coulson, R. A., Hernandez, T. and Dessauer, H. C. (1950). *Proc. Soc. exp. Biol. Med.* **74**, 866–869.
Crenshaw, J. W., Jr. (1962). *Physiol. Zool*, **35**, 157–165.
Crenshaw, J. W., Jr. (1965). *Evolution, N.Y.* **19**, 1–15.
Crescitelli, F. (1958). *Ann. N.Y. Acad. Sci.* **74**, 230–255.
Dawson, W. R. (1965). *In* 'Lizard Ecology, A Symposium' (W. H. Milstead, ed.), pp. 230–257. University of Missouri Press, Columbia, Missouri.
Dayhoff, M. O. and Eck, R. V. (1967/1968). 'Atlas of Protein Sequence and Structure'. National Biomedical Research Foundation, Silver Springs, Maryland.
Delpierre, G. R. (1968/1969). *Toxicon* **6**, 103–108.
Dessauer, H. C. (1961). *In* 'Blood and Other Body Fluids'. Biological Handbooks (D. S. Dittmer, ed.), pp. 45–46, *Fed. Amer. Soc. Exper. Biol.*, Washington, D.C.
Dessauer, H. C. (1966a). *Bull. serol. Mus., New Brunsw.* **34**, 4–8.
Dessauer, H. C. (1966b). *Bull. serol. Mus., New Brunsw.* **36**, 1–4.
Dessauer, H. C. (1967). *Herpetologica* **23**, 148–155.
Dessauer, H. C. (1968). *In* 'Systematic Biology' (C. Sibley, Organizer), pp. 325–357, National Academy of Sciences, U.S., Washington, D.C.
Dessauer, H. C. (1970). *In* 'Biology of the Reptilia' (A. d'A. Bellairs, C. Gans and T. Parsons, eds.), Vol 3. Academic Press, New York.
Dessauer, H. C. and Fox, W. (1956). *Science, N.Y.* **124**, 225–226.
Dessauer, H. C. and Fox, W. (1958). *Proc. Soc. exp. Biol. Med.* **98**, 101–105.
Dessauer, H. C. and Fox, W. (1959). *Am. J. Physiol.* **197**, 360–366.
Dessauer, H. C. and Fox, W. (1964). *In* 'Taxonomic Biochemistry and Serology' (C. A. Leone, ed.), pp. 625–647. The Ronald Press Co., New York.
Dessauer, H. C. and Nevo, E. (1969). *Biochem. Genets.* **3**, 171–188.
Dessauer, H. C., Fox, W. and Ramirez, J. R. (1957). *Archs Biochem. Biophys.* **71**, 11–16.

Dessauer, H. C., Hutton, K. E. and Musacchia, X. J. (1961). *In* 'Blood and Other Body Fluids'. Biological Handbooks (D. S. Dittmer, ed.), pp. 39–44. *Fed. Amer. Soc. Exper. Biol.* Washington, D.C.

Dessauer, H. C., Fox, W. and Hartwig, Q. L. (1962a). *Comp. Biochem. Physiol.* **5**, 17–29.

Dessauer, H. C., Fox, W. and Pough, F. H. (1962b). *Copeia* **1962**, 767–774.

Deutsch, H. F. and McShan, W. H. (1949). *J. biol. Chem.* **180**, 219–234.

DiMaggio, A. III and Dessauer, H. C. (1963). *Am. J. Physiol.* **204**, 677–680.

Dixon, G. H. (1966). *In* 'Essays in Biochemistry' (P. N. Campbell and G. D. Greville, eds.), Vol 2, pp. 147–204. Academic Press, London and New York.

Dobzhansky, T. (1968). *In* 'Evolutionary Biology' (T. Dobzhansky, M. K. Hecht and W. C. Steere, eds.), Vol 2, pp. 25. Appleton-Century-Crofts, New York.

Dowling, H. G. (1959). *Copeia* **1959**, 38–52.

Dowling, H. G. (1967). *Herpetologica* **23**, 138–142.

Dozy, A. M., Reynolds, C. A., Still, J. M. and Huisman, T. H. J. (1964). *J. exp. Zool.* **155**, 343–348.

Drujan, B. D., Segal, J. and Tovar, E. (1963). *Acta cient. venez.* **92** (Suppl. 1), 14.

Duguet, M., Chauvet, J. P. and Acker, R. (1971). *FEBS Lett.* (Federation of European Biochemical Societies). **18**, 185–188.

Ehrlich, P. R. and Holm, R. W. (1962). *Science, N.Y.* **137**, 652–657.

Ehrlich, P. R. and Raven, P. H. (1969). *Science, N.Y.* **165**, 1228–1232.

Eppenberger, M. E., Eppenberger, H. M. and Kaplan, N. O. (1967). *Nature, Lond.* **214**, 239–241.

Erspamer, V., Roseghini, M. and Cei, J. M. (1964a). *Biochem. Pharmac.* **13**, 1083–1093.

Erspamer, V., Vitali, T., Roseghini, M. and Cei, J. M. (1964b). *Archs Biochem. Biophys.* **105**, 620–629.

Erspamer, V., Vitali, T., Roseghini, M. and Cei, J. M. (1967). *Biochem. Pharmac.* **16**, 1149–1164.

Evans, E. E. (1963). *Fedn. Proc. Fedn. Am. Socs exp. Biol.* **22**, 1132–1137.

Evans, E. E., Kent, S. P., Attleberger, M. H., Sieberg, C., Bryant, R. E. and Booth, B. (1965). *Ann. N.Y. Acad. Sci.* **126**, 629–646.

Fine, J., Groulade, J. and Eyquem, A. (1954). *Annls Inst. Pasteur, Paris* **86**, 378–381.

Fine, J. M., Chalumeau-Le Foulgoc, M. T. and Amouch, M. P. (1967). *C. r. hebd. Séanc. Acad. Sci., Paris*, Ser. D. **265**, 1248–1250.

Fischer, F. B. and Neumann, W. P. (1954). *Hoppe-Seyler's Z. physiol. Chem.* **297**, 100.

Fisher, G. A. and Kabara, J. J. (1967). *In* 'Animal Toxins' (F. E. Russell and P. R. Saunders, eds.), pp. 283–292. Pergamon Press, New York and London.

Florkin, M. (1949). 'Biochemical Evolution' (S. Morgulis, translator). Academic Press, New York.

Fox, W. (1951). *Univ. Calif. Publs Zool.* **50**, 485–530.

Fox, W. and Dessauer, H. C. (1965). *Amer. Phil. Soc. Year Book* **1964**, 263–266.

Fox, W., Dessauer, H. C. and Maumus, L. T. (1961). *Comp. Biochem. Physiol.* **3**, 52–63.

Frair, W. (1962). *Bull. serol. Mus., New Brunsw.* **27**, 7–8.

Frair, W. (1963). *Science, N.Y.* **140**, 1412–1414.

Frair, W. (1964). *In* 'Taxonomic Biochemistry and Serology' (C. A. Leone, ed.), pp. 535–544. The Ronald Press Co., New York.

Frair, W. (1969). *Bull. serol. Mus., New Brunsw.* **42**, 1–3.

Frair, W. (1972). *Copeia* **1972**, 97–108.

Frieden, E. (1961). *Am. Zool.* **1**, 115–149.

Fuhrman, F. A., Fuhrman, G. J. and Mosher, H. S. (1969). *Science, N.Y.* **165**, 1376–1377.

Gans, C. and Elliott, W. B. (1968). *Adv. Oral Biol.* **3**, 45–81.
Gasser, M. F. and Cahet, P. (1967). *C. r. hebd. Séanc Acad. Sci., Paris*, Ser. D. **267**, 763–766.
George, D. W. and Dessauer, H. C. (1970). *Comp. Biochem. Physiol.* **33**, 617–627.
Gerzeli, G., Casati, C. and Gennaro, A. M. (1956). *Riv. Istochim. norm. Path.* **110**, 149–154.
Gillespie, J. H. and Crenshaw, J. W. (1966). *Copeia* **1966**, 889–893.
Goin, O. B. and Goin, C. J. (1968). *Am. Midl. Nat.* **80**, 289–298.
Goin, O. B., Goin, C. J. and Bachmann, K. (1968). *Copeia* **1968**, 532–540.
Gonçalves, J. M. (1956). *Anais Acad. bras. Cienc.* **28**, 365–367.
Gonçalves, J. M. and Vieira, L. G. (1950). *Anais Acad. bras. Cienc.* **22**, 141–150.
Gorman, G. C. and Atkins, L. (1969). *Bull. Mus. comp. Zool. Harv.* **138**, 53–80.
Gorman, G. C. and Dessauer, H. C. (1965). *Science, N.Y.* **150**, 1454–1455.
Gorman, G. C. and Dessauer, H. C. (1966). *Comp. Biochem. Physiol.* **19**, 845–853.
Gorman, G. C. and Shochat, D. (1972a). *Herpetologica* **28**, 106–112.
Gorman, G. C. and Shochat, D. (1972b). *Experientia.* **28**, 351–353.
Gorman, G. C., Licht, P., Dessauer, H. C. and Boos, J. O. (1971a). *Syst. Zool.* **20**, 1–18.
Gorman, G. C., Wilson, A. C. and Nakanishi, M. (1971b). **20**, 167–185
Graham-Smith, G. S. (1904). *In* 'Blood Immunity and Blood Relationship' (G. H. F. Nuttall, ed.), pp. 336–380. Cambridge University Press, London.
Granick, S. (1965). *In* 'Evolving Genes and Proteins'. (V. Bryson and H. J. Vogel, eds.), p. 82. Academic Press, New York.
Guttman, S. I. (1965). *Tex. J. Sci.* **17**, 267–277.
Guttman, S. I. (1967). *Comp. Biochem. Physiol.* **23**, 871–877.
Guttman, S. I. (1969). *Copeia* **1969**, 243–249.
Guttman, S. I. (1972). *In* 'Evolution of the Genus *Bufo*' (W. F. Blair, ed.), pp. 265–278, University of Texas Press, Austin, Texas.
Guttman, S. I. and Brown, L. E. (1968). *Am. Zool.* **8**, 805–806.
Hahn, W. E. (1962). *Comp. Biochem. Physiol.* **7**, 55–61.
Harpur, R. P. (1968). *Can. J. Zool.* **46**, 295–301.
Haslewood, G. A. D. (1967). 'Bile Salts,. Methuen, London.
Haslewood, G. A. D. (1968). *In* 'Chemotaxonomy and Serotaxonomy' (J. G. Hawkes, ed.), pp. 159–172. Academic Press, London and New York.
Hebard, W. B. (1964). *In* 'Taxonomic Biochemistry and Serology' (C. A. Leone, ed.), pp. 649–657. The Ronald Press Co., New York.
Hecht, M. K. (1969). *Ann. N.Y. Acad. Sci.* **167**, 74–79.
Herbert, J. D., Coulson, R. A. and Hernandez, T. (1966). *Comp. Biochem. Physiol.* **17**, 583–598.
Herner, A. and Frieden, E. (1960). *J. biol. Chem.* **235**, 2845–2851.
Hildemann, W. H. (1962). *Ann. N.Y. Acad. Sci.* **97**, 139–152.
Hinterstein, B. (1971). *Copeia* **1971**, 636–644.
Holmes, R. S., Masters, C. J. and Webb, E. C. (1968). *Comp. Biochem. Physiol.* **26**, 837–852.
Hubby, J. L. and Lewontin, R. C. (1966). *Genetics, Princeton* **54**, 577–594.
Hunsaker, D. II, Alston, R. E., Blair, W. F. and Turner, B. L. (1961). *Evolution, Lancaster, Pa.* **15**, 352–359.
Jeuniaux, C. (1961). *Nature, Lond.* **192**, 135–136.
Jimenez-Porras, J. M. (1963). 'Comparative Biochemical Studies on Venoms of Snakes of Costa Rica'. Dissertation, Louisiana State University, Baton Rouge, Louisiana.

Jimenez-Porras, J. M. (1964/1965a). *Toxicon* **2**, 155–166.
Jimenez-Porras, J. M. (1964/1965b). *Toxicon* **2**, 187–195.
Jimenez-Porras, J. M. (1967). *In* 'Animal Toxins' (F. E. Russell and P. R. Saunders, eds.), pp. 307–321. Pergamon Press, New York and London.
Johnson, B. D. (1968/1969). *Toxicon* **6**, 5–10.
Jouannet, M. (1968). *Comp. Biochem. Physiol.* **5**, 191–199.
Kabara, J. J. and Miceli, J. N. (1967/1968). *Toxicon* **5**, 133–134.
Kaplan, H. M. (1960). *Herpetologica* **16**, 202–206.
Kaplan, N. O. (1965). *In* 'Evolving Genes and Proteins' (V. Bryson and H. J. Vogel, eds.), pp. 243–277. Academic Press, New York.
Karig, L. M. and Wilson, A. C. (1971). *Biochem. Genet.* **5**, 211–221.
Kellaway, C. H. and Williams, F. E. (1931). *Aust. J. exp. Biol. med. Sci.* **8**, 123–132.
Khalil, F. and Abdel-Messeih, G. (1963). *Comp. Biochem. Physiol.* **9**, 75–79.
Kiortsis, V. and Kiortsis, M. (1960). *Revue suisse Zool.* **67**, 119–127.
Kluge, A. G. and Farris, J. S. (1969). *Syst. Zool.* **18**, 1–32.
Kmeťová, S. and Paulov, Š. (1966). *Acta Fac. Rerum nat. comen Bratisl. Zool.* **13**, 251–254.
Kochva, E. and Gans, C. (1967). *In* 'Animal Toxins' (F. E. Russell and P. R. Saunders, eds.), pp. 195–203. Pergamon Press, New York and London.
Kunz, Y. W. and Hearn, J. (1967). *Experientia* **23**, 683–686.
Kuwajima, Y. (1953). *Jap. J. exp. Med.* **23**, 21–25.
Lamb, G. (1902). *Lancet* **2**, 431–435.
Latifi, M., Shamloo, K. D. and Amin, A. (1965). *Can. J. Biochem. Physiol.* **43**, 459–461.
Lanza, B. and Antonini, F. M. (1955). *Bonap. Monit. Zool. Ital.* **63**, 293–299.
Lebherz, H. G. and Rutter, W. J. (1967). *Science, N.Y.* **157**, 1198–1200.
Leone, C. A. (1964). 'Taxonomic Biochemistry and Serology'. The Ronald Press Co., New York.
Leone, C. A. and Wilson, F. E. (1961). *Physiol. Zool.* **34**, 297–305.
Leviton, A. E., Myers, G. S. and Grunbaum, B. W. (1964). *In* 'Taxonomic Biochemistry and Serology' (C. A. Leone, ed.), pp. 667–671. The Ronald Press Co., New York.
Lewis, J. H. (1964). *In* 'Protides of the Biological Fluids' (H. Peeters, ed.), pp. 149–154. Elsevier Publishing Company, Amsterdam.
Licht, P. (1964). *Comp. Biochem. Physiol.* **12**, 331–340.
Lockwood, A. P. M. (1961). *Comp. Biochem. Physiol.* **2**, 241–289.
Low, B. S. (1968). *Comp. Biochem. Physiol.* **26**, 247–257.
Lutz, A. (1931). *Mem. Inst. Oswaldo. Cruz* **23**, 1–20.
Ma, P. F. and Fisher, J. R. (1968). *Comp. Biochem. Physiol.* **27**, 105–112.
Macgregor, H. C. and Uzzell, T. M., Jr. (1964). *Science, N.Y.* **143**, 1043–1045.
Maldonado, A. A. and Ortiz, E. (1966). *Copeia* **1966**, 179–182.
Malnate, E. V. (1960). *Proc. Acad. nat. Sci. Philad.* **112**, 41–71.
Manski, W. and Halbert, S. P. (1964). *In* 'Protides of the Biological Fluids' (H. Peeters, ed.), pp. 117–134. Elsevier, Amsterdam.
Manski, W., Halbert, S. P. and Auerbach, T. P. (1964). *In* 'Taxonomic Biochemistry and Serology' (C. A. Leone, ed.), pp. 545–554. The Ronald Press Co., New York.
Manski, W., Halbert, S. P., Javier, P. and Auerbach, T. P. (1967). *Int. Archs Allergy appl. Immun.* **31**, 529–545.
Manwell, C. (1966). *Comp. Biochem. Physiol.* **17**, 805–823.
Manwell, C. and Baker, C. M. A. (1966). *In* 'Phylogeny of Immunity' (R. T. Smith, P. A. Miescher and R. A. Good, eds.), pp. 3–15. University of Florida Press, Florida.

Manwell, C. and Schlesinger, C. V. (1966). *Comp. Biochem. Physiol.* **18**, 627–637.
Mao, S. H. and Dessauer, H. C. (1971). *Comp. Biochem. Physiol.* **40**, 669–680.
Marchlewski-Koj, A. (1963). *Folia biol. Kraków.* **11**, 167–172.
Margoliash, E. and Smith, E. L. (1965). *In* 'Evolving Genes and Proteins' (V. Bryson and H. J. Vogel, eds.), pp. 221–242. Academic Press, New York.
Markert, C. L. (1965). *In* 'Ideas in Modern Biology' (J. Moore, ed.), pp. 231–258. Natural History Press, Garden City, New York.
Masat, R. J. and Dessauer, H. C. (1968). *Comp. Biochem. Physiol.* **25**, 119–128.
Masat, R. J. and Musacchia, X. J. (1965). *Comp. Biochem. Physiol.* **16**, 215–225.
McBean, R. L. and Goldstein, L. (1967). *Science, N.Y.* **157**, 931–932.
McDowell, S. B., Jr. and Bogert, C. M. (1954). *Bull. Am. Mus. nat. Hist.* **105**, 1–142.
McKinney, C. O. and Turner, F. B. (1971). *Radiat. Res.* **47**, 530–536.
McLaughlin, E. T. and Conde del Pino, E. (1968). *Herpetologica* **24**, 251–252.
Mebs, D. (1968). Hoppe-Selye *Z. phys. Chem.* **349**, 1115–1125.
Michl, H. and Kaiser, E. (1962/1963). *Toxicon* **1**, 175–228.
Michl, H. and Molzer, H. (1964/1965). *Toxicon* **2**, 281–282.
Minton, S. A., Jr. (1967). *In* 'Animal Toxins'. (F. E. Russell and P. R. Saunders, eds.), pp. 211–222. Pergamon Press, New York and London.
Minton, S. A. Jr. (1967/1968). *Toxicon* **5**, 47–55.
Minton, S. A., Jr. (1968/1969). *Toxicon* **6**, 59–64.
Minton, S. A. and Salanitro, S. K. (1972). *Copeia* **1972**, 246–252.
Mirsky, A. E. and Ris, H. (1951). *J. gen. Physiol.* **34**, 451–462.
Mosher, H. S., Fuhrman, F. A., Buchwald, H. D. and Fischer, H. G. (1964). *Science, N.Y.* **144**, 1100–1110.
Neaves, W. B. (1969). *J. exp. Zool.* **171**, 175–184.
Neaves, W. B. and Gerald, P. S. (1968). *Science, N.Y.* **160**, 1004–1005.
Neaves, W. B. and Gerald, P. S. (1969). *Science, N.Y.* **164**, 557–559.
Neelin, J. M. (1963). *Can. J. Biochem. Physiol.* **41**, 1073–1078.
Newcomer, R. J. (1967). *Am. Nat.* **101**, 192–195.
Noble, G. K. (1931). 'The Biology of the Amphibia'. McGraw-Hill (Dover Publication's Edition), New York.
Ohno, S., Stenius, C., Christian, L. and Schipmann, G. (1969). *Biochem. Genet.* **3**, 417–428.
Parsons, T. S. and Williams, E. E. (1963). *Q. Rev. Biol.* **38**, 26–53.
Paulov, Š. and Kmeťová, S. (1964). *Folia Biol. Praha* **10**, 155–156.
Pearson, D. D. (1966). *Bull. serol. Mus. New Brunsw.* **36**, 8.
Petrakis, P. L. and Brown, C. W. (1970). *Comp. Biochem. Physiol.* **32**, 475–487.
Platz, J. E. (1972). *Copeia* **1972**, 232–240.
Porges, N. (1953). *Science, N.Y.* **117**, 47–51.
Porter, K. R. (1964). *In* 'Taxonomic Biochemistry and Serology' (C. A. Leone, ed.), pp. 451–456. The Ronald Press Co., New York.
Pough, F. H. (1969). *Comp. Biochem. Physiol.* **31**, 885–901.
Rapoport, S., Leva, E. and Guest, G. M. (1941). *J. biol. Chem.* **139**, 621–632.
Rapoport, S., Leva, E. and Guest, G. M. (1942). *J. cell. comp. Physiol.* **19**, 103–108.
Reichert, E. T. and Brown, A. P. (1909). 'The Differentiation and Specificity of Corresponding Proteins and Other Vital Substances in Relation to Biological Classification and Organic Evolution. The Crystallography of Hemoglobins'. Carnegie Inst., Washington, Publ. No. 116.
Remington, C. L. (1968). *In* 'Evolutionary Biology' (T. Dobzhansky, M. K. Hecht and W. C. Steere, ed.), Vol 2, 321–428. Appleton-Century-Crofts, New York.

Riggs, A., Sullivan, B. and Agee, J. R. (1964). *Proc. natn. Acad. Sci. U.S.A.* **51**, 1127–1134.
Robin, E. D., Vester, J. W., Murdaugh, H. V., Jr. and Millen, J. E. (1964). *J. cell. comp. Physiol.* **63**, 287–297.
Romer, A. S. (1956). 'Osteology of Reptiles'. University of Chicago Press, Chicago, Illinois.
Romer, A. S. (1966). 'Vertebrate Paleontology'. University of Chicago Press, Chicago, Illinois.
Romer, A. S. (1967). *Science, N.Y.* **158**, 1629–1637.
Rosenquist, J. W. (1969). 'Plasma Vitellin and Calcium Movement in *Anolis carolinensis*'. Ph.D. Dissertation, Tulane University, New Orleans, Louisiana.
Salthe, S. N. (1965). *Comp. Biochem. Physiol.* **16**, 393–408.
Salthe, S. N. (1969). *Biochem. Genet.* **2**, 271–303.
Salthe, S. N. and Kaplan, N. O. (1966). *Evolution, Lancaster, Pa.* **20**, 603–616.
Salthe, S. N. and Kitto, G. B. (1966). *Copeia* **1966**, 130–132.
Salthe, S. N. and Nevo, E. (1969). *Biochem. Genet.* **3**, 335–341.
Salthe, S. N., Chilson, O. P. and Kaplan, N. O. (1965). *Nature, Lond.* **207**, 723–726.
Sarich, V. M. and Wilson, A. C. (1966). *Science, N.Y.* **154**, 1563–1566.
Schenberg, S. (1959). *Science, N.Y.* **129**, 1361–1363.
Schenberg, S. (1962/1963). *Toxicon* **1**, 67–75.
Schmidt-Nielsen, K. (1962/1963). *Harvey Lect.* **58**, 53–93.
Schwantes, A. R., Schwantes, M. L. B. and Beçak, W. (1969). *Rev. Bras. Pesqui. Med. Biol.* **2**, 41–44.
Seal, U. S. (1964). *Comp. Biochem. Physiol.* **13**, 143–159.
Selander, R. K., Yang, S. Y., Lewontin, R. C. and Johnson, W. E. (1970). *Evolution, Lancaster, Pa.* **24**, 402–414.
Senió w, A. (1963). *Comp. Biochem. Physiol.* **9**, 137–149.
Shochat, D., Wyles, J. S. and Sarich, V. (1972). *Abstr. 52nd Meeting Am. Soc. Icthyol. Herp.* 83–84.
Shontz, N. N. (1968). *Copeia* **1968**, 683–692.
Shoppee, C. W. (1964). 'Chemistry of the Steroids'. Butterworths Inc., Washington D.C., pp. 371–385; 433–434.
Simkiss, K. (1967). 'Calcium in Reproductive Physiology'. Reinhold Press, New York.
Smith, H. W. (1932). *Q. Rev. Biol.* **7**, 1–26.
Smith, H. W. (1951). 'The Kidney: Its Structure and Function in Health and Disease'. Oxford University Press, New York.
Sorcini, M., Orlando, M. and Tentori, L. (1970). *Comp. Biochem. Physiol.* **34**, 751–753.
Strauss, W. G. and Grunbaum, B. W. (1961). *Copeia* **1961**, 488–490.
Sullivan, B. and Riggs, A. (1967a). *Comp. Biochem. Physiol.* **23**, 437–447.
Sullivan, B. and Riggs, A. (1967b). *Comp. Biochem. Physiol.* **23**, 449–458.
Sullivan, B. and Riggs, A. (1967c). *Comp. Biochem. Physiol.* **23**, 459–474.
Sullivan, B. and Riggs, A. (1967d). *Biochim. biophys. Acta* **140**, 274–283.
Sutton, D. E. (1969). 'Fingerprint Correspondence of Hemoglobin Applied to the Taxonomy of Reptiles'. Dissertation, Louisiana State University Medical Center, New Orleans, Louisiana.
Svedberg, T. and Hedenius, A. (1934). *Biol. Bull. mar. biol. Lab., Woods Hole* **66**, 191–223.
Tashian, R. E. (1965). *Am. J. hum. Genet.* **17**, 257–272.
Tu, A. T. and Ganthavorn, S. (1967/1968). *Toxicon* **5**, 207–211.
Tu, A. T. and Toom, P. M. (1967). *Experientia* **23**, 439–440.

Tu, A. T., James, G. P. and Chua, A. (1965/1966). *Toxicon* **3**, 5–8.
Tu, A. T. and Passey, R. B. and Tu, T. (1966/1967). *Toxicon* **4**, 59–60.
Tu, A. T., Toom, P. M. and Murdock, D. S. (1967). *In* 'Animal Toxins' (F. E. Russell and P. R. Saunders, eds.), pp. 351–362. Pergamon Press, New York and London.
Underwood, G. (1967). 'A Contribution to the Classification of Snakes'. British Museum (Natural History), London.
Uriel, J., Fine, J. M., Courcon, J. and Le Bourdelles, F. (1957). *Bull. Soc. chim., Paris* **39**, 1415–1427.
Ushakov, B. (1964). *Physiol. Rev.* **44**, 518–560.
Uzzell, T. M., Jr. and Goldblatt, S. M. (1967). *Evolution, Lancaster, Pa.* **21**, 345–354.
Vendrely, R. (1958). *Annls Inst. Pasteur, Paris* **94**, 142–166.
Volpe, E. P. (1960). *Evolution, Lancaster, Pa.* **14**, 181–193.
Voris, H. K. (1967). *Physiol. Zoöl.* **40**, 238–247.
Wake, D. B. and Özeti, N. (1969). *Copeia* **1969**, 124–137.
Wakeley, J. F., Fuhrman, G. J., Fuhrman, F. A., Fisher, H. G. and Mosher, H. S. (1965/1966). *Toxicon* **3**, 195–203.
Wald, G. (1960). *In* 'Comparative Biochemistry' (M. Florkin and H. S. Mason, eds.), Vol I, 311–345. Academic Press, New York.
Wallace, D. G., Maxon, L. R. and Wilson, A. C. (1971). *Proc. natn. Acad. Sci. U.S.A.* **68**, 3127–3129.
Webster, T. P. (1972). *Abstr. 52nd Meeting Amer. Soc. Ichtyol. Herp.* **93**.
Wells, M. R. (1964). *J. Tenn. Acad. Sci.* **39**, 50–53.
Williams, E. E. (1969). *Q. Rev. Biol.* **44**, 345–389.
Wilson, A. C. and Sarich, V. M. (1969). *Proc. natn. Acad. Sci. U.S.A.* **63**, 1088–1093.
Wilson, A. C., Kaplan, N. O. Levine, L., Pesce, A., Reichlin, W. and Allison, W. S. (1964). *Fedn. Proc. Fedn. Am. Socs exp. Biol.* **23**, 1258–1266.
Wittliff, J. L. (1962). *Evolution, Lancaster, Pa.* **16**, 143–153.
Wittliff, J. L. (1964). *In* 'Taxonomic Biochemistry and Serology' (C. A. Leone, ed.), pp. 457–464. The Ronald Press Co., New York.
Wittliff, J. L. (1968). *Toxicon* **6**, 73–74.
Wolstenholme, D. R. and Dawid, I. B. (1968). *J. Cell. Biol.* **39**, 222–228.
Wright, D. A. and Moyer, F. H. (1966). *J. exp. Zool.* **163**, 215–230.
Yunis, J. J. (1969). 'Biochemical Methods in Red Cell Genetics'. Academic Press, New York and London.
Zuckerkandl, E. and Pauling, L. (1965). *In* 'Evolving Genes and Proteins' (V. Bryson and H. J. Vogel, eds.), pp. 97–166. Academic Press, New York.
Zweifel, R. G. (1962). *Copeia* **1962**, 749–766.
Zweig, G. and Crenshaw, J. W. (1957). *Science, N.Y.* **126**, 1065–1067.

4 Fish

F. J. O'ROURKE

Department of Zoology,
University College,
Cork, Ireland

I. Introduction

As long ago as 1958, Crick pointed out that 'before long we shall have a subject called protein taxonomy' and Sibley (1963) added that the biochemical study of proteins was just comparative morphology at the molecular level although Wright (1966) correctly warned that 'amino acid sequences and molecular configurations are not directly observed and can only be studied by the indirect methods of chemical and physical analyses'. Yet Sibley (1967) felt able to suggest that within a few years the determination of the sequence of units 200-amino-acids long will become automated and take less than two weeks. Such determinations are still elaborate and tedious, involving the digestion of the protein molecule by enzymes which attack known points in the chain. Subsequent analysis of the peptide fractions by a combination of chromatography and electrophoresis gives the polypeptide 'finger-prints'. These can be used to determine the precise amino acid sequence in the protein. These sequences are currently known for only a limited number of proteins. Nevertheless Emanuel Margoliash and his team have determined the sequence of amino acids in cytochrome C from about two dozen species although Dayhoff's (1969) computer-analysed phylogenetic tree included only one teleost fish whereas the sequence of an elasm branch cytochrome C was also known at the time (Gladstone and Smith, 1967).

The heart cytochrome C has been fully studied in only two species of fish, a teleost, the tuna, species not given (Kreil, 1963) and an elasmobranch, *Squalis sucklii*, the Puget Sound dogfish (Gladstone and Smith, 1967). Tuppy and Paleus (1955) found that in the salmon the heme peptide positions 11 to 21 are identical with those of tuna fish. Whereas the tuna has only 103 amino acids *S. sucklii* has 104 as is the case with all other samples of cytochrome C from vertebrate hearts. The chymotryptic peptides from the dogfish material are similar to those in the tuna but the dogfish is unique in having valine at residue 9 (leucine, isoleucine and threonine have been found in this position in other species). Residue 33 is serine in the dogfish but tryptophan in the tuna and turtle, asparagine in the kangaroo and histidine in a number of other species. 'The variation of this locus suggests that the side chain of residue 33 probably plays no major role in the cytochromes and may be in the surface of the protein' (Gladstone and Smith, 1967). If the tuna lacks residue 104 the COOH terminal differences between it and the dogfish are very different but if it is assumed that residue 100 is deleted in the tuna the residues are more similar. (Residues 100 to 104 are known to vary considerably in the various species studied.) If it is assumed that residue 104 is missing in tuna cytochrome the proteins of tuna and dogfish differ in 18 residues and in missing one (=19) whereas if residue 100 is assumed missing in tuna then the two fish differ in 16 residues and the deletion (=17). Compared with various mammals the dogfish cytochrome shows 16 to 23 changes, 18 from the chicken and turtle and 23 from the rattlesnake.

Both fishes show a number of features in common in their cytochromes which are unknown in other groups (asparagine at residue 22 in place of lysine in all other vertebrates; serine at residue 54 in place of aspargine in all other vertebrates; at positions 22 and 54 other residues have been found in non-vertebrate cytochromes; at position 61 there is aspargine in tuna and glutamine in dogfish whereas all other species have either glutamic or aspartic acid there). Despite these similarities there are major differences between the two species. 'The total number of amino acid differences is as great between the two fish proteins as between dogfish cytochrome and the more recently evolved groups of vertebrates' (Gladstone and Smith, 1967). Thus the taxonomic distance between the Osteoichthyes and the Chondroichthyes is emphasized. There is clearly need for further studies on a variety of fishes as indeed the work of Gemeroy (1943) had earlier indicated.

Dayhoff (1969), in producing the best topological relation between species to correspond with the best phylogenetic tree, uses as a measure of branch length the number of mutations between nodes. This is underestimated by the observed changes since mutations can be superimposed so she corrects 'for superimposed mutations by applying factors based on the known probability that any amino acid will change to any other given amino acid. That provides our unit of branch length: accepted point mutations per 100 amino-acid positions (PAMs)'. To obtain a time scale in years 'The translation from PAMs to years is derived from geological evidence, the best of which dates the divergence of the lines of bony fishes and the mammals at about 400 million years ago. The cytochrome tree puts that divergence at 11·5 PAMs, on average. Therefore 11·5 PAMs corresponds to 400 million years'.

However Dayhoff (1969) admits that the rate of change of proteins varies greatly with each protein over geological time and that cytochrome C is the most slowly changing one so far studied in a wide variety of species (30 PAMs per 1,000 million years). Comparable figures for insulin is 40, glyceraldehyde-3-phosphate dehydrogenase 20, histones only 0·6, haemoglobin 120, ribonuclease 300 and fibrinopeptides 900.

It is clear much more data is required. Dayhoff (1969) suggested that slowly changing proteins will provide the best information on long-term evolution and rapidly changing ones 'will provide higher resolution for sorting out closely related species'.

The ever increasing concern with molecular biology has made many biologists consider species as groups of individuals sharing a gene pool with more or less similar sets of sequences in the purine and pyrimidine bases in their DNA and 'with a system of operators, controllers and repressors leading to the biosynthesis of similar sequences of amino acids, the integration of which in one cell, or in a number of variably differentiated cells leads to similar structural and functional characteristics, adapted to the ecological niche in which the species flourishes' (Florkin, 1966).

Commoner (1965) on the other hand, has drawn attention to some of the objections to the theory that identifies the biochemical specificity of DNA (and of the base nucleotide sequences which determine this) 'as the sole source of both the biochemical specificity of new DNA and of the cell's protein complement'. He has pointed out that detailed studies of ribonucleoprotein biosynthesis in TMV (tobacco mosaic virus), one of the best studied of all viruses, provide evidence of a dual origin of its biochemical specificity involving both its RNA and protein components.

> 'the total biochemical specificity of the virus is the resultant of a continued series of specificity transfers between segments of the RNA fibre and single protein subunits. In general, these considerations suggest strongly that the basic scheme for the generation of biochemical specificity in the cell is governed by epigenesis rather than by preformation . . . (the data supports) the view that biochemical and biological specificity is not derived from a single molecular constituent of the cell, but is rather a property of the entire cell and a manifestation of the living state'.

Mayr (1969) drew attention to the extraordinarily high information content carried by the chromosomal DNA and that, while a geneticist specializing on *Drosophila* or on mice, in a generous mood, might allow these species to have 50,000 genes, calculations show that the chromosomal DNA in a single mammalian cell nucleus contains enough material to enable the animal to have five million genes. Mayr posed the problem of what the 4,950,000 extra units are doing. The possible answers suggested by current research pose 'an entirely new set of problems' (Mayr 1969).

A partial answer to Mayr's (1969) query is provided by the work on repeated segments of DNA which has been reviewed by Britten and Kohne (1970). Repeated sequences of DNA have been found in mammals, birds, fish and amphibia as well as in lower animals, plants, bacteria and viruses.

The amount of DNA per cell (=genome size) is about the same in all mammals studied (between 10^9 and 10^{10} nucleotide pairs per haploid cell). In lower forms there is considerable variation, in fishes the number ranges from just less than 10^9 to nearly 10^{11}, while the amphibia range from just under 10^9 to just over 10^{11} and algae range from 10^8 to midway between 10^{11} and 10^{12}. If the time in hours on a logarithmic scale for half-reassociation of single copy DNA is plotted against genome size a straight line graph runs from T4 virus to salamander (*Amphiuma*). A plot of percentage DNA sequences held in common with the DNA of the *Rhesus* monkey (on a logarithmic scale) against time since divergence also gives a virtually straight line, with fishes having about 4% of repeated sequences and over 70% with primates. For the vertebrates, shown by Britten and Kohne (1970), 'the median age of the repeated families seems to be about 100 million years—the time in which half of the repeated DNA is replaced'. There appears to be little data on fishes but the total quantity of repetitive sequences is probably large 'In . . . Pacific salmon . . . more than 80% of the DNA shows repetition that is recognizable under conditions in which the calf shows only 45% repetition' (Britten and Kohne, 1970). Obviously this is a field in which rapid developments are likely and suggest that the current popular view that the 'Genetic Code' has been completely broken is rather simple minded.

Ohno and Atkin (1966) have shown that the placental mammals constitute a uniform group as regards DNA content, which level they term 100%. Birds are also uniform having 44–59% of the mammalian content. In the reptiles two distinct groups occur; the orders *Cocodylia* and *Chelonia* have a DNA value only slightly less than in mammals whereas the *Squamata* have a DNA value only slightly higher than that in birds. These authors then suggested 'that among ancient reptiles of the Mesozoic era, two different lineages with regard to total genetic content already existed, one eventually giving rise to placental mammals, the other the avian species'.

Gall, at Yale, in a personal communication to Ohno and Atkin showed that amphibians had very high values of DNA the lowest (*Bufonidae*) having 140% and *Necturus maculosus* having 2,789%. These amphibia have only 24–28 chromosomes which 'makes doubtful any propinquity of descent between present day amphibians and members of higher classes. It seems more reasonable to assume that the evolution of terrestrial vertebrates from aquatic vertebrates was polyphyletic'. Ohno and Atkin (1966) therefore made a preliminary survey of the chromosome complements of various fishes (representing eight species, two subclasses, six orders and eight families), where diploid chromosome numbers ranged from 38 in the Dipnoan *Lepidosiren paradoxa* to 102± in the Cyprinid *Carassius auratus*. The DNA values were obtained by measuring the Feulgen stain content of the red cell nuclei with the Deeley integrating microdensitometer (with a crushing condenser) (Deeley, 1955; Atkin *et al.*, 1965). Only a single *Lepidosiren* was available, an adult female 70 cm long. The 38 chromosomes were all metacentrics of enormous size and the DNA value extremely high—3,540%. The goldfish despite its 102± chromosomes only had a DNA value of 52%.

> 'The low diploid chromosome number, the absence of acrocentrics, the enormous size of the individual chromosomes, the very high DNA value found in the lung fish*, all these are precise characteristics of the genomes maintained by present-day members of the order Caudata of the class Amphibia. Although the chronology of evolution suggests that the lung fish could not have been the direct ancestor of the Amphibians, it is apparent that both belong to the same particular lineage of vertebrates which finally gave rise to the reptiles, birds and mammals of today'.
>
> 'Recent advances in knowledge of molecular structures of enzymes and other proteins revealed that multiplication of the ancestral gene and subsequent mutation in independent directions of these multiples played a very significant role in the evolution of vertebrates' (Ohno and Atkin, 1966).

These authors cited the five different kinds of polypeptide chain produced in ontogeny by man and other mammals and pointed out that Ingram (1963) believed that duplication of an ancestral gene first produced a gene for a polypeptide chain of myoglobin, on one hand, and a gene for the alpha chain of haemoglobin on the other. The genes for four other kinds of polypeptide chains of haemoglobin were, in turn, derived from multiples of the gene for the alpha chain.

Polypoidy can duplicate, even go as far as quadruplicate the entire genome but it

> 'is incompatible with the well established chromosomal sex-determining mechanism. For the tetraploid XXYY male is as likely to produce XY gametes as either XX or YY gametes . . . even among fishes of today normal hermaphroditism operates in certain members of the family Serranidae of the order Percomorphi (Atz, 1964). Thus, it is entirely conceivable that 300 million years ago, fishes ancestral to the terrestrial vertebrates of today did not have a firmly established chromosomal sex-determining mechanism, and there was no barrier against polyploid evolution then'.

These authors (Ohno and Atkin, 1966) propose that the three fish species representing two diverse orders with lowest DNA value (about 20% of mammals) and with identical diploid chromosome numbers made entirely of acrocentrics are the retainers of the original vertebrate genome. The green sunfish (*Lepomis cyanellus*) and the discus fish (*Symphysodon aequifasciata*), both Percomorphs, belong to the triploid lineage and the gold fish (*Carassius auratus*) of the Ostariophysi represents the pentaploid lineage from which are derived the birds and the Squamata. The rainbow trout (*Salmo irideus*) of the order Isospondyli may belong to the octoploid lineage to which the present day Crocodylia and Chelonia demonstrate close kinship: 'The base DNA value of 20% may, in turn, represent the polyploid lineage of the more ancient genome possessed by prevertebrate organisms'. This argument of Ohno and Atkin (1966) is supported by their references to LDH where the A, B and C

* Allfrey (1955) found a very high absolute value of DNA in *Protopterus* (the African lung fish) namely, 100×10^{-9} mg as compared with $7{\cdot}0 \times 10^{-9}$ mg in placental mammals.

polypeptides of mammalian LDH have been found in birds and many fishes. 'These findings are in conformity with the view that in vertebrates any DNA values above 20% of that of placental mammals indicate polyploid lineage. Flatfish of the order Heterosomata (with the lowest DNA measured) on the other hand, revealed the presence of the A polypeptide only'.

Rees (1964) pointed out that in the Salmonidae there were three groups with about 60,80 and 100 somatic chromosomes and that the numerical sequence had, in Svärdson's (1945) view, argued strongly in favour of a polyploid evolution. Svärdson suggested a basic haploid number of 10 for the group so that Atlantic salmon (*Salmo salar*) would be hexaploid, the brown trout (*Salmo trutta*) octoploid and the grayling (*Thymallus thymallus*) decaploid respectively. In support of this argument he showed that the 60–chromosome group contains 6 metacentrics and the 80–group 8 metacentrics which is what would be expected in a polyploid series based on a haploid set of 10. Rees (1964) found the similarity in DNA amount in salmon and trout nuclei difficult to reconcile with a polyploid evolution since if the 80 chromosomes of *S. trutta* were

> 'derived by the addition of two haploid sets of 10 from the 60 chromosome form we should have expected a corresponding increase in DNA. The case for polyploidy can be sustained only by postulating a chromosomal diminution in DNA with increasing chromosome number. Convincing evidence in favour of such diminution is exceptional although it may well occur in rare instances (see Hughes–Schrader and Schrader 1956)'.

Rees (1967) suggested that the evolution of the two species proceeded as follows:

> 1. The nuclear DNA content is similar in both trout ($2n=80$) and salmon ($2n=60$). The amount per chromosome is consequently greater in the latter and, as would be expected, the salmon chromosomes are larger than those of the trout. Indeed the nuclear DNA amount is approximately proportional to the total chromosome length which, it will be recalled, is much the same in the two species.
>
> 2. "Fusion" or "fragmentation" accounts for the change in number without appreciative change in the chromosome length or in nuclear DNA content. This accounts perfectly for the equivalence both with regard to DNA amount and to total chromosomal lengths in trout and salmon.
>
> 3. "Fusion" or "fragmentation" also accounts for the change in shape as well as in number that distinguish the chromosomes of the two species investigated and here it is worth pointing out that there is very good evidence to show that fusion or fragmentation of this kind does indeed occur in other salmonid species (Simon and Dollar, 1963).
>
> Both the cytological and cytochemical evidence are very satisfactorily explained by chromosome "fragmentation" or "fusion". While polyploidy is not completely ruled out by this evidence it would appear that polyploidy, at best, is unlikely.'

Bailey *et al.* (1969) have shown that salmonids (*Oncorhyncus tshawytscha* and *Salmo gairdnerii*), unlike higher vertebrates, contain duplicate genes for supernatant malate dehydrogenase, a dimeric enzyme, three forms of which exist in these fishes. Cory and Wold (1966) showed that enolase, another dimeric enzyme, occurs in three forms in salmonids and M_4 lactic dehydrogenase, a tetrameric enzyme, has duplicated genetic loci giving a series of five isoenzymes in this group of fishes. Bailey *et al.* (1969) pointed out that 'These multiple cases of gene duplication afford substantial confirmation of the hypothesis that salmonids are tetraploid' since it is unlikely that the duplicated enzymes represent 'four isolated and independent instances of regional chromosomal duplication'. These authors also note that both brown and brook trout have indications of extensive polymorphism at A and B loci in malate dehydrogenase.

Sibley (1967), discussing birds, conceded the fact that the work of the past decade on comparative protein studies has solved few taxonomic problems although it has thrown further light on many. Yet he remained optimistic that the complete amino acid sequence of proteins can provide an objective proof of taxonomic status provided that a sufficient number of proteins are compared. He believed that since 'it is highly improbable that two long sequences of amino acids in homologous proteins will result from convergent evolution' the number of proteins required to provide a high probability of a correct answer 'turns out to be possibly as few as two and certainly not more than five'. He does enter the caveat that 'there is a possibility of convergence in protein structure between closely related species'.

Zuckerkandl (1965) was not quite so optimistic as he said:

> 'In order to establish by chemical paleogenetics the evolutionary relationship between two different organisms it should not be necessary to know the sequential composition of thousands or even hundreds of homologous polypeptide chains. To require such knowledge would be discouraging indeed. It can reasonably be predicted, however, that a comparison of relatively few chains—perhaps a few dozen—should yield a large fraction of the maximum amount of information that polypeptide chains can provide. The reason is that even relatively few chains should yield a good statistical sample of the evolutionary behaviour of many chains'.

Zuckerkandl's high figure of a few dozen sequences seems to be rather pessimistic as Dayhoff (1969) pointed out that the probability of duplicating a sequence of a 100 amino acid chain is 20^{100} since each of the 20 amino acids can occur in each of the 100 positions in the sequence. On the other hand, her variation in the time required per PAM must be remembered.

Population genetic studies which have been carried out on many commercially important species of fish have enabled fisheries biologists to identify unit stocks and often confirm data produced by classical meristic studies.

The application of biochemical and especially serological techniques to fish studies has increased at a very rapid rate. In 1961 Sinderman said 'Immunogenetic studies of fishes have a remarkably brief history, mostly confined to the last ten years. At present I know of only eight groups in the world with major continuous research activity in this field'. Cushing's (1964) review listed 164 references while de Ligny's (1969) important and extensive review had no less than 269 references although many were to preprints, personal communications and Russian papers not easily available.

PREVIOUS REVIEWS

Apart from the two mentioned above there have been few general surveys of the use of all available biochemical and immunological techniques in the study of fish taxonomy. Urist and van de Putte (1966) have reviewed the chemical composition of the serum of Elasmobranchs and some other species in which urea and electrolyte levels are listed.

Buzzati-Traverso and Rechnitzer (1953) introduced the use of chromatographic techniques in taxonomic studies. Wright (1966) reviewed the chromatographic, electrophoretic and serological methods used by experimental taxonomists up to that time. In a section on 'Future Prospects' he made the important point 'that these methods are not substitutes for traditional morphological approaches but are auxiliary to them . . . . However, if a taxonomic problem cannot be readily resolved by morphological comparisons there is almost certainly some experimental approach which will throw light on the situation'. This is certainly even more true today when we have a very great variety of techniques increasing in number every day.

Wilkins (1967) has drawn attention to the innate plasticity of the morphometric and meristic characters traditionally used in the analysis of fish stocks and that this has led the fishery biologists to turn to newer techniques.

Nuttal (1904) wrote the first account of comparative serology although he mentioned few species of fish. Alan Boyden (1953, 1963) has summarized the trends in the field of systematic serology to which he and his school contributed so much.

Cushing's (1964) very valuable review of the literature on 'The blood groups of marine animals' was concerned mainly with fishes and marine mammals although it devoted a page to work on marine invertebrates and, despite its title, it gave two pages to soluble molecules under genetic control. Seven pages were devoted to blood serology techniques and key references were given to useful papers, especially reviews, on blood group research, general immunology and the genetics of blood group research.

While this paper was being prepared, Miss de Ligny (1969) produced her very comprehensive review 'Serological and Biochemical Studies on Fish Populations' in which she noted 'work on anadromous species is fully covered, studies on freshwater species are discussed if they are considered of value for the studies of marine fish, either with regard to the methods used or the results'. A number of other problems were considered also and she was kind enough to let me have a copy of her uncorrected manuscript in advance of publication for which I am very grateful and on which I have drawn.

In the present review I have tended to use freshwater fishes as examples whenever possible in order to extend de Ligny's review and because much of the work carried out in my laboratory has been on freshwater species. Thus it should be possible to regard this review together with those of Cushing (1964) and de Ligny (1969) as complementary to each other. It is hoped that these reviews cover all the important papers that have appeared on the relevant topics. Some overlap has of course been unavoidable.

The 55th Statutory meeting of the International Council for the Exploration of the Sea (ICES) decided that

> 'a special meeting will be held for one day in advance of the 1969 (57th) Statutory Meeting, for the discussion of methods for and application of biochemical identification (i.e. by sera, proteins etc.) of species and races of fish, with a special view to how these methods can be incorporated in the programmes for stock analysis'.

This meeting, which I had the privilege to chair, was held in Dublin on September 28th and 29th, 1969 and the Proceedings of that meeting have been published (de Ligny, 1971). Most of the authors provided preprints of their papers for the meeting and a number of them have been kind enough to allow me to refer to their papers. These references are cited in the text as Dublin (1969) and appear in the reference list under de Ligny (1971).

II. Taxonomic Problems in Fish

Despite the very large number of papers on comparative biochemistry and serology that have appeared in the last decade or so the fishes have not attracted as much attention proportionately as have other classes of vertebrates. It is very doubtful if we have comparable data for even one per cent of the 25,000 or so living species. Indeed Leone's (1968) list of fishes referred to in the immunotaxonomic literature was less than 120 species. Most of the species studied have been the limited number of commercially important types. Yet the fishes are by far the largest class of vertebrates and show a far greater diversity of morphology and possibly of origin than do the other classes of vertebrates.

In order to have a firm foundation for a biochemical and serological base to the taxonomy of fishes it is clear that studies are urgently needed of a number of groups. Among the Elasmobranchs the Holocephali especially need investigation, among the bony fish the Sarcopterygii have been very neglected and although the Dipnoi are not too difficult to obtain they have not been much studied. Of course the investigation of a *Latimeria* will give us invaluable information especially if we accept Zuckerkandl's (1965) postulate that

> 'Contemporary organisms that look much like ancient ancestral organisms probably contain a majority of polypeptide chains that resemble quite closely those of the ancient organisms . . . it is unlikely that selective forces would favour the stability of morphological characteristics without at the same time favouring the stability of biochemical characteristics, which are more fundamental'.

Fortunately there is a good deal of data available about the lampreys although the hagfishes have been less extensively studied.

The Actinopterygii are known largely from a limited number of the teleosts and there is little known about the Chondrostei and Holostei. Romer (1966) arranged the Teleosts in nine superorders and in the near future an effort should be made to cover the living orders systematically.

Work on haemoglobin and on the immunoglobulins and the evolution of immunological responses have thrown light on the higher taxonomic categories and some valuable work has been done on fishes which is covered in the 'Phylogeny of Immunity' by Smith, R. J. *et al.* (1966).

The fisheries biologist is primarily interested in intraspecific variation in species, the identification of fish stocks (non interbreeding populations) and possible race formation. The existence of sibling species defined by Mayr (1963) as 'morphologically similar or identical natural populations that are reproductively isolated' may also pose a problem to the economic icthyologist. An example is the case of the arctic *Sebastes* problem the solution of which required, not only morphological, but also, biochemical and serological techniques (p. 293). The Coregonid problem was one in which morphological data alone appear to have provided a satisfactory solution since the number of gill rakers were found to be unaffected by the environment (cf. the long series of papers of Svärdson from 1945 to 1965).

There would appear to be a very substantial body of evidence indicating that fishes may evolve in a greater variety of ways than there is data for in other animal groups.

Hybridization would appear to play a role in speciation in fish especially in freshwater. Some hybrids such as the salmon-trout produced by Piggins (1965) seem to continue to produce fertile and healthy fish at least as far as F_2 (Haen and O'Rourke, 1968). C. L. Hubbs (1955) showed that the occurrence of natural hybrids of freshwater species correlates well with suspected phylogeny based on morphological studies. Hubbs and Drewry (1960) agreed with this view and showed that artificially produced hybrids can be correlated with phylogeny and recommend the study of hybrids as a tool in phylogenetic studies. They studied artificial hybrids between *Crenichthys baileyi* and four species of *Fundulus*, *Lucania parva*, *Odina xenica* and *Cyprinodon variegatus* and concluded 'The marked difference in success of crossing *C. baileyi* with *Fundulus* and its allies, as compared with crosses with *Cyprinodon*, indicates that *Crenichthys* is phylogenetically closer to the Fundulines than the Cyprinodontines.' Hubbs and Strawn (1957) showed that the sperm of *Percina caprodes* (which produces many known natural hybrids) retains its fertilizing ability longer in heterospecific than homospecific matings than does that of *Etheostana spectabile* which has relatively few natural hybrids. 'The short temporal function of sperm of freshwater fishes in the natural environment is thought to be an isolation mechanism'. On the other hand C. L. Hubbs (1955) pointed out that hybrids in marine fishes are much rarer than those between north temperate freshwater fish.

Hubbs and Drewry (1958) believing that the duration of sperm function might be shorter in salt water and therefore, an effective isolating mechanism, tested this with two euryhaline species (*Fundulus grandis* and *F. similis*) but found that the sperm retains its fertilizing powers for heterospecific matings much longer than that of tested freshwater fishes. Hubbs (1955) suggested that the ephemeral and changing nature of the freshwater environment resulted in the high incidence of hybrids. Hubbs and Drewry (1958) suggest that, on the basis of their experiments, a lower survival rate than controls occurs in brackish water; whereas the freshwater hybrids tested seldom have lower survival rates than that of their controls (Hubbs and Strawn 1957).

Of special interest is the recent paper of Hagen and McPhail (1970) suggesting that in *Gasterosteus aculeatus* the two forms *leiurus* and *trachurus* are considered species in several streams and lakes in British Columbia whereas Miller and Hubbs (1969) have produced evidence of intergradation and introgression between *leiurus* and *trachurus* on the Pacific coast and regard them as subspecies. As Hagen and McPhail (1970) admit 'should conclusive data for introgression be forthcoming, it would then seem that the two sticklebacks act as species in places and as subspecies in others'. Such a situation is very similar to that described by Sibley (1961) in towhees.

III. Biochemical and Serological Techniques

While the rest of this review is divided into two main sections called Biochemical Methods and Serological Methods, the distinction is by no means clear cut. The literature on isoenzymes has been extensively reviewed as the application of their study to fish taxonomy is very recent and the enzymes present in serum, red blood cells and in a number of tissues have been studied by experimental taxonomists. The techniques used for isoenzymes are almost entirely biochemical although serological methods have been used to clarify certain problems in isoenzyme studies.

Amino acid studies are briefly reviewed as they have been somewhat underestimated in their potential. Special attention is devoted to mucus in view of my own interest in the potentialities of its study. Here, both electrophoretic and serological techniques have a future. The literature on eye-lens proteins is dealt with fully in view of the great potential they offer in taxonomic investigations involving immunoelectrophoretic and serological absorption techniques. Their study also indicates many of the problems that may arise in the use of all other biochemical methods. Serum proteins, haemoglobins and transferrins and myogens have been dealt with rather briefly as they have recently been extensively reviewed.

Serological methods reviewed are precipitin systems and the immunodiffusion and immunoelectrophoresis techniques that depend on the occurrence of precipitins. Complement fixation methods are mentioned in view of the great value of recently developed micro techniques and blood groups are very briefly referred to as Cushing (1964), de Ligny (1969) and Ridgeway (Dublin, 1969) have thoroughly covered the literature.

A. BIOCHEMICAL METHODS

1. Isoenzymes in Taxonomic Research in Fish

Many enzymes, even after crystallization, have been found to be made up of mixtures of closely related molecules. Two terms are used for these forms namely, isozymes and isoenzymes and in many cases they are used synonymously. Markert and Møller (1959) proposed the term isozyme 'to describe the different molecular forms in which proteins may exist with the same enzymatic specificity' and Latner and Skillen (1968) in their textbook 'Isoenzymes in Biology and Medicine' state that the word isoenzyme is the preferable one although they give no reason why this should be so. They attribute the origin of the word to Wroblewski and Gregory (1961) although reference to that paper shows this is not correct as these authors use both terms, apparently indiscriminately, and one of their references is to a paper entitled 'Isoenzymes and Myocardial Infarctation' written by Wroblewski *et al.* (1960). Augustinsson (1961) stated that the term isozymes 'should therefore be restricted to enzymes, the molecular structures of which differ only in those parts of the molecules that are not directly involved in the enzymatic reaction.' Fottrell (1967) stated that the term isoenzyme 'has been limited to multiple enzyme forms in a single species'. A subcommittee set up by the International Union of Biochemistry has recommended (Webb, 1964) that 'multiple enzyme forms in a single species should be known as isoenzymes although since either form is readily intelligible this recommendation is not to be interpreted as excluding the use of "isozyme" if any individual author prefers it'.

Giese (1968) says

> 'Isozymes (Isoenzymes) are enzymes in multiple forms, all performing the same general function at different rates. They are different to some extent in chemical composition so that they are separable electrophoretically. Thus lactate dehydrogenase is found in five electrophoretically distinct fractions. . . . The five lactate dehydrogenase (LDH) isozymes differ in many properties; catalytic activity, amino acid composition, sensitivity to heat and immunological responses'.

There is obviously a semantic problem in the use of the terms for these enzyme forms and this should be kept in mind when isoenzymes are being discussed.

The reader should be clear as to what a particular author means when he uses the term. Latner and Skillen (1968) admit that definition of an isoenzyme is difficult and they suggest it does not include the closely similar enzymes found in different tissues of the same individual or in related species.

This is such a rapidly growing field that Latner and Skillen (1968) listed one thousand references in the text and incorporated a further list of about 300 references which appeared while the book was going through the press.

Obviously in this situation of rapidly expanding information one must expect that much of the work done at present may have to be reassessed in the light of future developments especially in regard to the techniques used. The finding of multiple bands of enzymatically active elements after electrophoresis does not necessarily imply the existence of several isoenzymes since multiple bands may be produced by combination between serum proteins and enzymes (Latner, 1966). Extraction and purification techniques may produce artefacts, such as have been found in the multiple forms of ribonuclease (Shapira and Parker, 1960). Several forms of mammalian cytochrome C, when separated by ion exchange chromatography, were produced by the preliminary extraction with aqueous trichloracetic acid (Margoliash and Lustgarten, 1961).

The buffer used to extract the enzyme from the tissues may influence the isoenzyme pattern (Paul and Fottrell, 1961) and Fottrell (1967) emphasizes that

> 'These studies show that, for adequate proof of the existence of isoenzymes, several different extraction procedures should be tried. It is essential to provide good evidence that the various molecular forms do not in fact arise one from the other during the experimental procedures'.

Ontogenetic changes occur in the isoenzyme patterns in mammals and birds (Fottrell, 1967). In fishes these changes usually occur in the early phases of development (p. 264) and should be looked for before presuming that polymorphisms exist. The possibility that environmental factors may influence the patterns was suggested by the fact that low oxygen tension in tissue culture of monkey heart cells altered the type of LDH produced. (Dawson *et al.*, 1964). Changes also occurred in goldfish liver LDH patterns both during temperature adaptation (p. 264) and as a result of alterations in oxygen tension in the water (Hochachka, 1965). Possibly in both cases the altered level of dissolved oxygen caused the enzyme alterations.

Probably more is known about lactic acid dehydrogenase (LDH) than any other enzyme. It is proposed therefore to devote most of the available space to this enzyme which has been studied in the tissues, serum and red cells of a number of species of fish. The complexity in the distribution of the LDH types found and of the various genetic systems postulated to explain their inheritance in different species suggest that much work remains to be done on LDH before any certain conclusions can be reached about its phylogenetic significance in fishes. I have included enough data, especially on the Salmonidae, to indicate the difficulties involved and the reader is referred to de Ligny's (1969) extensive discussion of the major problems that have been encountered. Other enzymes which have been studied in fishes include, among others, esterases of serum and kidney, peroxidase in serum and aspartate amino transferase (AAT), malate dehydrogenase (MDH), phosphogluconate dehydrogenase, creatine kinase and phosphoglucomutase in tissues.

Pfleidorer and Jeckel (1957) found LDH from skeletal muscle and heart muscle in the rat had different kinetic properties. 'This was one of the first demonstrations of the occurrence of different molecular forms of functionally similar enzymes in the same animal species' (Fottrell, 1967). Haupt and Giersberg (1958) and Wieland *et al.* (1959) who investigated vertebrates, from fishes to mammals, were among the first to carry out comparative studies. Each LDH isoenzyme was found to be composed of four subunits of which there are two kinds, A and B, each a polypeptide chain with a specific amino acid sequence. Thus one can have the five possible tetramers (A_4, A_3B, A_2B_2, AB_3 and B_4)* commonly observed in different tissues. Markert's recent review (1968) has shown that the series demonstrate progressive differences in enzymatic activity, heat stability, amino acid composition and so on.

Vesell and Bearn (1962) found six bands of LDH in the red blood cells of carp and Bonavita and Guarneri (1963) studied the three LDHs found in the brains of a number of vertebrate species including fish and they showed that the mobility of the major fraction increases in species from the higher phylogenetic levels. Their study ranged from the Selachian *Mustelus mustelus* to the cat *Felis catus*.

Markert and Faulhaber (1965) found a very negatively charged LDH in eye extracts of 30 species of fish. They also studied the muscle LDH of these species and arranged them in four groups depending on whether the muscle had one, two, three or four major isoenzymes. A single fraction was found in the fluke and the hake (*Paralichthys dentalus*), two in the porgy, yellow perch (*Perca flavescens*) bream (*Diplodus argentens*), striped mullet (*Mugil cephalus*) and many other species. Three isoenzymes were found in some genera including sea-trout (*Cynoscion*), bass (*Roccus*) and pike (*Stizostedian*). In addition pyruvate inhibition was found to be more significant for LDH from mammals than fish and the brain LDHs of lower phyla had higher K_m values for lactate than those from higher phyla.

Kaplan *et al.* (1960) found higher APAD/TNAD activity ratios in fish heart muscle than in mammalian heart muscle LDH. This activity ratio has been studied in the heart muscle and in both the light and dark muscles of a number of fishes. Kaplan and Ciotti (1961) reported that flatfish (e.g. sole, flounder and halibut) have closely related LDHs which are strikingly different from those of other fish. Heart and skeletal muscle LDH of the flatfish appear to differ somewhat but share the ability to react rapidly with the acetylpyridine analogue. The malic and liver alcohol dehydrogenases of flatfish differ markedly from those of other vertebrates. Kaplan and Ciotti (1961) stated 'It has been reported that the flatfish group originated in the Eocene period'. This is about 50 million years later than the date of origin of most teleost groups.

* These have also been termed LDH 5, 4, 3, 2 and 1 and M_4H_0, M_3H_1, M_2H_2, M_1H_3 and M_0H_4 (the M and H standing for the component most abundant in mammalian heart (H) and skeletal muscle (M) respectively).

> 'As indicated in the table, the perch heart LDH when compared with other fish, shows characteristics somewhat similar to the flatfish enzymes. It has been suggested that the perch group might be the ancestor of the flatfish. Our data supports such a view and illustrates the usefulness of this type of enzyme comparison in evaluating the relationships of different organisms'.

This suggested phylogeny would agree with Romer's (1966) views as indicated in his family tree of the teleost orders (Fig. 91, p. 62). Recently Lush (1968) demonstrated multiple LDHs in the central nervous system of four families of flatfish examined, while Dando (Dublin, 1969) has independently studied LDH variation in the tissues of 17 species of flatfish. He found that LDH has in general 'a low frequency of variant alleles in the flatfish'.

Kaplan and Ciotti (1961) investigated the differences in LDH between light and dark muscles of some fish. In trout, mackerel, herring, scup, butterfish and dogfish there is

> 'a closer relationship between the heart and dark muscle dehydrogenases than between the dehydrogenases from the dark and light muscles. These results suggest that in these species the heart and dark muscles may have a common origin. In other fish, such as the sea robin, a similarity in properties of the enzymes from the dark and light muscles has been found'.

Whiting (*Merluccius bilinearis*) was the species that appeared to be most similar to mammals in its LDH patterns. Polymorphism at the B locus of whiting LDH has been found with one of the mutant alleles occurring as frequently as the normal (Markert and Faulhaber, 1965). Dissociation and recombination experiments with the whiting isoenzymes have shown that the five major LDHs are made up of tetramers (Latner and Skillen, 1968). Evidence of tissue LDH specificity was found in many, but not all, fish. Five LDHs were found only in three genera namely, whiting (*Merluccius*), herring (*Clupea harengus*) and the shad (*Alosa sapidissima*). Odense *et al.* (1966) suggested that in the herring the A and B subunits preferentially assemble in similar pairs before tetramers are formed hence AAAA, AABB and BBBB are the main combinations found.

Whitt (1969) has pointed out that in many vertebrates the LDH polypeptide subunits (A and B) are encoded in at least two codominant loci. Mammals and birds have three such loci, the C locus functioning only in primary spermatocytes. In teleosts a third locus, termed E by Markert and Faulhaber (1965) has been postulated; it has marked specificity. The E_4 isozyme normally has a very high negative charge and is localized in the nervous system, especially in the retina. Occasionally it is found in the non-neural tissues of teleosts as in the eye-lens. Whitt (1969) has proposed the E_4 isoenzyme may be more closely related to the B_4 than to the A_4 isozyme because the kinetic and physical properties are closely related. The genetic origin of the E subunit was investigated in a panmictic population of 245 adult killifish (*Fundulus heteroclitus*), from Wood's Hole, by studying the allelic isoenzyme variations at the LDH B locus. Three LDH phenotypes were found, slow (BB), fast (B′B′)

and the heterozygote hybrid (BB′) with intermediate mobility. The frequency of B_1 was 0·269 and of B, 0·731 which conforms with the Hardy–Weinberg formula. It was found that neither E_4 nor A_4 fractions changed their electrophoretic mobility even though the B subunit shows changed mobilities due to mutation.

> 'E_3B_1 and $E_4B'_1$, reflect in their number and mobility the subunit contributions of the two different alleles which makes it unlikely that E_4 can have arisen by the epigenetic alteration of the B subunit. The author rejects the idea that the amino acid substitution in the mutant B polypeptide occurs in a region that is eliminated or modified during epigenetic conversion of B subunits to E subunits'.

His conclusion that the subunits A, B and E are under separate genetic control is confirmed by immunochemical data using precipitin tests. Further in some species, as the weakfish (*Cynescion regalis*) A and B do not associate with each other, either *in vivo* or *in vitro*, although E will associate with either A or B subunits in molecular hybridization studies *in vitro*.

> 'This genetic, physical and immunochemical evidence suggests that that subunit is encoded in a separate LDH locus (the E locus)'.

The immunological studies show that B_4 and E_4 LDH isozymes are related and E_4

> 'is more closely related to the B_4 than the A_4 isozyme. This interpretation is also supported by the hierarchy of biochemical responses of $E_4 < B_4 < A_4$ observed for K_m with pyruvate as substrate, pyruvate concentration optimums, resistance to pyruvate inhibition and susceptibility to inactivation by heat and urea'.

Whitt (1969) suggests that the LDH loci A and B of vertebrates probably arose from a single ancestral gene represented by the A locus which is more active than the B gene in many vertebrate tissues, especially in flatfish, where the B gene function is restricted to only a few tissues. The A locus appears to be less variable in teleosts than the B locus. In some teleosts the A and B loci have each duplicated as a result of tetraploidization, (p. 260) and the duplicated genes still retain a high degree of homology.

Whitt (1969) believed the LDH gene E probably arose from a duplication of the B gene followed by mutation and selection resulting in divergence. 'The E locus is found only in fish, which suggests that it may have arisen after fish gave rise to the amphibians'. Amino acid sequence data and further immunochemical work along these lines is clearly warranted. Whitt (1969) drew attention to the possible similarity between vertebrate LDH genes and the haemoglobin system

> 'where the γ gene arose by gene duplication from the X locus, and later the B locus arose from the γ locus. The presence in teleosts of three LDH genes coding for related polypeptides affords an excellent opportunity to examine evolutionary relationships among genes, as well as providing a basis for investigation of the specificity of gene activation during cyto-differentiation with reference to gene homology'.

Mutants of the B subunit of LDH have also been found in the brook trout (*Salvelinus fontinalis*) by Morrison and Wright (1966) in the herring (*Clupea harengus harengus*) by Odense *et al.* (1966) whose work is discussed later (p. 262). in the hagfish (*Eptatretus stoutii*) by Ohno *et al.* (1967), and in the cod (*Gadus morhua*) by Odense *et al.* (1969).

Mutants at the A locus have been found to occur in trout (p. 259) and have also been found in the herring (Odense *et al.* 1966) and in the hagfish by Ohno *et al.* (1967). Ohno *et al.* (1968) showed that in the anchovy (*Engraulis mordae*) there was a similar homozygous three isoenzyme pattern as well as a heterozygote pattern due to an A locus variant.

Odense *et al.* (Dublin, 1969) added the information that further mutants had been found in four species of gadoid involving either the B locus (three species) or the A locus (in two species). Utter (1969) found two alleles at the A locus of LDH in the Pacific Hake (*Merluccius productus*). Wilkins (1968) has reported up to 10 LDH isoenzymes encoded at four separate loci in various tissues of Atlantic Salmon (*Salmo salar*); Duke and Kelly (1968) reported further LDH isoenzymes in the retina of this species although Wilkins (1968) noted that the eye-lens patterns contain only the tetramers found in other tissues. He found that the white muscle contained all the known isoenzymes of LDH found in the salmon although ontogenetic changes in quantity occurred.

Hodgins *et al.* (1969) have found LDH variants in the sera of the sockeye salmon (*Oncorhynchus nerka*). There are three phenotypes probably due to two alleles at the B locus. Ultracentrifugation showed that all variants have a molecular weight of about 150,000 and a similar number of subunits.

Using polyacrylamide gel electrophoresis of tissue extracts from speckled trout (*Salvelinus fontinalis*) collected in the Gaspe Hatchery, Quebec, an isolated selected inbred population, nine bands of LDH were reported in an early paper by Goldberg (1965) who obtained no evidence of tissue specificity but found evidence of a C gene locus in all tissues giving three homopolymers AAAA, BBBB and CCCC and 12 heteropolymers by the random combination of the three polypeptides. This theory he subsequently withdrew (Goldberg, 1966) and reported the occurrence of mutants at the A locus. He found that nine bands occurred only in hybrids (=splake) between *S. fontinalis* and *S. namaycush* and between female splake and male *S. fontinalis*. Five LDHs with species and tissue specificity occurred in the parental species from other areas in Canada and the U.S.A. He also found multiple LDHs in the lake trout (*S. namaycush*) which had LDH bands with mobilities similar to the closely related *S. fontinalis* while *Salmo gairdneri* had a different pattern. He proposed that the synthesis of the three polypeptide LDH subunits was controlled by three non-allelic gene loci and that there should be 15 different LDHs if the three subunits were arranged in all the possible subgroups of four. In the splake (*Salvelinus namaycush* female × *S. fontinalis* male) nine LDHs have been found and in the later paper (Goldberg, 1966) he suggested that

LDH[1] (BBBB) has a common gene locus in both species which would give the nine forms found. *In vitro* 'hybridization' of LDH from both parental species also gives nine LDHs on electrophoresis because some of the forms have identical mobilities. The Gaspé trout probably had a mutant allele in their genotype. Goldberg (1966) suggests that 'the B locus is common to all members of this group (Salmonidae) of fish and, from an evolutionary standpoint, that the B gene represents the ancestral type from which has evolved the A gene and, consequently, the multiplicity of LDH in the Salmonidae'. Breeding experiments by Morrison and Wright (1966) showed genetic control of the B subunits by codominant alleles on an autosomal gene. Crossing of brook and lake trout enabled backcrosses to be carried out which indicated linkage of the genes coding the A and B subunits (Morrison and Wright, 1966). At least two further LDH isoenzymes have been found in the eyes of trout which are probably under genetic control and are not found elsewhwere. The heterogeneity of LDH in fish is more complex than is the case in mammals (Goldberg, 1966).

Minor isoenzymes have been found in the eyes and gonads of many species of fish and not elsewhere in their tissues. A separate subunit C was proposed for the isoenzymes in the eye of all Salmonidae (Goldberg, 1966). This C appears to combine predominantly with B and because of this preferential association Morrisson and Wright (1966) suggested that the C subunit may have evolved from B by gene duplication and mutation. The muscle specific subunits D and E found by Hochochka (1966) may have arisen from A in a similar way but some hybridization may occur between all types of subunit. Cytogenetical evidence has been adduced by Ohno *et al.* (1968) and Klose *et al.* (1968) to back up the gene duplication theory. In the order Isospondyli (which includes the Salmonidae and the Clupeidae) formation of new LDH subunits took place in phylogeny by gene duplication as a result of polyploidy. There is a correlation between LDH isoenzymes and the chromosome number since additional LDH isoenzymes in muscle are only found in the tetraploid species of the order. Thus it was suggested that the specific muscle LDHs of Salmonidae (designated as D and E previously) which resemble those of a clupeid heterozygous for both LDH loci are two alleles of a single gene locus A in diploids while in tetraploids they would become two additional loci A^1 and A^2. A lengthy discussion of these problems has been included by de Ligny (1969) in her review.

The gonadal isoenzymes found in many species of fish are a group of five LDHs differing from those in somatic tissues and possibly controlled by two separate genes (Latner and Skillen, 1968). Hochachka (1966) who independently studied trout species suggested that five kinds of subunit occurred in these fish. Subunits A, B and C give the nine isoenzymes already mentioned. AC hybrids can be produced *in vitro* but have not been found in the fish themselves. Using muscle LDH, dissociation and recombination experiments have shown that 15 isoenzymes can be produced by randomly assembling the subunits in tetramers. *In vivo* however, a control mechanism is postulated which thus prevents certain possible combinations limiting subunit assembly to the nine found in nature. Two further subunits D and E (independent of A, B and C) give five further LDHs in the muscle of both species which migrate towards the cathode whereas the LDHs built from the A, B and C subunits migrate towards the anode. There are difficulties in all these hypotheses as is explained by de Ligny (1969).

Wilson *et al.* (1964) showed that all lower vertebrates including fish, amphibia and lower reptiles have heat unstable LDHs (half-lives of H_4-type LDH of 20 min at about 60°C) which migrate rapidly (electrophoretic mobility) whereas the higher reptiles have heat-stable and fast LDHs and mammals have unstable and very fast LDHs. The Palaeognathous birds have stable and fast LDHs while the Neognathous ones have stable and slow LDHs indicating that during the evolution of the reptiles there was a marked division into two groups one leading to the mammals the other to the higher reptiles and birds (Fig. 1). The activity of H_4 from man was reduced by 50%

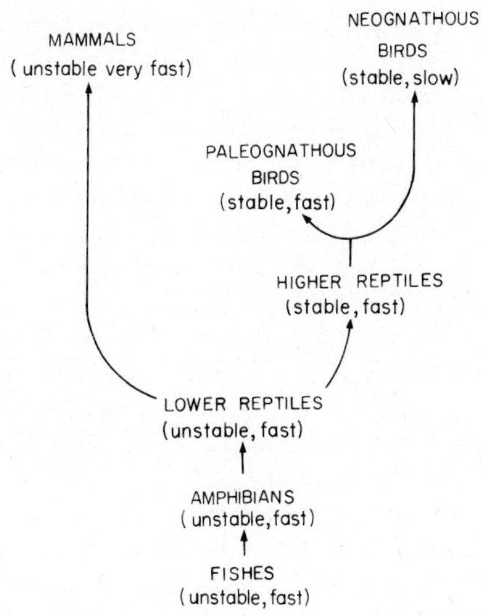

FIG. 1. Evolutionary tree showing the changes in heat stability and mobility of lactate dehydrogenase.

by exposure to 65°C for 20 min while the enzyme from the lizard *Iguana iguana* was reduced by 50% at 82°C for 20 min, that of the ostrich 50% at 80°C for 20 min. For a 50% reduction in the activity of LDH in the bullfrog the same effect was produced by a temperature of only 52°C. The electrophoretic mobility of H_4 LDH averaged less than 10 cm in 16 hr at 10 V/cm in 12 species of amphibia and fishes, 15 cm in 11 species of mammals and only about 2 cm in the 21 species of Neognathous birds examined.

Microcomplement fixation studies by Wilson *et al.* (1964) using anti-chicken H_4 serum, produced in rabbits, showed that H_4 isoenzymes from birds were closely related while those from reptiles, frogs and fish were less so, in that order. Studies with antisera to fish isoenzymes showed a marked change from the Agnatha to the bony fish.

Odense *et al.* (1966) were the first to describe polymorphisms in both LDH and aspartate aminotransferase (AAT) in the herring (*Clupea harengus harengus* L.) from the Western Atlantic. Five phenotypes of LDH occurred and represent two mutant alleles at the B locus and one at the A locus. They found that the frequency of the phenotypes fitted the Hardy-Weinberg hypothesis and that the three Canadian populations of herring had different allele frequencies. Ridgeway *et al.* (Dublin 1969) determined the LDH patterns of over 1,900 herrings from the Gulf of Maine and found some specimens of the mixed heterozygote phenotype AA^1BB^1 not found by Odense *et al.* (1966). These authors also used blood types and the esterase system described by Ridgeway *et al.* (1969) as Es-1 and the results with all three methods 'support the conclusion that the Georges Bank stock of herring is independent of the stocks of herring within the Gulf of Maine and in Southern Nova Scotia'. Their results support previous evidence from meristic studies by Anthony and Boyar (1968). Naevdal (Dublin, 1969) reported on the distributions of LDH and AAT isoenzymes showing that two rare mutant alleles were found in AAT. He also studied forms of serum esterase in herring from Norwegian waters. Simonarson and Watts (1969) studied white muscle esterase variants in herring stocks from the English Channel, Scotland and Ireland and found from two to ten bands of activity in different specimens.

Howlett and Jamieson (Dublin, 1969) have suggested that their preliminary electrophoretic study of white muscle esterase in the spratt (*Sprattus sprattus*) may prove a possible stock diagnostic characteristic since it is a simple system possibly involving 15 genotypes. Different frequencies of the five postulated allelomorphs in the two regions of the North Sea sampled, suggest that they hold two genetically isolated populations.

Scopes and Gosselin-Rey (1968) have reported polymorphism in both carp and *Tilapia mossambica* muscle and creatine kinase (CK) and phosphoglucose isomerase of carp and inolase of *Tilapia guineensis* show polymorphioms. A two-unit substructure for CK has been reported in mammals and birds but in carp only two bands were found in the presumed heterozygotes 'suggesting that either the normal homozygote enzyme consists of two non identical subunits produced at separate genes, or that random combination of subunits does not apply *in vivo* in the carp muscle'. Scopes and Hamoir (Dublin, 1969) reported on the glycolytic enzymes of the water soluble muscle extract from five species of *Tilapia*.

In 1961 Kaplan and Ciotti suggested that the evolutionary changes that have been observed with the mammalian and fish LDHs are not so apparent with the malate dehydrogenases (MDH). Some differences have been detected among a number of teleost fish and mammals however. They believed that with the data then available the MDHs were the result of a single evolutionary change whereas the LDHs of vertebrates, invertebrates and bacteria may have arisen independently in the course of evolution.

Augustinsson (1961a,b) found acetylcholinesterase (ChE) in the plasma of a variety of fishes (frogfish, wrasse, cod, eel, pike and shark). Aliesterases (AliE) were found in high concentration in the teleosts mentioned as well as in the hagfish in which it was the only plasma esterase detectable. Although in the AliE enzymes the specificity varied, frogfish and cod had acetyl AliE, while pike, shark, hagfish, wrasse and eel had propionyl AliE and the last two species also had butyl activity. The AliE of frogfish and pike are sensitive to eserine unlike that of the eel (*Anguilla anguilla*) whose plasma is the richest animal source of aliesterase (lipase) activity ever described and is highly resistant to eserine. While all mammalian plasmas studied have high concentrations of arylesterase this enzyme is absent in amphibian, reptilian and fish plasmas.

Nyman (1965c, 1965/66) studied the serum esterases of seven families and over 30 species of fish and found one to four bands were present in the electrophoretograms. These patterns were quite different from those found in the organ extract enzymes. Similar patterns were found in related species. Intraspecific variation was found in four families namely, Gadidae, Cyprinidae, Percidae and Salmonidae (*Salvelinus alpinus*). In *Gadus morhua* the variation in serum was not further studied although a length correlated variation was found in the liver esterase. This phenomenon was also detected by Nyman (1967a,b) in both the liver and muscle esterase patterns of *Salmo salar*.

In *Salvelinus alpinus* two bands migrating closely together were found and the distribution of the three phenotypes observed. The appearance of all liver and kidney parental bands in an interspecific hybrid showed that it was genetically controlled (Nyman, 1965c, 1967a). Kidney esterase in this species is possibly controlled by two alleles with three phenotypes. In the Cyprinidae (Nyman, 1967a) population genetic data on the roach (*Leuciscus rutilus*) supported the hypothesis that the two bands found were controlled by two alleles (Nyman, 1965/66).

In marine fish esterases have been used in the study of five species of Thunnidae (Sprague, 1967; Fujino and Kang, 1968), the herring (*Clupea harengus*) (Naevdal and Danielsen, 1967), plaice and flounder (*Pleuronectes platessa* and *P. flesus*) (de Ligny, 1966). This work is reviewed by de Ligny (1969). But it may be mentioned here that in the southern bluefin (*Thunnus maccoyii*) and the bigeye (*T. obesus*) Sprague (1967) found some individuals without serum esterase activity which is a feature yet to be explained. He also found that in the bigeye there was a sex-linked difference in esterase activity. De Ligny (1968) found quantitative differences in the esterase patterns of flounders of different lengths.

De Ligny (1969) considered that since 'the number of plasma specific enzymes is limited' enzymes are best studied in homogenates of various tissues and organs although she suggested that 'the variety of enzymes in red blood cells' might be profitably studied if the problem of preserving the red cells or their haemolysates could be overcome. She was 'not aware of any studies in this field so far'. She appeared to have overlooked Vesell and Bearn's (1962) study of the six bands of LDH in the red cells of the carp.

a. Ontogenetic changes. As in the case of serum proteins, haemoglobins etc. changes in isoenzymes occur during ontogeny although they have only been extensively studied in LDH, MDH and some esterases.

In the cyprinodont fish the medaka (*Oryzias latipes*) disc electrophoresis analysis by Nakano and Whitely (1965) showed only one LDH (AAAA) present in the egg (LDH5) but the four others, including the B subunit, appeared on hatching. In the adult one to three LDHs were found in all organs although LDH1 and LDH2 were found only in the retina.

Hitzeroth *et al.* (1968) showed, in the C subunit of the LDH of the retina (found only in Salmonids) that up to about 70 days after hatching the formation of subunits from maternal genes precedes the comparable process due to the paternal genes.

The different proportions of these isoenzymes found in various cells at different stages of development are probably related to the metabolic needs of the individual tissues. LDH isoenzymes with a high H (=B) content, such as are found in heart muscle, are most active at low concentrations of pyruvate. Such LDH forms tend to direct the lactate into the Krebs cycle rather than towards the formation of lactate. In the case of the M (=A) types they require energy sporadically and the lactate produced enters the blood and is metabolized elsewhere. This type of regulation is also of adaptive significance for heart cells where a constant energy supply is required.

Wilkins (1969) found in *Salmo salar*, that the LDH isoenzymes of the white muscle varied with age. In specimens less than about 15 cm the 'heart type' (most anodal tetramer) was present whereas fraction 5 (the muscle type) is relatively concentrated while fraction 1 occurs only in trace quantity. Ontogenetic changes in quantity were also found in the nonspecific liver esterases.

b. Changes caused by environment. Environmental factors also influence isoenzyme patterns. Thus in the goldfish alterations occur in the liver LDH pattern during adaptation from a 20–22°C water temperature to an environment at 4–5°C. In the liver the pattern was altered apparently by an increase in synthesis of the H subunit during cold adaptation. This was possibly associated with an increase in extra-mitochondrial metabolism known to occur in fish livers in these circumstances. The respiratory rate increases as adaptation develops and the energy for the activation of some of the reactions involved in metabolic activity is decreased. It was believed that the change in the respiratory rate might be associated with the appearance of isoenzymes with decreased energy of activation. It is of interest that in most other tissues the total LDH increased but the isoenzyme pattern was unaltered except in the liver (Hochachka, 1965).

In later papers Hochachka and Somero (1968), Somero (1969) and Somero and Hochachka (1969) produced a new view of what happens in temperature acclimation. They pointed out that *in vivo* substrate levels are usually so low that enzyme activities do not reach their maximum velocities and that under normal conditions in life the reaction velocities may be primarily dependent on the maximal enzyme–substrate affinity K_m. To test this hypothesis they examined LDH isoenzymes of the Salmonid *Salmo gairdneri*, whose relative activities are known to depend on acclimation temperatures, the South American lungfish (*Lepidosiren paradoxa*) which is highly warm-adapted (27–30°C) and the Antarctic fish *Trematomus borchgrevinki* which is markedly cold-adapted (−2°C). They also studied the tuna in which muscle temperature is regulated and held at a fairly high level whereas the temperature of its heart and other organs remain at the ambient level.

The experimental results showed that the K_m of a given LDH isoenzyme varied markedly with temperature and the highest affinity (minimum K_m) was at the habitat or acclimation temperature as indicated in Table I. Hochachka

Table I

Species	Temperature at which minimum K_m occurred with LDH
Trematomus borchgrevinki	0°C
[a] Tuna heart	10°C
[a] Tuna muscle	16°C
Lepidosiren paradoxa	30–35°C

[a] *Thunnus thynnus thynnus* the bluefin tuna.

and Somero (1968) suggest that in the case of enzymes from poikilotherms, temperature plays a role analogous to that of a positive effect on allosteric enzymes, so that a change in temperature directly promotes a compensatory change in the enzyme's affinity for the substrate. Later Somero (1969) showed the same phenomenon in the pyruvate kinase of the Alaskan king-crab where the 'warm' and 'cold' variants are formed by a temperature dependent interconversion of one protein species. This change takes place rapidly compared to the trout LDH change which takes weeks. Electrofocusing and electrophoresis revealed only one peak so that the two variants are said to appear to be isoenzymes in the conventional sense.

Further evidence that the K_m values of isoenzymes are temperature-dependent is provided by Baldwin and Hochachka (1970) who show that acetylcholine esterase from the trout brain occurs in two forms. One 'cold' variant appears after acclimatization to 2°C, another 'warm' one at 17°C. At the extreme temperatures only one form is present whereas both forms occur after acclimatization at an intermediate temperature. The K_m values of these variants are lowest at the acclimatization temperature. Therefore the authors conclude that the K_m-temperature relationship is adaptive and that the critical process during thermal acclimatization, when enzymes show sharp changes in K_m with temperature, is the result of the synthesis of a new enzyme variant that is better suited for catalysis and its control at the acclimatized temperature. Somero (1969) describes three types of temperature adaptation (1) *phylogenetic* which is the result of evolutionary selection and is illustrated by the antarctic fish and lungfish in Table I, (2) *acclimation* which takes place within days, weeks or months and is exemplified by the situation referred to above in the trout and (3) *immediate compensation* where in a matter of seconds the oxygen consumption of tissue slices can be changed in response to a temperature change.

Koehn (1969) has shown that temperature is responsible for maintaining serum esterase polymorphism, due to two alleles at a single locus, in the freshwater fish *Catostomus clarki* in the Colorado River of the western U.S.A. There is a north-south cline in this species and the enzyme most frequent in northern populations has maximum activity at the lowest experimental temperatures. The heterozygote showed the greatest activity over the widest range of temperature but was exceeded in activity at either end of the temperature scale by one of the two homozygotes.

Wilson *et al.* (1963) studied the forms of LDH in the pectoral muscles of birds and showed that LDH5 is the main form in species which fly sporadically and short distances (e.g. pheasant and grouse) while LDH1 was characteristic of the flight muscles of such species as the storm petrel and humming birds which undertake long flights. 'This survey provided a good correlation between the predominance of LDH5 and the ability to perform limited spells of severe exercise and the predominance of LDH1 and the capacity for sustained exercise' (Fottrell, 1967). Such correlations might be looked for in a wide variety of fishes, especially in salmon muscle whose energy utilization must be extremely high when during their spectacular upstream freshwater migrations they leap over waterfalls.

Since disease in man and domestic animals alters the isoenzyme patterns and is used in diagnosis (Fottrell, 1967; Latner and Skillen 1968) the occurrence of pathologically derived variation in the patterns of fishes should be anticipated. It is surprising that such changes have not yet been reported. Latner and Skillen (1968) refer to the fact that human liver and muscle isoenzyme changes occur in astronauts (presumably due to stress especially at blast-off and re-entry). Thus the stress of space flights must cause some cellular damage (Hawryhewicz and Blair, 1965) and it might be that osmotic stress may cause similar changes. Thus in the catadromous and anadromous fishes moving to and from marine to freshwater, environmental stress may well produce similar changes as indeed may spawning itself. These are obvious problems needing investigation.

It is clear that the investigation of isoenzyme patterns in fish blood and tissues is a promising and expanding line of study. Again one feels impelled to appeal to the workers involved to cover as wide a range of families of fish as possible in order to provide more fundamental taxonomic data which in turn will undoubtedly aid the applied worker.

2. *Amino Acids*

Buzzati-Traverso and Rechnitzer (1953) introduced the use of chromatographic methods for the study of the free amino acids of fishes. Dannevig (1956), Viswanathan and Krishna Pillai (1956) and Enge and McKee (1959) using tissue crushes reported finding species-specific patterns among various groups and Schaeffer's (1961) results on the amino acids in *Sebastes* muscles supported other evidence and seem to have stood the test of time. However, Farris (1957, 1958) showed that the chromatographic patterns of free amino acids observed in the Pacific sardine (*Sardinops caerulea*) could be influenced by environmental factors including diet. These findings, in the view of Cushing (1964), effectively blocked further work based on chromatographic techniques but I believe that, under certain circumstances, useful studies could still be made since the currently popular isoenzymes may also be subject to variation under the influence of environmental conditions.

3. *Mucus*

Hiscock (1949) extracted mucin from the foot of gastropod molluscs and found that antisera to it could be produced in rabbits. He showed that mucus from different species was immunologically species specific although they showed some cross-reactions. Wright *et al.* (1957) and Wright (1959) showed that the body surface mucus of a number of species of *Lymnaea* had species-specific chromatographic patterns. The chromatograms were examined under ultraviolet light and gave mixtures of fluorescent and absorption patterns. Later (Wright, 1964) showed that a number of chromatographically distinct forms of *Lymnea peregra* occurred. These forms had genetically controlled patterns and were differentially distributed geographically.

Although surface mucus is a characteristic of fishes, little is known about it as can be seen from Jakowska's (1963) review. Barry and O'Rourke (1959) showed that fish mucus was species specific but that ultraviolet analysis of the chromatogram was not, in itself, sufficient to distinguish closely related species which 'may require the use of several different solvents, and it may be necessary to examine, not alone the dry chromatogram, but also the wet one, under ultra-violet light. Furthermore, various location agents may be required'.

For example, *Gadus pollachius* mucus can be distinguished from the mucus of *G. virens* by the fact that the latter leaves no fluorescent material at the point of application after running with butanol acetic acid, whereas the former leaves a pink fluorescent material at the origin. On the other hand mucus from *Sebastes marinus* and *S. mentella* produced similar fluorescent patterns after running with butanol acetic acid and showed eight identical bands after treatment with ninhydrin. However, on viewing the chromatograms under ultra-violet light while still wet at the end of a run in phenol ammonia solvent, the diffuse outermost band (that is, the band with the highest R_f value) is split in the former species by an evanescent absorption band not seen in the latter. After drying, the split band of *marinus* is no longer detectable as it is replaced by a single diffuse band as in *mentella*. Ninhydrin-staining of the chromatograms produced by running in phenol ammonia gives ten identical bands in both species. Running in phenol water gives a fluorescent band in *marinus* with an R_f value of 0·21, not shown by *mentella*. The conclusions about the distinctness of the two species of *Sebastes* based on mucus differences were supported by immunological evidence (p. 293) (O'Rourke, 1961b).

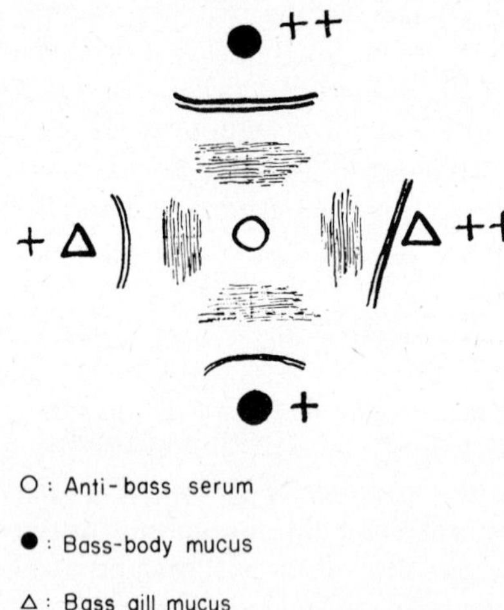

FIG. 2. Diagram of Consden-Kohn reactions between antibass serum R923 and bass mucus. Upper and lower reactions are with body mucus and left and right reactions are with gill mucus. Plus signs indicate approximate quantities applied. The tilt seen in the right hand reaction is an experimental artefact. The inner broad bands are nigrosine positive, the outer lines of precipitation are both ponceau-S and nigrosine positive. (From O'Rourke, 191a.)

The presence of blood serum antigens in both body mucus and gill mucus was shown in the bass *Morone labrax* (O'Rourke, 1960) and in *Sebastes* (O'Rourke, 1961b). In the case of anti-bass serum produced in a rabbit three AnAb reactions were given when the anti-bass serum was reacted with bass mucus using the Consden and Kohn (1959) immuno-diffusion method. It was also found, as can be seen in Fig. 2 that the gill and body mucus reacted somewhat differently. The reactions were stronger with body mucus than with that from the gill. This is perhaps surprising since gill mucus might be the more likely to be contaminated with blood. It should be pointed out that in the samples studied no contaminating blood was detected. Furthermore, the ponceau S-positive AnAb reactions were well separated from each other whereas with gill mucus the two lines of precipitate, although discrete, were very close to each other. This would indicate that the relative concentrations of the two antigens involved differ in gill and body mucus. Anticod serum (R 712) available from previous work on intrageneric relationships in the genus *Gadus* (O'Rourke, 1959) was used to study the body mucus of *Gadus morhua*. Only a single AnAb reaction, faintly ponceau S-positive, was found. Subsequent staining with nigrosine gave no further reaction bands, but the ponceau S-positive band stained more strongly.

These results showed that some of the serum protein antigens of fishes are also secreted in mucus. These species-specific antigens may be responsible, in part at least, for the ability of certain fish parasites to detect their specific hosts. Wright (1959) has pointed out that some parasites must detect their hosts by chemotactic methods, and the presence of species-specific serum proteins in the body mucus suggests a possible basis for this. This problem in parasitism has recently been further discussed by Rice (1968). Cushing (1964) referred to our mucus work as opening 'an interesting area of investigation' and calling

> 'attention to the potential value of mucus generally as a source of interesting immunological material (Hildemann's paper, 1959, on the nutritive and protective properties of parental mucus for the young of discus fish also is of interest in this connection). The papers above show that there is no question but that chromatographic procedures should be further investigated, providing that adequate measures are taken to learn the extent to which variations that are found are under genetic control'.

This proviso applies to every biochemical and serological technique.

O'Meara and O'Rourke (unpublished) have confirmed chromatographically the species specificity of the mucus of a wide variety of Irish marine fishes while Wilkins (1966a) has confirmed that proteins occur in mucus that are antigenetically similar, or identical to those of the homologous sera in whiting (*Gadus merlangus*), salmon (*Salmo salar*) and plaice (specific name not given but presumably *Pleuronectes platessa*). He reported that he was investigating the possibility of the secretion of soluble blood group substances in mucus and was interested in the possible effects of mucus on aggregation in social fishes.

Jakowska (1963) in her review of mucus secretion in fish noted only three reports on the chemical nature of fish mucus (Enomoto and Tomeyasu, 1960a,b; Enomoto *et al.*, 1961).

4. *Serum Proteins*

Sera from fishes have been extensively studied since the beginning of this century and Booke (1964) has given a good review of this work. Earlier investigations were concerned with the blood levels of protein, electrolytes and so on (Urist and van de Putte, 1966). The advent of electrophoresis brought a long series of studies of fish sera beginning with Deutsch and Goodhue (1945) who used the Tiselius (1937) method to report species specificity of the electrophoretic patterns of the blood proteins in fishes. This work was later extended by Deutsch and McShan (1949), Moore (1945) and after 1950 by many authors including Irisawa and Irisawa (1954) and Drilhon *et al.* (1966). Drilhon and Fine (1957) found that the blood of fishes shows a net dominance of globulins over albumins and that the concentration of the latter, is, as a rule, very low in Selachians. Other works that may be mentioned are those of Saito (1957a,b), who, among others, compared the individual components to the corresponding fractions in human blood. Such comparisons must be, at best, a preliminary attempt at characterization. Thus the transferrins, glycoproteins which bind iron, have been shown by Fine *et al.* (1965) and Møller and Naevdal (1966) to migrate to every possible position between the origin and the anodic end of the electrophoretogram in the Teleosts whereas in the lampreys and elasmobranchs they migrate towards the cathode!

Fine *et al.* (1964) using Fe_{59} showed that one of the electrophoretic fractions found by Drilhon and Fine (1960) in the European eel was Tf and that individual polymorphisms occurred. Møller and Naevdal (1966) found Tf polymorphism in the carp and four species of gadoid. The same holds true of many but not all species of fish studied.

De Ligny (1965) has detected over 15 transferrin variants in the plaice (*Pleuronectes platessa*) of the North Sea. She has shown by breeding experiments (de Ligny, 1968) that ontogenetic changes occur in these transferrins and that the differences between transferrins from young fishes and adults 'may represent the occurrence of transferrin molecules not yet possessing the full amount of sialic acid. This phenomenon not only interferes with the interpretation of data about the inheritance of transferrins, but also might impede identification of groups of very young fish'. In 1969 Miss de Ligny reviewed the transferrin situation very thoroughly and also covered other serum components such as the haptoglobins, albumins and prealbumins. Engle *et al.* (1958), Fine and Drilhon (1958), O'Rourke (1960), Sulya *et al.* (1961), Gunter *et al.* (1961) and Drilhon *et al.* (1961) all contributed to our knowledge of serum proteins.

Sanders (1964) was the first to study the serum of hybrids in the Salmonidae and further work was carried out by Haen and O'Rourke (1968, 1969a,b) and Nyman (1965a,b,c, 1966, 1967a,b).

Sex and age differences were first reported by Drilhon (1954) in the carp and confirmed in many species by other authors such as Robertson *et al.* (1961) Rall *et al.* (1961) (in *Petromyzon marinus*), Saito (1957d), Haen and O'Rourke (1969a,b) and by Booke (1964). Seasonal variation was first reported by Saito (1957c) and Sorvachev (1957) and subsequently the nutritional state was found to affect the serum in starved land locked sea lamprey (*P. marinus*) by Thomas and McCrimmon (1964).

Changes due to disease were reported in *Salmo trutta* kidney disease as far back as 1949 by Deutsch and McShan and by Phillips *et al.* (1957). Flemming (1958) reported changes in diseased carp, and Sinderman and Mairs (1958) found changes in herring with fungus disease. Mulcahy (1967, 1969) studied the serum protein changes in *Salmo salar* infected with UDN disease in Ireland and she suggested that the changes shown in the electrophoretograms may be used as a means of diagnosing the disease (Mulcahy, 1969).

Alterations in the serum protein pattern have been reported as responses to environmental factors such as falls in the oxygen content of the water (Bouch and Ball, 1964) and pollution (Fujiya, 1961). Thurston (1967) in a significant paper has shown that the serum protein electrophoretogram of rainbow trout may be affected by the method of capture and bleeding. Individual variations, which may be genetic, were reported by Deutsch and McShan (1949) in *Acipenser* sp. and by Creyssel *et al.* (1964) in the carp, while Wilkins (1967a) showed that in cod serum 6 of the 11 fractions detectable on electrophoresis were subject to individual variation. His technique involved immunoelectrophoresis which indicates how difficult it is to separate purely biochemical and serological approaches in experimental taxonomy.

To be sure that individual variation is genetically controlled, breeding experiments, if at all possible, are needed. Less satisfactory is population genetic data which may be explained in several different ways. De Ligny (1969) also suggested that 'the components which have been identified with proteins found to be polymorphic in higher Vertebrates' might be most profitably selected for study, although she wisely added the warning that 'it ought to be said, with the variety of the variation encountered in fish haemoglobins in mind, that more direct proof of the inheritance of the components in the fish species studied is to be preferred'. It is not surprising, therefore, that those attending the Dublin (1969) meeting recommended, that, wherever possible, breeding experiments should be carried out and that the applicability of the Hardy-Weinberg law to much population data must be very critically assessed.

5. *Eye-lens Proteins*

The soluble proteins of the eye-lens have great value in taxonomic studies. For comparative purposes these proteins may provide more data than the serum proteins because, whereas the latter are produced by a variety of cells the eye-lens proteins are synthesized by only one cell type present in the eye as a single layer. These proteins have an extremely wide range of serological cross reaction as was first shown by Uhlenhuth (1903); antiserum prepared against lenses of any of the vertebrate species give precipitin reactions with the lenses of all other vertebrates (Manski *et al.*, 1964). The possibility therefore exists of using these proteins to obtain data on the inter-relationships of higher taxa, since 'the lens proteins preserved and thus permit the study of phylogenetic relationships within the *whole* vertebrate subphylum' (Manski and Halbert, 1964).

Three main saline soluble eye-lens proteins are distinguishable by electrophoretic and immunological techniques. These are alpha-, beta-, and gamma-crystallin in order of decreasing electrophoretic mobilities, each of which constitutes a family of similar, but not identical proteins (Francois and Rabaey, 1959; Rabaey, 1959; Wood and Burgess, 1961; Maisel and Langman, 1961; Halbert *et al.*, 1961; van Heyningen, 1962). Protein with alpha-crystallin characteristics has been found in all vertebrate species studied and has come to be regarded as a classical organ-specific protein. The species specificity of the eye-lens protein pattern has been regarded as depending on beta- and gamma-crystallins (Maisel and Langman, 1961). Sibley and Brush (1967) have shown that these proteins may have a more limited taxonomic value in the study of birds than is the case with other groups, although Gysel's (1964, 1965) immunoelectrophoretic studies of avian lens proteins looked very promising. In fish we have found them very useful provided the material is used as soon as possible after collection.

In spite of the considerable amount of work done in recent years on lens proteins in many groups (Van Heyningen, 1962), only a few electrophoretic and immunoelectrophoretic analyses have been made of the lens proteins of fishes. Bon (1957) and Rabaey (1959) did some preliminary investigations and Smith (1962) undertook an electrophoretic study of eye-lens proteins using lenses of Pacific albacore (*Thunnus germo*—4 specimens), California bluefin (*T. saliens*—4 specimens) and kelp bass (*Palabrax clathratus*—16 specimens). The paper electrophoresis technique used was not sensitive and gave a two component curve for both bluefin and bass with slight pattern variations and the albacore had one component with two pattern variations.

Smith (1965) pointed out that 'the eye-lens core or nucleus is comprised of fibres that have sclerosed and died during formation because since they lacked intrinsic circulation they had been removed from sources of dietary protein and oxygen (Walls, 1942). Thus the nucleus contains only genetically-determined structural proteins—some readily soluble and ideal for studying genetic variation'. Smith (1965) used four techniques to study the eye-lens proteins of 108 yellowfin tuna *Thunnus albacares*, namely, (i) antilens serum; (ii) response to cooling; (iii) response to heating; and (iv) electrophoresis. The serological techniques did not produce particularly useful data nor did the turbidity produced by cooling. Heating eye-lens extracts produced rapid clear differential precipitations of proteins. This technique does not seem to have been further investigated. The paper electrophoretograms showed three patterns. 'The differences between them were presumably due to genetic differences within yellowfin tuna'. Two major components were common to all while the presence or absence of minor components produced the variations. De Ligny (1969) refers to further work by Smith (1966) which I have not seen and says that

> 'using cellulose acetate, electrophoretic patterns of the lens cortex and nucleus were compared and a higher resistance of the protein constituents of the latter during storage was reported. The pattern obtained from the isolated lens cortex was, moreover, found to show similarity to that of the aqueous humour of the eye, and was quite unlike that of the lens nucleus. In further investigations on intraspecific variation the author made use only of the lens nucleus. In a study of bluefin tuna (*T. thynnus*) from two different areas, the occurrence of seven different patterns was observed, heterogeneity being rare in one and large in the other population'.

In 1964 Rabaey published a comparative study of the lens and muscle proteins of 35 species of fishes. The sensitivity of his technique enabled him to stress the large number of protein fractions, up to 11, found in the eye-lens of fishes and he noted that this complexity was more pronounced than in most other vertebrate groups. Rabaey (1964) found that, as in the Amphibia, there was a very small amount of alpha-crystallin present in the eye-lenses of fishes as compared with those of reptiles, birds and mammals and that there was less in the *Selachii* than in the *Teleostei*.

Rabaey's (1964) study included investigation with gel filtration on Sephadex G-75 and immunoelectrophoresis. The low molecular weight lens proteins are responsible 'not only for the great complexity of the fish lens protein pattern but also for the marked variability even among two related species'. Rabaey found that in the electrophoretograms there are multiple protein fractions with different electrophoretic mobilities but with identical antigenic properties (as has been found also in birds). The proteins with higher molecular weights are more stable in the same family and even in the same order. Rabaey (1964) concludes 'this protein multiplicity pattern is highly specific and it is very probable that it will turn out to be one of the most reliable biochemical criteria in the study of animal relationship'. Bon *et al.* (1964) studied the eye-lens proteins of 32 species of fishes. They found that the lens proteins showed distinct species specificity with up to 13 possible fractions, but at the same time also exhibited certain family and order relationships. There is much work to be done in the identification of the nature of the various fractions. Most of the species studied by Bon *et al.* were marine, though the group did include two 'species' of salmonid, the brown and sea trout.

Bon *et al.* (1964) have shown that all crystallins investigated react with antiserum to α-crystallin of cow prepared in a rabbit. This organ specificity applies to vertebrates from fishes to mammals and in their investigation Bon *et al.* (1964) got positive reactions between the cow antiserum and *Scyliorhinus canicula*, *Crenilabrus cinereus*, *Gadus merlangus* and *G. morhua*. In all fish lenses investigated some proteins have a net positive charge at pH 8·6 and move towards the cathode. These proteins are absent from mammals. Although Bon *et al.* (1964) use the same terminology as is used for the mammals they do not intend this to imply that the components of similar groups are identical in the two classes. In the Euselachii there are pronounced α- and β-groups and a series of low peaks of the γ- and δ-groups. There is similarity between the patterns of *Carcharinus glaucus* (great blue shark) and *Squalus acanthias* (spurdog) and while the ichthyobibliographic card index (IBCI) puts these species in different orders both are surface swimmers. The other species *Scyliorhinus canicula*, *Eugaleus galeus* and *Raja clavata* are bottom-living and the first and third species show features in common such as α- and β-groups split and γ- and δ-groups show more differentiation. 'This resemblance could perhaps be related to the fact that both split off in the Jurassic Period' (Bon *et al.*, 1964).

These authors found minor differences in the subsidiary components of the sea trout (*Salmo trutta*) and river trout (*S. fario*) which are regarded by most authorities as conspecific. Of the six flatfishes investigated it was found that they fell into three groups each of which belonged to a separate family. In the Scombridae there was great differentiation in the lens proteins and a similarity in the split of the α- and β-peaks. There was a marked similarity between *Scomber scombus* and *S. japonicus*. 'Even more remarkable are the almost identical patterns of *Xiphas gladius* (Swordfish) and *Thunnus thunnus* (tunny)'. In this connection it is of interest that Barrett and Williams (1967) (cf. below) found virtually no interspecific differences in three species of *Thunnus*. Four species of Gadidae (cod, haddock, whiting and ling) were clearly separated from the hake (*Merluccius merluccius*) of the family Merluccidae. *Trachinus draco* (weaver) and *Lophius piscatorius* (sea-devil) were clearly separated since the former has

> 'a clear differentiation and a very complete set of components whereas the latter shows only an α–β-group and a γ–δ-group. As in the case of the Euselacchi there is perhaps an indication that this difference correlates with the difference in the power of vision. The weavers have highly developed eyes, the sea-devils have not. A similar correlation can be seen in mammals where the cow and horse have a simple pattern and carnivores a more differentiated one' Bon *et al.* (1964).

Goldberg (1966) and Morrison and Wright (1966) found an eye species-specific series of polymorphic forms of lactic acid dehydrogenase (LDH) in the extracts of the eyes of all the Salmonidae studied. Morrison and Wright (1966) undertook extensive genetic studies on *Salvelinus fontinalis* and the forms found in the eye are shown in Fig. 8 of de Ligny's (1969) review. Lush (1968) described the forms of LDH found in the retina of both cod (*Gadus morhua*) and coalfish (*G. virens*) and these too are figured in de Ligny's review (Figs 6 and 7). It is not clear whether these LDHs occur at all in eye-lens protein extracts. Barrett and Williams (1967) studied the eye-lens proteins of yellowfin tuna (*Thunnus albacares*), albacore tuna (*T. alalunga*), bluefin tuna (*T. thynnus*), skipjack tuna (*Kartsuwonus pelamis*) and bonito (*Sarda chiliensis*) storing the eyes at $-16°C$ until analysis was undertaken up to seven months later. They reported that the patterns were remarkably similar for the three

Table II
Lens-Protein Fraction Groups in Bonito, *Sarda chiliensis*

	Groups			Gene frequency	
	AA	AB	BB	qA	qB
Observed	17	12	6	0·657	0·343
Expected	15·1	15·8	4		
Length range of fish (cm)	51–72	47–54	43–44		

species of *Thunnus* especially on the cathodal side where the six or seven bands showed 'migration distances for each band almost identical for the three species'. Smith's (1965) report of polymorphism in the yellowfin tuna was not confirmed, the discrepancy may result from different techniques, since he found only four components, as contrasted to at least 11 fractions found by Barrett and Williams (1967). Smith (1962) failed to find polymorphisms in the albacore and bluefin soluble lens proteins but his work, using paper electrophoresis, was based on the separation of only one or two fractions. Barrett and Williams (1967) confirmed this although they examined only 32 specimens of albacore and 30 bluefin but in the 35 bonito they found marked species specificity and polymorphism in the cathodal fractions (Table II) where the fast fraction is A, the slower B. Assuming control by two codominant genes gave the result that

> 'The probability of a worse fit between the expected and observed frequencies is between 0·10 and 0·20 ($x^2 = 2{\cdot}153$, d.f. $= 1$), supporting the hypothesis of genetic control. But further analysis (Fig. 3) showed that B occurred in the smallest specimens, AB in the intermediate sizes and A only in the largest fish. Therefore, the conformity of the data to Hardy-Weinberg principles may only be fortuitous'.

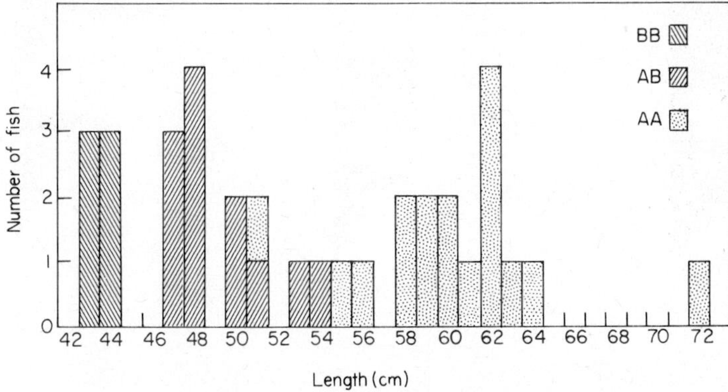

FIG. 3. Lengths of bonito, *Sarda chiliensis*, from which eye-lenses were collected, in relation to the distribution of the apparent phenotypes. (After Barrett and Williams, 1967.)

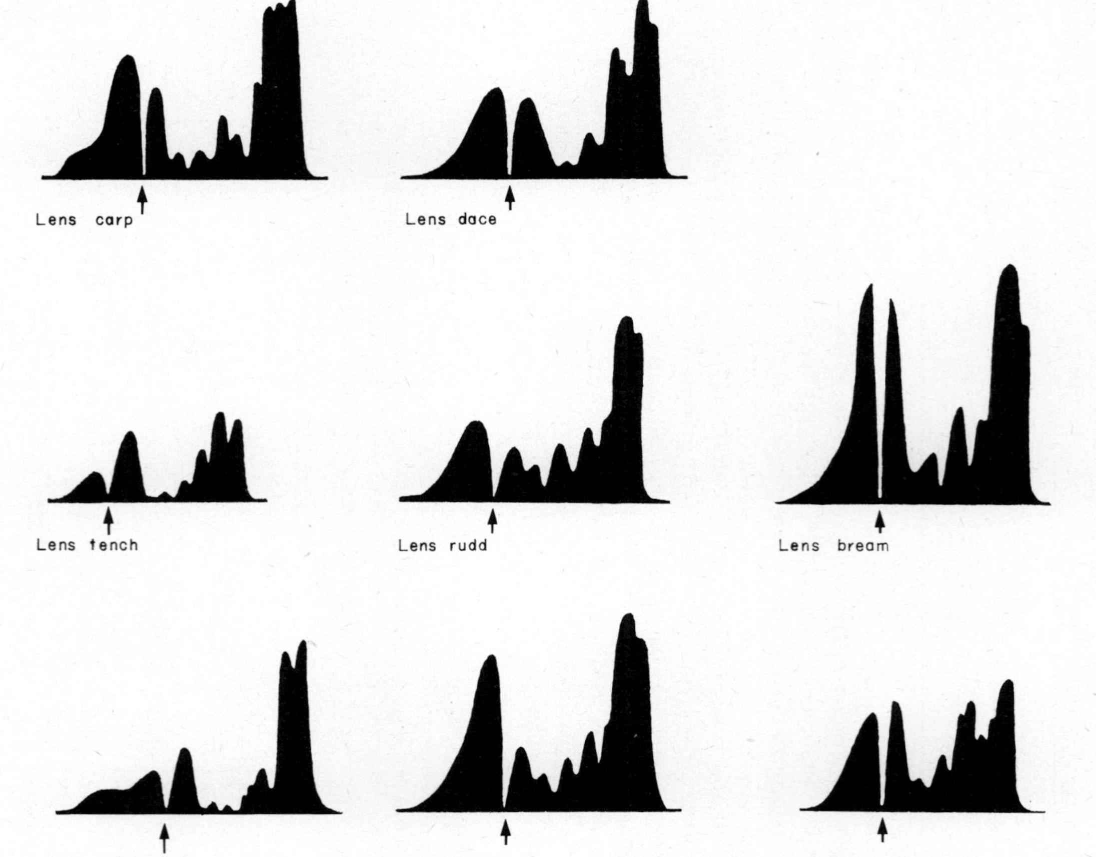

FIG. 4. Electrophoretic pattern in the eye-lens proteins of seven species of cyprinid and the rudd-bream hybrid. (From Haen and O'Rourke, 1969a.)

This is surely a salutary warning. The authors who found no differences between the patterns of eyes from each of 60 yellowfin concluded that gel-electrophoresis, *under the conditions used* (my italics) is not useful for population analysis and 'not even interspecific differences were apparent for the three species of *Thunnus*'. Initial studies of Pacific *Merluccius gayi gayi* and *M. polylepis* using paper and agar electrophoresis produced only three fractions and no interspecific differences were found (Schlotfeldt and Simpson, 1965). Six possible polymorphisms were found in a large sample of cod (*Gadus morhua*) using the starch gel technique (Odense *et al.*, 1966). Using nuclei and cellulose acetate, Smith and Goldstein (1967) reported polymorphism in the ocean whitefish (*Caulolatilus princeps*) in which three to six bands were found and four different patterns could be seen in two populations sampled.

Haen and O'Rourke (1969a) studied the lens proteins of 14 species of Irish freshwater fishes using electrophoresis in agar gel at pH 7·3 with phosphate buffer and each species showed complex characteristic patterns.

It can be seen (Fig. 4) that all seven cyprinid species studied, carp *Cyprinus carpio*, bream *Abramis brama*, tench *Tinca tinca*, goldfish *Carassius auratus*, rudd *Scardinius erythropthalamus*, roach *Rutilus rutilus*, and dace *Leuciscus leuciscus*, have their own specific lens protein patterns. No intraspecific variations, other than ontogenetic differences, could be found. Although each species shows its own characteristic pattern, certain relationships between all these lens separations can be detected. In each case the fastest moving group of fractions contains the highest proportion of the total protein and shows the highest peaks in the curves. Separation is best in the fractions nearest to the cathode. A second group of fractions with a high percentage of the total protein is found on both sides of the point of application. A third group of fractions is found in between the two former groups and usually contains a number of fractions present in low concentration. The carp and the goldfish are alike in having two anodic fractions.

The electrophoretic patterns of the eel *Anguilla anguilla*, the perch *Perca fluviatilis*, and the pike *Esox lucius*, lens proteins clearly demonstrated the specificity of these patterns. As in the case of the cyprinids, the fastest moving group of fractions are the most complex and contain the highest proportion of total protein. The main difference between these three species and the members of the cyprinid family that were studied can be found in the relative concentrations of the second and third group of fractions. In the cyprinids the group of fractions with the lowest mobility has a much higher percentage of the total protein than the middle group. In the eel, pike and perch, on the other hand, we found the reverse of this picture. Here the group of fractions near the origin is present in low concentration, while the middle group of fractions shows a much higher density.

Four members of the *Salmonidae* (salmon *Salmo salar*, rainbow trout *S. gairdnerii*, brown trout *S. trutta* and Cole's char *Salvelinus colii* now regarded as a form of *S. alpinus*) were studied. Each exhibits its own pattern, but at the same time similarities between the patterns are apparent (Fig. 5). Thus the total protein is divided fairly evenly between the main fractions. This contrasts with the other species investigated.

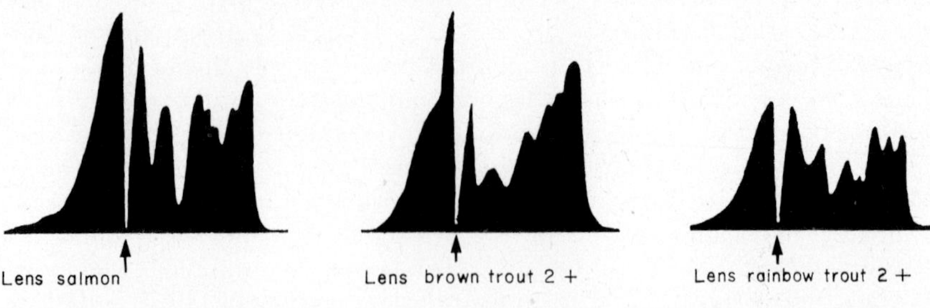

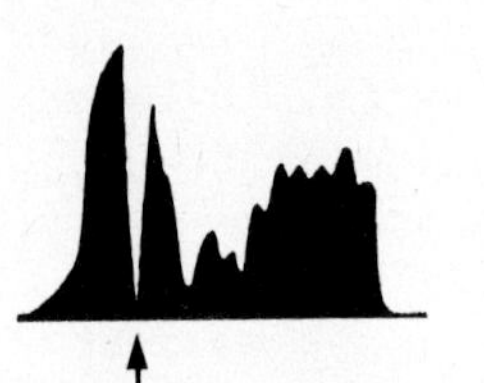

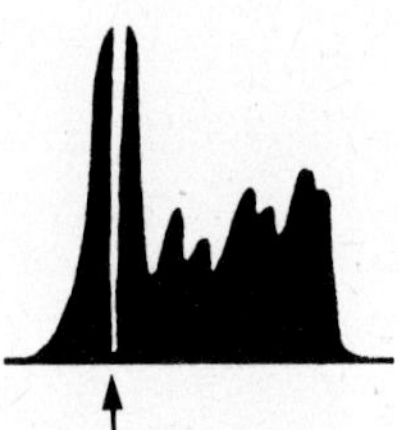

FIG. 5. Electrophoretic patterns of eye-lens proteins of four species of salmonid and one hybrid. (From Haen and O'Rourke, 1968.)

a. Hybrid fishes. The eye-lens protein patterns of hybrid fishes show some interesting features. The electrophoretic patterns of the lenses of the naturally occurring rudd-bream hybrid and its two parental species are shown in Fig. 4. The lens protein patterns of the rudd (with eight fractions) and the bream (with seven) are very alike, although they are often placed in separate genera, and the pattern of their hybrid resembles those of the parents and especially the rudd in having eight fractions. In contrast with the pattern in the rudd-bream hybrid, the pattern of the salmon-trout hybrid (F_2) is not very like that of either parental species (Fig. 5). The electrophoretograms show at least seven fractions in the brown trout, at least eight components in the salmon and at least ten fractions in the hybrid (Haen and O' Rourke, 1968). The hybrid pattern is, therefore, more complex than either of the two parental patterns. Furthermore, it does not seem to us to be a simple mixture of the two patterns.

b. Age changes. Francois and Rabaey (1957a; 1957b) reported changes in the composition and proportions of the lens proteins of mammals with age and Barrett and Williams (1967) reported age changes in the bonito. Ontogenetic differences in the lens proteins occurred in the two species of trout (brown and rainbow) we studied. The lens protein patterns of four different stages in the life-cycle of the rainbow trout are shown in Fig. 6. These two species may be considered together as the phenomena observed show a similar pattern in both. At the six-month stage the lens protein pattern in both species shows a high proportion of the total protein in the fastest moving group of fractions. The middle group of fractions is very weak, and some proteins found in older fish are either completely absent or show only traces.

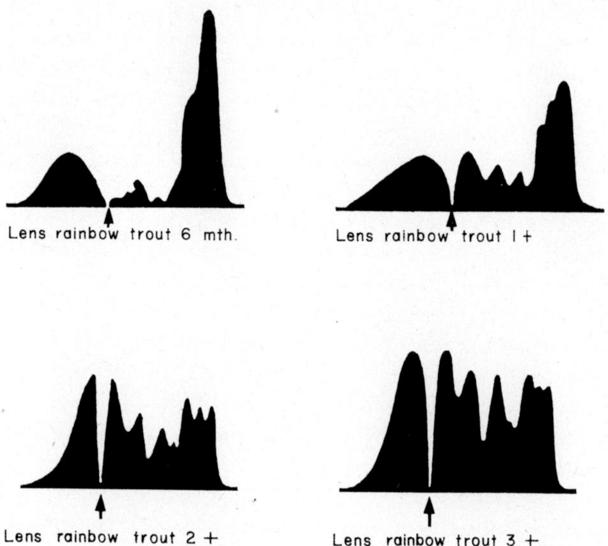

Fig. 6. Age changes in eye-lens protein electrophoretic patterns of rainbow trout. (From Haen and O'Rourke, 1968.)

Near the point of application an anodic band is present, while on the cathodic side protein is either completely absent or present in very low concentration. It may also be observed that separation of the lens fractions in the fastest moving group of the six-month-old fish is fairly poor, as most bands seem to merge into one another. In the 1+ stage differentiation of this group of fractions has already improved. The middle group of fractions become more prominent and the slowest moving cathodic fraction is now present as a weak band. In the 2+ lens protein pattern the fastest moving fractions separate out more clearly, the middle group of fractions increase considerably in density, and the two fractions on either side of the point of application become quite pronounced. The 3+ lens protein patterns show the final stage of development. There are no basic differences between the 2+ and 3+ stages, except that all the fractions increase in concentration and show up as completely separated bands. No further differentiation occurs after the 3+ stage.

Some eye-lenses of rainbow trout, bream, rudd and rudd-bream hybrids found in Ireland were infected with the metacercariae of the fluke *Diplosto-*

mulum spathaceum which gave a peculiar opaqueness to the lenses (Haen and Ryan, 1967). Such infected eye-lenses were discarded. The possibility of such infections affecting results should be investigated.

The results of studies on eye-lens proteins in fishes show great promise. The eyes of fishes can be easily collected even from commercially valuable species and large numbers of specimens can be studied and readily stored. Our samples, stored at 4°C were used within a week but Barrett and Williams (1967) stored the eyes at −16°C and used them up to seven months later and stated 'Increasing the time of frozen storage of the supernatant solution caused no differences in the electrophoretic patterns of the lens proteins of the five species'. Bon *et al.* (1964) found striking similarities between *Salmo trutta* and *S. fario* (now regarded generally as conspecific) where the only differences in the lens protein patterns occurred in the subsidiary components. These differences may be due to environmental factors. Haen and O'Rourke's (1969a) study included four salmonid species, which all showed a close similarity in eye-lens pattern. The differences which occurred were found only in the protein components present in low concentrations.

Wright *et al.* (1966) used acrylamide gel to study the eye-lens proteins of two freshwater salmonids (*Salmo gairdnerii* and *Salvelinus fontinalis*) and reported polymorphism in the latter species. We understand that one of Professor Wright's team, Dr Eckroat has undertaken an important study of the inheritance of this polymorphism by breeding experiments and these results will be awaited with interest.

Smith (1968) reported that an 0·018g% saline extract produced both albumins and globulins from the eye-lens nucleus in both yellowfin tuna and a desert wood rat (*Neotoma lepida*) and produced no deleterious 'salt effects'. Physiological saline extracts gave simpler patterns on cellulose acetate electrophoresis due to the fusion of some proteins into single bands and the failure to extract others. Smith (1968) adds a note that G. L. Peterson of the University of Hawaii got similar results with 0·018% saline extracts of the nuclear lens proteins of the three elasmobranchs studied namely, the ray *Urotrygon davies* and two sharks *Hexanchus griseus* and *Carcharhinus milberti*. This modification of the usual technique of extraction would appear to mark a very significant advance in the study of eye-lens proteins.

The only study that seems to have been made of the eye-lens proteins of hybrid fishes were those of Haen and O'Rourke (1969a) whose results suggest that the lens proteins pattern of hybrid fishes may or may not resemble that of the parents.

The discovery that ontogenetic changes take place in the eye-lens proteins of some species studied, is a most interesting result. Very little work seems to have been done on intraspecific variation of fish eye-lens proteins due to age. Characteristic changes take place both in the differentiation of already existing components and in the appearance of new fractions as the fish grows older.

It is obvious that further investigations must be undertaken to determine the limiting factors in the use of eye-lens proteins in fishes. The insensitivity of the techniques used by Smith (1962, 1965) make his findings unreliable while apparent conformity with the Hardy-Weinberg Law does not necessarily prove a genetic hypothesis. There seems to be a tendency to forget all the assumptions that are presumed in the Hardy-Weinberg hypothesis. The absence of immigration and emigration, the occurrence of random mating and the absence of selection are by no means easy to exclude in fish populations. In the last analysis only breeding studies can provide certain answers.

Manski and Halbert (1964) in their fascinating study of the phylogeny of eye-lens proteins, by using immunoelectrophoretic techniques have been able to plot out a satisfying sequence of changes during the phylogeny of the whole vertebrate subphylum. They related the proteins appearing at different stages of evolution with the time scales of palaeontologists assuming that proteins common to modern vertebrates must have originated with their common ancestor. By comparing vertebrates from different classes in relation to the presence or absence of additional lens components it was possible to establish the order of the appearance of these proteins independently of palaeontological knowledge.

Their method can be summarized by a quotation

> 'In a relatively simple situation, the procedure can be exemplified by the reactions of anti-lamprey and anti-shark lens serum. Absorption of anti-shark lens serum with lamprey lens abolished the reaction with this species, but left positive reactions with the lenses of all other vertebrates tested. On the other hand, absorption of anti-lamprey lens serum with shark lens abolished the cross reactions with all other vertebrate lenses. It follows from this that the shark and other vertebrates share components not present in the Agnatha'.

They are thus able to refer to 'Devonian', 'Permian' etc. components in avian and mammalian lenses.

Manski and Halbert (1964) suggest that the lens proteins show slow rates of evolutionary changes whereas the better known serum proteins (more limited in their cross reactions) have relatively faster rates of evolution. 'In addition, serum proteins are synthesized by a variety of cells which may have different rates of evolution'. These authors used lenses excised rapidly after death and 'Precautions were taken to carefully dissect, clean and freeze the lenses rapidly and all specimens were preserved by lyophiliziation'. These precautions may explain the difficulties encountered by Sibley (1967) especially on the rapid denaturation found in bird lenses.

Thus the eye-lens proteins seem to offer an extremely useful source of data about the relationships of both higher taxa and interspecific differentiation. The uniformity of the species of *Thunnus* suggests that different buffers may be required and for subspecific analyses it would appear that other methods may be more appropriate although Smith's (1966) reports of seven polymorphic forms of the lens nucleus proteins of *Thunnus thynnus* is encouraging as is the paper by himself and Goldstein (1967) on ocean whitefish.

Pichot (Dublin 1969) reported that the 'cristallin' proteins of *Merluccius merluccius*, *M. senegalensis* and *M. cadenati* were the best way of separating the species although this division of taxa was supported by meristic characters. Examination of electrophoretograms of cristallins in more than 40 species of fishes enabled him to say that *M. cadenati* seems to be more distant systematically than the two other species are from each other. The transferrins also confirm the belief that *M. cadenati* is a well defined species. Morphological evidence indicates that it has only five cervical vertebrae as compared with six in the other two species.

While it would be foolish to rely on one technique it does appear to me that there is a strong case to be made for much more extensive investigations into the possibilities afforded by the eye-lens proteins.

6. *Muscle Proteins (Myogens)*

There are relatively few studies of the water soluble muscle proteins of fishes. Connell (1953a,b), who used the Tiselius technique of electrophoresis, was the first to analyse the muscle proteins of fishes for comparative purposes.

Dingle *et al.* (1955) studied muscle extracts of freshly killed and post-rigor cod using the Tiselius method and found no qualitative differences between them. Hamoir (1955) reviewed extensively studies on fish proteins, including those of the muscle. Hewitt *et al.* (1963) studied the water-soluble muscle proteins of 20 species of poeciliid fishes, using paper electrophoresis, and found constant, marked and reproducible differences between various tribes and genera in this family. Rabaey (1964) investigated the lens and muscle proteins of 35 fish species using agar-gel electrophoresis. He reported marked differences between the protein patterns of white muscle and of the red *musculus lateralis superficialis*. Greenberg and Kopac (1965) studied the water-soluble muscle proteins of nine species and subspecies as well as hybrids of the genus *Xiphophorus*, using paper electrophoresis.

Tsuyki *et al.* (1965) studied the muscle proteins of approximately 50 species of fishes using starch gels and found 'virtual constancy and species specific nature of these myogens' and suggested that they could be used in phylogenetic studies.

Haen and O'Rourke (1969b) studied the water-soluble muscle proteins of the white muscle of 14 species and two hybrid forms of Irish freshwater fishes using electrophoresis in polyacrylamide gel. Each species showed a characteristic electrophoretogram pattern. It was found that in some species the protein pattern of white muscle varied with the part of the body from which it was taken. Changes in the muscle proteins with age, reported by other authors in the cod and salmon, were found in brown trout and rainbow trout.

The muscle protein patterns of five cyprinids studied (Haen and O'Rourke, 1969b) are shown in Fig. 7.

Among hybrids, the rudd-bream hybrid patterns showed only quantitative differences between the fastest moving fraction of the shoulder and tail muscle proteins. In the salmon–trout hybrids marked differences between the faster moving fractions of the specimens from the different muscle areas were

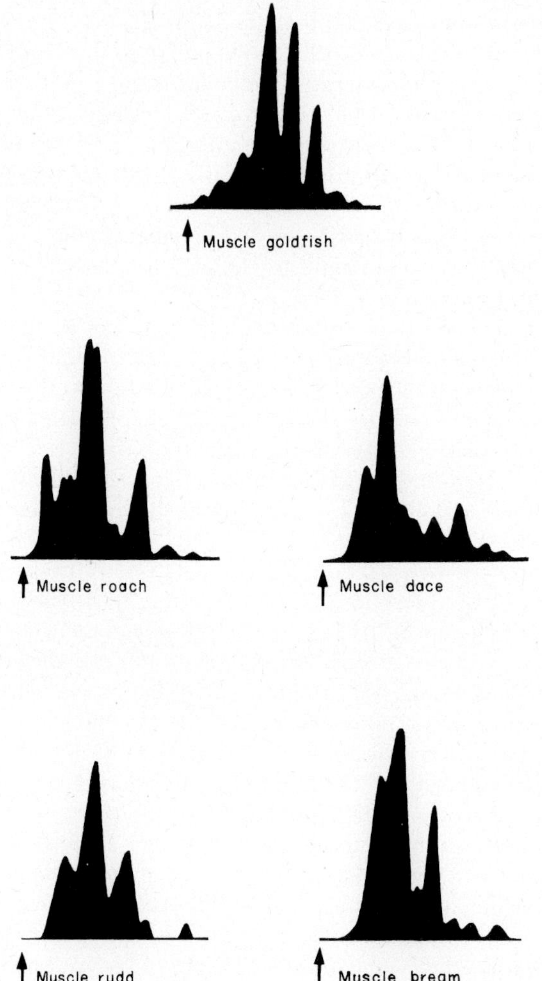

FIG. 7. Electrophoretic patterns of the mid muscle protein of five cyprinids. (From Haen and O'Rourke, 1969b.)

apparent. In the brown trout, the salmon and the salmon–trout hybrid relatively little difference can be detected between the patterns of the two parental species, while the hybrid is also very similar (Haen and O'Rourke, 1968).

The mid-muscle protein patterns of the rudd and bream hybrid seem to have inherited muscle proteins from both parents, as the slower moving fractions are more like those of the bream, while the fastest moving fraction is more like that in the rudd. The fastest moving fraction of the bream is not represented in the hybrid.

Our findings on the water-soluble muscle proteins of the two hybrid forms studied (Haen and O'Rourke, 1968) are similar to those of Greenberg and Kopac (1965) in that the patterns of the hybrids show characteristics of both parental species. Nyman (1965), using starch gel electrophoresis could not distinguish between the muscle protein patterns of rudd and roach but found their hybrid showed a quantitative increase in the density of one of the major bands. However, we found quite marked differences between the parental species (Fig. 7) using polyacrylamide gel. Nyman found eight fractions in both species while we found nine fractions in the roach but only five (or possibly six) in the rudd.

We know of no study that stresses the importance of using muscle extracts from the same region of the body in comparing water-soluble muscle proteins from different species. Our results indicate, however, that this is of the utmost importance as in most of the species studied there existed an appreciable degree of differentiation between shoulder and tail muscle within a single specimen of a given species (Haen and O'Rourke, 1969b). The relative amounts of muscle and fibrous tissue change along the length of the fish and so may account for these results.

The only intraspecific variations that were detected (provided muscle from the same area of the body was taken) were due to difference in age. Our studies showed a change in the protein composition of the white muscle of two species of trout with increase in age. Connell (1953b) who studied the cod (*Gadus morhua*) also found some evidence that age has an effect on the proportions of the protein components of muscle. Our findings agreed with Nyman (1967) who studied the *musculus lateralis superficialis* of the salmon and found marked ontogenetic changes with gradual transitions of the protein patterns and a marked increase in the number of bands when sexual maturity is reached. He described four muscle protein systems (A to D designated from the most rapidly moving system to the slowest) giving a maximum of twelve bands. The newly hatched salmon has a total of nine bands, five of which are in the A system as compared with three in older specimens. The two fastest moving fractions in the C system only occur in the newly hatched forms and are absent from the one-year-old stage onwards. The one-year-old stage has only five bands (3A and 2C) while the maximum of twelve is reached at the two-year-old stage and from then on only quantitative changes occur. Connell (1953b) did not specify the quality or degree of change that occurred with age.

It may be concluded from these studies that, in some fishes at least, there occur changes in the muscle proteins as the animals get older. These changes in the Salmonidae consist of a decrease of fast moving fractions and an increase in slow moving components in all three species (the salmon, the brown trout and the rainbow trout) investigated. Further studies on muscle proteins seem well worth pursuing.

7. Haemoglobins

a. Multiple haemoglobins. The first reported electrophoretic studies on fish haemoglobins (Hbs) were published in 1959 by workers in America, Japan and India. Each group believed that it was the first in the field. In America, Buhler and Shanks (1959) examined the Hbs of 14 freshwater and saltwater fish species using paper and moving-boundary electrophoresis. Good electrophoretic separations were achieved in sodium barbital (veronal) buffer at pH 8·8, and it was found that many of the Hbs had several components. The isoelectric points of these Hbs were always rather high as compared with those found in higher vertebrates. Five species were found to have electrophoretically homogeneous Hbs including the carp (*Cyprinus carpio*), two Hb components were present in five other species while three fractions, all in very similar ratios, were found in both steelhead and rainbow trout, and the brook trout (*Salvelinus fontinalis*). In India, Chandrasekhar (1959) made comparable analyses using agar gel electrophoresis at pH 8·6 on the Hbs of ten species of freshwater fish. Again one to three components were demonstrated in different species. In Japan, Hashimoto and Matsuura (1959a,b) in their studies of fish Hbs used a Tiselius electrophoresis apparatus with a veronal buffer at pH 8·6. Among the five species investigated were the rainbow trout and the carp. Again from one to three fractions were found. All their analyses were made on fresh haemolysates containing only oxyhaemoglobin, except that in the case of the carp the haemolysate was also oxidized with ferricyanide to give methhaemoglobin. Whereas the oxyhaemoglobin of the carp showed two components, the methhaemoglobin showed only one. Hashimoto and Matsuura (1960) published a further report on the heterogeneity and species specificity of fish Hbs. This time, using the same technique, a number of sea-water, brackish-water and freshwater fishes were investigated. Apart from a number of closely related Salmonids, the fishes again were shown to have one to three Hb components on electrophoresis. The Salmonid fishes showed from two to four fractions and the authors distinguished clearly two groups. One group had Hb which was made up of two components; this group included the chum salmon (*Oncorhynchus keta*), red salmon (*O. nerka var. nerka*), and hime-masu (*O. nerka var. adonis*), which all showed closely related patterns and were named by the authors as 'Salmon type' fishes. The second group, consisting of the pink salmon (*O. gorbuscha*), king salmon (*O. taschawytscha*), rainbow trout and brook trout (*Salvelinus fontinalis*) had three of four fractions, and were called trout type fishes. Very little difference was noted between the pink and king salmon on the one hand, and the rainbow and brook trout on the other.

Schumann (1959) investigated the Hbs of a large number of freshwater fishes by means of agar gel electrophoresis using a borate buffer system at pH 8·6. He confirmed the work of previous authors showing that heterogeneity of blood Hb is widespread in fishes. He studied 13 members of the salmon family, eight Cyprinid species, three Percidae, and one species each of the families Gadidae, Esocidae, and Cottidae. He concluded that the Salmonidae have multiple Hbs with at least two or three fractions, and that many species showed strong family and generic similarities. Of all the species studied at most four species had Hb fractions that migrated with identical velocities.

Since these first reports on the multiple Hbs of fishes and the excellent review of Gratzer and Allison (1960) discussing the Hbs of the vertebrates in general, several publications have extended the number of species in which the phenomenon is observed. Allison *et al.* (1960) studied the electrophoretic fractions of the brook lamprey (*Lampetra fluvialilis*) and found that all specimens investigated showed a similar pattern of one minor and two major components. Rumen and Love (1963) studied the Hb pattern of the sea lamprey (*Petromyzon marinus*) by means of free boundary electrophoresis and found six distinct fractions. Buhler (1963) using moving boundary electrophoresis analysed the Hbs of the chinook salmon (*Oncorhynchus tschawytscha*) and the rainbow trout and demonstrated the occurrence of two electrophoretically distinct Hbs in the chinook salmon and three in the rainbow trout. Tsuyuki and Gadd (1963) investigated the Hb patterns of the sockeye salmon (*Oncorhynchus nerka*) and rainbow trout by means of starch gel electrophoresis. They found Hb patterns showing 12–16 fractions, distributed on either side of the point of application. This is quite in contrast to the two or three fractions recorded up to that time. The overall Hb patterns of the two species were similar but species specificity was evident. The same authors also studied the spring salmon (*O. tschawytscha*) and the brook trout and again extreme complexity of the Hb patterns was observed. It was noted that this complexity appeared to be a feature of all salmonid fishes.

The agar electrophoresis technique was used by Sick *et al.* (1962) to study the haemoglobin patterns of the Atlantic representatives of the genus *Anguilla*, which according to Schmidt (1925) constitute two different species, namely, the American eel (*A. rostrata*) and the European (*A. anguilla*). In addition, the Japanese eel (*A. japonica*) was included in the investigations for comparison. The experiments showed that each species proved to be monomorphic as to their Hb pattern. The pattern of the American eels could not be distinguished from that of the European ones, while the Japanese eels showed a different pattern.

b. Intraspecific variation. Further refinement of technique resulted not only in better separations of multiple Hbs, but also showed the existence of intraspecific and ontogenetic variations in fish Hbs. Sick (1961a) was the first to describe the phenomenon of intraspecific variation. His studies on whiting (*Gadus merlangus*) and cod (*G. morhua*) showed geographical variation in the Hb patterns, these variations reflecting gene frequency. This polymorphism was detected by a refined agar gel electrophoresis technique, using phosphate buffer at pH 7·3. In whiting (*G. merlangus*) three different electrophoretic patterns were found. Each pattern has a slow moving component and in addition may have either one or both of two faster moving fractions. No correlation of Hb type with sex or size could be detected. In cod (*G. morhua*) three distinct patterns were again observed. Sick (1961b) mentioned that he had observed the same phenomenon in the sole (*Solea solea*) and the eel-pout (*Zoarces viviparus*) and suggested that this may be of rather common occurrence in fishes. He further suggested that zone electrophoresis of Hbs might become a valuable tool in taxonomic investigations at subspecific level.

Sindermann and Honey (1963) studied the Hbs of five Clupeoid fishes from the Western Atlantic by means of agar gel electrophoresis for possible variant patterns that could be used to determine discreteness of subpopulations of the alewife (*Alosa pseudoharengus*), blueback herring (*A. aestivalis*), American shad (*A. sapidissima*), Atlantic herring (*Clupea harengus*) and Atlantic menhaden (*Brevoortia tyrannus*). However, they found a remarkable similarity in numbers and mobilities of the fractions. One species, the Atlantic herring was sampled throughout much of its area of abundance in the western North Atlantic Ocean, and at various stages and maturity conditions without any variant Hb patterns being found. Wilkins (1963) and Wilkins and Iles (1966) on the other hand did find distinct pre-adult patterns in the Atlantic herring on the European side of the ocean.

c. Ontogenetic differences. The occurrence of ontogenetic variations in the Hb patterns of lower vertebrates was first reported by Adinolfi and Chieffi (1958) who investigated the electrophoretic behaviour of the cyclostome (*Petromyzon planeri*) in the adult and ammocoetes stage. Paper electrophoresis did not detect any difference between adult and larval blood, both of which showed a single band migrating towards the anode. On the other hand, starch gel electrophoresis provided evidence of the presence of two distinct bands with different mobilities at each stage. In a further paper Adinolfi *et al.* (1959) elaborated on their earlier findings, and showed that in the process of metamorphosis three or four Hb components are found.

The first demonstration of distinct foetal or embryonic Hbs in elasmobranchs was made by Manwell (1958a,b), using oxygen equilibria as a criterion. In a later study, Manwell (1963) made use of zone electrophoresis to compare the adult and foetal Hbs of the spiny dogfish (*Squalus suckleyi*). Paper electrophoresis showed that *S. suckleyi* foetal Hb moved approximately 20% more slowly than adult Hb. The latter tends to form two broad zones, resolving quite poorly, but each moving more rapidly than the foetal Hb.

Wilkins (1963) and Wilkins and Iles (1966) showed that there are distinct preadult (larval, foetal or juvenile) and adult Hb types in the atlantic herring on the European side of the Atlantic Ocean. They found a positive correlation between the size of the herring and the Hb pattern. Wilkins and Iles (1966) found corresponding results when they investigated the developmental stages and the corresponding Hb patterns in the sprat (*Sprattus sprattus*).

Koch, *et al.* (1964) investigated the developmental patterns of the Atlantic salmon (*Salmo salar*) using a new micro-starch gel electrophoresis technique. This technique gave a very high degree of separation. Each pattern could be divided into two groups of components, one group migrating to the anode, the other to the cathode. During development, the number of components in the anodic group increased from two to five. In the cathodic group five components are already present in the two-year-old animals while during subsequent development and growth the number of components increases to a maximum of eight. By adding thyroid tissue to the food of two-year-old salmon, the differentiation of the Hb pattern is accelerated without corresponding influence on the body growth. Size-dependent changes were also found by Vanstone *et al.* (1964) in starch gel electrophoretgrams of the Hb patterns of several populations of the Pacific salmon (*Oncorhynchus spp.*).

d. Haemoglobins of hybrid fishes. The occurrence of different Hb patterns in hybrid fishes was demonstrated by Sick *et al.* (1963). They studied the plaice (*Pleuronectes platessa*) and the flounder (*Platichthys flesus*) and their hybrids which seem to occur frequently in nature. They found that both species were monomorphic with regard to their Hb pattern, and that there exists a distinct difference between the two Hb pictures. The supposed hybrids all showed the same Hb pattern which was different from those of the parent species. The pattern also differed clearly from that obtained by running an *in vitro* mixture of Hb solutions from the two parent species. The range of mobilities found in the supposed hybrid pattern is narrower than that of the *in vitro* mixture. In the more compact hybrid pattern seven distinct bands can be recognized.

Manwell *et al.* (1963) studied the Hb patterns of approximately 1,000 centrarchid fishes of the species *Lepomis cyanellus* (green sunfish, *L. macrochirus* (bluegill), *L. microlophus* (redear sunfish), and *Chaenobryttus gulosus* (warmouth sunfish), all possible F_1 hybrids of these four species, several different F_2 populations, and one tetrahybrid cross. In addition, hybrids between the centrarchid fishes *Pomoxis nigromaculatus* (black crappie) and *P. annularis* (white crappie) were analysed. The authors concluded that within a given species or F_1 hybrid no intraspecific variations could be found, and that each species and each hybrid had its own unique Hb pattern. Furthermore, it was found that in most cases the F_1 hybrids yielded a Hb pattern that was identical to that obtained by simply mixing Hbs of the two parental species. However, in some of the crosses (bluegill × warmouth sunfish, warmouth sunfish × green sunfish, and green sunfish × warmouth sunfish) the F_1 hybrids had in every case 25–40% of their Hb with electrophoretic properties different from those of the parental species.

Nyman (1966) reported that the F_1 salmon-brown trout hybrid serum protein pattern is a simple combination of the proteins of both parents. He later (1967a,b) noted that some of these proteins seem to be determined by the

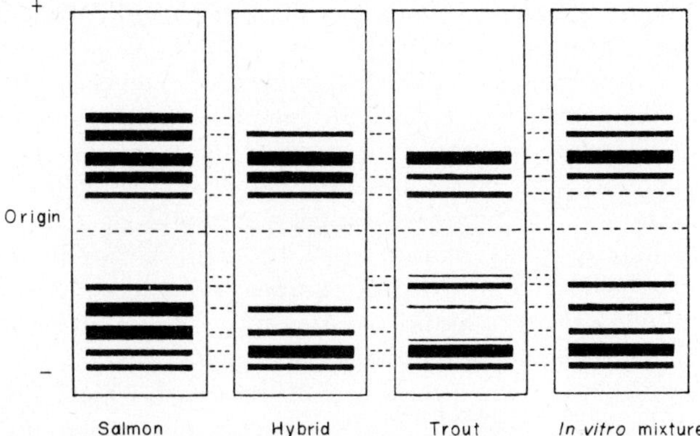

FIG. 8. Haemoglobin patterns of salmon, trout, salmon-trout hybrid and an *in vitro* mixture of salmon and trout blood in starch-gels. (From Haen and O'Rourke, 1968.)

same genes in both parental species and only one band is present in the F_1 hybrid in a concentration equal to that in the parents. He found that in the F_2 generation, selection occurs against the salmon proteins of the serum and the kidney, liver and serum esterases.

Haen and O'Rourke (1968) studied the Hbs of F_2 salmon–sea trout hybrids. It may be seen from Fig. 8 that no new hybrid haemoglobin was found using the starch gel technique; the hybrid pattern is not, however, just a mixture of the two parental patterns. On both the anode and the cathode side, only four of the five salmon fractions are present in the hybrid and the C_1, C_2 and C_4 fractions of the trout do not occur in the hybrid. The micro-agar gel technique also shows that there is an overall reduction in the number of fractions in the hybrid, as seen in Fig. 9. Koch *et al.* (1964) have shown that the relative

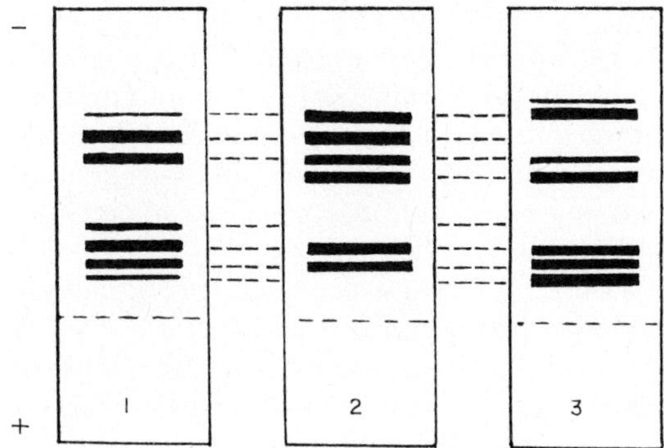

FIG. 9. Haemoglobin patterns of salmon (1), trout (3) and salmon-trout hybrid (2) in agar gels. C_1 and C_2 of the hybrid pattern correspond to the C_2 and C_3 of the salmon and trout patterns. The C_3 component of the hybrid has a migration rate similar to the C_4 of the trout, while the C_4 fraction of the hybrid has comparable components in C_5 of the salmon and C_5 of the trout. (From Haen and O'Rourke, 1968.)

quantities of the components, which at pH 8·3 migrate to the anode, decrease as the salmon and sea trout grow.

Manwell *et al.* (1963) noted that there was little evidence for the formation of any new hybrid protein in various animals. These authors studied the Hbs of some hybrid birds and fishes to investigate the possibility that new Hbs might arise in hybrids from recombinations of the different polypeptide chains and be significant in explaining some aspects of hybrid vigour.

'Artificial hybridization' of Hbs from the small mouth bass *Micropterus dolomieu* and the large mouth bass *M. salmoides* was achieved by mixing haemolystates and leaving the mixture in pure carbon monoxide for several days. The new Hbs which were produced seem to be identical with those of the red-eyed bass *M. cooosae* and the spotted bass *M. punctulatus* which, it is suggested, may have evolved from hybrids of the large-mouth and small-mouth bass. Hybrid fishes do not seem to have any unique Hb peptides, which again suggests that polypeptide chain recombination rather than biosynthetic interaction is responsible for the appearance of new hybrid Hbs. Manwell *et al.* (1963) have shown that in some hybrid fishes there is better blood gas transport by the new Hb, thus providing an understanding of the nature of hybrid vigour at the molecular level.

The recent development of isoelectric focussing has produced a tool with remarkable analytical powers. Quast and Vesterberg (1968) have detected 11 Hb zones in the blood of *Myxine glutinosa* and Koch and Backx (1969) have found that a mature salmon may have in its blood more than 25 Hbs with different isoelectric points. Obviously this is a most important development in the study of Hbs and we may expect to learn much from this tool.

B. SEROLOGICAL METHODS

1. Precipitin Reactions

Kraus (1897) first discovered the occurrence of precipitins in antibacterial sera. Tchistovich (1899) (in Nuttal, 1904) was apparently the first to produce precipitins by the injection of the blood of one animal (horse and eel serum and corpuscles) into that of another. He showed that long continued treatment of rabbits (with antigens) leads to the disappearance of precipitins from blood of rabbits rendered immune to eel serum.

Tchistovich obtained eel serum which would precipitate the red blood cells of goats, dogs, rabbits and guinea pigs. De Lisle (1902) (in Erhardt, 1929) was the first to claim to have obtained antisera from eels treated with rabbit serum. Nuttal (1902) mentioned that 'Tests upon different fish bloods are being made the subject of study by Dr Graham Smith at my suggestion. The results should be of considerable interest'. I have been unable to trace Graham Smith's work on fishes.

Nuttal (1904), the founder of Systematic Serology, reported the earliest series of tests of antisera against fish blood sent to his laboratory on dried filter paper. An exception was the blood from the dogfish *Scyllium canicula* which was fluid. Seven elasmobranchs, 12 species of Teleosts (including salmon and pike blood from Ireland) and five unclassified fishes were tested. None of these reacted with any of the 25 antisera used (including anti-mammal, anti-bird, anti-reptile, anti-frog and anti-lobster sera produced in rabbits). An anti-ammocoetes serum was produced by Nuttal (1904) by the injection into a rabbit of ammocoetes serum which had been dissolved out of filter paper with saline. This reacted by medium clouding followed by well marked precipitation with the one cyclostome species tested (not named). Two of the 26 species of Teleost and one of the 11 elasmobranchs tested gave only a doubtful clouding with this antiserum of 'only moderate power'.

Boyden (1964) pointed out that Nuttal 'Commonly used only a single arbitrarily selected and often unknown concentration of antigen in his tests. Actually his dilutions would correspond roughly with protein concentrations ranging from 1/2,000 to 1/8,000 which would fall in the mid region of many precipitin systems'. Boyden (1964) pointed out that Erhardt (1929) had said that the results of some of the precipitin testing by others agreed well with the then current views of systematic relationships, particularly in fishes but that he believed that serological tests would only have a subordinate value in taxonomy. Boyden's own work and that of his school has shown how valuable serological studies can be in taxonomic work.

The next author to study fishes was Gemeroy (1943) who worked in Boyden's laboratory. He used both the ring test and the photronreflectometer (photroner) which was devised by Libby (1938) and introduced into comparative serology by Boyden and De Falco (1943). He investigated the relationships of 31 species of freshwater and marine fishes including *Petromyzon marinus*, 6 elasmobranchs, *Acipenser rubicundus*, *Lepidosteus osseus*, *Amia calva* and 21 teleosts.

Antishark serum (Figs 2 and 3 in Gemeroy) did not react with the lamprey or sturgeon 'The inability to get any reaction between these three classes is an indication of how far apart they really are . . . there does not seem to be any justification for placing three such widely separated groups, as determined by serological methods, into one class'. The results also indicated that the sharks and rays 'have also had a long period of evolutionary divergence. It seems clear then that orders and classes may have very different values as regards their degree of relationship when different vertebrate groups are being discussed'. The differences found between the three ganoids 'were not as wide, however, as found with the species of Elasmobranchii tested'.

In the Teleosts cross reactions were found between anti-brook trout serum and rainbow trout and brown trout but not with carp, bowfin, catfish and tarpon. Pike and muskallunge cross reacted. Cross reactions also occurred with the Percoidea (Fig. 5 in Gemeroy, 1943). All these results agreed with the systematic relationships established by classical morphological methods. The cross reaction between catfish and bowfin was 7·7% as compared with

2·0% catfish-carp and 1·6% for catfish-trout. However the antiserum used was produced by a multiple series of injections which as Wolfe (1933) had shown decreased specificity.

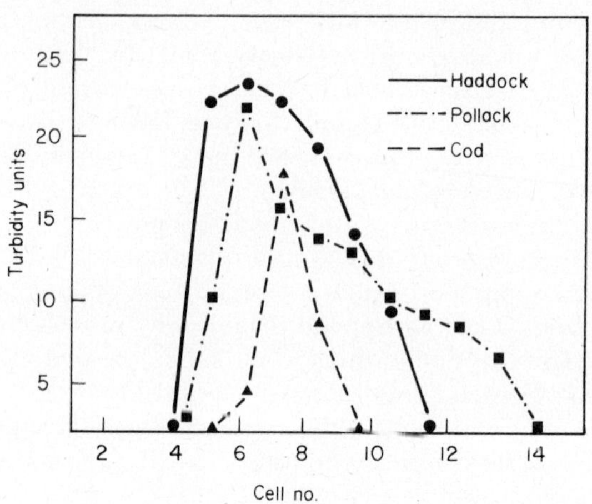

Relationship ratios

	Antiserum used	Ratio	Mean
Cod/Haddock	*R* 712	100:24	
	R 72	100:37	25·3
	R 71	100:25 *	
Cod/Pollack	*R* 712	100:21	24·0
	R 71	100:27 *	
Haddock/Pollack	*R* 71	100:94	90·5
	R 712	100:87 *	

* Reciprocal figures.

FIG. 10. Relationship ratios of haddock, pollack and cod. (From O'Rourke, 1959.)

Gemeroy (1943) pointed out some possible sources of error. Lipoids may decrease the specificity of the reaction. When the ring test is used comparable amounts of antigen and antibody must be present as only the end point is determined. The photroner has the advantage that it covers the full range of AnAb-reactions from antigen to antibody excess. Dilution of antisera produced by multiple series of injections (and therefore with decreased specificity) eliminated doubtful cross-reactions with very distantly related species. Gemeroy (1943) wisely suggested that it may prove 'necessary to compare only specific fractions of blood serum rather than antigenic mixtures' and he found that the biochemical gulf which separates species and orders among fishes is far wider than in birds. This of course raises the question of how comparable are the various taxonomic categories in different animal groups.

The next photroner study of fishes would appear to be the work of O'Rourke (1959) on four species of the genus *Gadus* (family Gadidae) (Fig. 10) and later work (O'Rourke, 1961b) on two species of *Sebastes* which were shown to differ serologically to a greater extent (19%) than did the two clearly defined gadoids *G. aeglefinus* L. and *G. pollachius* L. which differed by 9·5%. Although it was not intended to suggest that any particular percentage difference should be taken to mark off the species level, it is of interest, that considering the current lack of interest in photroner studies of fish, that the separation of *Sebastes marinus* and *S. mentella* on photroner grounds was supported by immuno-diffusion studies using both Ouchterlony and Consden-Kohn techniques and also by chromatographic differences in their mucus using two different chromatographic techniques (O'Rourke, 1961b). Schaefer (1961) added further biochemical evidence of the distinctness of the two species by comparing the amino acid content of muscle from the two forms.

The problem of the taxonomic status of the North Atlantic *Sebastes* form was not solved by traditional methods discussed over a week's symposium in 1959 at Copenhagen and the fish are almost impossible to tag. Recent work of Altukhov and Nefyodov (1968) on the *Sebastes* problem using agar gel electrophoresis showed that there are statistically significant differences between the sera from both species both in regard to α_1-globulins and to albumin phenotypes B and AB. Thus these authors agree with 'the results obtained by Schaeffer and O'Rourke who demonstrate the possibility of distinguishing both redfish types'. The 'Giant' type is shown by Altukhov and Nefyodov (1968) to be a hybrid occurring between the two species. 'In albumin A they approach the *marinus* type and in the α globulin the *mentella* and they occupy an intermediate position in albumin B frequency'. A later paper by Altukov *et al.* (1968) showed that the thermostability of isolated muscles, shown to be a species specific criterion by Ushakov (1959, 1964), from the '*marinus* and *mentella* types are quite different'.

The latter authors criticise my proposal to regard the two forms as 'sibling species' and taking Mayr's (1963) definition of the term I would accept this as a valid criticism especially as I agree with their view 'that these fishes appear quite different from one another in most parts of their areal distribution'. My view is that these forms are recently evolved species and I would agree with Altukhov *et al.* (1968) that two of Kotthaus's (1961) suggestions offer the most probable solution to the situation namely, that

> 'they are independent species connected by intermediary fish of hybrid origin and that the *marinus* type is a phylogenetically more ancient type from which the *mentella* type has detached itself. Here the process of the speciation has not been terminated yet, hence, the emergence of intermediary forms'.

It is significant that the West Greenland population is represented by sterile F_1 hybrids whose hybrid vigour enables them to tolerate the cold water there. De Ligny (1969) however seems to disagree as she says

> 'It may be noticed that the samples of both the *marinus* and the "giant" type were collected over a wider geographical range than those of the *mentella* type and could conceivably represent different geographical populations of which the existence has also been suggested by the variability of morphological and meristic characters observed within the whole range of *Sebastes* sampled at different locations (Kelly, Barker and Clarke, 1961)'.

It is also of interest to note that Tucker's (1959) suggestion that the European and American eels were merely ecophenotypes of the same species spawning in one area, was first disproved by the photroner study of Gemeroy and Boyden (1961) who compared the sera from the two populations. Sick *et al.* (1962) initially found no haemoglobin differences between the two forms and found the chromosome numbers were identical. Later Sick *et al.* (1967) using much larger numbers of eels did find that a rare haemoglobin allele occurred in American eels. Sinderman and Krantz (1968) who said that as regards the differences in the frequencies of three transferrin types in European and American eels. Sinderman and Kranz (1968) who said that as regards the eel problem the photroner evidence 'does not entirely eliminate possible environmental differences' showed the extraordinary complexity of the iso-antigen–isoagglutinin system of American eels which 'seems to exceed that presently known for any other vertebrate'. They have not apparently investigated the European eels although work on Japanese eels does not indicate such extreme complexity.

Mairs and Sinderman's (1962) work on the eastern Atlantic clupeoids is one of the most detailed and thorough serological studies on fish species yet carried out since they used photroner, agar diffusion, red cell agglutination and paper electrophoresis of the blood. The forthcoming work of Cross, Haen and O'Rourke on a series of Irish freshwater Cyprinidae including the study of naturally occurring hybrids is comparable in the variety of techniques used.

The study by the same authors of some Irish Salmonidae (Haen and O'Rourke, 1968, 1969a,b) including the investigation of the F_2 generation of an artificially produced hybrid between the salmon and the sea trout (Piggins, 1965) covered a range of soluble proteins (serum, muscle, eye-lens and haemoglobin) and compared different techniques but unfortunately did not include the serological reactions which have been studied in the European salmon by Wilkins (1967).

Ridgeway (1963) using the Ouchterlony technique with antisera produced in rabbits showed interspecific differences in the adult sera of six species of tuna. He showed that the soluble antigens of the muscle could not be used to distinguish three of the six species. He found that the discriminatory qualities of the best antisera to three species were not great and remarks 'This points up the fact that there is considerable variability between individual rabbit sera, so that one may need to immunize 6 to 20 rabbits in order to produce an antiserum of the specificity necessary to make a particular distinction. It is this fact that makes ultramicro methods such as the Consden-Kohn and the microcomplement fixation technique appeal so much to us all. In the paper quoted (Ridgeway, 1963) he makes the interesting suggestion that the results 'indicate that there are definite possibilities in the application of immunochemical methods to the problems of identifying the species of larval forms of tuna'.

I believe that photroner studies may still be profitable although I have (O'Rourke, 1960) pointed out that 'Unfortunately the photroner requires 3·5 ml or more antiserum per estimation and as we find it hard to produce good antisera this is very wasteful since at least centilitres are required to produce useful data'. The Consden-Kohn (1959) technique requires only microlitre quantities of both serum and antigen and should be further studied as a precipitin system. Further studies should show that the multiple lines of precipitation may be analysed either by projection or examination under a microscope.

References and Bibliography*

Adinolfi, M. and Chieffi, G. (1958). *Nature Lond.* **182**, 730.

Adinolfi, M., Chieffi, G. and Siniscalo, M. (1959). *Nature Lond.* **184**, (4695), 1325–1326.

Allfrey, V. G. (1955). *Adv. Enzymol.* **16**, 411–500.

Allison, A. C. *et al.* (1960). *Biochim. biophys. Acta* **42** (1), 43–48.

Altukhov, J. P. and Nefyodov, G. N. (1968). *Res. Bull. Int. Comm. Northw. Atlant. Fish* **5**, 86–90.

Altukhov, J. P., Nefyodov, G. N. and Payusova, A. N. (1968). *Res. Bull. Int. Comm. Northw. Atlant. Fish.* **5**, 130–136.

Alm, G. (1964). *Rep. Inst. Freshwat. Res. Drottningholm* **36**, 13–59.

Anthony, V. C. and Boyar, H. C. (1968). *Res. Bull. Int. Comm. Northw. Atlant. Fish.* **5**, 91–98.

Archer, R. K. (1965). 'Haematological Techniques for Use on Animals'. Blackwell Scientific Publications, Oxford.

Ascoli, M. (1902). *Münch. med. Wschr.* **49**, 398–401.

Atz, J. M. (1964). *Q. Rev. Biol.* **27**, 366–377.

Atkin, N. B. *et al.* (1965). *Chromosoma* **17**, 1–10.

Augustinsson, K. B. (1961a). *Ann. N.Y. Acad. Sci.* **94**, 844–860.

Augustinsson, K. B. (1961b). *Proc. natn. Acad. Sci. U.S.A.* **45**, 753.

Bailey, G. S. *et al.* (1969). *Biochem. biophys. Res. Commun.* **34**, (5), 605–609.

Baldwin, J. and Hochachka, P. W. (1970). *Biochem. J.*, **116**, 883.

Barrett, I. and Williams, A. A. (1967). *Copeia* **2**, 468–471.

Barry, J. M. and O'Rourke, F. J. (1959). *Nature Lond.* **184**, 2039.

Bon, W. F. (1957). *Pubbl. Staz. zool. Napoli* **30**, 373.

Bon, W. F. *et al.* (1964). *Pubbl. Staz. zool. Napoli* **34**, 59–65.

Bonavita, V. and Guarneri, R. (1963). *J. Neurochem.* **10**, 743.

Booke, H. E. (1964). *N.Y. Fish Game J.* **11**, (1), 47–57.

Börtitz, S. (1964). *Biol. Zbl.* **83**, (6), 225–238.

Bouch, G. R. and Ball, R. C. (1965). *Trans. Am. Fish. Soc.* **94**, (4), 363–370.

* This survey of the literature was concluded in June 1970.

Boyden, A. A. (1953). *Syst. Zool.* **2**, 19–30.
Boyden, A. A. (1963). *Syst. Zool.* **12**, 1.
Boyden, A. A. (1964). *In* 'Taxonomic Biochemistry and Serology'. (C. A. Leone, ed.). The Ronald Press Co., New York.
Boyden, A. A. and De Falco, R. J. (1943). *Physiol. Zool.* **16**, 229–241.
Britten, R. J. and Kohne, D. E. (1970). *Scient. Am.* **222**, 24–31.
Buhler, D. R. (1963). *J. biol. Chem.* **238** (5), 1665–1674.
Buhler, D. R. and Shanks, W. E. (1959). *Science N.Y.* **129**, 899–900.
Buzzati-Traverso, A. A. (1953). *Proc. natn. Acad. Sci. U.S.A.* **39** (5), 376–391.
Buzzati-Traverso, A. A. and Rechnitzer, A. B. (1953). *Science N.Y.* **117**, 58.
Chandrasekhar, N. (1959). *Nature Lond.* **184**, 1652–1653.
Commoner, B. (1965). *Protides biol. Fluids* **13**, 36–51.
Connell, J. J. (1953a). *Biochem. J.* **54**, 119–126.
Connell, J. J. (1953b). *Biochem. J.* **55**, 378–388.
Consden, R. and Kohn, J. (1959). *Nature Lond.* **183**, 1512.
Cory, R. A. and Wold, F. (1966). *Biochemistry, N.Y.* **5**, 3181.
Creyssel, R. *et al.* (1964). *Bull. Soc. Chim. biol.* **46** (1), 149–159.
Crick, F. H. C. (1958). *In* 'The Biological Replication of Macromolecules'. Symposium No. 12, Soc. Exptl. Biol. New York, pp. 138–163.
Cushing, J. E. (1964). *Adv. mar Biol.* **2**, 85–131.
Dannevig, E. H. (1956). *FiskDir. Skr.* **11** (6), 1–12.
Davis, B. J. (1964). *Ann. N.Y. Acad. Sci.* **121**, 404.
Dawson, D. M., Goodfriend, T. L. and Kaplan, N. O. (1964). *Science N.Y.* **143**, 929.
Dayhoff, M. O. (1969). *Scient. Am.* **221** (1), 86–95.
Deelder, C. I. (1960). *Nature Lond.* **185**, 589–591.
Deeley, E. M. (1955). *J. Sci. Instrum.* **32**, 263–267.
Deutsch, H. F. and Goodhue, M. E. (1945). *J. biol. Chen.* **161**, 1–20.
Deutsch, H. F. and MacShan, W. H. (1949). *J. biol. Chem.* **180**, 219–234.
Dingle, J. R., Eagles, D. E. and Neelin, J. M. (1955). *J. Fish. Res. Bd. Can.* **12** (1).
Dray, S. *et al.* (1962). *Nature Lond.* **195**, 785–786.
Drilhon, A. (1954). *C.r. Séanc. Biol.* **153**, 1532–1535.
Drilhon, A. and Fine, J. M. (1957). *C.r. hebd. Séanc. Acad. Sci., Paris* **245**, 1676.
Drilhon, A. and Fine, J. M. (1960). *C. r. hebd. Séanc. Acad. Sci. Paris* **250**, 4044–4045.
Drilhon, A. *et al.* (1966). *C. r. hebd. Séanc. Acad. Sci. D, Paris,* **262**, 1315–1318.
Drilhon, A., Fine, J. M. and Magnin, E. (1961). *C. r. Séanc. Soc. Biol.* **155**, 451.
Dublin (1969)—see de Ligny, (1971).
Duke, E. J. and Kelly, M. (1968). *Isoenzyme Bull.* Jan. 1968.
Enge, K. and McKinley McKee (1959). *Nytt Mag. Zool.* **8**, 34–36.
Engle, R. L. *et al.* (1958). *Proc. Soc. exp. Biol. Med.* **98**, 905–909.
Enomoto, N. and Tomiyasu, Y. (1960a). *Bull. Jap. Soc. scient. Fish.* **26**, 745–748.
Enomoto, N. and Tomiyasu, Y. (1960b). *Bull. Jap. soc. scient. Fish.* **26**.
Enomoto, N., Nagao, R. and Tomiyasu, Y. (1961). *Bull. Jap. Soc. scient. Fish.* **27**, 143–146.
Erhardt, A. (1929). *Ergeb. Fortschr. Zool.* **7**, 279–377.
Farris, D. A. (1957). *U.S. Fish Wildlife Serv. Spec. Sci. Rept., Fisheries* **208**, 35.
Farris, D. A. (1958). *J. Cons. perm. int. Explor. Mer.* **23**, 235–244.
Fine, J. M. and Drilhon, A. (1958). *C. r. hebd. Séanc. Acad. Sci., Paris.* **246**, 3183.
Fine, J. M. *et al.* (1964). *C. r. hebd. Séanc. Acad. Sci. Paris* **258**, 753–759.
Fine, J. M. *et al.* (1965). *Protides biol. Fluids* **12**, 165.
Flemming, H. (1958). *Z. Fisch.* **7**, 91–152.

Florkin, M. (1963). *ICSU Rev.* 5, 202–209.
Florkin, M. (1964). *In* 'Colloquium on the Protides of Biological Fluids'. (H. Peeters, ed.). Elsevier, Amsterdam. pp. 17–28.
Florkin, M. (1966). 'A Molecular Approach to Phylogeny', Elsevier, Amsterdam.
Fottrell, P. F. (1967). *Sci. Progr.* **55**, 543–559.
Fox, A. C. (1956). *Science N.Y.* **321**, 343.
Francois, J. and Rabaey, M. (1957a). *Am. J. Opthamol.* **44**, 347–357.
Francois, J. and Rabaey, M. (1957b). *Am. Med. Assoc. Arch. Opthamol.* **57**, 672–680.
Francois, J. and Rabaey, M. (1959). *A.M.A. Archs Ophthal.* **62**, 991.
Fujino, K. and Kang, T. (1968). *Copeia* **7**, 56–63.
Fujiya, N. (1961). *J. Wat. Pollut. Control Fed.* **33** (3), 250–257.
Gemeroy, D. G. (1943). *Zoologica, N.Y.* **28**, 109–123.
Gemeroy, D. and Boyden, A. (1961). *Bull. serol. Mus., New Brunsw. N.J.* **26**, 7–8.
Giblett, E. R. *et al.* (1959). *Nature Lond.* **183**, 1589–1590.
Giese, A. C. (1968). 'Cell Physiology'. W. B. Saunders, Philadelphia 3rd. ed.
Gladstone, A. and Smith, E. L. (1967). *J. biol. Chem.* **242**, 4702–4710.
Goldberg, E. (1965). *Science N.Y.* **148**, 391–392.
Goldberg, E. (1966). *Science N.Y.* **151**, 1091–1093.
Gratzer, W. B. and Allison, A. C. (1960). *Biol. Rev.* **35**, 459.
Grabar, P. and Williams, Jr. C. A. (1953). *Biochim. biophys. Acta* **10**, 193.
Greenberg, S. S. and Kopac, M. J. (1965). *Physiol. Zool.* **38** (2), 149–157.
Gunter, G., Sulya, L. L. and Box, B. E. (1961). *Biol. Bull. mar. biol. Lab., Woods Hole* **121** (2), 302–306.
Gysels, H. (1964). *Experientia* **20**, (145), 1–4.
Gysels, H. (1965). *J. Orn. Lpz.* **106** (2), 208–217.
Haen, P. J. and Ryan, B. (1967). *Ir. Nat. J.* **15** (9), 270–271.
Haen, P. and O'Rourke, F. J. (1968). *Nature Lond.* **217** (5123), 65–67.
Haen, P. and O'Rourke, F. J. (1969a). *Proc. R. Ir. Acad.* **68**B (4), 67–75.
Haen, P. and O'Rourke, F. J. (1969b). *Proc. R. Ir. Acad.* **68**B (7), 101–108.
Hagen, D. W. and McPhail, J. D. (1970). *J. Fish. Res. Bd. Can.* **27** (1), 147ff.
Halbert, S. P., Manski, W. and Auerback, T. (1961). *In* 'The Structure of the eye'. (G. K. Smelser, ed.) Academic Press, New York.
Hamoir, G. (1955). *Adv. Protein Chem.* **10**, 227–288.
Harris, H. (1969). *Proc. R. Soc.* B, **174**, 1–31.
Hashimoto, K. and Matsuura, F. (1959a). *Bull. Jap. Soc. scient. Fish.* **24** (9), 719–723.
Hashimoto, K. and Matsuura, F. (1959b). *Nature Lond.* **184**, 1418.
Hashimoto, K. and Matsuura, F. (1960). *Bull. Jap. Soc. scient. Fish.* **26** (3), 354–360.
Haupt, I. and Giersberg, H. (1958). *Naturwissenschaften* **45**, 268.
Hawryhewicz, E.J. and Blair, W. H. (1965). *Aerospace Med.* **36**, 369.
Hewitt, R. E. *et al.* (1963). *Copeia* **2**, 296–303.
Heyningen, R. van (1962). *In* 'The Eye', (H. Davson, ed.). Academic Press, New York.
Hickman, C. G. and Smithies, O. (1957). *Proc. Genet. Soc. Can.* **2**, 39.
Hildemann, W. H. (1959). *Am. Nat.* **96**, 193–246.
Hiscock, I. D. (1949). *Aust. J. Sci.* **2**, 209.
Hitzeroth, M. *et al.* (1968). *Biochem. Genet.* **1**, 287–300.
Hochachka, P. W. (1965). *Archs Biochem. Biophys.* **111**, 96–103.
Hochachka, P. W. (1966). *Comp. Biochem. Physiol.* **18**, 261–269.
Hochachka, P. W. and Somero, G. N. (1968). *Comp. Biochem. Physiol.* **27**, 659–668.

Hodgins, H. O., Ames, W. E. and Utter, F. M. (1969). *J. Fish. Res. Bd Can.* **26**, 15–19.
Howe, A. F. (1964). *Analyt. Biochem.* **9** (4), 443–453.
Hubbs, C. L. (1955). *Syst. Zool.* **4** (1), 1–20.
Hubbs, C. (1967). *Bull. natn. Inst. Sci. India* **34**, 48–59.
Hubbs, C. and Drewry, G. E. (1958). *Am. Nat.* **867**, 378–380.
Hubbs, C. and Drewry, G. E. (1960). *Publ. Inst. mar. Sci. Univ. Texas* **6**, 81–91.
Hubbs, C. and Strawn, K. (1956). *Evolution Lancaster, Pa.* **10** (4), 341–344.
Hubbs, C. and Strawn, K. (1957). *J. exp. Zool.* **134**, 33–62.
Hughes-Schrader, S. and Schrader, F. (1956). *Chromosoma* (Berl.) **8**, 709.
Hunn, J. B. (1967). 'The Chemistry of Fish Blood'. Res. Rep. U.S. Fish. Wildl. Serv. Vol 72, p. 32.
Ingram, V. M. (1963). 'The Haemoglobins in Genetics and Evolution'. Columbia University Press, New York.
Irisawa, W. and Irisawa, A. F. (1954). *Science N.Y.* **120**, 849–851.
Jakowska, S. (1963). *Ann. N.Y. Acad. Sci.* 458–462.
Kabara, J. J. *et al.* (1964). *Clin. Chem.* **10** (12), 950–959.
Kaplan, N. O. and Ciotti, M. M. (1961). *Ann. N.Y. Acad. Sci.* **94** (3), 701–722.
Kaplan, N. O. *et al.* (1960). *Science N.Y.* **131**, 392.
Kawerau, K. (1956). *Chromat. Meth.* **1** (2).
Klose, J. *et al.* (1968). *Humangenetik* **5**, 190–196.
Koch, H. J. A. and Backx, J. (1969). *Sci. Tools* **16**, 44–47.
Koch, H. J. A., Bergström, E. and Evans, J. (1964). *Laxforskningsinstitutet Meddelande* **6**, 1–7.
Koch, H. J. A. *et al.* (1967). *Meded K. vlaam Acad.* **29** (7), 1–16.
Koehn, R. K. (1969). *Science N.Y.* **163**, 943–944.
Koehn, R. K. and Perez, J. E. (1967). *Biochem. Genet.* **1**, 131–144.
Kotthaus, A. (1961). *Rapp. P.-v. Reun. Cons. perm. int. Explor. Mer.* **150**, 42–44.
Kraus, R. (1897). *Wein Kein. Wechr.* **10**, 736.
Kreil, G. (1963). *Z. Physiol. Chem.* **334**, 154–160.
Latner, A. L. (1966). 'The Binding of Circulating Enzymes by Plasma Proteins'. Proc. 5th West. Europ. Symp. Clin. Chem. Paris.
Latner, A. L. and Skillen, A. W. (1968). 'Isoenzymes in Biology and Medicine'. Academic Press, London.
Leone, C. A. (1964). 'Taxonomic Biochemistry and Serology'. The Ronald Press Co., New York.
Leone, C. A. (1968). *Bull. serol. Mus. New Brunsw.* **39**, 1–23.
Libby, R. L. (1938). *J. Immunol.* **34**, 71–73.
Ligny, W. de (1965). *Proc. IXth European Conference on Animal blood groups and Biochemical Polymorphisms.*
Ligny, W. de (1966). Inst. natn. Rech. Agron. Paris.
Ligny, W. de (1968). *Genet. Res. II*, 179–182.
Ligny, W. de (1969). *Oceanogr. Mar. Biol. Ann. Rev.* **7**, 411–513.
Ligny, W. de (Ed.) (1971). Proceedings of the special meeting on the biochemical and serological identification of fish stocks ICES, Dublin, 1969. *Rapp. P-v. Réun. Cons. Perm. int. Explor. mer.* **161**.
Love, R. M. (1957). *In* 'The Physiology of Fishes'. Vol 1. (M. E. Brown, ed.), pp. 401–418, Academic Press, New York.
Love, R. M. (1970). 'The Chemical Biology of Fishes'. Academic Press, London.
Love, R. M., Lovern, J. A. and Jones, N. R. (1959). *Spec. Rep. Fd. Invest. Bd. D.S.I.R.* **69**, 62 pp.

Lush, I. E. (1967). 'The Biochemical Genetics of Vertebrates except Man'. J. Wiley and Sons, New York 1967.
Lush, I. E. (1968). 'Proc. XI Conf. on Animal Blood Groups Polish Acad. Sci.'
Mairs, D. F. and Sinderman, C. J. (1962). *Biol. Bull. Woods Hole* **123**, 330–343.
Maisel, H. and Langman, J. (1960). *Anat. Rec.* **140**, 185–198.
Maisel, W., Auerback, T. P. and Holberts, P. (1964). *J. opt. Soc. Am.* **50**, 985–990.
Manski, W. and Halbert, S. P. (1964). *In* 'Colloquium on the Protides of Biological Fluids'. (H. Peeters, ed.), pp. 117–134. Elsevier, Amsterdam.
Manski, W. *et al.* (1960). *J. opt. Soc. Am.* **50**, 985–995.
Manski, W., Halbert, S. P. and Auerbach, T. P. (1964). *In* 'Taxonomic Biochemistry and Serology'. (C. A. Leone, ed.), pp. 545–554, The Ronald Press Co., New York.
Manwell, C. (1958a). *Physiol. Zool.* **31**, 93.
Manwell, C. (1958b). *Science N.Y.* **128**, 419–420.
Manwell, C. (1963). *Archs. Biochem. Biophys.* **101** (3), 504–511.
Manwell, C. and Baker, A. (1970). 'Molecular Biology and the Origin of Species: Heterosis, Protein Polymorphism and Animal Breeding'. Sidgwick and Jackson, London.
Manwell, C., Baker, A. and Childers, W. (1963). *Comp. Biochem. Physiol.* **10**, 103–120.
Margoliash, E. and Lustgarten, J. (1961). *Ann. N.Y. Acad. Sci.* **94**, 731.
Markert, C. L. (1968). *Ann. N.Y. Acad. Sci.* **151**, 14–48.
Markert, C. L. and Faulhaber, I. (1965). *J. exp. Zool.* **159**, 319–332.
Markert, C. L. and Møller, F. (1959). *Proc. natn Acad. Sci. U.S.A.* **45**, 473.
Masat, R. J. and Musacchia, J. (1965). *Comp. Biochem. Physiol.* **16**, 215–225.
Mayr, E. (1963). 'Animal Species and Evolution'. London University Press, Oxford.
Mayr, E. (1969). *J. Linn. Soc. Zool.* **1**, 311–320.
Mayr, E. and Amadon, D. (1951). *Amer. Mus. Novit.* **1406**, 1–42.
Møller, R. C. and Hubbs, C. (1969). *Copeia*, **8**, 52–69.
Miller, D. and Naevdal, G. (1966). *Nature, Lond.* **210**, 317.
Moore, D. H. (1945). *J. biol. Chem.* **161**, 21–32.
Morrison, W. J. and Wright, J. E. (1966). *J. exp. Zool.* **163**, 259–270.
Mulcahy, M. F. (1967). *Nature, Lond.* **215**, 143–144.
Mulcahy, M. F. (1969). *J. Fish. Biol.* **1**, 333–338.
Naevdal, G. and Danielsen, D. S. (1967). Int. Counc. Explor. Sea Counc. Meeting 1966 J7. (Mimeo 7 pp.) quoted by de Ligny (1969).
Nakano, E. and Whitely, A. H. (1965). *J. exp. Zool.* **159**, 167–179.
Nuttal, G. H. F. (1904). 'Blood Immunity and Blood Relationship', Cambridge University Press, London.
Nyman, L. (1965/66). *K. Vetensk. Soc. Arsb.* **9–10**, 84–102.
Nyman, L. (1965a). *LantbrHögsk. Annlr.* **31**, 225–230.
Nyman, L. (1965b). *Hereditas* **53**, 117–126.
Nyman, L. (1965c). *Swed. Salmon Res. Inst. Rep. LFI Medd.* **13**, 1–11.
Nyman, (1965c). *Swed. Salmon Res. Inst. Rep. LFI Medd.* **13**, 1–11.
Nyman, L. (1966). *Swed. Salmon Res. Inst. Rept. LFI Medd.* **3**, 1–6.
Nyman, L. (1967a). *Rep. Inst. Freshwat. Res. Drottningholm* **47**, 5–38.
Nyman, L. (1967b). *Swed. Salmon Res. Inst. Rep. LFI Medd.* **8**, 1–11.
Odense, P. H., Allen, T. M. and Leung, T. C. (1966). *Can. J. Biochem.* **44**, 1319–1326.
Odense, P. H. *et al.* (1969). *Biochem. Genet.* **3**, 317–334.
Ohno, S. and Atkin, N. B. (1966). *Chromosoma* **18**, 455–466.
Ohno, S. *et al.* (1967). *Science N.Y.* **156**, 93–98.
Ohno, S., Wolf, U. and Atkin, N. B. (1968). *Hereditas* **59**, 169–187.

O'Rourke, F. J. (1959). *Nature Lond.* **183**, 1192.
O'Rourke, F. J. (1960). *Proc. R. Ir. Acad.* **61***B* (9), 167–176.
O'Rourke, F. J. (1961a). *Nature Lond.* **189**, 943.
O'Rourke, F. J. (1961b). ICES/ICNAF Redfish Symposium, 1961, pp. 100–103.
Ouchterlony, O. (1949). *Acta path. microbiol. scand.* **25**, 507–515.
Oudin, J. (1948). *Annls Inst. Pasteur, Paris* **75** (30–31), 109–129.
Paul, J. and Fottrell, P. (1961). *Biochem. J.* **78**, 418.
Pederson, K. O. (1933). *Kolloidzeitschrift* **63**, 286.
Phillips, J. *et al.* (1957). *Fish Res. Bull. N.Y.* **20**, 61 pp.
Pfleiderer, G. and Jeckel, D. (1957). *Biochem. Z.* **329**, 371.
Piggins, D. J. (1965). *Ann. Rep. Salm Res. Trust Ire.* 1964, 27–37.
Prie, C. E. (1968). *Q. Jl Fla. Acad. Sci.* **31** (3), 190–196.
Quast, R. and Vesterberg, O. (1968). *Acta chem. scand.* **22**, 1499.
Rabaey, M. (1959). University of Ghent, Ghent. 'Onderzoek der lenseiwitten met behulp van micro-electrophoresis in agar'.
Rabaey, M. (1964). *Protides biol. Fluids* **12**, 273–277.
Rall, D. P., Schwab, P. and Zubrod, C. G. (1961). *Science N.Y.* **133**, 279–280.
Rees, H. (1961). *Chromosoma* **15**, 275–279.
Rees, H. (1967). *Chromosoma* **21**, 472–474.
Ridgeway, C. J. (1963). *Fish Wldlife Service, U.S.A. Fishery Bull.* **63** (1), 205–211.
Ridgeway, C. J., Sherburne, S. W. and Lewis, R. D. (1969). *Spec. Publs Am. Fish Soc.*
Robertson, O. H. *et al.* (1961). *Endocrinology* **68** (5), 733–746.
Romer, A. S. (1966). 'Vertebrate Palaeontology'. University of Chicago Press, Chicago (3rd. ed.).
Rumen, N. M. and Love, W. E. (1963). *Archs Biochem. Biophys.* **103**, 24–35.
Rutter, L. (1948). *Nature Lond.* **161**, 435–436.
Saito, K. (1957a). *Bull. Jap. Soc. Scient Fish.* **22** (12), 752–759.
Saito, K. (1957b). *Bull. Jap. Soc. scient. Fish.* **22** (12), 760–766.
Saito, K. (1957c). *Bull. Jap. Soc. scient. Fish.* **22** (12), 768–772.
Sanders, B. G. (1964). *In* 'Taxonomic Biochemistry and Serology'. (C. A. Leone, ed.), pp. 673–679. The Ronald Press Co., New York.
Sarich, J. (1969). *Triangle* **9** (2), 55–60.
Sarich, J. and Wilson, A. C. *Science N.Y.* **154**, 1563–1566.
Schaeffer, H. (1961). ICES/ICNAF Redfish Symp. 1961, pp. 104–110.
Scheidegger, J. J. (1955). *Int. Archs Allergy appl. Immun.* **7**, 103.
Schlotfeldt, H. S. and Simpson, J. G. (1965). *Investnes zool. chil.* **12**, 45–65.
Schmidt, J. (1925). *Rep. Smithson Instn.* **1924**, 279–316.
Schumann, G. O. (1959). *Rep. Inst. Freshwat. Res. Drottningholm* **40**, 176–197.
Scopes, R. K. and Gosselin-Rey, C. (1968). J. Fish. *Res. Bd. Can.* **25** (12), 2715.
Shapira, R. and Parker, S. (1960). *Biochem. biophys. Res. Commun.* **3**, 200.
Sibley, C. G. (1960). *Ibis* **102**, 215–284.
Sibley, C. G. (1961). *In* 'Vertebrate Speciation' (W. F. Blair, ed.), pp. 69–88, University of Texas Press, Austin.
Sibley, C. G. (1963). *System. Zool.* **11** (3), 108–118.
Sibley, C. G. (1967). *Discovery, New Haven, Conn.* **3** (1), 5–20.
Sibley, C. G. and Brush, A. H. (1967). *Auk* **84** (2), 203–219.
Sibley, Corbin, K. W. and Haavie, J. H. (1968). *Condor* **71**, 155–179.
Sick, K. (1961a). *ICES Gadoid Fish Comm.* No. 28. Quoted by de Ligny (1969).
Sick, K. (1961b). *Nature Lond.* **192**, 894–896.
Sick, K., Westergaard, M. and Frydenberg, O. (1962). *Nature Lond.* **193**, 1001–1002.

Sick, K., Frydenberg, O. and Nielson, J. T. (1963). *Nature Lond.* **198**, 712.
Sick, K. *et al.* (1967). *Nature Lond.* **214**, 1141–1142.
Simonarson, B. and Watts, D. C. (1969). *Comp. Biochem. Physiol.* **31**, 309–318.
Simson, R. C. and Dollar, A. M. (1963). *Can. J. Genet. Cytol.* **5**, 43–49.
Sinderman, C. J. (1961). *Trans. N. Am. Wildl. Conf. no. 26*, 298–309.
Sinderman, C. J. and Honey, K. H. (1963). *Copeia* **3**, 534–537.
Sinderman, C. J. and Krantz, G. E. (1968). *Chesapeake Sci.* **9**, 94–98.
Sinderman, C. J. and Mairs, D. F. (1958). *Anat. Rec.* **131** (3), 599–600.
Sinnot, E. W., Dunn, L. C. and Dobzhansky, T. (1958). 'The Principles of Genetics' (5th ed.). McGraw Hill, New York.
Smith, A. C. (1962). *Calif. Fish Game* **48**, 199–201.
Smith, A. C. (1965). *Calif. Fish Game* **51**, 163–169.
Smith, A. C. (1966). *Am. Zool.* **6**, 577.
Smith, A. C. (1967). *Calif. Fish Game* **53**, 197–202.
Smith, A. C. (1968). *Comp. Biochem. Physiol.* **27** (2), 543–549.
Smith, A. C. and Goldstein, R. A. (1967). *Comp. Biochem. Physiol.* **23**, 533–539.
Smith, A. C. *et al.* (1963). *Calif. Fish Game* **49**, 44–49.
Smith, R. J., Miescher, P. A. and Good, R. A. (1966). 'The Phylogeny of Immunity'. University of Florida Press, Gainesville.
Smith, I. (1968). 'Chromatographic and Electrophoretic Techniques', 2nd Ed. 2 vols. W. Heinemann, London.
Smithies, O. (1957). *Nature Lond.* **180**, 1482.
Smithies, O. and Hiller, O. (1959). *Biochem. J.* **72**, 121.
Somero, G. N. (1969). *Biochem. J.* **114**, 237.
Somero, G. N. and Hochachka, P. W. (1969). *Biochem. J.* **114** (2), 237.
Sorvachev, K. (1957). *Biokhimiya* **22** (5), 822–827.
Sprague, L. M. (1967). *Hereditas* **57**, 198–205.
Stormont, C. (1968). 'Proc. XI European Conference on Animal Blood Groups and Biochemical Polymorphism'. W. Junk, The Hague, pp. 53–66.
Sulya, L. L., Box, B. E. and Gunter, G. (1961). *Am. J. Physiol.* **200** (1), 152–154.
Svärdson, G. (1945). *Rep. Inst. Freshwat. Res. Drottningholm* **23**, 151 pp.
Svärdson, G. (1965). *Rep. Inst. Freshwat. Res. Drottningholm* **46**, 95–123.
Thomas, M. L. and McCrimmon, H. R. (1964). *J. Fish Res. Bd. Can.* **21** (2), 239–246.
Thurston, R. V. (1967). *J. Fish. Res. Bd. Can.* **24** (10), 2169–2188.
Tiselius, A. (1937). *Trans. Faraday Soc.* **33**, 524–531.
Tsuyki, H. and Gadd, R. E. A. (1963). *Biochim. biophys. Acta* **71**, 219–221.
Tsuyki, H., Roberts, E. and Vanstone, W. E. (1965). *J. Fish. Res. Bd. Can.* **22**, 203–213.
Tsuyki, H., Roberts, E. and Kerr, R. H. (1967). *J. Fish. Res. Bd. Can.* **24**, 299–304.
Tucker, D. W. (1959). *Nature, Lond.* **183**, 495–501.
Tuppy, H. and Paleus, S. (1955). *Acta. chem. scand.* **9**, 353.
Uhlenhuth, P. T. (1903). 'Robert Koch Festscript'. Fischer, Jena.
Urist, M. R. and van de Putte, K. A. (1966). *Amer. Inst. Biol. Sciences. Wash. D.C.*
Utter, F. M. (1964). Thesis, University of California (Santa Barbara).
Utter, F. M. (1969). *J. Fish. Res. Bd. Can.* **26**, 3268–3271.
Utter, F. M. and Hodgins, H. O. (1969). *J. exp. Zool.* **172**, 59–68.
Utter, F. M. *et al.* (1969). *J. Fish. Res. Bd. Can.* **26**, 15–19.
Ushakov, B. P. (1959). *Zool. Zh.* **38** (9), 1292–1302.
Ushakov, B. P. (1964). *Physiol. Rev.* **44** (3), 518–560.
Vanstone, N. E. *et al.* (1964). *Can. J. Physiol. Pharmacol.* **42**, 697–703.
Vesell, E. S. and Bearn, A. F. (1962). *J. gen. Physiol.* **45**, 553–565.

Viswanathan, R. and Krishna Pillai, V. (1956). *Proc. Indian Acad. Sci.* **43** b (6), 334–339.
Walls, G. L. (1942). *Bull. Cranbrook Inst. Sci.* **19**.
Walsh, K. A., Ericcson, L. H. and Neurath, H. (1966). *Proc. natn. Acad. Sci. USA* **56**, 1339.
Webb, E. C. (1964). *Experientia* **20**, 592.
Whitt, G. S. (1969). *Science N.Y.* **166**, 1156–1158.
Wieland, I. *et al.* (1959). *Biochem. Z.* **332**.
Wieme, R. J. (1965). 'Agar Gel Electrophoresis', Elsevier, Amsterdam.
Wilkins, N. P. (1966a). 'Proc. X European Conference on Animal Blood Groups and Biochemical Polymorphisms', *Inst. natn. Rech. Agron. Paris* 355–359.
Wilkins, N. P. (1966b). *Comp. Biochem. Physiol.* **17**, 1141–1158.
Wilkins, N. P. (1967). *J. Cons. perm. int. Explor. Mer* **31** (1), 77–88.
Wilkins, N. P. (1968). *J. Fish. Res. Bd. Can.* **25** (12), 2651–2653.
Wilkins, N. P. (1970). 'Proc. XI European Conference on Animal Blood Groups and Biochemical Polymorphism', W. Junk, The Hague, 539–543.
Wilkins, N. P. and Iles, T. D. (1966). *Comp. Biochem. Physiol.* **17**, 1141–1158.
Williams, N. P. (1963). North Sea Herring Working Group 61, M.M. 63.
Wilson, A. C. and Kaplan, N. O. (1964). *Fedn. Proc. Fedn. Am. Socs exp. Biol.* **23**, 1255.
Wilson, A. C., Cahn, R. D. and Kaplan, N. O. (1963). *Nature Lond.* **197**, 331.
Wilson, A. C. *et al.* (1964). *Fedn. Proc. Fedn. Am. Socs exp. Biol.* **23**, 1258–1266.
Wolfe, H. R. (1933). physiol. zöol. **6**, 55–90.
Wood, D. C. and Burgess, L. (1961). *Am. J. Ophthal.* **51**, 305–314.
Wright, C. A. (1959). *J. Linn. Soc. Zool.* **44**, 222–237.
Wright, C. A. (1964). *Proc. zool. Soc. Lond.* **142** (2), 371–378.
Wright, C. A. (1966). *Int. Rev. gen. exp. zool.* **2**, 1–42.
Wright, C. A., Harris, R. H. and Claugher, D. (1957). *Nature Lond.* **180**, 1498.
Wright, J. E. (1955). Penn. State Univ. Contrib. No. 195 Dept. Bot. Plant Path. 172–175.
Wright, J. E. (1955). *Progre. Fish. Cult.* **17**, 172–176.
Wright, J. E. *et al.* (1966). *Proc. Pacif. Sci. Congr.* **7**, 11.
Wroblewski, F. and Gregory, K.F. (1961). *Ann. N.Y. Acad. Sci.* **94**, 912.
Wroblewski, F., Ross, C. and Gregory, K. F. (1960). *New Engl. J. Med.* **263**, 531–536.
Zuckerkandl, E. (1965). *Science, N.Y.* **212** (5), 110–118.

5 Insects

W. P. STEPHEN

Department of Entomology, Oregon State University, Corvallis, Oregon, U.S.A.

I. Introduction

Through the use of comparative biochemistry and immunology the taxonomically oriented biochemist and the biochemically oriented taxonomist are seeking 'objective' criteria on which to base genetic distances between organisms. The selection of proteins, or their constituent polypeptide chains, as these criteria, is based on the premise that they are manifestations of the nucleotide sequence of the gene. Although this premise has been widely held for a number of years, it was only recently that Yanofsky and his colleagues first clearly demonstrated the colinearity of protein amino acid sequence and DNA nucleotide sequence in their studies of tryptophan synthetase in *E. coli* (Yanofsky *et al.*, 1964; Carlton and Yanofsky, 1965; Guest and Yanofsky, 1965).

It is largely on the basis of the colinearity premise that Florkin (1964) defined the species in biochemical terms as 'consisting of groups of individuals with more or less similar combinations of sequences of purine and pyrimidine bases in their macromolecules of DNA, and with a system of operators and repressors leading to the biosynthesis of similar amino acid sequences'. Perhaps the most significant segments of this definition are the uses of the term 'similar' in relation to base sequence combinations and amino acid sequences, for just as other portions of the phenotype have evolved at different rates and in different ways, intrinsic structural variability is equally evident at the macromolecular level. The amino acid sequencing studies that have been done, reveal that amino acid replacements resulting from spontaneous mutations which affect the homologous DNA apparently are maintained in a population if they do not impair enzyme function. Henning and Yanofsky (1963) have shown that seven out of nine mutationally-induced single amino acid substitutions in tryptophan synthetase modified the macromolecule sufficiently so as to be detectable by their different rates of migration in polyacrylamide gels. The significant feature of these observations is that substitutions apparently have not occurred at those sites which affect the substrate-binding capacity or its steric fit, nor are there residue losses which impinge on catalytic functions. Relatively few amino acid sequences are known for the primary proteins of animals, i.e. hemoglobin, cytochrome C, insulin, etc., but the structural similarities of a number of these taken from different organisms suggest a relative stability in certain of the DNA base sequences through time (see Dayhoff, 1969). These observations that proteins vary in the degree to which they can be modified without loss of function has prompted Anfinsen (1959) and Florkin (1966) to propose a protein spectrum concept in which the process of mutation and selection may alter or impart considerable structural variability to proteins at one end of the spectrum (=violable) and virtually none to those at the other (=inviolable). If phylogenetic interpretations can be drawn from this concept, then from among the multiplicity of animal proteins there should occur those that have retained structural and functional primitiveness. The verification of the postulate is now contingent on the sequencing of an array of primary structures and to trust the probability that similarities in amino acid sequences will permit us to infer protein homology.

In the absence of data from insects that can be used for comparative purposes we need refer only to a few of the sequencing studies that have been done on vertebrates to determine to what degree they may be in accord with taxonomic relationships. Hemoglobin of man consists of two pairs of identical chains, an α- and a β-chain. In sequencing analysis conducted on hemoglobin of normal and sickle-cell anemics, a disease which is lethal in its homozygous state, Ingram (1958) found that the two hemoglobins differed only by the substitution of a valine for the glutamic acid residue at one position in the β-chain. Sequence analysis of hemoglobin from man and gorilla has shown that differences in amino acid residues occur at one to two percent of the positions of the nearly 300 amino acid residue chain (Zuckerkandl and Pauling, 1962), whereas differences in the α-chain of hemoglobin between man and horse occurred at approximately 12% of the positions and between man and rabbit at 20%. Comparable similarities were found among amino acid sequences of ribonuclease from cattle and sheep, in which the amino acids differ at only 3 out of the 124 residue positions. Sequence data available for cytochrome C indicate a difference at approximately 12% of the positions when man and horse are compared, but at well over 90% of the positions when horse and *Pseudomonas*, or human and *Pseudomonas* sequences are compared. These studies have prompted Wilson and Kaplan (1964) to speculate on the potential applicability of amino acid sequence differences of enzymes for taxonomic use. They postulate that 'enzymes from different families of mammals differ in sequence by about 2%, enzymes from different cohorts of mammals by about 12–20%, and enzymes from different kingdoms by more than 90%.

It is abundantly clear that proteins represent a means by which the genetic proximity of organisms may ultimately be quantified, but it is equally clear that the relationship has often been over-simplified. Of the enormous number of proteins which characterize each of the more than a million insect species, the only studies on primary structure are those of the amino acid sequences of cytochrome C of four species (Dayhoff, 1969). The overwhelming nature of the task in insects has thus necessitated the development of alternative approaches to the problem, some with greater potential than others.

It is the purpose of this chapter to provide the non-biochemically oriented biologist with a review and appraisal of the various immunological and biochemical techniques that have been applied to insect taxonomy. Comparative studies in invertebrates have lagged considerably behind other animal groups, primarily because of the difficulty in obtaining sufficient amounts of protein for sequence analysis. Although the alternative approaches to the evaluation of protein structure lack the definitive quality of amino acid sequencing they hold a potential for the acquisition of objective data, by means of which taxa at various levels may be phenetically compared.

II. Zone Electrophoresis

A. PROTEIN ANALYSIS

The principles and techniques of electrophoresis are outlined elsewhere in this volume and presented in considerable detail in Williams and Chase (1968). Suffice it to say that zone electrophoresis is the movement of charged particles on a supporting medium under the influence of an external electric field. The use of paper as the supporting medium in which electrophoresis could occur provided the first rapid, economical means of obtaining gross comparisons of macromolecular structure such as the proteins. Paper, however, because of its high molecular adsorptive qualities, variable pore size, and high electro-endosomotic buffer flow was replaced as a separating medium by agar gel, thence starch and more recently acrylamide gel, each of which provided an increase in macromolecular differentiation. The introduction of starch and acrylamide as media introduced a second dimension to protein separation. The physical impediments presented by starch or acrylimide matrices in the gel can be predictably altered by their concentration, thus superimposing a sieve effect on macromolecular migration.

A multitude of descriptive studies on insects appeared employing the new techniques, but few were more than exploratory studies of the soluble proteins extracted from various tissues at various stages of development. The majority of analytical studies on insects conducted during the period in which paper electrophoresis was widely employed, used hemolymph as the tissue source, primarily because of its ready accessibility, its quantity, and the incomplete information on the potential sources of protein variability. Wyatt (1961) provides a comprehensive review of the literature on hemolymph biochemistry to this date. In the few comparative studies in which attempts were made to utilize protein pherograms as taxonomic characters, the hemolymph was again the principal tissue used (Clark and Ball, 1957; Brezner and Enns, 1958; Stephen, 1961; Marty and Zalta, 1967). The protein pherograms resulting from paper electrophoresis of cockroach hemolymph were characterized by the presence of three or four bands whose patterns showed species and/or species-group specificity (Stephen, 1961). With the marked increase in the discriminating ability among protein fractions through the use of starch and acrylamide as the electrophoretic medium, from 20 to 36 fractions are discernible in the hemolymph from these same cockroaches, and the specificity, while less apparent, is just as real. It is from these pherograms that we have been able to draw conclusions parallel to those of Florkin (see above) as to the species- or population-protein 'violability' or 'inviolability'. Rejection of starch or acrylamide as an electrophoretic medium because of the complexity of the resulting protein patterns, reflects an *a priori* logic and undoubtedly accounts for the inability of these authors to distinguish among and between a number of vertebrate species (Johnson, 1968; Brown, 1964).

The past several years have witnessed a substantial decline in the appearance of papers whose taxonomic orientation is based on the comparison of protein pherograms. This is in part attributable to the increase in interest in specific enzyme systems but also relates to the relative difficulty in determining inter- and intrapopulation variability in protein patterns. There have been few comparative studies on either specific enzyme systems or total proteins of insects that meet even the most liberal interpretations of what might be considered to be acceptable taxonomic criteria. Phenetic comparisons and phylogenetic conclusions drawn on the basis of protein extract analysis of a single species from each of several genera lying in diverse families, comparisons of insects merely categorized as 'fly', 'wasp', 'beetle', or comparisons of protein extracts from larvae of one species with adult extracts of a second represent flagrant violations of the taxonomic ethic, but nevertheless because of their biochemical basis such papers have appeared in reputable journals. In addition to the customary amenities accorded any taxonomic character, that is comparing specimens of the same stage and same sex, additional precautions must be observed when proteins are being used in biochemical phenetics. The intrinsic qualitative and quantitative variability in insect tissue proteins has been alluded to earlier and the ubiquitous nature of this variability among all insect tissue is extrapolated to a large extent from the comparative biochemical and physiological studies on the hemolymph (see Wyatt, 1961; and below). Analysis of the soluble esterases from the thoracic muscle of the American cockroach, *Periplaneta americana*, support this generality (Beckendorf and Stephen, 1970).

An assessment of inter-individual variation in a population is implicit in taxonomic studies, be they biochemically or morphologically based, but biochemistry has provided characters capable of being homologized at much lower levels. Thus segments of a phenotype of a pherogram representing soluble proteins from the thoracic muscle of a species, may not be considered as homologous to comparable segments of pherograms derived from hemolymph, testis, or whole body extracts of the same species. In short, the biochemist in addition to inter-individual variation must recognize and account for intra-individual variability particularly as it is reflected in biochemical tissue specificity. The importance of these sources of biochemical character variation cannot be over-emphasized and the following list, by no means definitive, is drawn largely from studies made on insect hemolymph, and is presented as an indication of the type of lability that may be anticipated in other insect tissues.

1. *Sexual Effects*

Sex-specific or sex-linked proteins have been reported from the hemolymph of *Musca domestica* (Bodnaryk and Morrison, 1966), from the cockroach, *Nauphoeta cinerea* (Adiyodi, 1967), and detectable sexual differences in protein patterns have been reported from the hemolymph of the last larval stages of *Hyalophora* and *Samia* (Laufer, 1964). Hudson (1966) however, found no sex-specific proteins in the hemolymph of larvae, pupae, or adults of the tomato hornworm, *Protoparce quinquemaculata*.

2. *Development Stage*

Distinct quantitative and qualitative changes in proteins have been associated with the developmental processes of both holometabolous and hemimetabolous insects. The developmentally associated changes are elaborately documented in the studies by Denucé (1958) on *Bombyx mori* and *Galleria mellonella*, by Laufer (1960) on *Hyalophora cecropia* and *Samia cynthia*, and by Chen (1959) on *Culex pipiens* among the holometabola; as well as those by Wang and Patton (1968) on *Acheta domesticus* and by Marty and Zalta (1966) on *Gryllus campestris* among the hemimetabola.

3. *Tissue Specificity of Proteins*

Tissue specificity in soluble proteins has been reported from the tomato hornworm, *Protoparce quinquemaculata* (Hudson, 1966) and the saturniid moths, *Hyalophora cecropia* and *Samia cynthia* (Laufer, 1960). The tissue specificity is not always apparent for certain tissues at certain times may contain most proteins in common with the insect hemolymph. The reason for the similarity between tissue and hemolymph proteins may relate to their point of synthesis, for example, proteins synthesized in the fat body of saturniid moths (Laufer, 1960; Shigematsu, 1958) and of *Drosophila melanogaster* are liberated directly into the blood, as are certain proteins synthesized in the gut of *Samia* (Laufer, 1960) and in the blood cells of *Prodenia* (Wigglesworth, 1959). Recently, it has been shown that even isolated fat body of *D. melanogaster* was capable of *in vitro* synthesis of proteins into an incubation medium, certain of which had electrophoretic mobilities identical with those of several hemolymph proteins (Rüegg, 1968). Loughton and West (1965) detected hemolymph proteins in fat body, midgut wall, heart muscle, and midgut lumen, and concluded that the pattern of incorporation of hemolymph proteins into tissues of *Malacosoma americana* was specific and the absorption process was selective. Doyle and Laufer (1969) found that 10 of the 11 soluble proteins from the salivary glands of *Chironomus tentans* were also present in the hemolymph, and that the glands synthesized protein *de novo*, took up hemal proteins selectively, and transported and secreted them as well.

4. *Moult-associated Changes*

Significant qualitative and quantitative changes in proteins have been shown to occur in the moth, *Hyalophora cecropia* (Telfer and Williams, 1953), in the cockroach, *Periplaneta americana* (Steinhauer and Stephen, 1959; Fox and Mills, 1969), as well as in the foot pads of *Sarcophaga* (Goldberg *et al.*, 1969) at the time of moulting. On the other hand no detectable changes in either protein patterns or in amino acids were detectable in *Rhodnius* at the time of moulting (Coles, 1965).

5. *Diet Induced Changes*

Hemolymph proteins of the adult housefly, *Musca domestica*, are profoundly altered by diet (Bodnaryk and Morrison, 1966), but no dietary effect on the protein composition of the hemolymph of *Pieris brassicae* larvae was observed by van der Geest (1968). However it should be noted that the latter studies were quantitative rather than qualitative.

6. *Wound Induced Changes*

Alteration of protein patterns, particularly those of the esterases, have been shown to occur in Lepidopterous larvae within two hours after being punctured by a needle (Laufer, 1960). Marek (1969) reports that injury proteins in prepupae of the moth, *Galleria mellonella* are synthesized both in the midgut and the hemolymph and are detectable as esterases in the hemolymph but not in the fat body. Insects are capable of recovering rapidly from considerable changes in volume or composition of the hemolymph, at least in simple sugars such as glucose (Treherne, 1958), but on removal of large amounts of hemolymph, the new proteins that are formed may well represent those common to a developmental stage succeeding that in which surgery occurred (Laufer, 1964).

7. Age-induced Changes

Protein concentration values in the hemolymph of the adult house cricket, *Acheta domesticus* attains a plateau in 30-day-old adults, following which there is a rapid senescence-related decline (Nowosielski and Patton, 1965). Significant increases in the succinoxidase activity in thoracic muscle of *Leucophaea maderae* occur after its emergence as an adult (McShan *et al.*, 1954), and cytochrome oxidase activity increased up to 200% during the first three days of the life of the adult housefly (Sacktor, 1950). Acid phosphatase activity has been shown to increase 90% above the level at the time of emergence of adult honeybee workers, whereas alkaline phosphatase falls approximately 44% during that same period (Rockstein, 1953). Comparable post-emergence changes have been reported for a variety of compounds, indicating that maturation of holometabolous insects in particular, which may be initiated in the pupal stage, is not really complete with adult emergence. There appears to be an obligatory period of post-emergence maturation during which quantitative and qualitative protein changes may occur.

8. Genetically Based Variation

One of the most exciting applications of protein analysis has been its use in detecting genetic polymorphism. Most of the genetic studies to date have involved the use of esterase isoenzymes as allelic tags particularly in *Drosophila* (Ogita, 1962; Wright, 1963; Hubby and Lewontin, 1966). The abundant work that has been done on the genetics of organophosphate resistance in houseflies has revealed that many of the resistant strains are characterized by strain-specific esterase patterns (van Asperen and van Mazijk, 1965; van Asperen, 1962; Menzel *et al.*, 1963). Marty and Zalta (1967) have reported that two isolated populations of the orthopteran, *Cophopodisma pyrenaea* could be distinguished on the basis of population specific hemolymph proteins.

In an open circulatory system characteristic of the invertebrates, in which all tissues are bathed in hemolymph we should anticipate that tissue acting not only in a buffering capacity, but also as a hormonal messenger, nutritional reservoir and metabolic waste repository. Thus the extent of biochemical variability in the hemolymph may be considerably greater than might be anticipated in other tissues, and may limit its use in biochemical phenetics.

As indicated earlier, the literature is resplendent with reports on the proteins or enzymes extracted from various tissues of various insect species, and although many of these studies are comparative in nature, few would qualify as taxonomic studies. Much of the difficulty in meeting even the most basic acceptable taxonomic standards is associated with the inaccessibility of living specimens of species in those genera or families which are to be studied. A notable exception is the Hubby and Throckmorton (1965) analysis of over 360 electrophoretically separated proteins extracted from whole body homogenates of nine sibling species of the *virilis* group of *Drosophila*. This group is separated into two phylads on the basis of cytological data, one including the species *virilis*, *nova mexicana*, *americana* and its subspecies *a. texana*; and the second consisting of five species of the *montana* phylad which is cytologically subdivided into two subphylads. The study is based on the electrophoretic analysis of four saturated ammonium sulfate precipitable fractions of pooled samples of each species separated on acrylamide gels. Proteins of species with mobilities differing from those of all other proteins of all other species were designated as *unique*; proteins of species that had identical mobilities with proteins from one or more species belonging to the same cytological phylad were designated *phylad*; and those proteins that were common to species of both major cytological phylads were termed *ancestral*. The distribution of proteins from the nine species is presented in Fig. 1, which includes all except five

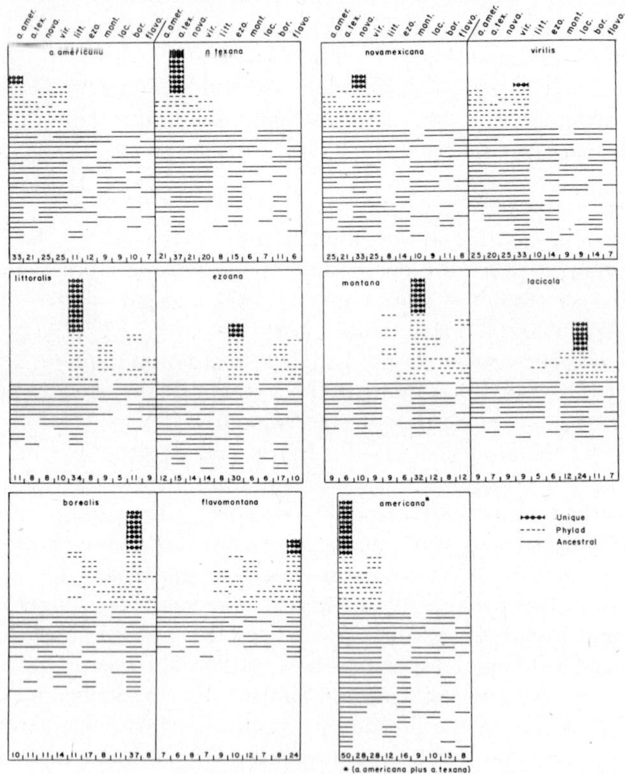

FIG. 1. The distribution of proteins from species of the virilis group of *Drosophila*. Proteins that are unique to a species appear at the top, those restricted to members of a phylad are in the middle, and those occurring in members of both phylads are at the bottom. Proteins in adjacent diagrams are not necessarily in the same order. (After Hubby and Throckmorton, 1965.)

fractions that are shared by all species. The sequence of proteins in adjacent diagrams is not necessarily the same and the numbers at the bottom of each column are those that each species has in common with the lead species in that particular diagram. Thus, *a. americana* has two fractions which are unique, nine which it shares with one or more members of the *virilis* phylad, and 22 in common with at least one of the species in both cytological phylads.

The authors conclude that each of the species within this group retained from 45 to 60% of the genic material from the common ancestor for the species group, 25% from the common ancestor for the phylad, and 15% that is unique. It is evident from Fig. 1 that the members of the *virilis* phylad have almost 80% other protein fractions in common, thus suggesting that the divergence of species of this phylad may have been more recent than the

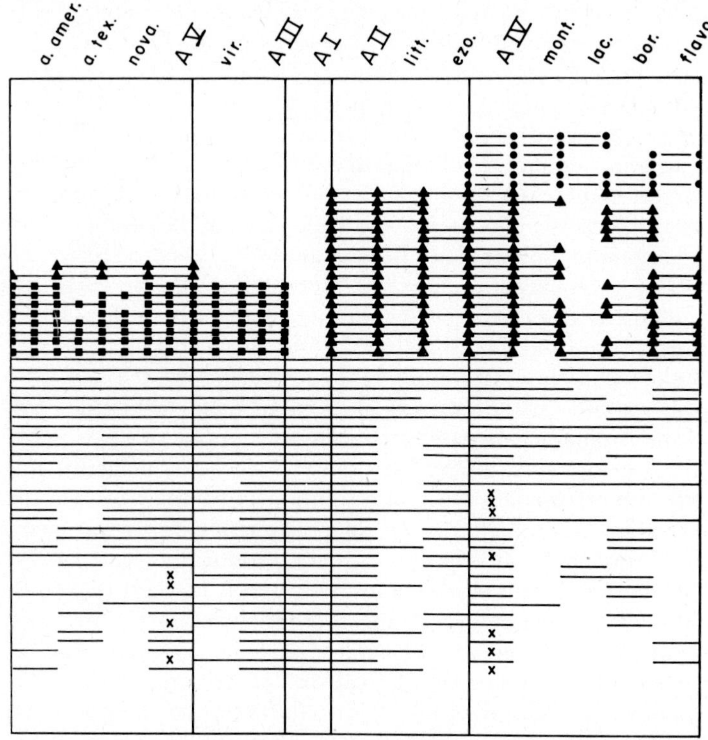

FIG. 2. Reconstruction of ancestral gene pools for species in the virilis group of *Drosophila*. Ancestral columns are headed by roman numerals preceded by A. Thus, AI indicates the ancestor for the group as a whole, AIII the ancestor for the virilis phylad, and so on. Unadorned lines toward the bottom of the figure indicate proteins derived from the ancestor for the species group. Lines adorned with solid circles indicate proteins present only in *montana*, *flavomontana*, *borealis*, and *lacicola*. Lines adorned with squares indicate proteins restricted to the *virilis* phylad. Lines adorned with triangles to the left of AI indicate proteins restricted to *americana* and *novamexicana*. Lines adorned with triangles to the right of AI indicate proteins restricted to the *montana* subphylad. With the exception of Ancestrals, proteins to the left and right of AI are not identical. Uniques are not included in the figure. (After Hubby and Throckmorton, 1965.)

divergence of the species of the *montana* phylad. In addition the authors have reconstructed a model of the ancestral gene pool of the species of this group which is illustrated in Fig. 2. The data were derived from information in Fig. 1. By pooling the fractions of the four species of the *montana* group, a composite hypothetical gene pool (A. IV) is presented as the ancestor of these species. A second hypothetical gene pool (A1. II) is derived by pooling the fractions of the species *littoralis* and *ezoana* with the hypothetical gene pool A. IV, which would represent the ancestral stock of the entire *montana* phylad. The hypothetical gene pools are then treated as species and the percent identity between and among all of the species in the group is calculated so as to estimate divergence that may have occurred.

The study indicates, in an admirable fashion, the manner in which electrophoretically separated proteins may be used in phenetic and phylogenetic taxonomy. Perhaps the difficulties involved in securing live material and the ready calculation of the relative mobility of the protein fractions accounts for the paucity of studies of comparative quality.

Aside from the geographic variability in hemolymph proteins shown to occur in disjunct populations of the orthopteran, *Cophopodisna pyrenaea* (Marty and Zalta, 1967) there is little or no other comparable data taken from natural populations of insects. The strain specificity among laboratory cultures of DDT and organophosphate resistant houseflies would suggest this as a fruitful area for study, as would the quantitative and qualitative differences in plasma proteins reported from different geographical races of turtles, snakes and amphibians (Zweig and Crenshaw, 1957; Dessauer and Fox, 1958).

Because of the intrinsic variability in tissue proteins it is felt that when total protein pherograms are to be used in phenetics that techniques providing maximal discriminatory properties be routinely employed. Variability must be identified before even preliminary macromolecular similarities might be estimated. Maximal separation and resolution of complex protein mixtures has been achieved through the use of polyacrylamide as the medium in routine electrophoresis, for not only is it inert, optically clear, lacks macromolecular absorption and electroendosmosis, but it has been shown that the ratio of absolute protein mobilities in two different acrylamide gel concentrations permitted estimation of simple molecular weights (Zwaan, 1967). Yet in all media including acrylamide, the calculation or computation of relative mobilities of the protein fractions of two samples to be compared is at best difficult and often impossible. The use of gel sheets will often facilitate estimat-

---

FIG. 3. A: Pherograms of soluble protein extracted from thoracic muscle of three species of *Periplaneta*: (1) *P. fuliginosa* (Serville); (2) 'Split-gel' with *P. fuliginosa* on left and *P. australasiae* (Fabr.) on right; (3) and (4) duplications of *P. australasiae*; (5) 'Split-gel', *P. australasiae* on left and *P. brunnea* Burm. on right; (6) *P. brunnea*. Similarities in most fractions is evident. B: Pherograms of soluble proteins from two species in different cockroach genera in which the quantitative and qualitative dissimilarities are striking: (1) *Panchlora nivea* (L); (2) 'Split-gel', *P. nivea* on left and *Phortioeca phoraspoides* (Walker) on right; (3) *P. phoraspoides*.

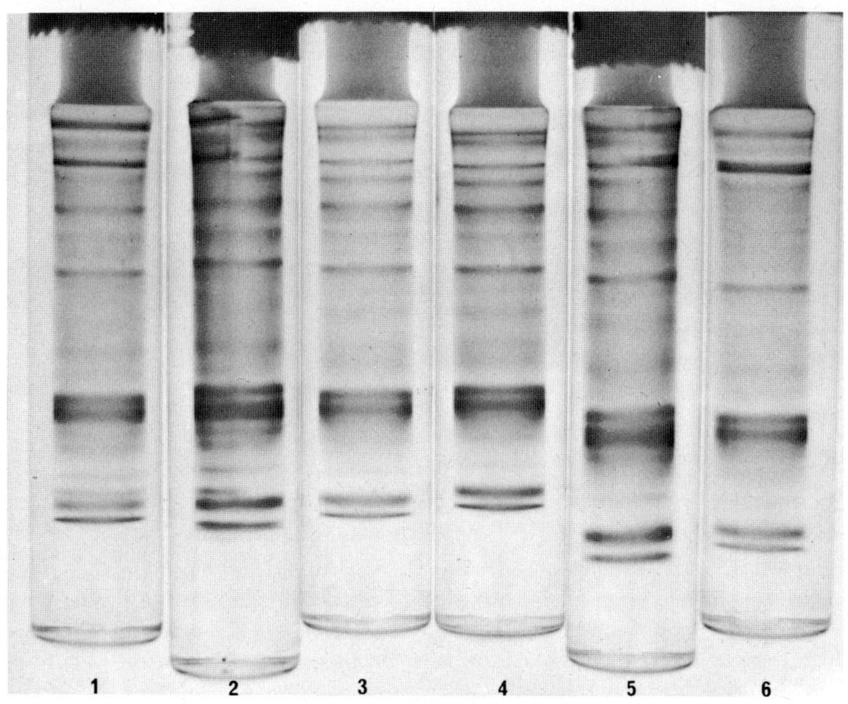

A

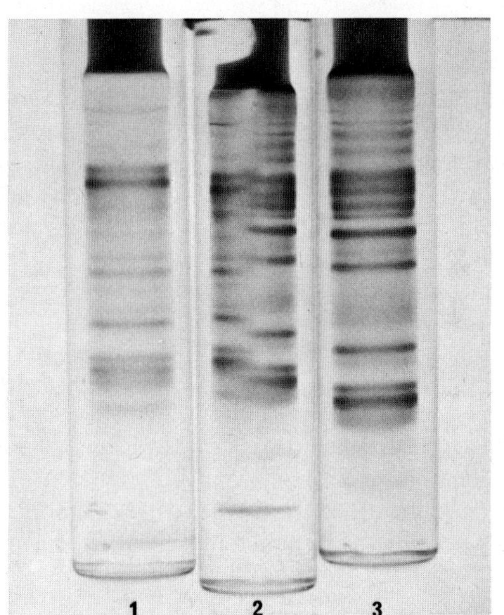

B

ing the comparative mobility of fractions, but even with samples separated by 5 mm, comparable mobilities can only be estimated. Fraction mobilities can be objectively compared through the use of a method such as the 'split-gel' technique developed in this laboratory in which the samples to be compared lie immediately adjacent to each other. The technique is used routinely in studies being conducted in this laboratory wherein a representative of each species or population is run against a standard so that comparative fraction mobility might be readily obtained. In addition, whenever possible the tissue from individual specimens is used in each analytical run, so as to be able to assess at least some of the genetic variability of the populations with which we are working. An indication of the inter- and intra-generic similarity in protein fractions is readily apparent through the use of 'split-gels' (Fig. 3) in the analysis of tissue extracts from several species of cockroaches (Stephen, unpublished).

B. GDH—GLYCEROPHOSPHATE DEHYDROGENASE (L-Glycerol-3-phosphate: NAD Oxidoreductase, EC 1.1.1.8)

The enormous energy demands made upon insect flight muscle to permit them to contract from several hundred to 1,000 contractions per second over extended periods of time resulted in the evolution of a non-glycolytic, aerobic oxidase system in which the oxidation is complete and immediate without the accumulation of metabolic intermediates such as lactate in the LDH cycle. This is accomplished in insects, and apparently in other invertebrates through the glycerol–phosphate cycle during which the extramitochondrial NADH is rapidly oxidized without the accumulation of intermediaries. As might be anticipated GDH activity is greatest in those tissues that show little or no LDH activity, that is, the well tracheated flight muscle of the insect thorax (Sacktor and Cochran, 1957; Zebe and McShan, 1957; Sacktor and Dick, 1962). Although the enzyme is exceedingly high in concentration in insect flight muscle it is by no means restricted to flight muscle or to insects, having been reported from all mammalian and amphibian tissues thus far studied (Rosen *et al.*, 1968).

Earlier studies on amino acid composition of honeybee and rabbit GDH led Jukes (1965) to postulate a common origin for both molecules, and the suggestion that many NAD+-dependent dehydrogenases are oligomeres with subunits of about 35,000 small m.w., prompted Kaplan (1965) to postulate a common ancestral protein for all such enzymes. Critical structural analysis of purified GDH from rat, rabbit and honeybee muscle have not resolved the problems on the primary, conformational and subunit structural properties of the enzyme from these sources (Brosemer and Kuhn, 1969; Fondy *et al.*, 1968, 1969). The many similar structural properties such as amino acid composition, NADH binding turnover number, similarities in hydrophobicity, optical rotary dispersion spectrum, reaction with NAD+

analogs, etc., led Fondy *et al.* (1969) to conclude that these were features to be expected of genetically homologous enzymes. Other unresolved structural questions, such as differences in C-terminal sequence, electrophoretic behavior, solubility, molecular weight, and tryptic peptide maps suggest that the GDH molecule has undergone many primary structural changes in the course of evolution, leading to distinct functional specificity (Brosemer and Kuhn, 1969). The many similar features exhibited by GDH from each of the sources, coupled with its specific function in the glycolytic cycle makes it suspect as a primitive enzyme system and an intriguing enzyme in biochemical phenetics (Brosemer *et al.*, 1963).

There have been no extensive comparative studies on this enzyme system in insects although generic and/or subgeneric stability have been reported in

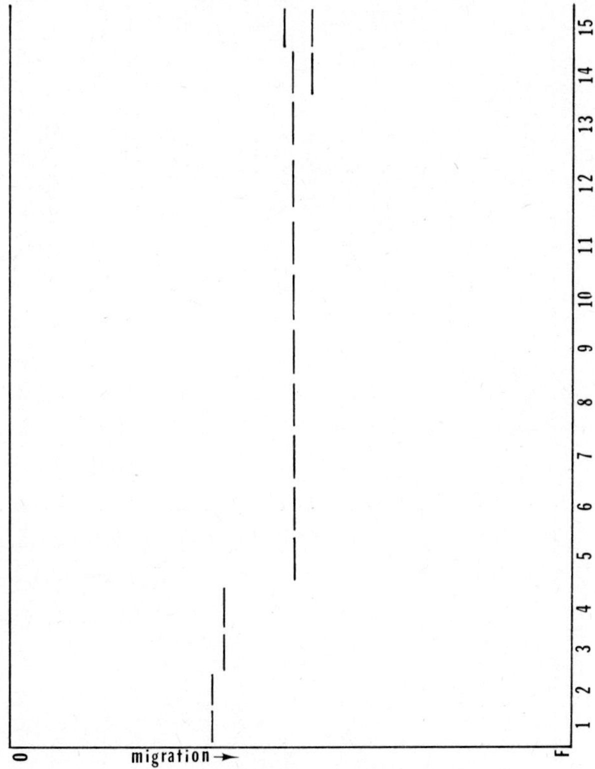

FIG. 4. Diagrammatic representation of the R_f values of GDH fraction(s) extracted from muscle of 15 species of cockroaches. O = origin of sample; F = front marker. Species are numbered 1–15 on the abscissa: (1) = *Eurycotis floridana* (Walker); (2) *E. decipiens* (Kirby); (3) *Blatta orientalis* L.; (4) *Periplaneta americana* (L.); (5) *Leucophaea maderae* (Fabr.); (6) *Nauphoeta cinerea* (Oliver); (7) *Gromphadorhina portentosa* (Schaum); (8) *Pycnoscelus surinamensis* (L.); (9) *Panchlora nivea* (L.); (10) *Capucina patula* (Walker); (11) *Blaberus giganteus* (L.); (12) *B. craniifer* Burm.; (13) *Eublaberus posticus* (Erichson); (14) *Phortioeca phoraspoides* (Walker); and (15) *Diploptera punctata* (Eschsch.).

Table I
Comparative Classification of Cockroach Species. (After Stephen and Cheledin, 1970b.)

Based on GDH	Based on genitalia, proventriculus (McKittrick, 1964)	Based on wing venation (Rehn, 1957)	Based on external morphology (Princis, 1960)
Group I	Group I	Group I	Group I
Subgr. 1. *Eurycotis*	Subgr. 1. *Eurycotis*	*Eurycotis*	*Eurycotis*
Subgr. 2. *Periplaneta*	Subgr. 2. *Periplaneta*	*Periplaneta*	*Periplaneta*
Blatta	*Blatta*	*Blatta*	*Blatta*
Group II	Group II	Group II	Group II
Subgr. 1. *Leucophaea*	Subgr. 1. *Leucophaea*	Subgr. 1. *Leucophaea*	Subgr. 1. *Leucophaea*
Nauphoeta	*Nauphoeta*	*Nauphoeta*	*Nauphaeta*
Gromphadorhina	*Gromphadorhina*		*Gromphadorhina*
Pycnoscelus	Subgr. 2. *Pycnoscelus*	*Pycnoscelus*	Subgr. 2. *Pycnoscelus*
Panchlora	Subgr. 3. *Panchlora*	*Panchlora*	Subgr. 3. *Panchlora*
Capucina	*Capucina*		Subgr. 4. *Capucina*
Blaberus	Subgr. 4. *Blaberus*	Subgr. 2. *Blaberus*	*Blaberus*
Eublaberus	*Eublaberus*	*Eublaberus*	*Eublaberus*
Subgr. 2. *Phortioeca*	Subgr. 5. *Phortioeca*		Subgr. 5. *Phortioeca*
Group III		Group III	
Diploptera	Subgr. 6. *Diploptera*	*Diploptera*	Subgr. 6. *Diploptera*

GDH isoenzymes among the bees (Stephen and Cheldelin, 1971). Electrophoretic stability of GDH at the species level has been reported by Sims (1965) and Hubby and Lewontin (1966) for *Drosophila*, and Hubby and Throckmorton (1965) found different GDH patterns between species of this genus. Intraspecific variation in GDH was determined in inbred and wild populations of *D. melanogaster* and the electrophoretic variants traced to two alleles located on the second chromosome (Grell, 1967). A similar observation was reported by Berger and Milkman (in Fondy *et al.*, 1968) with the suggestion that GDH exists as a dimer in *Drosophila*.

Comparative electrophoretic analysis of thoracic tissue extracts of cockroaches and bees undertaken in this laboratory, reveal no sex- or age-associated qualitative changes in GDH and suggests that its presumed primitiveness may make it valuable as a super-specific criterion. As in the protein analysis, tissue extracts from individual specimens were analyzed, and because of the similarity in mobility of the fractions, extracts of several specimens of each species were run against one or more species used as standards in each of the groups in which studies were conducted. Among the cockroaches, *Periplaneta americana* and *Leucophaea maderae* were used as standards in 'split-gels' against each of the 28 species in the 21 genera of cockroaches that have thus far been analyzed. A composite of the electrophoretically separated GDH fractions from 15 species in 10 genera of cockroaches are diagrammed in Fig. 4. The diagram is characterized by two major groups of GDH, a slower moving GDH fraction (species 1–4), and a fast moving fraction(s) (species 5–15). Arbitrary super-generic groupings based solely on GDH of the 15 species are presented in Table I, in which comparisons are made with super-generic classifications proposed by McKittrick (1964), Rehn (1951), and Princis (1960). Two major groups are proposed on the basis of the fast and slow moving GDH, each of which is further subdivided. The separation of the slow moving GDH-characterized species is straightforward and not inconsistent with other acceptable classifications. On the basis of migration alone, the 11 species characterized by fast moving GDH fraction(s) are separated into two groups at the same hierarchal plane. *Phortioeca* is characterized by a basal band with a mobility comparable to that of the *Leucophaea* group and the apical band with a mobility identical to that of the apical band of *Diploptera*, and on this basis is considered to be a member of a sub-group of the *Leucophaea* complex. Both GDH fractions of *Diploptera* have mobilities distinct from those of the *Leucophaea* group and on the basis of this electrophoretic uniqueness it is given full group status. The proposed classification, while not in complete accord with any of the three other classifications appears to incorporate features of all three (Stephen and Cheldelin, 1970b).

The stability of the GDH enzyme system, in cockroaches at least, is further evidenced in the analysis of thoracic muscle extracts of mature female specimens of *Pycnoscelus surinamensis* and *P. indicus*. The karyology of the group has been thoroughly documented by Roth and Cohen (1968). *P. indicus* is restricted

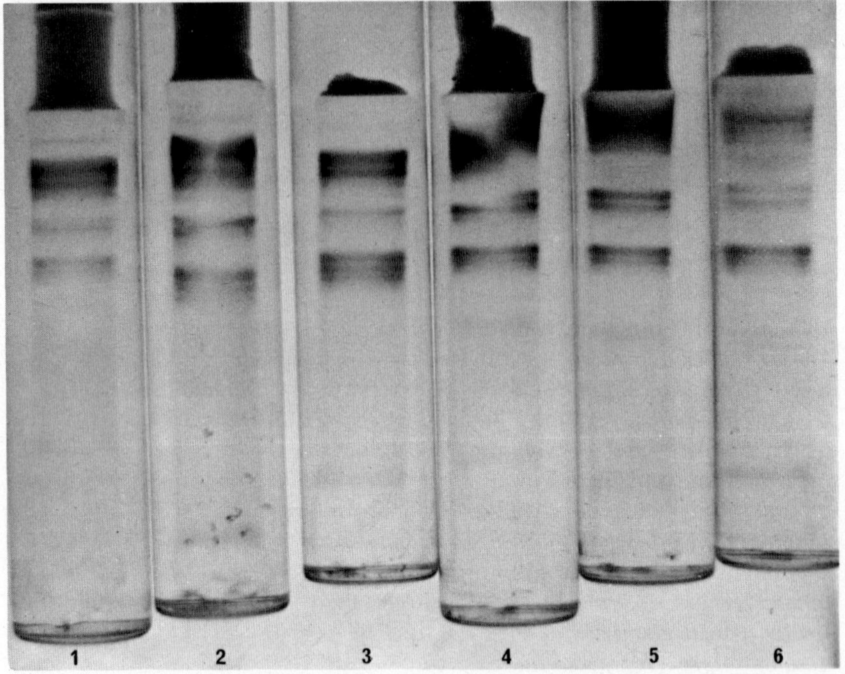

FIG. 5. Zymograms of GDH from 3 species of *Bombus* representing 2 subgenus-specific patterns: (1) *B. californicus* Smith; (2) 'Split-gel' with *B. californicus* on left and *B. fervidus* (Fabr.) on right; (3) *B. fervidus*; (4) 'Split-gel' with *B. fervidus* on left and *B. occidentalis* Greene on right; (5) and (6) *B. occidentalis*. Note the difference in mobilities of the 3 fractions in the basal group and 2 in the mid-group in (5).

to Hawaii where it is wholly bisexual and characterized by a ♀ chromosomal complement of 2N = 36 or 38. Throughout the rest of its range, it is largely or exclusively parthenogenetic (*P. surinamensis*) with ♀ chromosomal complements of 34 (Brazil and Australia), 35 (Thailand), 37 (Indonesia), 53 (Florida, Brazil) and 54 (Panama, Jamaica and Uganda). 'Split-gel' analysis of 5 to 23 specimens from cultures of each of 11 geographically disjunct populations of *surinamensis* reveal that all are characterized by a single GDH fraction with the same R_f value. Thus diploidy and triploidy has no electrophoretically discernible effect on GDH in this genus.

Among the bumblebees the electrophoretic patterns of GDH are considerably more complex, characterized in most species by the presence of eight fractions grouped in a 3–2–3 arrangement (Fig. 5). However, in the genus several distinct mobility patterns are evident, two of which are illustrated, which coincide with subgeneric groupings based on morphological and behavioral criteria (Stephen, 1957). Very recent electrophoretic studies on GDH from 10 bumblebee species is reported by Brosemer and his colleagues (Brosemer *et al.*, 1970). Using cellulose acetate as a separating medium, they were able to discern five, and in one case six GDH activity sites which served as criteria to subdivide the genus into four general groups. The proposed groups are not entirely consistent with morphologically based criteria, and this may be attributable to the limitations imposed by the separating medium used.

Purified GDH from the honeybee, *Apis mellifera*, was prepared and microcomplement fixation studies performed against antigens from six races of this species (Brosemer *et al.*, 1967). The exceptional specificity of the microcomplement fixation test is outlined below but it is an analytical method which is sufficiently sensitive to detect differences between hemoglobins which differ by as little as two amino acid residues out of a total of 574 (Reichlin *et al.*, 1964). No immunological differences in the GDH of the six races or of the three castes of honeybees were detected, a result which supports conclusions on the stability of the GDH in the insect groups being studied in this laboratory. Similarly, the antigenic properties of GDH of newly emerged bees was determined to be the same as those in older adults, and the increase in enzymic activity is attributed to a greater number of enzyme molecules rather than to a more active enzyme moiety (Brosemer *et al.*, 1967). Contrary to the conclusions reached by Brosemer *et al.* (1967), studies in this laboratory would suggest that the electrophoretic GDH patterns are not only highly stable at the super-specific level but they warrant further study on their potential as a taxonomic character.

The major, or slowest moving GDH fraction has been crystallized from four species of bumblebees and used for the preparation of antisera for microcomplement fixation studies on other bumblebee species and the honeybee (Brosemer *et al.*, 1970). The results of these studies and the interpretations of the authors are outlined in Section VI.

There are a few references to analytical data on this enzyme system from other insects but none are comparative. Muus (1968) reports a single band from pooled thoraces of the housefly, *Musca domestica*, and a single band from the pooled thoraces of the tsetse fly, *Glossina*. A single GDH band was found in muscular extracts of *Leucophaea maderae*, and possibly also from the fat body of the same species (Gilbert and Goldberg, 1966). The latter observation corresponds with the studies from this laboratory.

C. MDH—MALATE DEHYDROGENASE (L-Malate: NAD Oxidoreductase, EC 1.1.1.37)

This enzyme like GDH has a major role in aerobic metabolism and is an essential component of the Kreb's cycle. It is thus present in most tissues of all animals and is invariably present in multiple molecular forms. MDH is unusually high in insect muscle and is believed to be one of the principal components of the energy cycle in insects where carbohydrates serve as a major energy source.

In spite of its near universality in animal tissue there have been virtually no comparative studies conducted among the animals. This enzyme system has been used as a tool in studies relating to developmental changes in hemi- and holometabolous insects (Laufer, 1961; Gilbert and Goldberg, 1966; Goldberg *et al.*, 1969). Generally the activity has been found to be greatest during the period of tissue synthesis when the energy demand is at its peak (Goldberg *et al.*, 1969).

Gilbert and Goldberg (1966) reported the same two MDH isoenzymes in all tissues of *Leucophaea maderae* (ovaries, embryo, testes, fat body) of both the nymphs and the adults, but suspected that one may be mitochondrial and the other extra-mitochondrial. MDH, as three forms, was isolated from foot pad extracts of *Sarcophaga bullata* in the late larval, pupal and adult stages (Goldberg *et al.*, 1969). Muus (1968) found three MDH fractions from extracted thoraces of the tsetse fly *Glossina*, and two from whole thoracic extracts of *Musca domestica*. Two MDH bands were resolved from hemolymph extracts of the cynthia moth (*Samia cynthia*) and three from the hemolymph of cecropia (*Hyalophora cecropia*).

Preliminary data from electrophoretic analysis of thoracic extracts from the cockroaches indicate that there are five distinct MDH fractions in extracts of *L. maderae*, and from three (*Diploptera punctata*) to six (*Blaberus gigantus*) in other cockroach species (Stephen, unpublished). It is interesting to note that the closely related genera *Leucophaea* and *Nauphoeta* have four of the five MDH fractions in common and the fifth (the fastest migrating fraction) present in only one of eight specimens of *N. cinerea* analyzed. Further analyses of this enzyme system in cockroaches and bees is underway in this laboratory even though existing studies are indecisive as to its potential role in insect phenetics.

D. LDH—LACTATE DEHYDROGENASE (L-Lactate: NAD Oxidoreductase, EC 1.1.1.27)

Another incompletely known enzyme system in insects is lactate dehydrogenase which has proven to be an extremely useful enzyme system in genetic and developmental studies among vertebrates (Markert, 1963; Wilkinson, 1965; Vessell, 1965). The enzyme is responsible for glycolytic energy production through oxidation or partial oxidation of glycogen or glucose under conditions of limited oxygen supply. Its ubiquitousness in vertebrate tissues, which are much more poorly oxygenated than many invertebrate tissues, has been one of the prime reasons for these investigations (Zebe and McShan, 1957; Gilbert and Goldberg, 1966).

The available information on LDH isoenzymes in insects is too fragmentary to permit conclusions as to its taxonomic value. Gilbert and Goldberg (1966) report LDH isoenzymes from most tissues of *Leucophaea maderae*, except the thoracic muscle, with the greatest concentration in the adult ovaries. Four fractions were isolated from adult ovaries, three from larval ovaries, two from the fat body of larvae and adults, and one from the accessory glands, testis, and Malpighian tubules. Only two isoenzymes were detected in the testis of the grasshopper, *Melanoplus differentialis*, but these were absent in the muscle (Aronson *et al.*, 1968); and a single fraction was isolated from the foot pad of *Sarcophaga* during the late larval, the pupal, and adult stages (Goldberg *et al.*, 1969). Thoracic muscle analysis of the tsetse fly (*Glossina*), and the house fly, (*Musca domestica*) (Muus, 1968), as well as the analysis of thoracic muscle extracts from females of 17 species of *Bombus* and representative species of 14 genera of cockroaches (Stephen, unpublished) suggest that LDH is either absent or present in such low quantities as to make its identity difficult.

LDH exists in two essentially different forms in most vertebrates, designated as M (muscle type), and H (heart type). These forms differ in a number of catalytic properties including their electrophoretic mobility, amino acid composition, rates of reaction with the same coenzyme analogs, and in their immunological properties. It is noteworthy that in a few vertebrates, such as the adult lamprey, only the M form of the enzyme exists (Kaplan and Ciotti, 1961).

Two comparable forms have been isolated from invertebrate tissue, and these also differ catalytically. One of the forms is common to highly aerobic tissue and is similar in many respects to the H form of vertebrates (Wilson and Kaplan, 1964). This form of the enzyme reacts more strongly with the acetyl pyridine analog (AcPyAD) and is found in the muscle tissue of crustaceans, centipedes, insects, millipedes, and many other invertebrates. However, neither of the two isoenzymes of LDH that have been isolated from either the tissue of arachnids or the horseshoe crab, (*Limulus*) have been shown to exhibit this property. This difference has been utilized to support a commonly accepted view that two very distinct invertebrate subphyla exist, the Mandibulata, which include the former, and the Chelicerata, which include the latter, of the two groups.

E. ESTERASES—(The Carboxylic Ester Hydrolases of the Subgroup EC 3.1.1)

The soluble esterases differ from most other animal enzyme systems in the number and heterogeneity of the fractions. Multiple molecular forms of the soluble esterases exhibit a high degree of tissue specificity among vertebrates (Holmes and Masters, 1968). This heterogeneity, associated to some extent with genetic heterozygosity, imparts to the esterases a value in characterization of specific tissues, in following the mechanics of tissue development, as well as offering promise in comprehending more of their genetic basis.

Since the primary purpose of this chapter is to apprise taxonomists of what has been done in the Insecta and the significance of those results, the myriad of related papers on the use of esterase isoenzymes as indicators of genetic polymorphism, or as a reflection of developmental change will not be considered. Variability in the esterase isoenzyme patterns have been attributed to both of these factors as well as to a high degree of tissue specificity. Nevertheless it is studies in these areas in which the use of esterase isoenzyme analysis has been the most fruitful and potentially most applicable to taxonomy (Hubby and Narise, 1967; Lewontin and Hubby, 1966; Prakash and Lewontin, 1968). It has been shown that several alleles at one locus may be identified electrophoretically, from which minimal estimates of genetic heterozygosity within and among populations may be made (Lewontin and Hubby, 1966). On the basis of this study it has been suggested that since homologous loci in disjunct populations of *Drosophila* are identifiable by electrophoretic and histochemical means, that these methods could be extended to determine the occurrence of homologous loci in sibling species.

There has been no clear indication of the potential use of soluble esterases as taxonomic criteria above the population level, although underlying species characteristics in the soluble esterases have been recognized in man, guinea pig, rat and mouse (Holmes and Masters, 1967). Interpretive differences have arisen as a result of the considerable amount of work that has been done on the association of esterase activity with organophosphate resistance in houseflies. van Asperen and van Mazijk (1965) concluded that fly-strain patterns of esterase isoenzymes are highly specific and constant over long periods of time, whereas Collins and Forgash (1968) found considerable inter- and intra-strain differences in isoenzyme patterns of the same species. Further, any meaningful synthesis of the published studies on the multiple molecular esterase forms in insects is fruitless, for the differences in techniques and materials are matched by equal or greater differences in methods by which samples were prepared and extracted. Thus the general applicability of this enzyme system in biochemical phenetics is contingent upon the development of a practical means of obtaining replicable data with very limited material usually available.

Using an isogenic population of the American cockroach, *Periplaneta americana*, restricting sampling to thoracic and appendage muscle, and conducting analysis under rigidly controlled conditions, Beckendorf and Stephen (1970) found no qualitative differences in the patterns of soluble esterases taken from the thoracic or appendage muscle, nor was any sex- or age-associated qualitative difference observed. Of the 19 different esterase isoenzyme fractions extracted from the thoracic muscle of the American cockroach, ten were found to be common to all of the specimens analyzed, four of which were carboxylesterases, three were lipases, and the others included an arylesterase, pseudocholinesterase and an acetyl esterase (Stephen and Cheldelin, 1970a). The esterases are unique among the insect enzyme systems thus far studied in their heterogeneity and numerical abundance, but because of these characteristics this enzyme group could well prove to be one of the more

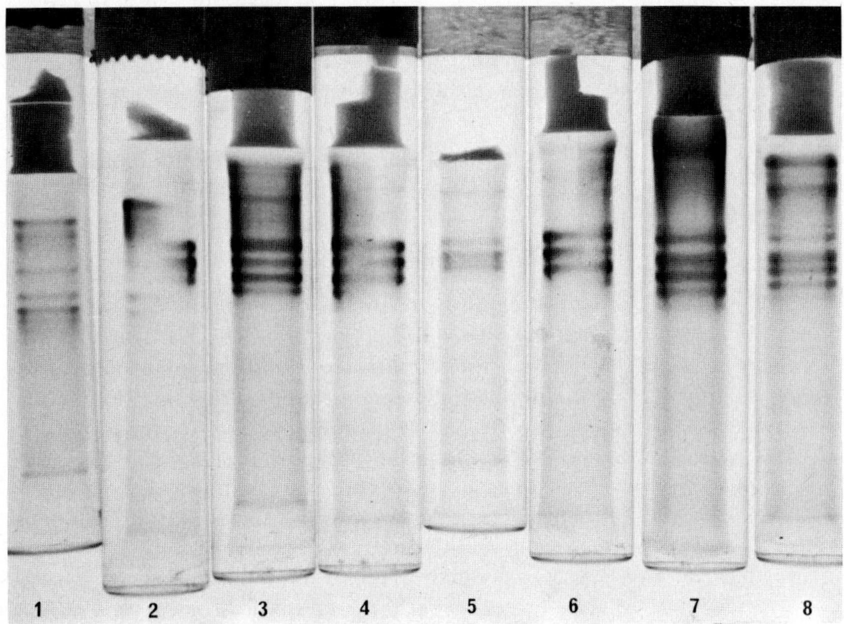

FIG. 6. Esterase zymograms of 5 *Bombus* species indicating pattern similarities among members of the same subgenus. (1) *B.* (*Pyrobombus*) *vosnesenskii* Rad; (2) 'Split-gel', *B.* (*P.*) *vosnesenskii* on left and *B.* (*Bombus*) *occidentalis* Greene; (3) *B.* (*B.*) *franklini* (Frison); (4) Split-gel', *B.* (*B.*) *franklini* on left and *B.* (*B.*) *occidentalis* on right; (5) *B.* (*B.*) *occidentalis*; (6) 'Split-gel', *B.* (*B.*) *occidentalis* on left and *B.* (*Bombus*) *terricola* Kirby on right; (7) *B.* (*B.*) *terricola*; (8) *B.* (*Bombus*) *affinis* Cresson.

valuable in biochemical phenetics particularly after preliminary characterizations and population specificity have been established for the species groups in question.

The accompanying figure illustrates esterase zymograms prepared solubilized from the thoracic muscle extracts of queens of several bumblebee species of the subgenus *Bombus* (Fig. 6). Subgeneric similarities in the patterns (primarily carboxylesterases) are consistent and the patterns are distinct from the species with which a comparison is made, as well as from those of all other subgenera (Stephen, unpublished).

There are, however, certain biochemical and biological properties of esterases which demand stringent control and adequate replication when insect materials are being used. Aside from tissue specificity and developmental changes, three to four times the amount of tissue extracts satisfactory for analysis of other enzyme systems are required for quality esterase zymograms. In addition a single gene has been reported as influencing the migration of a group of esterases in *Tetrahymena* (Allen, 1961) and mammalian tissues wherein each group represents a set of isoenzymes whose structure can be altered by a mutation at a single locus. More recently Marek (1969) indicated that wounding of the prepupae of *Galleria mellonella* resulted in the synthesis of two 'injury proteins' both of which were esterases and both of which were detected only in the hemolymph. However, with proper precautions and rigid replicability the heterogeneity in the esterase enzyme systems from insects are a distinct asset in genetic and, potentially, in taxonomic research.

III. Chromatographic Analyses

Insect hemolymph is characterized by an unusually high complement of free amino acids, ranging from 5 to 30 times higher than that possessed by other animal groups (Florkin, 1960). An enormous amount of quantitative and qualitative information on the free amino acid composition of insect hemolymph has been made available through the use of microbiological assays (Duchateau and Florkin, 1958), paper chromatography (see reviews by Wyatt, 1961; Chen, 1962) and more recently through the use of various automatic analyzers (Chen, 1966). Comparatively little work has been done on the amino acid composition of other insect tissues. Generally, hemolymph analysis has revealed a preponderance of the aliphatic amino acids. Certain amino acids, or their derivatives, have been reported as being unique to a single species (thyroxine in *Rhodnius*, Harrington, 1961) and a number of amino acids such as β-alanine, α-aminobutyric acid and ornithine have been identified which are not known to occur in any insect protein molecule. The analyses suggest that the free amino acids are characteristic of any given stage of any given insect species, but that considerable qualitative and quantitative hemolymph variability is evident between stages of the same species.

The available analytical data on free amino acids has prompted Florkin and Jeuniaux (1964) to conclude that the more primitive Hemimetabola have free amino acid ranges from 293 to 636 mg/100 ml, considerably lower than the free amino acid values found in the Holometabola, and that the relative proportions of the free amino acids in the hemolymph differ between the two major groups.

Factors that influence amino acid quality and quantity, such as diet and starvation, development, silk production, disease and parasitism as well as the genetics of the organisms, are reviewed by Wyatt (1961) and Chen (1962). Changes in the free amino acid pool of hemolymph and tissue extracts associated with the development of hemimetabolous and holometabolous insects are thoroughly and completely reviewed by Chen (1966). In the decade beginning in 1953, a number of papers appeared in which it was suggested that the degree of similarity in free amino acid composition of the hemolymph and/or tissue could be interpreted as a direct indication of the relationship among those forms (Ball and Clark, 1953; Buzzatti-Traverso and Rechnitzer, 1953; Micks, 1956; Micks and Gibson, 1957). Fortunately taxonomically oriented papers based on chromatographic analysis of free amino acids in the hemolymph of insects have appeared infrequently since the early 1960s. This is not meant to suggest that the complement of free amino acids in specific insect tissues may not be species- or tissue specific, but rather that the rigid control necessary to reduce or eliminate the causes of variation would be difficult to achieve.

A variety of chemicals including primary, secondary and tertiary metabolic products have been suspect of chemotaxonomical value. Generally, however, the secondary and tertiary metabolic products are proving to be rather unreliable characters in this area. The yellow pigment, 3-hydroxykynurenine, an amino acid in the metabolic pathway from tryptophan to nicotinic acid, is stored in the wings of certain groups of butterfles. Analysis of a distribution of this pigment in the Nymphalidae, however has suggested taxonomic groupings which are contradictory to those based on other morphological characters (Brown, 1967).

Micks *et al.* (1966) were able to separate morphologically indistinguishable strains of *Aedes aegypti* on the basis of chromatographic patterns of pteridines, and Throckmorton (1962) found several pteridines could be correlated with taxonomic position in the Diptera. Bhalla (1968) on the other hand was unable to draw any conclusion regarding the taxonomic value of three of the major pteridines extracted from the heads of eye color mutants of *Aedes* and *Culex*.

Long-chain fatty acid triglycerides are a major lipid component of insect fat body and the major energy source where fatty acids are used for flight. The isolation of triglycerides from several insect species and the determination of their carbon number distribution by gas-liquid chromatography has shown that each of the species sample was characterized by a specific triglyceride carbon number distribution (Harlow *et al.*, 1969).

IV. Finger-printing

The influence of nongenetic parameters on the free amino acid composition of various animal tissues presents a major limitation to the utility of that technique in taxonomy. Our goal is essentially to enable us to critically and accurately examine the amino acid sequences as they present an image of the nucleotide sequences of the DNA–RNA genetic materials. Since proteins or enzymes consist of polypeptides having their origin at one or several gene loci and since certain areas of the molecules may relate specifically to enzyme function while the remainder may vary in structure and charge without impinging on that function, an analytical procedure has been developed which is aimed at a stage in protein metabolism lying somewhere between amino acid and electrophoretic analysis. This technique referred to as finger-printing involves the partial digestion of related proteins in an attempt to detect the subtle differences in their product peptides. After partial digestion of the protein by one or more proteolytic enzymes, the digests are separated in two dimensions, using both chromatography and/or electrophoresis.

Tryptic digests of hemoglobins from several species have yielded two dimensional peptide patterns which support the taxonomic positions of the animals used (Zuckerkandl *et al.*, 1960). Finger-printing of myoglobins from seven animal species including two whales, a porpoise, seal, walrus, penguin and horse reveal peptide patterns whose taxonomic interpretations are at odds with those based on conventional morphological and immunological data. Comparable differences with accepted taxonomies resulted from the comparison of finger-prints of pancreatic ribonuclease. This has lead Wilson and Kaplan (1964) to conclude that both of these compounds are unreliable for the measurement of structural similarity.

In the absence of any comparable work on insects the inference can be drawn that the technique may be of value in determining relationships when based on functionally similar proteins. However, preliminary evaluation as to their potential taxonomic value must be highly discriminating and the ultimate selection based on *a priori* taxonomic conclusions.

V. Immunological Analysis

Other more rapid means of determining structural correspondence among proteins are continually being sought to circumvent the laborious, albeit objective, amino acid sequencing. One such area, for which expectation has regularly exceeded result, is the area of comparative immunology or systematic serology (Boyden, 1958, 1964). Partly because of the extensive complexity of the immune phenomena and partly because of poor experimental and taxonomic methods, the data from most of the comparative immunological studies on insects are incomplete, inconclusive or at odds with taxonomies based on other characters. However, the abundance of new theoretical information on the immune process and refinements in analytical techniques has been reflected by a recent resurgence of impressive works on the qualitative and quantitative aspects of comparative immunochemistry (Williams and Chase, 1968, 1967; Ouchterlony, 1968; Campbell *et al.*, 1964; Kabat and Mayer, 1961; Leone, 1964; Basford *et al.*, 1968). Among the more recent studies, several have used insects as their test organisms: at the specific and infraspecific level (Simon, 1969; Downes, 1963); and at the generic and supergeneric level (Butler and Leone, 1967; Basford *et al.*, 1968).

The vocabulary peculiar to immunology and immunochemistry is outlined in some detail in other parts of this volume and in ultimate detail in Kabat and Mayer (1961), but a skeletal outline of the terms used in succeeding paragraphs is presented below for convenience sake.

The injection of a foreign material, usually a protein (*antigen*) into an animal will stimulate that animal to generate compounds (*antibodies*) which will react with a high degree of specificity to the material that was injected.

Animals in which this process of antibody formation has occurred are considered to be *immunized*. The process of immunization may in turn be detected by the formation of a precipitate (the *precipitin*) when the soluble antigen is mixed with its immune serum in optimal proportions. It is the quantitative and qualitative analysis of the precipitin reaction which serves as a basis for the studies in systematic serology.

Antigens: Until recently only proteins were considered to be able to stimulate the production of antibodies, but it is now recognized that a number of polysaccharides, even those composed of a single simple sugar such as glucose (dextran) or fructose (levan), synthetic polypeptides (Sela and Fuchs, 1967), lipids, especially glycosphingolipids (Rapport and Graf, 1967), as well as a variety of conjugated proteins have this property. It has also been noted that there is a differential specificity on the part of the antigens, for example the polysaccharides mentioned above are antigenic in man and mice but are not so in rabbit or guinea pig. Thus antigenicity can be a function of the species immunized (Kabat and Mayer, 1961).

Antibodies: The antibodies synthesized in response to an antigen are all serum globulins and it is thought by some that the entire gamma globulin fraction may be a composite of antibodies that are formed in response to a variety of antigens to which the organism has been exposed. Many of the physico-chemical properties, including the molecular weights of antibody-gamma globulins and normal gamma globulins, are identical and the difference, if any, between the two lies in the possession by the anti-gamma globulin of several areas which have a configuration complementary to that of the antigen with which it will react. This area of complementarity is exceedingly small. For human antidextran it has been calculated to be equivalent to an open chain of six or seven glucose units, which when considering the size of the gamma globulin molecule is less than one percent of its surface—assuming a molecular weight of approximately 160,000 (Kabat, 1960). Complementary areas, however, are not uniform even for simple glucose chain units, for each animal synthesizes a population of antibody molecules that have varying combining site sizes (Kabat, 1958). Further, the use of a purified antigen rarely results in the formation of an antibody which is unique to that protein. It has thus far rarely been possible to prepare purified antigens that are not mixtures of numerous molecular species and assumptions as to antibody purity can never be made without prior testing. Even more frustrating is the inability to determine differences in serological activity of the antigens, particularly in the not infrequent cases where the precipitin reactions resulting from heterologous, or cross reacting, antigens are greater in optical density than those resulting from homologous, or reference, antigens. When this phenomenon is observed in a mixed antigen series it may well be that the serological activity of one or possibly even several of the antigens may differ, in which case it would be impossible to control.

Nevertheless, in spite of the number and complexity of problems involved in the immune reaction, immunological correspondence of cross reactions among antigens of closely related species are high, and the level of cross reaction decreases and disappears as the relationships between and among the species from which the antigens are taken become more remote. There is, in turn, no known criterion on which to elaborate on primitiveness in immunological reactions, and phylogenetic interpretations so based are no less subjective. Boyden (1964) concludes that because of the sliding scale of specificities of antisera of the same kind, it is not possible to give percent relationships to indicate precipitin parameters for species, genera, families, etc. It is however, a general rule that interordinal reactions are usually weak, interclass reactions are rare and interphylar reactions exceptional.

There have been fewer than a dozen papers on the immunotaxonomy of insects published in the past 20 years, and less than half of these can be considered to be anything other than survey in nature. The reason for the paucity of comparative immunological studies on insects are twofold: the inability to quantitatively and qualitatively evaluate immunological data obtained; and difficulties in obtaining readily available insect species to serve as an antigen source. The latter can be resolved by purely mechanical and manipulative means but it is the former on which the ultimate applicability of systematic serology will depend.

The ultimate analysis and comparison of antigens with specific antibodies is effected by bringing together optimal concentrations of each so as to form visible precipitin bands. This may be accomplished in a fluid (interfacial or ring test) or in a semisolid medium (a gel diffusion test). The gel- or immunodiffusion method has excellent resolving powers and is more qualitative in that it permits the identification of more than a single antigen–antibody reaction in a single test (Campbell *et al.*, 1964). While providing some qualitative and essentially quantitative data immunodiffusion, or for that matter all immunological analysis, is limited in that it can only provide minimal information on the quality of the determinate groups, and no information on the size, shape or structure of the reacting molecules.

This shortcoming has led to the development of immunoelectrophoresis (Graber and Williams, 1953) in which the physico-chemical and immunological analytical methods are combined. Immunoelectrophoretic analysis (IEA) may be conducted in a variety of ways employing a variety of separating media (Hirschfeld, 1968). In all of the published immuno-taxonomic papers on insects in which IEA has been used, the method has involved electrophoretic separation of a test antigen on a microscope or lantern slide using a 0·75–1% Ionagar medium. Following electrophoresis, troughs are cut on either side of the electrophorated antigen and filled with either a homologous or heterologous antiserum. Immunodiffusion is then permitted to proceed for 24 to 48 hours and homologous precipitin arcs form in the intervening area between the trough and the separated antigen (Fig. 9). In its present stage, the science requires a combined use of single AnAb systems, multiple AnAb

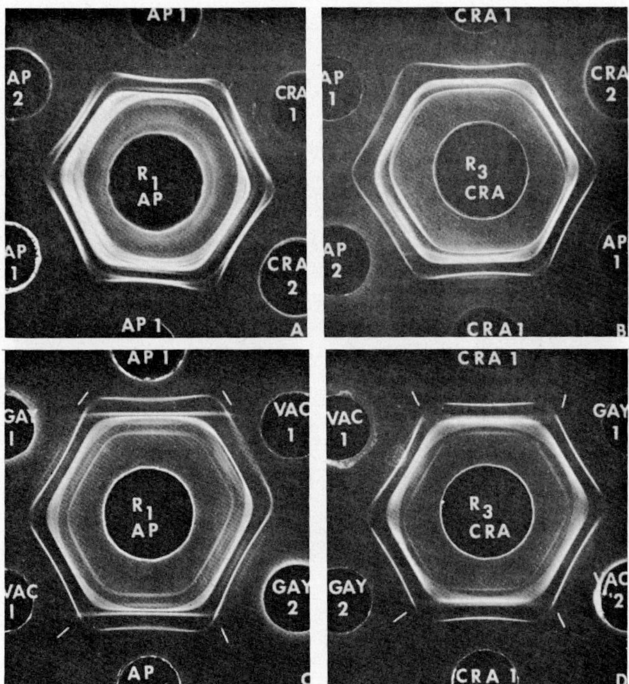

FIG. 7. Immunodiffusion patterns produced on Ouchterlony plates from reactions between antisera produced from 2 populations of *Rhagoletis pomonella*, anti-AP (A, C) and anti-CRA (B, D), and 8 antigen samples. Serological differences between homologous and some heterologous reactions are indicated by white pointers. (After Simon, 1969a.)

systems, physiochemical and histochemical analyses, all rigidly controlled and all with heirarchic equivalence (*sensu* Throckmorton, 1968), in order to provide acceptable and potentially valuable contributions to taxonomy.

Reference is here made to two papers both of which appeared recently and both of which satisfy many of the above criteria. Both of the studies utilize several immunological methods in the development of their theses, and/or supplement these with numerical taxonomic methods. The first paper deals with the comparative serology of eight populations of the tephritid fly, *Rhagoletis pomonella*, taken from four fruit hosts: apple; *Crataegus* (2 species); *Vaccinium* (2 species); and *Gaylussacia* (2 species). The forms have been variously treated as host races, sympatric subspecies, or sibling species (Simon, 1969). Each of the populations were compared by serological methods, quantitative fluid antigen–antibody serological correspondence; gel immunodiffusion; immunoelectrophoresis; and absorption of antisera. The latter technique has proven to be exceedingly useful in removing those antigens which are shared by two taxa to be compared so that only those unique to one

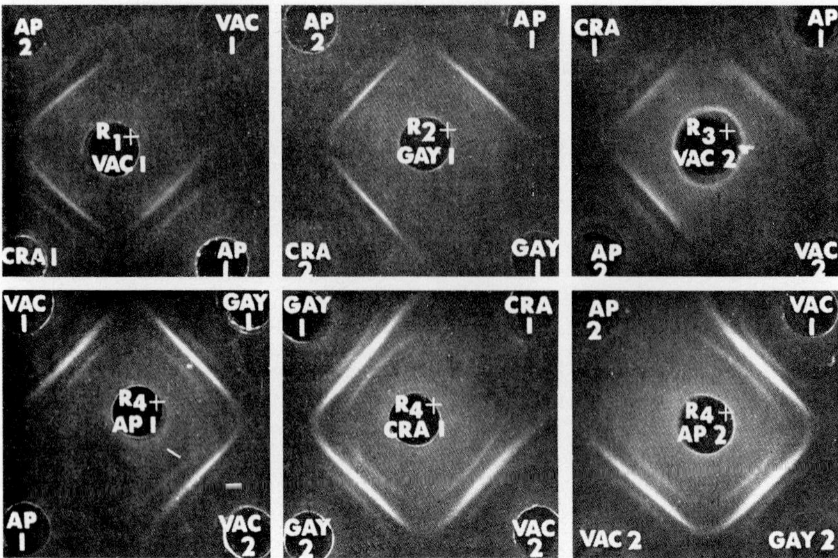

FIG. 8. Immunodiffusion patterns produced on Ouchterlony plates from reactions between absorbed antisera (center well) and antigen samples: (A) R_1 + VAC_1 = anti-AP serum (R_1) is absorbed by VAC_1 antigen; (B) R_2 + GAY_1 = anti-AP serum (R_2) absorbed with GAY_1 antigen; (C) $R_3 + VAC_2$ = anti-CRA serum absorbed with VAC_2 antigen; (D) $R_4 + AP_1$ = anti-VAC_1 serum absorbed with AP_1 antigen; (E) $R_4 + CRA_1$ = anti-VAC_1 serum absorbed with CRA_1 antigen; and (F) $R_4 + AP_2$ = anti-VAC_1 serum absorbed with AP_2 antigen. (After Simon, 1969.)

or other of the taxa may be evaluated. This is accomplished by precipitating the antigens common to both homologous and heterologous reactions through combining the antiserum with an appropriate amount of extract from a species showing all of the reaction arcs in common. The resulting antigen–antibody precipitate is discarded and antigens common to both are eliminated. Except for the optical density curve of the fluid antigen–antibody reactions, representative data are illustrated in Figs. 7, 8, and 9. The immunodiffusion patterns of whole body pupal extracts are illustrated in Fig. 7, in which the center wells in each of the four plates contain antisera against populations from apple (R_1-AP) or antisera against populations taken from *Crataegus* (R_3-CRA). Antigen of each of the eight populations of *Rhagoletis pomonella* were placed in appropriate antigen wells about the central antiserum well and diffusion was permitted to occur. Nine precipitin bands are observed in the homologous reactions in the four plates illustrated, and the cross reactions between AP and CRA reveal complete fusion of all of the precipitin bands in the zone of overlap (corners of the precipitation hexagon) between the different reactions. However, the reactions between anti-AP or anti-CRA sera and antigens from VAC and GAY populations produced all but one of the bands present in the

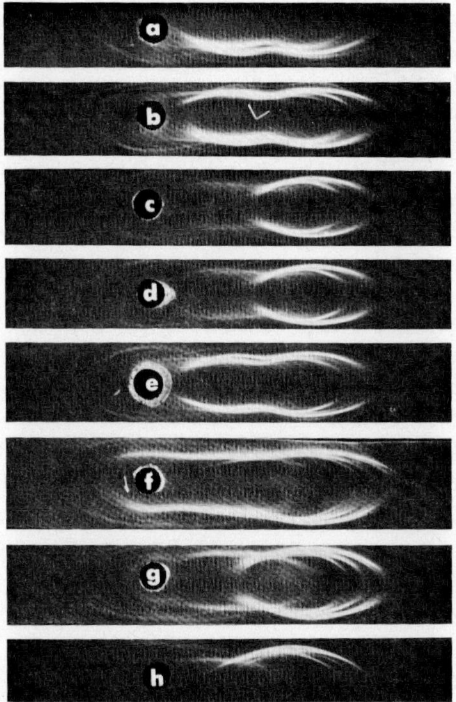

FIG. 9. Stained immunoelectrophoresis patterns resulting from reactions between the 8 electrophorated antigens and anti-AP_1 serum. (After Simon, 1969.)

homologous reactions, separating the populations phenetically into two groups.

The data from the absorption experiments corroborate those using unabsorbed antisera. No precipitin bands resulted when AP or CRA antisera were absorbed with either AP or CRA antigens, confirming that all populations of AP and CRA shared common antigens. However, when AP and CRA antisera were absorbed with either VAC or GAY antigens, two precipitin bands resulted when these were permitted to diffuse against AP or CRA antigens (Fig. 8a, b, c). Similarly, anti-VAC serum absorbed with AP or CRA antigen yielded three precipitin bands when the resulting serum was permitted to diffuse against either VAC or GAY antigen (Fig. 8d, e, f).

In the final phenetic comparison, Simon analyzed each of the eight populations immunoelectrophoretically. Proteins, separated electrophoretically on Ionagar were permitted to diffuse toward the diffusing antisera located in troughs along either side. Complementarity was reflected in the formation of a precipitin arc. The reactions between one of the antisera (anti-AP-1) and the eight electrophorated antigens are illustrated in Fig. 9. From the test of that study '—AP-1, AP-2 and CRAT-2 antigens show identical IE patterns. The reaction with CRAT-1, however, differs from the three others in that one of the heavy bands (possibly a composite of four antigen–antibody reactions) continues toward the cathodic side of the immunoelectropherogram (arrow Fig. 9f). The spectra produced by VAC and GAY antigens are indistinguishable from each other but are differentiated from those of AP and CRAT in that, while the latter show a group of at least four wavy precipitin bands produced over the entire length of the anodal reactive zone (arrow, Fig. 9b), the VAC and GAY spectra show bands with only one area of precipitation toward the anode (Fig. 9c, d, g).' Simon suggests that the two continuous arcs over the anodal zone in the AP and CRA pherograms may indicate electrophoretic heterogeneity of the proteins carrying homologous antigenic determinants which in turn are located in two different areas.

On the basis of the remarkable agreement among the quantitative and qualitative immunological tests, the eight host specific populations of the

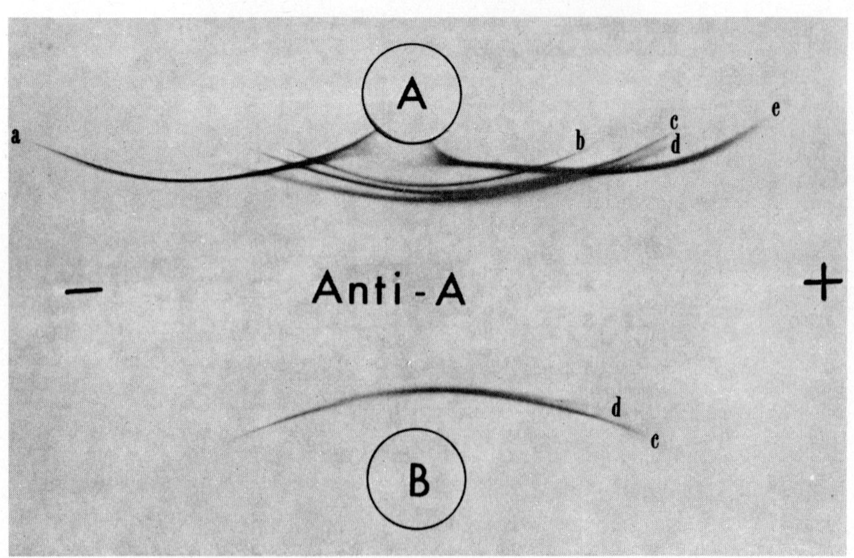

Fig. 10. Quantification of immunoelectrophoretic pattern. A is the reference species and B is a cross reacting species. Macromolecules in A and B migrate toward both poles in an electric field. When antiserum A is added, precipitin zones form as shown above. A, with 5 zones at 5 units each = 25 or 100%. B, with one zone (c) at 5 and one (d) at 4 = 9 or 36%. (After Basford *et al.*, 1968.)

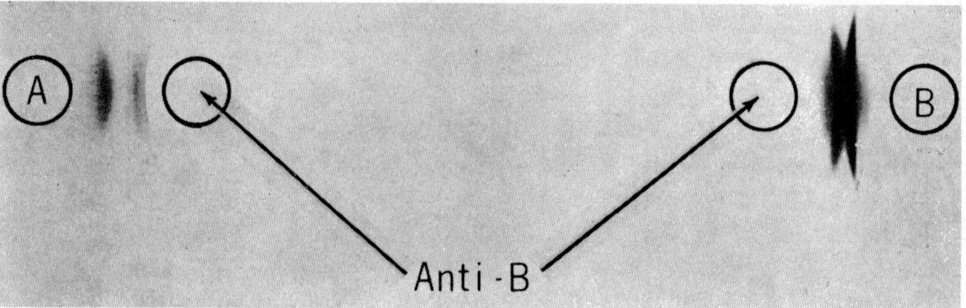

FIG. 11a. Immunodiffusion study patterned for quantification. B is the reference species and A is a cross reacting species. Both are reacted with antiserum against B which results in formation of precipitin zones between the wells as shown above. (After Basford *et al.*, 1968.)

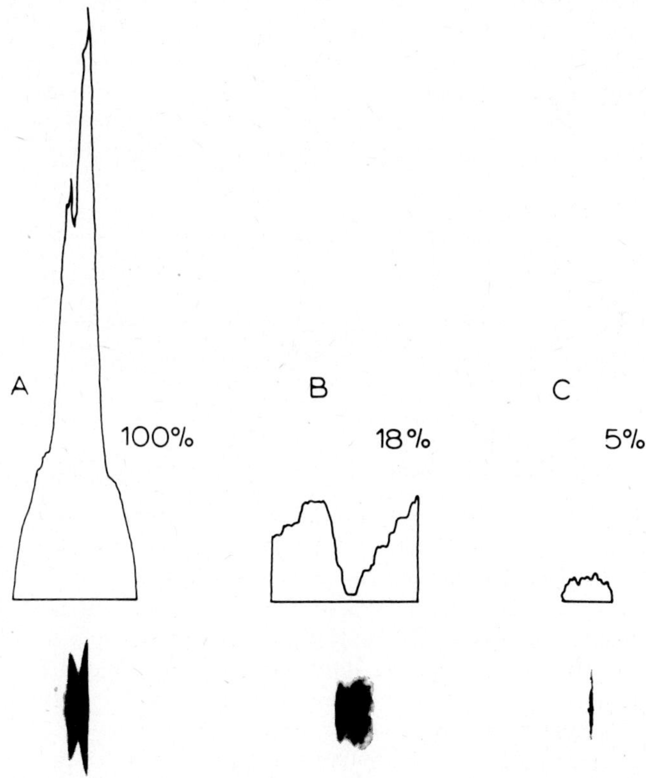

FIG. 11b. Densitometric scans from immunodiffusion reactions similar to those illustrated in Fig. 11a. In these scans A is the reference species and is given a value of 100%, and the values of the cross reacting species B and C are calculated relative to A by comparing the total areas under the respective curves. (After Basford *et al.*, 1968.)

species are separated into two serological groups: AP and CRA; VAC and GAY. The magnitude of serological differences that occur between the two population groups is so great that it could not be maintained were free gene interchange to occur between them (Simon, 1969).

In this study the methods are exclusively phenetic and the clustering of equivalent macromolecular characters has permitted the author to define a next higher step in the taxonomic hierarchy.

In the second of the two comprehensive studies, pherograms are generated from numerical data collected from immunoelectrophoretic and immunodiffusion studies on species representing eight families of Coleoptera (Basford *et al.*, 1968). This study is unique in that the authors elaborate on means of precipitin quantification from which they are able to generate numerical data from biochemical characters of the same hierarchic level. Both immunoelectrophoretic patterns and immunodiffusion patterns are obtained through cross reactions between and among all the species involved. The immunoelectrophoretic tests were quantified by giving a value of five to each of the precipitin arcs of the reference test, and a value of five to each corresponding precipitin arc formed in the cross reaction zone if it was as dense as that in the reference zone, and four if it was less dense (Fig. 10). Species cross reactions as determined by this method, were computed using the following formula:

$$\text{Percent cross reactivity} = \frac{\sum \text{values assigned to zones in cross reaction}}{\sum \text{values assigned to zones in reference reaction}} \times 100$$

Precipitin arcs resulting from the immunodiffusion tests were quantified (Butler and Leone, 1968) and the cross reaction values computed as follows:

$$\text{Percent cross reactivity} = \frac{\text{average densitometric value of cross reaction}}{\text{average densitometric value of reference reaction}} \times 100$$

(See Figs 10 and 11). The values obtained from the two serological analyses were subjected to cluster analysis of correlation and distance coefficient matrices and to principal components analyses. The conclusions of the study are of lesser significance to this review than is the application of the taxonomic method to immunological analysis of insects. Nevertheless, the authors conclude that the immunological data gave the highest correlations with established morphological classification when processed as unstandardized logarithms of the percent, cross reaction values.

The development of analytical techniques, many of which are mentioned above, has enabled the comparative biochemist to discriminate between and among macromolecules with relative ease, particularly where purified materials are available as a standard of reference. One of the major limiting factors to the conduct of detailed, broadly based studies in insects has been the limited amounts of tissue available to the researcher. Thus, most of the studies on biochemical phenetics in insects have been based on complex mixtures of antigens derived from extracts of whole bodies, or in a few cases specific tissues. The development of antibodies against such extracts has made the interpretation of the resulting antigen–antibody complexes exceedingly difficult. Objective turbidity measurements of antigen–antibody precipitates in fluid systems have been used as a basis of establishing multidimensional relationships (Boyden, 1964) but provide no opportunity to evaluate the individual antigen components of that particular system. Precipitin arcs resulting from two dimensional immunodiffusion tests can be quantified and occasionally do give minimal data of qualitative value (see Figs 11a, b).

At present however it is immunoelectrophoretic analysis (IEA) that provides us with maximal qualitative and quantitative discriminating powers for evaluating similarities of mixed antigen extracts. Improvement of IEA techniques either through better spatial separation of the antigens electrophoretically has permitted us to associate some precipitin arcs with specific pherogram antigens. In most stained pherograms obtained from separation of insect tissue extracts on a standard 7·5% separating acrylamide gel, there invariably appears one or more centers at which the majority of protein fractions are concentrated. These are discernible on the gel, but when subjected to immunodiffusion yield a mass of precipitin arcs many of which are superimposed or in such close proximity that qualification and quantification is impossible. The increase in discriminating power of acrylamide disk electrophoresis prompted us to use this medium in a two-step IEA process. The size of the tube was increased in both length and diameter (92 × 4·5 mm id) so as to increase the

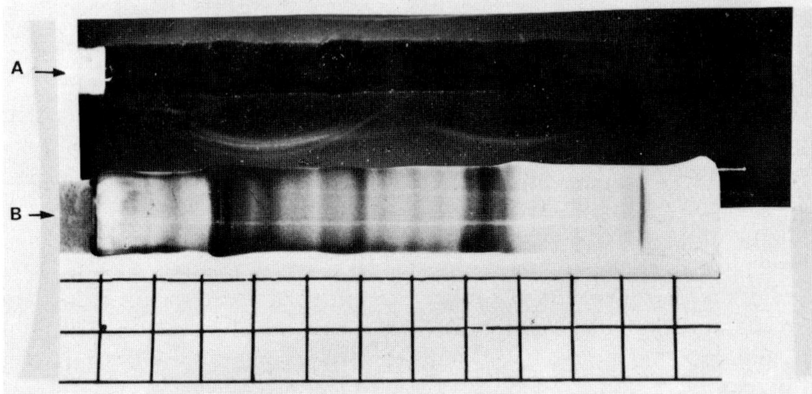

Fig. 12. Composite photographs of electrophorated *Leucophaea maderae* antigen cross reacted with anti-*Periplaneta americana* serum. The two halves of a single acrylamide gel are illustrated, one-half (A) embedded in Ionagar for double diffusion, and one-half (B) stained and aligned to associate precipitin arcs with protein fractions. The 4 acrylamide concentrations making up the separating gel can be distinguished by the abrupt diameter changes in the gel.

total amount of antigen electrophorated. A more uniform distribution of the protein fractions in the gel was achieved by using four different concentrations of acrylamide with the most concentrated at the anodic end. The graded series in each tube consisted of 5·5, 6·0, 6·5 and 7% acrylamide and comparable migration was effected through rigid control of the amount of gel of each concentration added to each tube. Upon electrophoresis each of the gels was bisected longitudinally, one-half embedded in Ionagar for immunodiffusion studies and the other half stained with Coomassie Blue so as to associate the precipitin arc with its antigen (unpublished, Fig. 12). The method has promise in that it moves us one step closer in our ability to associate electrophoretic with immunological data in a more objective manner.

The student of immunology and immunochemistry is fortunate indeed for the several comprehensive treatises that have appeared on these subjects in the past few years. By far the most comprehensive, is the four volume series 'Methods in Immunology and Immunochemistry' edited by Williams and Chase (1967, 1968), which is an open-end treatise dealing with methodology in this area. The four volumes are concerned with preparative methods for antigens, antibodies and the care of laboratory animals; details of analytical methods involved in immunological research; techniques and methods for analysis of antigen–antibody reactions; and an evaluation of the approaches to the study of the immune response. In addition, the long term standards of Kabat and Mayer (1961) in which the theoretical aspects received due attention, Campbell *et al.* (1964) in which immunological methods are presented in ultimate detail, and Ouchterlony (1968) in which the techniques of performing and interpreting immunodiffusion are outlined in minute detail, would provide a complete reference library on immunological methods and techniques.

VI. Microcomplement Fixation

Microcomplement fixation is one of the more recent developments that appears to hold promise in providing close estimation of phenetic similarity between and among macromolecules without amino acid sequencing. The technique, developed by Wasserman and Levine (1961) is used to assay very small quantities of highly purified protein immunologically. One of the principal advantages of the technique, particularly when working with invertebrates, is that the quantities of antigen and antiserum required for testing are up to 100 times less than that required for conventional complement fixation tests, and up to 1,000 times lower than required for quantitative precipitation evaluation.

The MCF technique has proven to be sufficiently sensitive to develop distance relationships of the primate species, including chimpanzees, on the basis of serum albumin (Sarich and Wilson, 1966), to permit the distinction between hemoglobins A, S, and C which have been estimated to differ by only two out of 574 amino acid residues (Reichlin *et al.*, 1964), and to detect minor differences in primary amino acid sequences (Arnheim *et al.*, 1969). The measure of cross reactivity is quantified through calculation of a number to represent the immunological distance (ID) between a homologous and a heterologous reaction (Sarich and Wilson, 1966).

Taxonomically oriented MCF studies in insects have been restricted to two dehydrogenase enzyme systems, α-glycerophosphate dehydrogenase (GDH) (Brosemer *et al.*, 1967, 1970) and glyceraldehydephosphate dehydrogenase (GAP-D) (Marquardt *et al.*, 1968). In the first of the two studies, purified GDH was extracted from honeybees and the antisera prepared in rabbits. MCF comparisons with five honeybee races, three other bee species, a wasp, and two flies confirmed Kaplan's conclusion (1965) that it is necessary to utilize antibodies against antigens of one of the members of the same or a similar taxon. The calculated IDs among the higher Hymenoptera coincided well with conventional classifications. However the cross reactions using the Diptera as the antigen sources were very weak and no reaction was observed with other non-hymenopterous insects. Using the GDH-based ID values from insects and comparing these with comparable ID values from MCF studies on vertebrate dehydrogenases, the authors suggest that the rate of GDH speciation in insects may not be more rapid than that of vertebrate dehydrogenases. The absence of detectable differences among the six races of honeybees or their castes, plus cross reaction based IDs which are consistent with classification of the higher Hymenoptera implies that the technique and the enzyme system warrant attention as tools in biochemical phenetics and phylogeny.

Brosemer *et al.* (1970)* crystallized the major (i.e. slowest) flight-muscle specific GDH isoenzyme from four *Bombus* species and prepared antisera for MCF studies against one of them. They observed that there was measurable cross reaction only with the major flight-muscle specific fraction in each of 12 Apinae, and that six of the eight bumblebees all cross reacted to the same extent (ID = 1·2 compared to an ID = 1·0 for the homologous reaction). The most distant cross reaction among the bumblebees against the *B. nevadensis* antisera was shown by *B. occidentalis* (ID = 1·4). The authors interpret those data to indicate that identical electrophoretic patterns of the 'slow' GDH isoenzyme in these two species, does not reflect the considerable structural differences between the two shown by the MCF study. Although an ID value of 2·4 was obtained for the honeybee—*B. nevadensis* cross reaction, suggesting a more distant relationship, the cross reaction occurred *only* with the major GDH isoenzyme of flight-muscle. The identical cross reaction with anti-honeybee GDH in all of the species in the test suggests to the authors that little or no evolutionary change has occurred in this variant since *Apis* and *Bombus* diverged. This conclusion is based on the premise that *Bombus* species

* I am indebted to R. W. Brosemer for providing me with a copy of this paper presented at a symposium at the AAAS meetings in Boston in December, 1969.

share among themselves a more recent common ancestor than do *Apis* and *Bombus*, thus 'the time since divergence from the honeybee lineage is identical for all bumblebee species.' Unfortunately the latter part of the premise in particular, is open to considerable question. This does not, however, reflect on their subsequent conclusion, which is consistent with their data, that anti-*Apis* GDH should cross react to the same extent with the corresponding protein from all *Bombus* species if the rate of evolutionary change of this *Bombus* protein has been constant. The findings of this study suggest that MCF studies probably provide valid estimates of primary structural similarities.

In the second study from Brosemer's lab (Marquardt *et al.*, 1968), anti-honeybee glyceraldehydephosphate dehydrogenase was prepared from crystallized glyceraldehyde-P dehydrogenase (GAP-D) extracted from honeybee thoraces. Cross reactions with extracts of various other insects were evaluated using the MCF technique quantified in terms of immunological distance (ID). As in the MCF studies on GDH no differences were detectable in the GAP-D from the six races of honeybees tested but unusually high ID values, indicating great taxonomic dissimilarity, were found to exist between honeybees (1·0) and bumblebees (10 to 14), and between the honeybee and *Tetralonia* (13), whereas an ID value of 2·5 was obtained for the leafcutting bee, *Megachile rotundata.* The relationship as derived from the MCF comparisons are at odds with the accepted classification schemes, but the same relative order was realized when GAP-D was tested using three other immunological techniques. Further, cross reactions of anti-honeybee GAP-D with bumblebee and lobster extracts gave essentially identical ID values, and the cross reaction of the flesh-fly enzyme was lower than that of the lobster enzyme. The authors conclude that 'unless one is ready to completely challenge classical taxonomy, it is obvious that the microcomplement fixation test with anti-honeybee glyceraldehyde-P dehydrogenase does not give meaningful taxonomic data.' Allison (in Marquardt *et al.*, 1968) on the basis of the similarity in the primary structure of GAP-D from a mammal and a crustacean, suggests that the protein is extremely conservative, which is reflected in the limited structural changes that have occurred during evolution. Thus, the MCF comparisons of a number of arthropods against anti-lobster GAP-D which yielded cross reactions consistent with classical taxonomy could only be attributed to chance, if the above hypothesis were to be accepted (Allison and Kaplan, 1964). This too, appears highly improbable.

Of the half-dozen insect enzymes that have been crystallized only the two dehydrogenases have been compared using microcomplement fixation procedures. That anti-honeybee GAP-D analysis should result in such anomalous taxonomic interpretations, whereas MCF comparisons made using a variety of vertebrate and invertebrate enzymes, including anti-lobster GAP-D have yielded IDs which were consistent with conventional classifications, is unsettling. It is, however, no cause to denigrate the technique or its potential applicability, but should caution against the use of a single character, be it biochemical or morphological as a sole criterion for phyletic or phenetic conclusions.

VII. Summary

This review has been directed at the biologist with an interest in taxonomically oriented comparative biochemistry, but whose duties or background have not permitted him to remain abreast of developments in that field. With the thrust at a presentation and evaluation of the methodologies, the limited space precludes any in depth consideration of problems of a conceptual nature. Fortunately Throckmotron's (1968) concern for the philosophical problems involved in integrating biochemistry and taxonomy make further comments in this area superflous.

A certain aura of glamour has evolved each of the 'new' tools that have offered promise in taxonomy, but none has retained its 'newness' or promise as long as comparative biochemistry. From the data in the text it is apparent that an enormous potential still exists, recurring with our ability to objectively evaluate each of the lower constituent levels of the molecular species.

Generalizations, be they praise or criticism serve no useful scientific purpose, and so to denigrate existing comparative biochemical studies as sanctification of a typological taxonomy is a misrepresentation (Rasmussen, 1969). A distinction is essential between typology as a concept and preliminary characterization of a taxon through the evaluation of the phenotype. This is not easily done for the rudiments of the concept pervades all logic, the difference being essentially one of degree. Perhaps much of the energy expended by Neo-Darwinian purists over the degree of objectivity accorded the lowest of the nomenatorial taxa in the evolutionary continuum could be conserved were some attempt made to distinguish between exploratory and definitive studies. The evaluation of each new taxonomic tool is of necessity rigid and generally proceeds through a series of steps: (1) identification of characters believed to hold taxonomic promise; (2) verification of their taxonomic value at the population level, and (3) application or rejection in taxonomic studies. Molecular taxonomy is about to make the transition from Step (1) to (2).

VIII. Postscript

A number of papers pertinent to biochemical taxonomy have appeared since this chapter was prepared in 1969. Several of these have stressed the need for a constant re-evaluation of improvements, limitations and biases in the analytical techniques employed. Thus current reviews such as those of Mattenheimer (1970) and Chrambach and Rodbard (1971) are invaluable to the biologically oriented student. The former provides a rather complete survey of the micromethods in biochemical analyses and the latter consists of a comprehensive review of the physico-chemical characteristics of polycrylamide gel as an electrophoretic medium.

There is a group of molecules which are difficult to separate on the conventional electrophoretic media. Many are more readily fractionated on the basis of their molecular net charge at a 'zero' gel concentration, that is, in a medium lacking molecular seiving properties. This characteristic has been utilized in the development of *isoelectric focusing* where fractionation procedes in a free, sucrose density-gradient, or in very low concentration polyacrylamide electrophoresis (Dale and Latner, 1968; Riley and Coleman, 1968; Finlayson and Chrambach, 1971). The method holds considerable promise where sieving properties of the gel matrix impede separation.

Several significant contributions have been made to insect biochemical systematics in the past two years. These have been confined largely to two major areas: development associated variation; and, biochemically based genetic studies of population differentiation.

A. DEVELOPMENT-ASSOCIATED VARIATION

Egg protein patterns from 23 insect species from 7 different orders were found to be species specific and the patterns reflected generic and family relationships (Salkfield, 1969). Koch (1968) reported that he was able to distinguish between two geographical races of *Acheta domesticus* L. on the basis of egg protein pherograms. In both studies eggs of comparable age were used, an apparent necessity to avoid the complications arising from protein synthesis in their development (Brookes and Dejmal, 1968).

Contradictory conclusions were reached in analytical studies on the esterase isozymes present in the developmental stages of holometabolous insects. Pantelouris and Downer (1969) report that in *Drosophila immigrans* electrophoretic fractions that are strong in the larva are identical to those that are strong in the adult and thus that metamorphosis is a continuous process in the ontogeny of this species. They conclude that there is no evidence to support the view that metamorphosis involves a massive switchover in gene action in the transition from the larval to the adult stages. Mestriner (1969) likewise states that esterase patterns in the honey bee, *Apis mellifera ligustica* are constant during the entire cycle of development. Simon (1969b) in an analysis of the esterase isozymes in the mosquito, *Culex pipiens fatigans*, reports distinctive qualitative differences in the zymograms of each of the major developmental stages of the species. Although patterns were consistent within each stage the vast majority of isozymes were unique to each stage analyzed. The esterases from the adults were specific to that stage although sexual differences were evident. It is unlikely that the metabolic pathways in these three species are as different as these studies would suggest. Rather the differences observed may be a reflection of the techniques employed by each of the authors.

Combre *et al.* (1971) found little or no difference in LDH or GDH activity between larvae and adults of the mosquito *Aedes aegypti*. Schneiderman (1967) cautions against the misinterpretation of isozyme patterns taken from whole body homogenates. He reports that a pupal alkaline phosphatase previously described in *Drosophila melanogaster* is indeed a modified larval skin phosphatase which has been altered as a result of exposure to larval gut enzymes during pupation.

Although it has been reported that the protein composition of the hemolymph of several insects undergoes cyclic changes during vitellogenesis (Hill, 1962) it has been only recently that such proteins have been characterized. Dufour *et al.* (1970) partially characterized a protein found in the hemolymph of the locust, *Schistocerca gregaria*, which was present in females only during the period of ovarian development and the sexually active stage. deLoof and deWilde (1970a) described a vitellogenic female protein (VFP) from the hemolymph of the Colorado beetle, *Leptinotarsa decemlineta*, also associated exclusively with the period of ovarian maturation. In addition, deLoof and deWilde (1970b) found that the hemolymph of adult Colorado beetles reared on a short day regimen (10:14 LD) displayed three unique proteins. All three fractions were found in the hemolymph of both males and females and were unrelated to the VFP protein of the female. Such sources of variation must be accounted for in any comparative biochemical study.

A fluctuation in the concentration of total hemolymph protein during a 24-hr period was reported from males of the Madeira cockroach, *Leucophaea maderae*, held on a 12:12 LD cycle (Hayes *et al.*, 1970). Maximum protein concentrations were recorded shortly after 'dawn', near 'sunset' and late in the dark period. Since photoperiod phenomena affect membrane permeability and water balance in general, these changes need not necessarily reflect changes in the rate of protein synthesis. No attempt was made to assess associated qualitative changes.

It has been reported that the enzyme, tryptophan pyrrolase, shows a monophasic circadian rhythm in rat liver, and the increase is due to *de novo* enzyme synthesis (Hardeland, 1969). This observation suggests that further studies on photoperiod effects on protein synthesis in insects are warranted.

The value of the electrophoretic technique as a tool in genetic studies is reflected in the number of recent papers in this area (see below). The reports of extensive population genetic polymorphism has prompted some question as to how close these estimates represent the actual variability in the genome. As available histochemical methods permit the detection of only a restricted number of the enzymes in any system, the non-random choice of the enzyme system itself may be a factor in biasing the variability reported. Gillespie and Kojima (1968) and Kojima *et al.* (1970) compared the electrophoretically detectable allozyme variations (variant proteins produced by allelic forms) in glucose-metabolizing enzymes with those exhibiting broad substrate specificities. They separate the enzymes into two groups: Group I enzymes (GDH, MDH, G6PDH, etc.) with very restricted *in vivo* substrates resulting from previous enzyme reactions which do not come directly from the environment; and Group II enzymes (esterases, alkaline and acid phosphates, ADH, etc.) usually having broad substrate specificity which act on a class of molecules, such as esters and alcohols, many of which come directly from the external environment. Gillespie and Kojima (1968) in an analysis of a number of populations of *Drosophila ananasse* found that Group I enzymes were but 9 to 24% as variable as Group II enzymes. Kojima *et al.* (1970) extended and confirmed these findings in an analysis of 5 different *Drosophila* species. These differences should be anticipated, for Group I enzymes are components of pathways in which most of the intermediates are necessary in other reactions, thus demanding a rigid standard for the kinetic parameters so as to retain optimal intermediate pool sizes. On the other hand, Group II enzymes are peripheral to the major energy producing and anabolic pathways and as such would not have the rigid kinetic standards so that variation could proceed over a wider range of activities. With the less rigid kinetic standards we would expect a more diverse array of molecular forms subject to the more diverse selective pressures of the environment.

Recent evidence supports the tenets of differential polymorphism among the various enzyme systems. In three of four sibling species of *Drosophila*, GDH was found to be monomorphic and fixed for the same allele (Ayala *et al.*, 1970), and in one of the nine species triads investigated by Hubby and Throckmorton (1968) the monomorphic appearance of GDH lead the authors to suggest that the enzyme has retained its structure over a long period of time. An analysis of GDH pherograms of thoracic muscle extracts of 24 species of *Bombus* has shown not only species specificity but suggests groupings which are consistent with morphological and behavior interpretations (Stephen and Cheldelin, 1973).

MDH in *Drosophila* is under the control of two loci. O'Brien and MacIntyre (1969) report that MDH-1 is monomorphic in several populations of *D. melanogaster* whereas Ayala *et al.* (1970) find it polymorphic in all four sibling species of the *willistoni* group. The former authors warn that MDH-1 may often stain as two bands in the homozygous state and that the two or more 'molecular' species that are detected, may be a result of different levels of autopolymerization, differential saturation with co-factors, conformational states, or alteration of the gel components. They stress that such artifacts may also appear with GDH, alcohol dehydrogenase, and acid phosphatase in the homozygous condition and appropriate determinations as to polymorphism can only be determined through knowledge of the genetic and functional autonomy of the system being studied. MDH-2 has been shown to be monomorphic in three of the four species of the *D. willistoni* group but in each of the three species fixation has occurred at different alleles. Two isozymes of MDH were found in each of four species of *Drosophila* (Red'kin, 1970). There was both inter- and intraspecies variability in the activity of each isozyme, and the ratio of their activities varied among the different species.

G-6-PDH shows dosage compensation in *D. melanogaster* (Steele *et al.*, 1969) and is stable in the homozygous state in most of the populations of *D. pseudoobscura* populations examined (Prakash, 1969). Hubby and Throckmorton

(1968) record that G-6-PDH appears to be monomorphic in eight of the nine triads of *Drosophila* examined but that it is represented by a unique form in eight of these nine species groups. This led the authors to suggest that the molecule is more labile than GDH over longer periods of evolutionary time.

Among the non-glucose metabolizing enzymes, (Group II, Gillespie and Kojima, 1968) ALKALINE PHOSPHATASE (APH) was found to be highly species specific in three species of *Anopheles* (Bianchi and Pirodda, 1968). They report three different patterns from the three species, each monomorphic and the APH apparently present as dimers. Both *in vivo* and *in vitro* tests indicated that that APH from the palearctic species *A. atroparvus* and *A. labranchiae* hybridized freely, whereas no hybridization occurred between *A. labranchiae* and the Nearctic *A. freeborni.* Banerjee (1967) found three different APH patterns from the worker, soldier, and reproductive cast of the termite *Odontotermes redemenni*, along with three distinct larval APH patterns which resemble those of the adults.

The apparent conflicting observations may in fact be real but they should serve to underscore a need for a critical examination of all aspects of the technique and especially to ascertain the relative importance of the limitations and biases associated with it (O'Brien and MacIntyre, 1969).

The esterases are the most complex of the enzyme systems thus far studied among insects. Esterase zymograms are usually the result of multiple gene loci in insects although Mestriner (1969) suggests that the esterase of *Apis mellifera* is a result of a single codominant locus with monofactorial inheritance. This limited number of esterases, may be a result of the technique employed by the latter investigator, for four activity zones were recorded from *Apis indica* and *Apis mellifera* (Tanabe *et al.*, 1970) and 9–11 have been detected from *A. mellifera* muscle in this laboratory. Esterase activity among the haplodiploid Hymenoptera is four to eight times greater in the females than in the males (Tanabe *et al.*, 1970). However since the somatic tissue of male *Apis* is diploid with a high degree of polysomaty, the authors conclude that the reduced activity may be due to the homozygous state of the male tissue rather than to haploidy. Trebatoski and Craig (1969) were able to distinguish between single gene and double gene dosage effects in their esterase-6 fraction of *Aedes aegypti*, the homozygote yielding a band double in intensity to that of the single codominant allele at that site. Burns and Johnson (1967) in an analysis of the polymorphic esterases in several populations of *Colias eurytheme* were unable to demonstrate any regional association with allelic frequencies. Johnson *et al.* (1969) investigating 14 alleles of two esterase loci in the harvester ant, *Pogonomyrmex barbatus*, found a significant correlation between esterase patterns and weather conditions and suggests that natural selection is an important process in determining the allozyme patterns. Subsequent studies by Burns and Johnson (1971) on 9 to 14 alleles of a single dimeric esterase in the butterfly, *Hemiargus isola*, found a near uniformity in allelic frequency in ecologically variable time and space. They suggest that the polymorphism is a result of the heterozygous advantage in an adaptation to an as yet undetected environmental heterogeneity in each locality.

B. GENE-ENZYME VARIABILITY IN POPULATIONS

The most extensive use to which comparative biochemistry has been put in the past two years has been in the area of genetic variability among natural populations. The major studies have been directed at one or more of four questions: What is the level of genetic polymorphism in populations of widespread and restricted range species? Is the level of polymorphism similar in all populations of a given species? Do sibling species show lesser or greater genetic divergence than non-sibling pairs? and, Does speciation require a massive, or only minor, changes in the gene pool? Investigations have been confined largely to species of *Drosophila* because of the extensive genetic information available on species of the genus.

There is general agreement that the level of genetic polymorphism in populations of a number of species of *Drosophila* is exceedingly high, with 38 to 70% of all loci polymorphic (Prakash *et al.*, 1969; Ayala *et al.*, 1970; O'Brien and MacIntyre, 1969) and from 10 to 22% of the loci heterozygous in each individual. O'Brien and MacIntyre (1969), analyzing 10 gene-enzyme systems in *Drosophila simulans* and *D. melanogaster*, found that in 8 natural populations of the latter species polymorphism ranged from 30 to 80% with an average of 54%. Comparable analysis made on a laboratory population of *melanogaster* held in culture for 20 years indicated the level of polymorphism to be equivalent to those of the field populations. Prakash *et al.* (1969), in a study of 24 different loci of central and peripheral populations of *D. pseudoobscura* noted very little difference in the levels of polymorphism. They found, however, that the isolated population of the species in Bogota, Colombia had but 25% of its loci polymorphic and less than 5% heterozygosity per individual.

Contradictory interpretations are made of the findings relating to the changes in gene pool required in speciation. Hubby and Throckmorton (1968) investigated the amount of genetic similarity in 9 pairs of sibling species of *Drosophila*, comparing each pair to a third morphologically distinct species of the same species group. Their analyses based on an average of 18 loci showed that sibling species had 50% of their loci in common whereas non-sibs shared an average of 18%. They report, however, that among the sibling species tested, one of the pairs, *D. victoria* and *D. lebanonensis*, had 85·1% of their proteins in common whereas the sibling pair *paulistorum* and *willistoni* shared only 22·5% of their proteins. This led the authors to conclude that speciation did not necessarily require a change in a large number of gene loci. Ayala *et al.* (1970) on the other hand, studied 14 structural genes of four sibling *Drosophila* species, *willistoni*, *paulistorum*, *equinoxialis*, and *tropicalis*. Their results indicated that the main proportion of polymorphic loci per species is 81% at the 1% criterion or 67% at the 5% level for polymorphism, and that individuals belonging to the different species differ from each other in approximately 50 of the genes sampled. This high level of genic difference between the four sibling species led the authors to conclude that a major reorganization in the gene pool is the most likely explanation of the speciation process. There is however, general agreement in all of these studies that extensive changes in structural genes may or may not be associated with comparable morphological change, that is, similarities in morphology and genetics do not necessarily go hand in hand. In addition, it can be concluded that the observations generally support the concept of a widespread balancing selection at most of these polymorphic loci.

Acknowledgements

I am grateful to the editor of 'Genetics' for permission to reproduce Figs 1 and 2; and to the editors of 'Systematic Zoology' to reproduce Figs 7 to 11. Studies in this area were supported in part by National Science Foundation Grant GB-25587 and U.S. Public Health Service Grant No. 00040.

References and Bibliography

Adiyodi, K. G. (1967). *J. Insect Physiol.* **13**, 1189–1195.

Allen, S. L. (1961). *Ann. N.Y. Acad. Sci.* **94**, 753–773.

Allison, W. S. and Kaplan, N. O. (1964). *J. biol. Chem.* **239**, 2140–2152.

Anfinsen, C. B. (1959). 'The Molecular Basis of Evolution'. J. Wiley and Sons, New York.

Arnheim, N., Prager, E. M. and Wilson, A. C. (1969). *J. biol. Chem.* **244**, 2085–2091.

Aronson, J. N., Fish, J. R. and Muckenthaler, F. A. (1968). *Comp. Biochem. Physiol.* **24**, 657–659.

Asperen, K. van (1962). *J. Insect Physiol.* **8**, 401–416.

Asperen, K. van and Mazijk, M. E. van (1965). *Nature* **205**, 1291–1292.

Ayala, F. J., Maurao, C. A., Perez-Salez, S., Richmond, R. and Dobzhansky, T. (1970). *Proc. natn. Acad. Sci., U.S.A.* **67**, 225–232.

Ball, G. H. and Clark, E. W. (1953). *Syst. Zool.* **2**, 138–141.

Banerjee, B. (1967). *Insectes soc.* **14**, 51–56.

Basford, N. L., Butler, J. E., Leone, C. A. and Rohlf, F. J. (1968). *Syst. Zool.* **17**, 388–406.

Beckendorf, G. W. and Stephen, W. P. (1970). *Biochim. Biophys. Acta* **201**, 101–108.

Bhalla, S. C. (1968). *Genetics* **58**, 249–258.

Bianchi, U. and Pirodda, G. (1968). *Riv. Parassit.* **29**, 297–303.

Boyden, A. (1958). *In* 'Serological and Biochemical Comparisons of Proteins' (W. H. Cole, ed.). Rutgers University Press, New Brunswick, N.J.

Boyden, A. (1964). *In* 'Taxonomic Biochemistry and Serology' (C. A. Leone, ed.), pp. 75–99. The Ronald Press Co., New York.

Bodnaryk, R. P. and Morrison, P. E. (1966). *J. Insect Physiol.* **12**, 963–976.

Brezner, J. and Enns, W. R. (1958). *J. Kansas ent. Soc.* **31**, 241–246.

Brookes, V. J. and Dejmal, R. (1968). *Science* **160**, 999–1001.

Brosemer, R. W. and Kuhn, R. W. (1969). *Biochemistry, N.Y.* **8**, 2095–2105.

Brosemer, R. W., Vogell, W. and Bucher, Th. (1963). *Biochem. Z.* **338**, 854–910.

Brosemer, R. W., Grosso, D. S., Estes, G. and Carlson, C. W. (1967). *J. Insect Physiol.* **13**, 1757–1767.

Brosemer, R. W., Fink, S. C. and Carlson, C. W. (1970).
Brown, K. S. (1967). *Syst. Zool.* **16**, 213–216.
Brown, L. E. (1964). *Syst. Zool.* **13**, 92–95.
Burns, J. M. and Johnson, F. M. (1967). *Science* **156**, 93–96.
Burns, J. M. and Johnson, F. M. (1971). *Proc. natn. Acad. Sci., U.S.A.* **68**, 34–37.
Butler, J. E. and Leone, C. A. (1967). *Syst. Zool.* **16**, 56–63.
Butler, J. E. and Leone, C. A. (1968). *Comp. Biochem. Physiol.* **25**, 417–426.
Buzzati-Traverso, A. A. and Rechnitzer, A. B. (1953). *Science* **117**, 58–59.
Campbell, D. H., Garvey, J. S., Cremer, N. E. and Sussdorf, D. H. (1964). 'Methods in Immunology'. W. A. Benjamin, Inc., New York.
Carlton, B. C. and Yanofsky, C. (1965). *J. biol. Chem.* **240**, 690–693.
Chen, P. S. (1959). *Revue suisse Zool.* **66**, 280–289.
Chen, P. S. (1962). *In* 'Amino Acid Pools' (J. T. Holden, ed.), pp. 115–138.
Chen, P. S. (1966). *In* 'Advances in Insect Physiology' Vol 3 (J. W. L. Beament, J. E. Treherne and V. B. Wigglesworth, eds.), pp. 53–132. Academic Press, London and New York.
Chrambach, A. and Rodbard, D. (1971). *Science* **172**, 440–451.
Clark, E. W. and Ball, G. H. (1957). *Physiol. Zool.* **29**, 206–212.
Coles, G. C. (1965). *J. Insect Physiol.* **11**, 1317–1323.
Collins, W. J. and Forgash, A. J. (1968). *J. Insect Physiol.* **14**, 1515–1523.
Combre, A., Beytout, D., Mouchet, J. and Bastide, P. (1971). *J. Insect Physiol.* **17**, 527–543.
Dale, G. and Latner, A. L. (1968). *Lancet* 1968–*I*, 847.
Dayhoff, M. O. (1969). 'Atlas of Protein Sequence and Structure' 4th edit. National Medical Research Foundation, Silver Springs, Md.
Denucé, J. M. (1958). *Z. Naturf.* **13b**, 215–218.
Dessauer, H. C. and Fox, W. (1958). *Proc. Soc. exp. Biol. Med.* **98**, 101–105.
Downes, A. E. R. (1963). *Science* **139**, 1286–1287.
Doyle, D. and Laufer, H. (1969). *J. Cell. Biol.* **40**, 61–78.
Duchâteau, G. and Florkin, M. (1958). *Archs int. Physiol. Biochim.* **66**, 573–591.
Dufour, D., Taskar, S. P. and Perron, J. M. (1970). *J. Insect Physiol.* **16**, 1369–1377.
Finlayson, R. and Chrambach, A. (1971). *Analyt. Biochem.* **40**, 296.
Florkin, M. (1960). 'Unity and Diversity in Biochemistry'. Pergamon Press, London.
Florkin, M. (1964). *In* 'Protides of the Biological Fluids' (H. Peeters, ed.), Elsevier, Amsterdam.
Florkin, M. (1966). 'A Molecular Approach to Phylogeny'. Elsevier, New York.
Florkin, M. and Jeuniaux, C. (1964). *In* 'The Physiology of Insects' (M. Rockstein, ed.), pp. 110–152. Academic Press, New York and London.
Fondy, T. P., Levin, L., Sollohub, S. J. and Ross, C. R. (1968). *J. biol. Chem.* **243**, 3148–3160.
Fondy, T. P., Ross, C. R. and Sollohub, S. J. (1969). *J. biol. Chem.* **244**, 1631–1644.
Fox, F. R. and Mills, R. R. (1969). *Comp. Biochem. Physiol.* **29**, 1187–1195.
Geest, L. P. S. van der (1968). *J. Insect Physiol.* **14**, 537–542.
Gilbert, L. I. and Goldberg, E. (1966). *J. Insect Physiol.* **12**, 53–63.
Gillespie, J. H. and Kojima, K. (1968). *Proc. natn. Acad. Sci., U.S.A.* **61**, 582–585.
Goldberg, E., Whitten, J. and Gilbert, L. I. (1969). *J. Insect Physiol.* **15**, 409–420.
Grabar, P. and Williams, C. A. (1953). *Biochim. Biophys. Acta* **10**, 193–194.
Grell, E. H. (1967). *Science* **158**, 1319–1320.
Guest, J. R. and Yanofsky, C. (1965). *J. biol. Chem.* **240**, 679–689.
Hardeland, R. (1969). *Z. vergl. Physiol.* **63**, 119–136.

Harrington, J. S. (1961). *J. ent. Soc. sth. Afr.* **24**, 216–217.
Harlow, R. D., Lumb, R. H. and Wood, R. (1969). *Comp. Biochem. Physiol.* **30**, 761–769.
Hayes, D. K., Mensing, E. and Schechter, M. S. (1970). *Comp. Biochem. Physiol.* **34**, 733–737.
Henning, U. and Yanofsky, C. (1963). *J. molec. Biol.* **6**, 16–21.
Hirschfeld, J. (1968). *In* 'Handbook of Immunodiffusion and Immunoelectrophoresis', pp. 163–182. Ann Arbor Science Publishers, Inc., Ann Arbor, Michigan.
Hill, L. (1962). *J. Insect Physiol.* **8**, 609–619.
Holmes, R. S. and Masters, C. J. (1967). *Biochim. Biophys. Acta* **132**, 379–399.
Holmes, R. S. and Masters, C. J. (1968). *Biochim. Biophys. Acta* **151**, 147–158.
Hubby, J. L. and Lewontin, R. C. (1966). *Genetics* **54**, 577–594.
Hubby, J. L. and Narise, S. (1967). *Genetics* **57**, 291–300.
Hubby, J. L. and Throckmorton, L. H. (1965). *Genetics* **52**, 203–215.
Hubby, J. L. and Throckmorton, L. H. (1968). *Am. Nat.* **102**, 193–205.
Hudson, A. (1966). *Can. J. Zool.* **44**, 541–555.
Ingram, V. M. (1958). *Biochim. Biophys. Acta* **28**, 539–545.
Johnson, F. M., Schaffer, H. E., Gillaspy, J. E. and Rockwood, E. S. (1969). *Biochem. Genet.* **3**, 429–450.
Johnson, M. L. (1968). *Syst. Zool.* **17**, 23–30.
Jukes, T. H. (1965). *Am. Scient.* **53**, 477.
Kabat, E. A. (1958). *In* 'Serological and Biochemical Comparisons of Proteins' (W. H. Cole, ed.), pp. 92. Rutgers University Press, New Brunswick, N.H.
Kabat, E. A. (1960). *J. Immunol.* **84**, 82–85.
Kabat, E. A. and Mayer, M. M. (1961). 'Experimental Immunochemistry'. Charles C. Thomas, Springfield, Illinois.
Kaplan, N. O. (1965). *In* 'Evolving Genes and Proteins' (V. Bryson and H. J. Vogel, eds.), p. 243. Academic Press, New York.
Kaplan, N. O. and Ciotti, M. M. (1961). *Ann. N.Y. Acad. Sci.* **94**, 701–720.
Koch, P. (1968). *Proc. R. ent. Soc. Lond.* (C), **33**, 13–20.
Kojima, K., Gillespie, J. and Tobari, N. (1970).
Laufer, H. (1960). *Ann. N.Y. Acad. Sci.* **89**, 490–515.
Laufer, H. (1961). *Ann. N.Y. Acad. Sci.* **94**, 825–835.
Laufer, H. (1964). *In* 'Taxonomic Biochemistry and Serology' (C. A. Leone, ed.), pp. 171–189. The Ronald Press Co., New York.
Leone, C. A. (1964). 'Taxonomic Biochemistry and Serology' (C. A. Leone, ed.). The Ronald Press Co., New York.
Lewontin, R. C. and Hubby, J. L. (1966). *Genetics* **54**. 595–609.
Loughton, B. G. and West, A. S. (1965). *J. Insect. Physiol.* **11**, 919–932.
deLoof, A. and deWilde, J. (1970a). *J. Insect Physiol.* **16**, 1455–1466.
deLoof, A. and deWilde, J. (1970b). *J. Insect Physiol.* **16**, 157–169.
McKittrick, F. A. (1964). *Mem. Cornell Univ. agric. Exp. Stn.* **389**, 197 pp.
McShan, W. H., Kramer, S. and Schlegel, V. (1954). *Biol. Bull. mar. biol. Lab., Woods Hole* **106**, 341–352.
Marek, M. (1969). *Comp. Biochem. Physiol.* **29**, 1231–1237.
Markert, C. L. (1963). *Science* **140**, 1329–1330.
Marquardt, R. R., Carlson, C. W. and Brosemer, R. W. (1968). *J. Insect Physiol.* **14**, 317–333.
Marty, R. and Zalta, J. P. (1966). *C. r. hebd. Séanc. Acad. Sci., Paris* **263**, 180–182.
Marty, R. and Zalta, J. P. (1967). *C. r. hebd. Séanc. Acad. Sci., Paris* **264**, 643–646.

Mattenheimer, H. (1970). (Micromethods for the Clinical and Biochemistry Laboratory'. Ann Arbor Science Publishers Inc., Ann Arbor, Michigan. 232 pp.
Menzel, D. B., Craig, R. and Hoskins, W. M. (1963). *J. Insect Physiol.* **9**, 479–493.
Mestriner, M. A. (1969). *Nature, Lond.* **223**, 188–189.
Micks, D. W. (1956). *Ann. ent. Soc. Am.* **49**, 576–581.
Micks, D. W. and Gibson, F. J. (1957). *Ann. ent. Soc. Am.* **50**, 500–505.
Micks, D. W., Rehmet, A. and Jennings, J. (1966). *Ann. ent. Soc. Am.* **50**, 602–606.
Muus, J. (1968). *Comp. Biochem. Physiol.* **24**, 527–536.
Nowosielski, J. and Patton, R. L. (1965). *J. Insect Physiol.* **11**, 263–270.
O'Brien, S. J. and MacIntyre, R. J. (1969). *Am. Nat.* **103**, 97–113.
Ogita, Z. (1962). *Jap. J. Genet.* **37**, 518–521.
Ouchterlony, O. (1968). 'Handbook of Immunodiffusion and Immunochemistry' pp. 215. Ann Arbor Science Publishers, Inc., Ann Arbor, Michigan.
Pantelouris, E. M. and Downer, R. G. H. (1969). *J. Insect Physiol.* **15**, 2357–2362.
Prakash, S. (1969). *Genetics* **62**, 778–784.
Prakash, S. and Lewontin, R. C. (1968). *Proc. natn. Acad. Sci., U.S.A.* **59**, 398–405.
Prakash, S., Lewontin, R. C. and Hubby, J. L. (1969). *Genetics* **61**, 841–858.
Princis, K. (1960). *Eos. Madr.* **36**, 427–449.
Rapport, M. M. and Graf, L. (1967). *In* 'Methods in Immunology and Immunochemistry' (C. A. Williams and M. W. Chase, eds.), pp. 187–196. Academic Press, New York.
Rasmussen, D. I. (1969). *Bioscience* **19**, 418–420.
Red'kin, P. S. (1970). *Zh. obshch. Biol.* **31**, 458–463.
Rehn, J. W. H. (1951). *Mem. Am. ent. Soc.* **14**, 1–134.
Reichlin, M., Hay, M. and Levine, L. (1964). *Immunochem.* **1**, 21–30.
Riley, R. F. and Coleman, M. K. (1968). *J. Lab. clin. Med.* **72**, 714.
Rockstein, M. (1953). *Biol. Bull. mar. biol. Lab., Woods Hole* **105**, 154–159.
Rosen, M. H., Fried, G. H. and Schaefer, C. W. (1968). *Bioscience* **18**, 302–305.
Roth, L. and Cohen, S. H. (1968). *Psyche* **75**, 53–76.
Rüegg, M. K. (1968). *Z. vergl. Physiol.* **60**, 275–307.
Sacktor, B. (1950). *J. econ. Ent.* **43**, 832–838.
Sacktor, B. and Cochran, D. G. (1957). *Biochim. Biophys. Acta* **25**, 649.
Sacktor, B. and Dick, A. R. (1962). *J. biol. Chem.* **237**, 3259.
Salkfeild, E. H. (1969). *Can. Entomol.* **101**, 1256–1265.
Sarich, V. M. and Wilson, A. C. (1966). *Science* **154**, 1563–1566.
Schneiderman, H. (1967). *Nature* **216**, 604–605.
Sela, M. and Fuchs, S. (1967). *In* 'Methods in Immunology and Immunochemistry' (C. A. Williams and M. W. Chase, eds.), pp. 167–175. Academic Press, New York.
Shigematsu, H. (1958). *Nature, Lond.* **182**, 880–882.
Simon, J. P. (1969a). *Syst. Zool.* **18**, 169–184.
Simon, J. (1969b). *Ann. ent. Soc. Am.* **62**, 1307–1311.
Sims, M. (1965). *Nature* **207**, 757–758.
Steele, M. W., Young, W. J. and Childs, B. (1969). *Biochem. Genet.* **3**, 359–370.
Steinhauer, A. L. and Stephen, W. P. (1959). *Ann. ent. Soc. Am.* **52**, 733–738.
Stephen, W. P. (1957). *Oregon State College Tech. Bull.* **40**, 163 pp.
Stephen, W. P. (1961). *Syst. Zool.* **10**, 1–9.
Stephen, W. P. and Cheldelin, I. H. (1970a). *Biochim. Biophys. Acta* **201**, 109–118.
Stephen, W. P. and Cheldedin, I. H. (1970b). *Comp. Biochem. Physiol.* **37**, 361–373.
Stephen, W. P. and Cheldelin, I. H. (1971). *13th Int. Congr. Ent.* **1**, 442–444.

Stephen, W. P. and Cheldelin, I. H. (1973). *Biochem. System.* **1**, 69–76.
Tanabe, Y., Tamaki, Y. and Nakano, S. (1970). *Jap. J. Genet.* **45**, 425–428.
Telfer, W. H. and Williams, C. M. (1953). *J. gen. Physiol.* **36**, 389–413.
Throckmorton, L. H. (1962). *Univ. Texas Publ.* **6206**, 415–487.
Throckmorton, L. H. (1968). *A. Rev. Ent.* **13**, 99–114.
Trebatoski, A. M. and Craig, G. B. (1969). *Biochem. Genet.* **3**, 383–392.
Treherne, J. E. (1958). *J. exp. Biol.* **35**, 611–625.
Vessell, E. S. (1965). *In* 'Progress in Medical Genetics' (A. G. Steinberg and A. G. Bearn, eds.), pp. 128–175. W. Heinemann, London.
Wang, C. and Patton, R. L. (1968). *J. Insect Physiol.* **14**, 1069–1075.
Wasserman, E. and Levine, L. (1961). *J. Immunol.* **87**, 290–295.
Whittaker, S. R. and West, A. S. (1962). *Can. J. Zool.* **40**, 655–671.
Wigglesworth, V. B. (1959). *A. Rev. Ent.* **4**, 1–16.
Wilkinson, J. H. (1965). *In* 'Isoenzymes' (C. Long, ed.), pp. 43–141. Spon, London.
Williams, C. A. and Chase, M. W. (1967). 'Methods in Immunology and Immunochemistry' Vol I. Academic Press, New York.
Williams, C. A. and Chase, M. W. (1968). 'Methods in Immunology and Immunochemistry' Vol II. Academic Press, New York.
Wilson, A. C. and Kaplan, N. O. (1964). *In* 'Taxonomic Biochemistry and Serology' (C. A. Leone, ed.), pp. 321–346. The Ronald Press Co., New York.
Wright, T. R. F. (1963). *Genetics* **48**, 787–801.
Wyatt, G. R. (1961). *A. Rev. Entomol.* **6**, 75–102.
Yanofsky, C., Carlton, B. C., Guest, J. R., Helsinki, D. R. and Henning, U. (1964). *Proc. natn. Acad. Sci., U.S.A.* **51**, 266–272.
Zebe, E. C. and McShan, W. H. (1957). *J. gen. Physiol.* **40**, 779–790.
Zuckerkandl, E., Jones, R. T. and Pauling, L. (1960). *Proc. natn. Acad. Sci., U.S.A.* **46**, 1349–1360.
Zuckerkandl, E. and Pauling, L. (1962). *In* 'Horizons in Biochemistry' (M. Kasha and B. Pullman, ed.), pp. 189–225. Academic Press, New York.
Zwaan, J. (1967). *Analyt. Biochem.* **21**, 155–168.
Zweig, G. and Crenshaw, J. W. (1957). *Science* **126**, 1065–1066.

6 Biochemical and Immunological Taxonomy of the Mollusca

C. A. WRIGHT

British Museum (Natural History), London, England

I. Introduction

The application of experimental techniques to molluscan taxonomy has both a pragmatic and an empirical foundation. Pragmatic because most of the studies have centred upon species which have some economic significance to man, either as transmitters of disease or as sources of food. Empirical because, in the early stages, techniques tended to be applied at random in attempts to supplement the not wholly satisfactory approaches of traditional morphology. A strong stimulus to some of the biochemical studies has been the parasitologists' desire to take a worm's eye view of the Mollusca. For these reasons this review is limited to a few molluscan groups and the conclusions which have been reached are restricted in their application. The emphasis has been upon species discrimination and the characterization of infra-specific categories rather than on the relationships of the phylum as a whole or even of classes and orders within the phylum. In many cases the techniques employed have been tested almost to destruction and this allows some assessment to be made of the utility of the methods themselves. It must be remembered that any such assessments are confined to the context of the application of the techniques and that direct extrapolation to other groups may not be justified. This point becomes clear in this review as it is seen that some methods which work well within one family of gastropods are simply not applicable to other families.

The material is arranged under method headings. Such a presentation leaves a certain amount to be desired but the results from most of the early exploratory investigations do not readily lend themselves to any other treatment. Fortunately some of the recent contributions have been of a more precise nature both with respect to the methods used and to the taxonomic problems involved. Many of the methods have become sufficiently well established for them to be applied as routine adjuncts to some taxonomic studies and where several techniques have been applied to a single problem it is possible to make appropriate evaluations of their respective merits. Although such mixed approaches will involve a certain amount of cross-reference within the text they also serve to draw together current progress in the experimental field.

II. Chromatographic Methods

The earliest applications of chromatographic methods to taxonomic work were directed towards the study of amino acids in the hope that basic differences in the protein composition of different species of organisms might be revealed. In the preparation of material for analysis two general approaches were used, either pieces of tissue or small, whole animals were squashed directly onto paper or extracts were prepared from which soluble proteins were precipitated so that the resultant solution contained only amino acids and small peptides. The second method gave more precisely quantifiable results and clearer separations of the ninhydrin-positive materials but the squash technique, because of the presence of a variety of substances in the tissues, provided the opportunity for investigating other biochemical differences. Of these other substances the compounds which fluoresce in ultra-violet light have attracted most attention and it was this approach which was first used in molluscan studies by Kirk *et al.* in 1954.

A. TISSUE EXTRACTS

In their pioneer contribution Kirk and his co-workers examined a number of species of stylommatophoran land snails from Western Australia (including several species of the native genus *Bothryembryon*) and found that each species yielded a characteristic pattern of fluorescent bands. They examined samples of *Theba pisana* and *Helix aspersa* from widely separate localities and were unable to find any signs of geographical variation but since neither of these species is native to Australia and must, therefore, be relatively recent introductions, one would not expect to find marked population differences. The technique used by Kirk *et al.* (1954) was a horizontal disk method using improvised apparatus. Results obtained in this way were not easily reproducible but development of the Kawerau dish with mechanically cut slotted disks led to great improvements (Wright *et al.*, 1957). A further modification came with the discovery that all of the fluorescent materials found in chromatograms made from squashed fragments of molluscan foot are present in the body-surface mucus. Using mucus only Wright (1959a) carried out a survey of the British species of *Lymnaea* and was able to demonstrate the existence of species-specific patterns and of generalized patterns characteristic for species groups corresponding to some of the older generic divisions of *Lymnaea*. Thus *L. peregra* and *L. auricularia*, members of the former genus *Radix*, have in common a brilliant blue band between Rf 0·4–0·45 while *L. palustris*, previously included in *Stagnicola*, has two bands, one pink and the other lilac between Rf 0·5–0·65. No fluorescent substances were found in the mucus of *L. stagnalis* or *L. truncatula*.

Michejda and his co-workers in Poland (Michejda, 1958; Michejda and Turbanska, 1958; Michejda and Urbanski, 1958) made fairly extensive investigations into the uses of chromatography in molluscan taxonomy. They compared the ninhydrin-positive and fluorescent patterns obtained from aqueous tissue homogeneates of five species of freshwater prosobranchs and five basommatophoran pulmonates. They found that there was a general agreement between the chromatographic patterns and the accepted relationships of the species concerned. The agreement appeared to be more marked among the prosobranchs than among the pulmonates. They made quantitative comparisons between the ninhydrin-positive patterns obtained from skin, muscle, digestive gland and 'visceral sac' of two species of the prosobranch genus *Viviparus*, *Lymnaea stagnalis* and three species of helicids. Their findings showed that skin and muscle gave similar patterns and that there was close agreement between the patterns given by the other two organ systems. A generalized group relationship emerged from this investigation in that curves of the density of the pattern plotted against Rf value were bi-modal for the helicids, unimodal for the viviparids and 'parabolic' for *L. stagnalis*. The third part of the study consisted of a survey of fifty species of gastropods, mostly land snails among which the Helicidae were particularly well represented, using aqueous homogeneates of both whole snails and foot muscle tissues. These analyses gave indications of family relationships but the conclusions drawn were somewhat tentative due to a certain amount of conflict in the results from different groups.

Subsequent applications of chromatography to molluscan taxonomy have been largely concentrated on two aspects of the freshwater basommatophora but before dealing with these it is appropriate to look at three isolated but interesting contributions from the marine field. Collyer (1961) used foot-muscle squash chromatograms to investigate the significance of minor morphological differences between two British populations of the gastropod

Nassarius. Consistent differences in the ninhydrin-positive patterns from the two populations were obtained. Size of the individuals did not appear to affect the pattern but their nutritional state did, certain fractions being lost in starved specimens. These nutritional differences however proved to be reversible. In the light of these results Collyer was inclined to uphold the distinction between *Nassarius reticulatus* and *N. nitida*. Working with the eastern North American oyster *Crassostrea virginica* Hillman (1964) was able to demonstrate differences in the ninhydrin-positive pattern between populations from Long Island Sound and the James River in Virginia. The materials used in this case were direct squashes and alcoholic extracts of adductor muscle tissue. Rucker (1965) compared the amino acid patterns derived from hydrolysis of conchiolin of eight species of marine molluscs. He found striking differences between the species but, with one exception, the patterns given by *Crassostrea virginica* from six localities were similar. The exceptional sample consisted of dead weathered shells in which some degradation of the conchiolin appeared to have occurred. Thus, in these three examples differences were found at the species level in *Nassarius* and at the population level in *Crassostrea* while examination of different material from *Crassostrea* failed to reveal population differences but suggested the existence of specific characteristics.

B. FREE AMINO ACIDS

Taxonomy has not been the prime objective of most investigations of the amino acids of freshwater gastropods. However, studies concerned with the host as an environment for developing trematode larvae have yielded a certain amount of comparative data. Targett (1962b) compared the free amino acids from various tissues in *Biomphalaria glabrata*, *Lymnaea stagnalis* and *Planorbarius corneus* and showed that there were few significant qualitative differences between comparable tissues from the three species. An interesting exception was methionine which, although present in the digestive glands of all three was found in the 'anterior tissues' of only *B. glabrata*. Also of interest was the absence of arginine and histidine in the anterior tissues of laboratory bred *L. stagnalis*, both of these substances being found in wild-caught individuals of the same species. In an earlier paper Targett (1962a) reported on free amino acids in the blood of several species and some of these results are shown in Table I together with those of other workers for comparison.

To what extent the differences which appear in this table are a true reflection of differences between species and populations or simply result from technical variations it is diffcult to say. Gilbertson *et al.* (1967) examined four strains of *B. glabrata* and although they found some quantitative differences between the strains they got almost comparable levels of variation from replicate tests on a single population. There are discrepancies between the findings of Targett (1962a) and Gilbertson *et al.* for *B. glabrata* but the most surprising report is that of Dusanic and Lewert (1963). They claimed that

Table I
Free amino acids in the blood of some freshwater basommatophoran pulmonates

	Biomphalaria glabrata			*Lymnaea stagnalis*		*L. palustris*
	Targett (1962a)	Dusanic and Lewert (1963)	Gilbertson *et al.* (1967)	Targett (1962a)	Friedl (1961) (*L.s. jugularis*)	Morrill (1963)
Alanine	+		+	+	+	+
Aminobutyric acid					+	
Arginine			+			+
Asparagine					+	+
Aspartic acid	+		+	+		+
Cystine	+			+	+	+
Glutamic acid	+		+	+	+	+
Glutamine					+	+
Glycine	+	+	+	+	+	+
Histidine			+		+	+
Hydroxyproline						+
Isoleucine	+		+	+	+	+
Leucine	+	+	+	+	+	+
Lysine	+	+	+	+	+	+
Methionine			+			
Ornithine					+	+
Phenylalanine	+		+	+	+	+
Proline		+	+		+	+
Serine	+		+	+	+	+
Threonine	+	+	+	+	+	+
Tryptophan					+	+
Tyrosine	+		+	+	+	+
Valine	+		+	+	+	+

threonine and leucine only appeared after exposure of the snails to miracidia of *Schistosoma mansoni* and that leucine subsequently disappeared 14 hours after exposure and threonine after 24 hours. Both of these amino acids have been found in uninfected snails in all of the other investigations but proline which Dusanic and Lewert report in uninfected *B. glabrata* was not found by Targett either in this species or in *L. stagnalis*.

The conflicting evidence of these reports and the results of our own investigations (Wright and Ross, unpublished) tend to confirm the doubts expressed by Burch (1960) concerning the value of amino acid studies in the taxonomy of freshwater gastropods. It is perhaps significant that only one recent paper (Stiglingh and van Eeden, 1970) has appeared on the subject.

C. MUCUS

The other application of chromatographic techniques to taxonomic problems in the basommatophora, the study of fluorescent substances in mucus, has had slightly more success. Following on Wright's (1959a) basic survey of the British species of *Lymnaea* collections were made of some of the well-known variants of *L. peregra* from mountain lakes in Southern Ireland (Ballantine and Bradley, 1963). Chromatograms of the body-surface mucus from this material revealed fluorescent patterns which differed from that shown by the seven populations previously examined from various localities in England. In the light of these results an intensive survey of *L. peregra* was undertaken throughout western Europe with particular attention concentrated in Scotland and Ireland. Over 700 populations were sampled and seven distinctive pattern types were found. Of these, three were very rare, three have wider but restricted distributions and the seventh is widespread throughout the species' range (Wright, 1964, 1966b). Of the three rare forms one was found only in streams draining from hot springs in Iceland, one in a single lake in the extreme north of Scotland and the third in a Scottish lake and a stream in the north of England. Of the three restricted forms one appears to be largely confined to small lakes on granite rock in the west of Ireland and north-west Scotland (*patternless*), the second is also lacustrine but has a wider distribution in Scotland and Ireland and extends into lakes in the north of England and on high ground in Wales (*no-delta*) and the third is common in streams in Scotland, Ireland and Wales and a few localities in the west of England (*split-beta*); this form also occurs in some lakes in Scotland. The seventh form (*normal*) occurs in all kinds of habitats and has been found throughout the British Isles, all over France and the Iberian peninsula, northern Italy, Turkey and the Atlas mountains in Morocco. The Icelandic, *patternless*, *no-delta*, *split-beta* and *normal* forms have all been bred in the laboratory on standard diets and their patterns have remained unchanged through many generations. Hybrids between some of the possible crosses have been achieved although *patternless* has not yet been successfully mated with *normal*. In nature hybrids between *split-beta* and *normal* occur commonly and specimens with intermediate patterns have been found in streams draining from lakes containing different forms. There is no apparent correlation between the fluorescent mucus pattern and shell shape, the true-breeding involute forms from Lough Crincaum in Ireland and Loch Fleodach Coire in Sutherland both have the *normal* pattern type while the large spireless form originally described as *L. burnetti* from Loch Skene in Dumfriesshire has a *no-delta* pattern like that of many spired Scottish lake populations.

Two very useful lessons can be learned from this study. The first is that some biochemical variation between populations of a single species may be greater than the differences between populations of different species. The second is that the degree of intra-population variation within the range of a species is far from uniform. In the case of *L. peregra* and its close relative *L. auricularia* Wright (1964) suggested that high levels of intra-population variation occurred in the regions in which the two species probably evolved but this may be a special case and not necessarily of general application.

The work on *Lymnaea* was developed to provide background information for comparable studies on the planorbid snail hosts of African *Schistosoma* spp. Preliminary mention of the existence of population differences in mucus chromatograms of *Bulinus truncatus* was made by Wright (1962). Three populations from Iraq and one each from Iran and Israel showed a basically similar three-banded pattern but the innermost of these bands was lacking in a sample from near Cairo and was scarcely visible in a Sudanese population from near Khartoum. In contrast to the fluorescent materials in lymnaeid mucus, which show up clearly in ultra-violet light (365 nm) on freshly developed chromatograms, those of *Bulinus* are usually difficult to detect until the chromatogram is exposed to ammonia fumes. No fluorescent substances have so far been found in the mucus of any species of *Biomphalaria*. Within the genus *Bulinus* the members of the *B. africanus* species group (=*Physopsis*) lack fluorescent substances, species in the *B. forskali* complex have a single fast-moving band (Rf about 0·85 in a butanol-acetic acid-water solvent) which fluoresces a pale greenish-blue colour and members of the *B. truncatus* and *B. tropicus* groups have the basic three-banded pattern between Rf 0·45 and 0·65 previously mentioned for *B. truncatus*. The inner of these bands (1) (Rf 0·45–0·49) is very pale blue or yellowish, the middle (2) is also blue (Rf 0·55–0·58) and the outer (3) is yellow (Rf 0·60–0·65). A population of *B. liratus* from Madagascar and several populations of *B. tropicus* from the South African highveldt have a wide bright blue band (5) which appears only on exposure to ammonia fumes between Rf 0·80 and 0·95 (Wright, 1971). In a recent study of a polyploid complex of *Bulinus* from the Ethiopian highlands Brown and Wright (1972) found that there was considerable variation between different populations in the presence, absence or intensity of the three bands in the basic pattern but that generally speaking the pattern for any particular population was characteristic. In four diploid populations band 5 was found, in one case it was present in four out of six individuals from

FIG. 1. Paper disk chromatograms of the body-surface mucus of *Bulinus* species, viewed under ultra-violet light after exposure to ammonia fumes (except (i)). (a), (b) and (c) octoploid populations from the highlands of Ethiopia; (d) *B. truncatus* (tetraploid) from Sardinia; (e) *B. truncatus* (tetraploid) from Morocco; (f) *B. natalensis* (diploid) L. Sibayi, South Africa; (g) *B. liratus* (diploid) Madagascar; (h) diploid population from Ethiopia; (i) *B. bavayi* (*forskali* group) from Madagascar.

The patterns in (a)–(c) are typical for most populations in the *Bulinus truncatus*/*B. tropicus* complex. In (a) bands 1–3 are all well developed and roughly equal in intensity, in (b) band 1 dominates and bands 2 and 3 are less well developed while in (c) it is band 2 which dominates. In (d) band 2 is virtually alone, there being only slight traces of bands 1 and 3 while in (e) band 2 is dominant, 3 is stronger than 1 and band 4 is present although weak. In (f) all five bands are present but only band 4 is well developed while in (g) and (h) bands 1, 2, 3 and 5 are all well developed with 5 dominant. In both of these populations band 1 is yellowish in contrast to its blue colour in the other patterns and band 1 may not be homologous throughout. All specimens of *B. liratus* examined have band 5 well developed also all specimens of *B. tropicus* from the highveldt near Potchefstroom in South Africa but it has only been found in three diploid populations from Ethiopia and is not present in all individuals. Specimens in these Ethiopian samples showing band 5 have given rise to offspring lacking the band but selection of those with it has increased its frequency in laboratory culture. The greenish-blue band in (i) is characteristic for all members of the *B. forskali* group and it is visible (as in this specimen) without exposure to ammonia fumes.

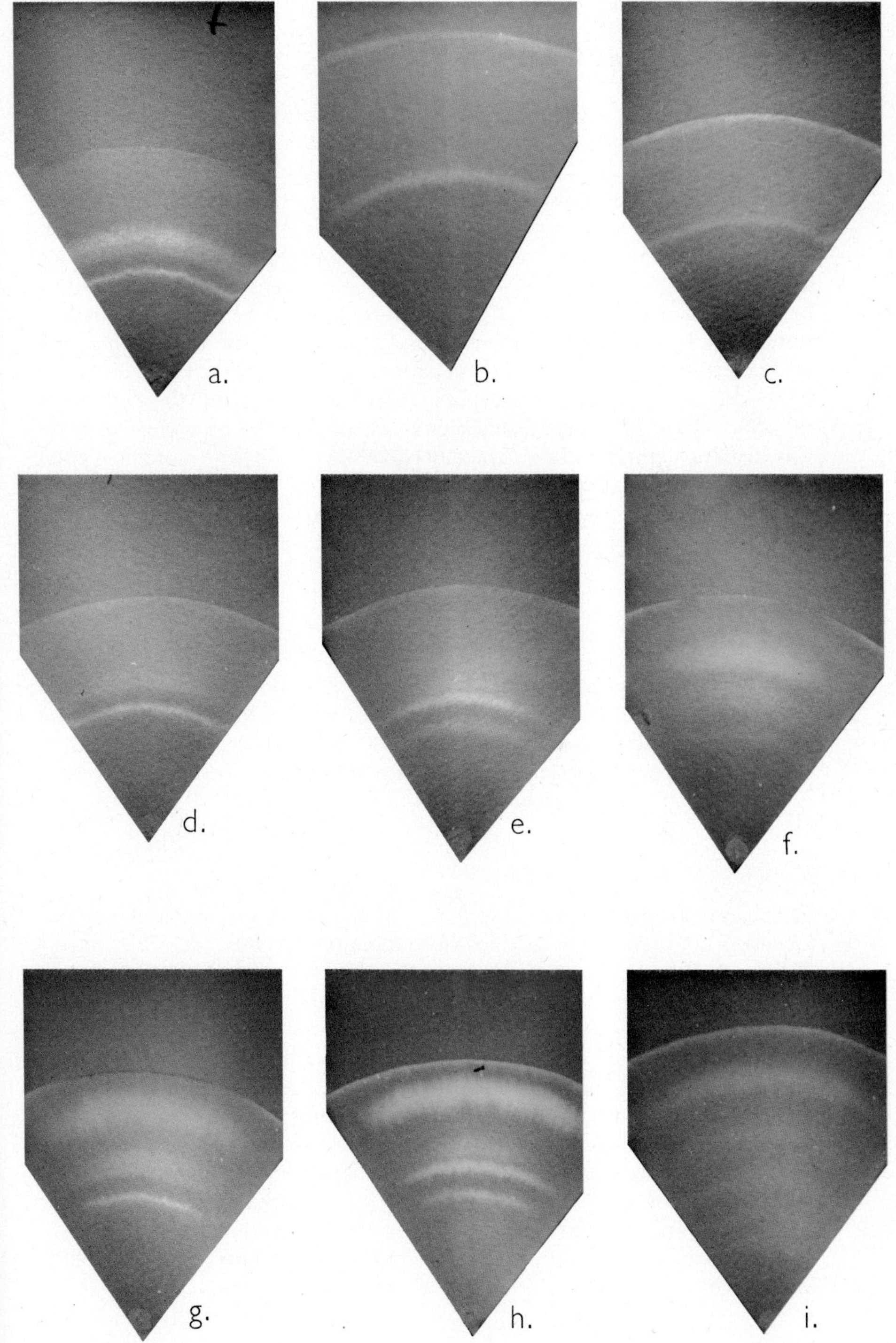
a.
b.
c.
d.
e.
f.
g.
h.
i.

a sample and from another locality in 12 out of 49 specimens examined. Individuals with band 5 produced offspring both with and without the band and when present it was always intense, no apparent intermediates being seen. Among tetraploid populations a similar range of variation in the basic three bands occurred but specimens from Lake Margherita show a diffuse band, visible before exposure to ammonia fumes in the region of bands 4 (Rf 0·70–0·80) and 5. Octoploid samples also showed the usual pattern of basic variation and some parents with a three-banded pattern gave rise to offspring in which band 3 was lacking while a single individual from one locality had bands 1–3, a pale blue band 4 and a diffuse blue in the position of band 5 (unlike the bright colour of typical band 5) and gave rise to offspring with similar patterns. A selection of these patterns is illustrated in Fig. 1 and gives some idea of the range of inter-population variation. Until the fluorescent substances have been chemically characterized and the physiological pathways leading to their synthesis are understood we can do little more than point to them as potentially useful taxonomic characters. Below the species level they allow discrimination between populations and races of snails and they also provide a basis for drawing together species at the sub-generic or species-group level.

III. Electrophoretic Methods

Electrophoresis of proteins has been widely used in molluscan taxonomy and has made appreciable contributions of value. Most of the techniques giving improved resolution have been employed (moving boundary and paper electrophoresis gave inadequate resolution) and there has been some diversity in the choice of protein systems selected for examination.

A. GENERAL BLOOD PROTEINS

Although blood proteins would seem to be the most obvious source of material they have not been widely used for various reasons. Woods *et al.* (1958) included blood of the oyster *Ostrea virginica* and a squid (*Loligo pealeii*) in a general survey of marine invertebrate sera on starch gel and Wright and Ross (1959) made a preliminary study of the blood of the planorbid *Biomphalaria glabrata* using cellulose acetate as the supporting medium. However, Wright and Ross (1963) found that considerable depletion of the blood protein fractions of *B. glabrata* occurred with increasing age and Targett (1963) noted that the blood proteins other than haemoglobin in a variety of planorbid species were of irregular occurrence. Targett also found that the electrophoretic mobility of the haemoglobin fractions was similar in ten planorbid species (from three separate subfamilies) thus confirming his earlier observations on the similarity of the absorption spectra of various planorbid haemoglobins (Targett, 1962c). More recently Richards (1970) has shown that haemoglobin synthesis in

B. glabrata is subject to genetic variation. In the course of his breeding experiments clones were established in which both juvenile and adult snails were deficient in haemoglobin, others in which the juveniles were deficient but the adults were normal and others in which haemoglobin levels were normal at all ages. Davis and Lindsay (1964, 1967) used polyacrilamide gel disc electrophoresis and found that the 12 fractions which they were able to separate in the blood of *Helix aspersa* decreased quantitatively with increasing size of the snails while there were no qualitative or quantitative changes associated with size in the 20 fractions present in foot muscle extracts of the same species. Furthermore they found that the foot-muscle proteins were capable of indicating differences at the population level in the freshwater prosobranch genera *Oncomelania* and *Pomatiopsis*. In contrast to these observations some workers have reported much less variation in molluscan blood proteins. Michelson (1966b) used disc electrophoresis to study the blood of *B. glabrata* and of the 9–12 fractions which he found he mentioned that there was some variation in only a few of the minor components and that these variations were not correlated with age or sexual maturity. On cellulose acetate at the high pH used by Wright and Ross (1963) and Targett (1963) the haemoglobin fraction is the fastest moving component (about half as fast as baboon serum albumin) but in Michelson's system using a pH of 8·2 the haemoglobin appeared to resolve into as many as four subfractions, all of low mobility with the major one not migrating at all. The fastest moving fraction in Michelson's system stained strongly with bromphenol blue and alcion blue and was also PAS positive suggesting that it may be a protein associated with a muco-polysaccharide. More recently Stadnichenko (1970) has studied blood proteins in *Lymnaea stagnalis*. Refractometer measurements showed an increase in total blood protein associated with increasing size of the snails but electrophoresis in agar gel at pH 8·6 revealed only five fractions, two strongly staining and moving to the anode and three weaker cathodal components. All five of these fractions were present in the three size groups examined (mean shell lengths 17·7 mm, 27·5 mm and 39·2 mm) but some variation in the relative concentration of fractions 2 and 3 (the two on either side of the start line) was noted. Cheng's (1964) claim that the electrophoretic profiles of molluscan blood proteins show specific characteristics was not justified by his results because of the distant relationships between the nine species he examined. Rosenfield and Sindermann (1965) in a preliminary report mentioned that there were differences in the blood proteins of oysters at the generic level but specific distinctions were only apparent in some of the isoenzyme systems. Grossu and Tesio (1971) were able to confirm the distinct status of the morphologically similar slug species *Limax zilchi* and *L. cineroniger* by finding constant differences in the blood proteins and tissue esterases of individuals of comparable states of maturity.

The suspect nature of blood proteins has led to the adoption of other systems for detailed taxonomic work. A basic dichotomy has arisen in the choice of material, dictated largely by the characteristics of the snail groups studied. Wright and Ross (1963) working with planorbid basommatophora turned to egg proteins, while Davis and Lindsay (1964) adopted extracts of foot muscle of freshwater prosobranchs. Both systems have their merits and their disadvantages. To obtain egg proteins in a suitably early state of development necessitates maintenance of living colonies of snails while foot muscles can be collected from wild-caught material without the need for laborious breeding work. Electrophoretic analysis of egg proteins does not damage the individual which produced them and laboratory colonies of known constitution can be established for further work but many freshwater prosobranches are viviparous and in other species eggs are difficult to harvest.

B. EGG PROTEINS

Morril (1961) and Morril *et al.* (1964) made an intensive electrophoretic study of the egg proteins of *Lymnaea palustris*. Using cellulose acetate as the supporting medium they were able to separate nine fractions, four of which were PAS positive and one which proved to be an α-napthyl esterase. In starch gel at the same pH (8·6) 11 protein fractions and two esterases were separated. Extracts of the albumen gland gave protein patterns identical with those of the egg albumen but showed ten esterase fractions two of which were identical with those in the egg contents. The greater resolving power of the starch gel showed slight differences in the patterns between different egg masses from the same individual snail but these were presumed to be due to different total protein concentrations with loss of some of the weaker fractions. There was apparently no selective depletion of particular protein fractions by the developing embryos, a fact also noted by Wright and Ross (1963) for *Biomphalaria glabrata*. Mean values for the total protein content of individual *L. palustris* eggs given by Morril (1964) are 16·6 μg when first laid and 2·6 μg at 7 days.

FIG. 2. Egg proteins of *Bulinus* species separated by electrophoresis on cellulose acetate. (a) *B. truncatus*, Sebha Oasis, Libya. (b) *B. truncatus*, Hedera, Israel. (c) *B. tropicus*, Potchefstroom, South Africa. (d) *B. liratus*, Basybasy, Madagascar. (e) *B. reticulatus*, Kisumu, Kenya. (f) *B. reticulatus*, Masingire, Mozambique. (g) *B. wrighti*, Marbum, South Arabia.

(d) and (b) show the typical sub-divided main fraction characteristic for members of the *B. truncatus* (tetraploid) group. Despite the geographical distance between the origins of these two samples the egg protein patterns are virtually identical. (c) and (d) show the clearly defined major fraction of the *B. tropicus* (diploid) group which contrasts with that of *B. truncatus*. These two populations are closely similar in most respects but there are differences in their egg proteins, particularly the slow-running fractions between the origin and the main fractions. (e) shows the egg protein pattern for topotype *B. reticulatus* and a similar pattern has been obtained from specimens of Tanzanian origin while (f) illustrates the pattern found in material from Mozambique, Rhodesia and South Africa. These results indicate that the populations of *B. reticulatus* from East Africa are distinct from those of southern central Africa. (g) shows the pattern given by *B. wrighti*, a species which was at one time confused with *B. reticulatus* but now shown to be distinct from it.

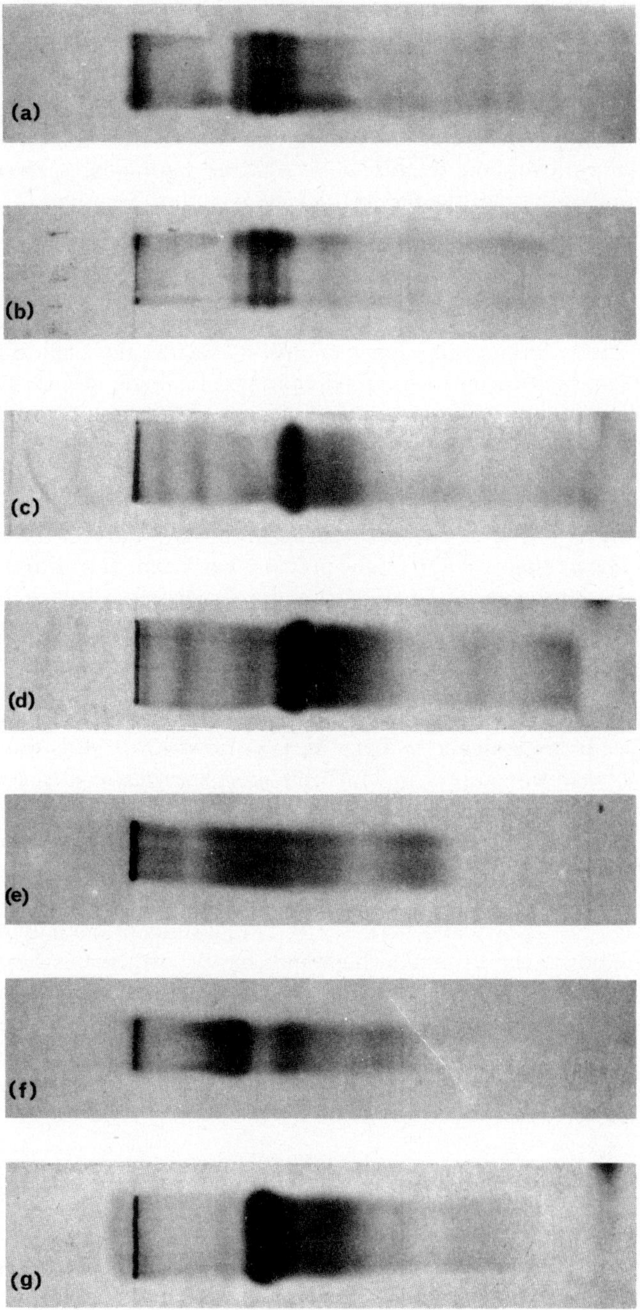
(a)
(b)
(c)
(d)
(e)
(f)
(g)

Wright and Ross (1965) described and figured the egg protein patterns obtained by electrophoresis on cellulose acetate at pH 11·78 of twelve species of the African planorbid genus *Bulinus* and seven species of *Biomphalaria*. Many of the species were represented by several populations and egg protein patterns at the population, species and species-group level were characterized. The most useful contribution in this work was the finding of a characteristic species-group pattern for the *Bulinus truncatus* complex. Members of this group are intermediate hosts for *Schistosoma haematobium* in northern and western Africa and the Middle East but morphologically they are not easy to separate from the *B. tropicus* complex which do not act as hosts for *S. haematobium*. The egg protein pattern is one of the most reliable features for distinguishing these groups. Another important aspect of this study was the demonstration of differences in egg proteins between populations of a single species. This reflection of physiological differences at the intermediate host population level provided further support for hypotheses on the evolution of trematode strains in parallel with their molluscan hosts (Wright, 1960; 1966a). Wright and Ross (1966) continued the survey of bulinid egg proteins with a detailed study of all of the members of the *B. africanus* species group and most of the *B. forskali* complex. These results reinforced the previous work and, in addition, revealed the close relationship between *B. nasutus* and *B. abyssinicus* and their position slightly apart from other members of the *africanus* group as well as demonstrating a distinction between topotype *B. reticulatus* from Kenya and the form from southern Central Africa. Egg protein patterns have also been used to confirm the unique nature of *B. nyassanus* and the relationship of *B. succinoides* (both of these species are endemic to Lake Malawi) to the *B. tropicus* group rather than the *B. truncatus* complex to which the morphological evidence suggested an affinity (Wright *et al.*, 1967). More recently (Wright, 1971) egg proteins have provided evidence of the diversity of populations of *B. cernicus* on Mauritius, a fact confirmed by divergent characters of the radular teeth, also they have shown the close similarity between populations of *B. bavayi* on Aldabra and Madagascar. Brown and Wright (1972) in a study of the polyploid complex of bulinids in the Ethiopian highlands have shown the existence of two distinct egg protein patterns in octoploid populations as well as the typical *B. truncatus* pattern among tetraploids and two further patterns among diploid populations. Possible intermediates between the two diploid pattern types were found in some mixed populations but no intermediates between the two octoploid patterns have been seen.

C. MUSCLE PROTEINS

It has already been mentioned that the majority of electrophoretic studies of foot muscle proteins have concentrated on the freshwater prosobranchs but two brief notes relating the application of the technique to basommatophoran pulmonates are interesting. Pace and Lindsay (1965) examined thirteen populations of *Bulinus* belonging to three species complexes as well as three other species in different planorbid genera. They found strikingly little overall variation but the level of variation between two populations of a single species of *Bulinus* was as great as that between the three planorbid subfamilies represented in their material. They concluded that there appeared to be little prospect of the technique being of any value in the Planorbidae. Burch and Lindsay (1969) in examining the relationships of two Indian species of *Lymnaea*, *L. luteola* and *L. acuminata* compared the foot muscle proteins of both species with those of European *L. auricularia* and the holarctic *L. stagnalis*. Existing systems of classification had associated *L. luteola* with *L. stagnalis* on the basis of morphological similarities in the structure of the prostate gland and *L. acuminata* had been included in the synonomy of *L. auricularia rufescens*. The electrophoretic results showed that *L. luteola* and *L. auricularia* were so similar that they could not be distinguished while *L. acuminata* was distinct from them and all three of these species were clearly different from *L. stagnalis*. The distinction between *L. stagnalis* and the other three species was upheld by chromosome studies.

In a major morphological and biological review of the relationships between the hydrobiid genera *Oncomelania* and *Pomatiopsis* Davis (1967) included characteristics of the foot muscle proteins. All *Oncomelania hupensis* subspecies had one or two dense, fast-moving fractions beyond Rf 0·75 and these were lacking in *Pomatiopsis* species. The problem of the relationships of the snail from Taiwan originally described as *Tricula chiui* was partly resolved by study of the foot muscle proteins. The species had been transferred from *Tricula* to *Oncomelania* and Davis (1968a) subsequently showed that the electrophoretic pattern of its foot muscle proteins was identical with that of some populations of *O. hupensis formosana*. Immunological studies supported the conclusion that *O. chiui* is no more than a local race of *O. h. formosana* (see p. 375). Comparison by disc electrophoresis of foot muscle proteins from field-collected and laboratory-bred populations of the Japanese *Oncomelania hupensis nosophora* from the same locality showed some variation in two of 24 major fractions. Of these two fractions one was never found in laboratory-bred snails and it was suggested that this might represent a genetic difference (Davis and Takada, 1969).

Some snails of the pleurocerid genus *Semisulcospira* in the Far East act as hosts for the lung fluke *Paragonimus*. In Japan various species groups of *Semisulcospira* have been established on the basis of chromosome numbers. Two species belonging to the same cytological group ($n = 18$), *S. libertina* and *S. trachea* are distinguished by shell characters, the first being smooth and living in rivers while the other is lacustrine and has prominent ribs. No significant biological characters could be found to distinguish between the two, and the foot muscle proteins differed only in that eight out of the 24 fractions resolved had slightly higher Rf values in *S. trachea* than their counterparts in *S. libertina*. Differences of this order are only considered significant at the population level and immunological work (see p. 376) supported the treatment of *S. trachea* as a synonym of *S. libertina*, the few differences found not justifying subspecific status (Davis 1968b). Davis (1969) also used foot muscle proteins to help in establishing the identity of a Japanese population of *Semisulcospira* found to be naturally transmitting *Paragonimus westermani*. The lack of significant differences between this population and topotype *S. libertina* led to the inclusion of the host population in that species and the prediction that the topotype population would probably also be capable of transmitting the parasite. Davis' most recent contribution (1971) in this field concerns another host for *Paragonimus*, the thiarid *Brotia costula episcopalis* from Malaysia. Two populations from separate drainage systems were compared and found to agree with respect to 23 reproducible foot muscle protein fractions but three bands of low reproducibility were found in only one of the populations. When 12 key fractions were compared with the pattern given by *Semisulcospira libertina* it was found that only three peaks were common to both patterns. The most fruitful part of this study is the discussion of relationships between the families to which these two species belong. However, that discussion depends in a large part on the immunological results and will be dealt with later in this review (see p. 376).

To conclude this section on electrophoresis of largely unspecified proteins extracted from molluscan tissues an interesting study on species of the intertidal marine prosobranch genus *Littorina* provides a change from the rather intense air of medical malacology. Wium-Andersen (1970) has used polyacrylamide gel electrophoresis to compare the proteins extracted from the buccal mass of three species, *L. littorea* from Denmark, *L. striata* from Tenerife and *L. rudis* from Denmark and Greenland. The patterns for *L. striata* and *L. littorea* were constant, three bands in *L. striata* of which the two outer are haemoglobin fractions and five bands in *L. littorea*, the two outer fractions again being haemoglobin but considerably faster moving than those of *L. striata*. *L. rudis* by contrast showed great variability both between individuals and populations. A total of 14 bands were found in this species of which eight were common and six rare. Of the eight common bands two are haemoglobin with similar mobility to the two fractions in *L. littorea* but in *L. rudis* they may occur together or separately. Wium-Andersen suggests that these two haemoglobin fractions are determined by two allelic genes and that the individuals in which both fractions occur are therefore heterozygotes. The actual occurrence of the various phenotypes agrees closely with the predicted frequency on this hypothesis and the occurrence of the fast moving HbI on the west coast of Jutland suggests the existence of a gene frequency cline. The rest of the protein bands in *L. rudis* present a confusing picture with high levels of individual variation within populations with the exception of *L. r. groenlandica* from Heligoland and Greenland. From both of these localities the patterns are constant with four bands in the Heligoland population and five in that from Greenland. It is suggested that this may justify consideration of *L. r. groenlandica* as a distinct species. The high levels of variation in *L. rudis* are explained by the fact that it is a brackish-water species and such habitats in Denmark are geologically young so that it may perhaps be a relatively recent colonizer. In contrast to the other two species, both of which have pelagic larvae, *L. rudis* is ovo-viviparous and its populations are likely to be somewhat parochial while in *L. littorea* and *L. striata* there is a much greater chance of gene exchange throughout the species' range.

D. ISOENZYMES

Although a few minor reports of molluscan enzyme systems had appeared before 1964 it is the work of Norris and Morrill (1964) which can be regarded as the basis for subsequent detailed studies. In this work they examined the organ specificity and embryological development of nine hydrolytic enzymes in *Lymnaea palustris*. In all they resolved 42 fractions, 16 aromatic esterases, 6 alkaline phosphatases, 4 acid phosphatases, 4 leucine aminopeptidases, 4 alanine aminopeptidases, 4 β-galactosidases, 2 α-glucosidases and 1 each of β-glucosidase and β-glucuronidase. The maximum number of fractions found in one organ was 32 in the digestive gland and the minimum was 10 in the muciparous and oothecal glands. In the first two days of development five bands are present in extracts of the embryo, 2 acid phosphatases and 1 each of leucine and alanine aminopeptidase and galactosidase. Esterases do not appear in the embryo until the third day and at the time of hatching 29 of the total of 42 adult fractions can be resolved. In a comparable study on hydrolytic enzymes of developing embryos of the marine prosobranch *Ilyanassa obsoleta* the same authors (Morrill and Norris, 1965) found that some of the bands present in the earliest stages disappear between the second and fourth days of embryonic development while others persist throughout and still others do not appear until the third to the seventh days. Similar observations on the sequential development of lactic dehydrogenase enzymes in the marine prosobranch *Argobuccinum oregonense* were reported by Goldberg and Cather (1963).

As a prelude to taxonomic application of isoenzyme studies Wright *et al.* (1966) carried out a comparative survey of esterases and acid and alkaline phosphatases in the organs of one species each in the planorbid genera *Bulinus* and *Biomphalaria*. Alkaline phosphatases proved to be weak and difficult to resolve, the acid phosphatases showed relatively little variation either between the species or in different organs with the exception of the digestive gland but the esterases were more abundant and gave more fractions, especially in the kidney and digestive gland. In the light of these results it was decided to concentrate for taxonomic purposes on the α-napthyl esterases of digestive gland extracts. Burch and Lindsay (1967) used disc electrophoresis to compare the esterases of foot muscle extracts in 17 populations of *Bulinus* belonging to the *B. tropicus* and *B. truncatus* groups. They found 4–10 fractions and although there was some variation in the slow-moving bands there were always two heavily-staining fast bands in all diploid samples but only one in polyploids.

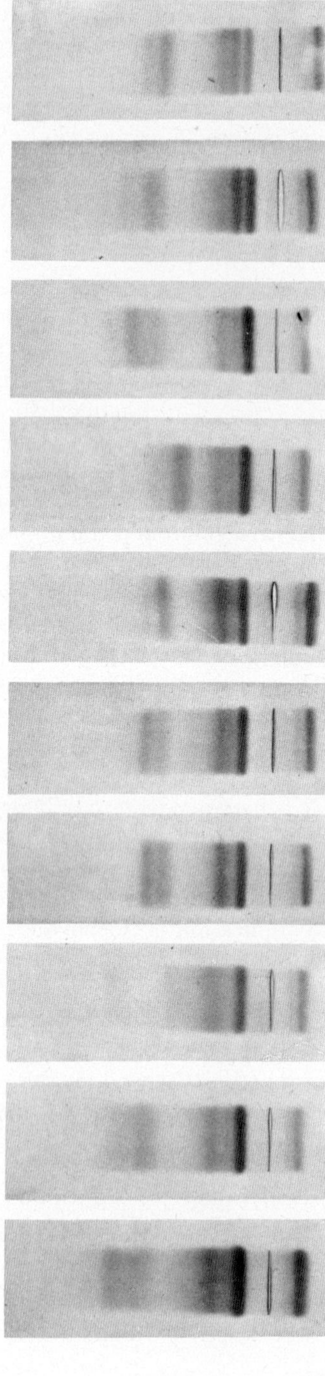

Fig. 3. Digestive gland esterase isoenzymes of a random sample of ten adult *B. tropicus* from a laboratory-bred population derived from Lake Duluti near Arusha, Tanzania. Most populations of *Bulinus* show fairly uniform patterns of esterase isoenzymes with seldom more than two bands varying within a population. The colony from which this sample was taken has been bred in the laboratory for over nine years and with the exception of the fourth and fifth specimens in this series no two out of the ten give the same pattern.

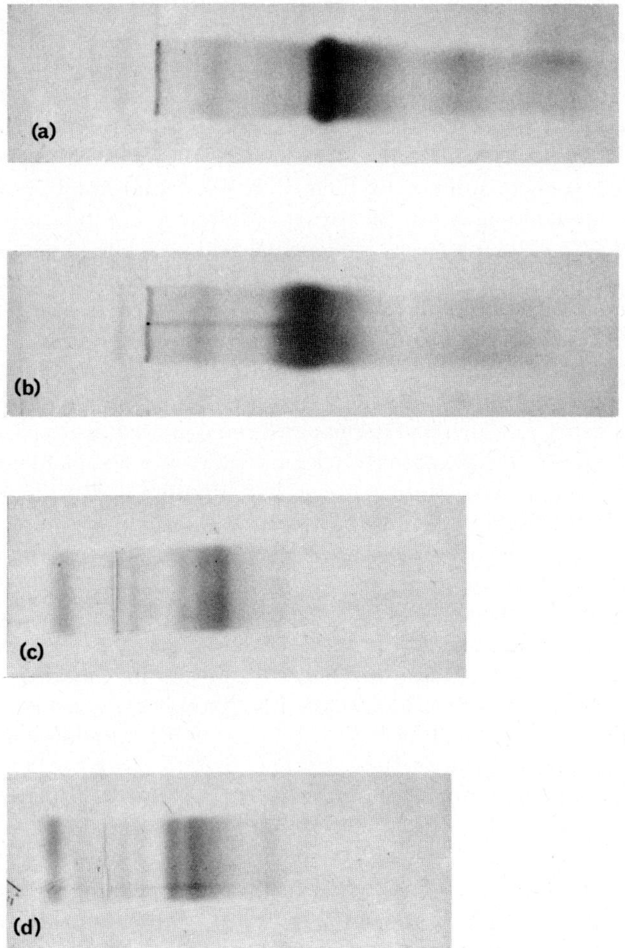

FIG. 4. Egg proteins and digestive gland esterase isoenzymes of octoploid *Bulinus* from Ethiopia. (a) 'b' type egg protein. (b) 'd' type egg protein. (c) and (d) two of the three esterase isoenzyme patterns found in octoploid populations. 'b' type egg proteins were found in 17 octoploid populations and 'd' type in three. Both types were found together in a single sample but no intermediate patterns were seen and isolated individuals of each type gave rise only to progeny with the same type. The esterase pattern shown in (c) was found in seven populations and that in (d) in four, in neither case did the esterase pattern appear to segregate with the egg protein pattern.

Wright and File (1968) pursuing their earlier work described the digestive gland esterases from about 50 populations of 17 nominal species of *Bulinus* distributed throughout the four species groups into which the genus was at that time divided. The range of variation encountered between populations was so great as to make the characterization of species impossible and even the definition of species groups was difficult. A certain amount of individual heterogeneity within populations was encountered, particularly in two populations of the *B. tropicus* complex. These belong to the assemblage of diploid populations investigated by Burch and Lindsay (1967) and, although the two fast running bands mentioned by them were not found in all of our diploid samples, in those in which they did occur individuals which lacked the faster moving fraction of the pair were present. Wright and File (1968) concluded that for most taxonomic purposes digestive gland esterases are only suitable for comparisons at the population level. Wright (1971) showed that an Aldabran and a Madagascan population of *B. bavayi* differed by only a single slow-moving esterase fraction but the marked differences in both esterases and acid phosphatases served to distinguish *B. obtusispira* clearly from *B. liratus*, two Madagascan species which had in the past been confused. Brown and Wright (1972) found three esterase pattern types among twelve octoploid populations of Ethiopian *Bulinus*. One of the patterns had affinities with a group of tetraploid populations and the other two with two groups of diploids. These differences in esterase patterns were not correlated with the differences in egg proteins already discussed (see p. 364).

Coles in a series of papers (1969a, b; 1970) has studied a wide range of enzyme systems in the digestive glands of a few species of *Bulinus* and *Biomphalaria*, *Lymnaea natalensis* and the African freshwater prosobranch *Pila ovata*. The enzymes studied were acid and alkaline phosphatases, acetyl and butyl esterases, leucine aminopeptidase, β-glucosidase, β-glucuronidase, peroxidase and the following dehydrogenases: lactate, malate, β-hydroxybutyrate, α-glycerophosphate, α-alanine, glutamate, isocitrate, glucose-6-phosphate and 6-phosphogluconate. Unfortunately the range of material covered in this survey is inadequate to assess the taxonomic value of most of the systems. Coles suggests that the leucine aminopeptidases have the greatest potential use. However, although Coles separated four fractions of this enzyme in *Bulinus nasutus* and three in *B. africanus*, Wright and Moule (unpublished) were unable to resolve more than one in several species of *Bulinus*. On the other hand Coles obtained only three acid phosphatase fractions and one β-glucuronidase in these two bulinid species, while Wright and Moule resolved seven acid phosphatases and three β-glucuronidases.

The work of Manwell and Baker (1968) on two populations of the helicid land snail genus *Cepaea* has provided some interesting background ideas for molluscan isoenzyme studies. Working with two population samples from localities in the south-west of England they compared *iso*citrate, malate and 6-phosphogluconate dehydrogenases. One of the populations (Tintagel) consisted entirely of *C. nemoralis* while the other (Erme valley) was made up of about two-thirds *C. hortensis* and one-third *C. nemoralis*. All three of these enzyme systems (and others mentioned in less detail) showed considerable polymorphism. The NADP-dependent *iso*citrate dehydrogenases form two distinct groups of fractions, one slow-moving and the other fast. The fast-moving fractions in *C. hortensis* show a triallelic variation and the six possible phenotypes, three homozygous and three heterozygous were found, the heterozygotes showing both parental fractions together with a usually more intense 'hybrid' zone of intermediate mobility. Polymorphism of the slow NADP-dependent *iso*citrate dehydrogenases was also found but the variation was less easily analysed and did not coincide with that of the fast-moving enzyme. In *C. nemoralis* only a single fast NADP-*iso*citrate dehydrogenase was found in the pure population and thus corresponded to one of the *C. hortensis* forms but in the mixed population a number of *C. nemoralis* showed a heterozygous form comparable to some *C. hortensis*. Similar patterns of variation were found to occur in both the 6-phosphogluconate and malate dehydrogenases and in both cases the *C. nemoralis* from the mixed population exhibited at least some affinities with *C. hortensis* which were not found in the pure *nemoralis* sample. The most striking example of this was shown by the glucose-6-phosphate dehydrogenase which occurred as a single fraction in both species. However, in the pure population of *C. nemoralis* from Tintagel this fraction has a mobility 10–12% faster than that found in the mixed population from Erme valley, where the enzyme appears to be identical in both *C. nemoralis* and *C. hortensis*. The authors mention that other samples of *C. nemoralis* from Devon also have the slower moving fraction but they do not make it clear whether these specimens were from pure or mixed populations. In discussing their results Manwell and Baker offer four explanations.

1. The presence of variants in common between closely related species may be the result of retention of variation present in the ancestral form.
2. The common variants may actually represent different mutations.
3. Similar selection pressures on sympatric individuals of the two species may preserve variants which are lost in populations subject to different ecological conditions.
4. Introgressive hybridization which may allow the occasional transfer of some genes from one species to the other.

On the whole the authors appear to favour the theory of introgressive hybridization but detailed discussion on the basis of results from only two populations must remain highly speculative.

In the marine field taxonomic enzyme studies are so far limited. Reid (1968) compared the distribution of esterases and endopeptidases in the bivalve genera *Lima*, *Mya*, *Chlamys*, *Pecten*, *Glycymeris*, *Modiolus*, *Crassostrea*, *Arcopagia*, *Tresus*, *Arctica* and *Cardium* and showed differences between related genera and between two species of *Mya*, both in extracts of the digestive diverticula and in the stomach fluid. The patterns for three groups of these lamellibranchs showed characteristic features which could be related to gut morphology and methods of digestion. Changes due to starvation were observed in some species and these included an increase in several of the esterase fractions. In a subsequent study of the esterases of eight species of the bivalve genus *Macoma* Reid and Dunnill (1969) examined 40 specimens of each of two species and 10 each of the other six. In some species they found a high level of individual variation but this was less marked in other species. In *M. secta* the gastric contents and diverticular extracts gave identical patterns but in the other species there were differences in the esterases from these two sources. Multiple arylesterases were present, there was a single aliesterase common to all of the species but no cholinesterases were found. Differences in diet surprisingly did not appear to affect the distribution of these digestive system esterases. The authors were cautious in the taxonomic conclusions which they drew in view of the relatively limited amount of material examined.

IV. Immunological Methods

The application of immunological techniques to problems of molluscan systematics antedates the use of all other experimental methods. A preliminary contribution was made by Makino (1934) and ten years later Wilhelmi (1944) published a study on the relationships of the Mollusca to other phyla. Wilhelmi prepared antisera to tissue extracts of an annelid (*Nereis virens*), an arthropod (*Limulus polyphemus*), an echinodern (*Asterias forbesi*), two gastropod molluscs (*Busycon carica* and *B. canaliculatum*) and a bivalve (*Pecten irradians*), using both 'whole' and lipid-free antigens. Comparisons were made by the interfacial ring-test method and the conclusion was reached that the molluscs had some ancestral affinity with the annelids since this was the only reciprocal interphylar reaction which was obtained. Cross reactions between the molluscs used were rather weak, the heterologous titres between the two species of *Busycon* being only 36 and 27% of their respective homologous titres and between the gastropods and the bivalve the heterologous reactions were at only about 2% of the homologous titres. Wilhelmi found that the reactions obtained with antisera made from 'whole' extracts gave unreliable results and advocated the use of lipid-free antigens for future work. Hiscock (1949) tried to overcome this problem by using the mucins extracted from chopped foot tissue of a number of Australian marine prosobranchs. It seems likely that these extracts contained more than just mucins and in the only preliminary results which were published the homologous titre for an antiserum to *Austrocochlea torri* was 1/1,024 and the heterologous titres with *Bembicium melanostoma* and *Cellana tramoserica* were respectively 1/8 and 1/2 using a ring-test procedure. From this time onwards the progress of immunological taxonomy of molluscs follows the familiar pattern of domination by studies on the freshwater snails of medical importance.

In a preliminary note Wright (1959b) mentioned attempts to immunize rabbits with foot muscle proteins of *Bulinus*. This project was unsuccessful, in part because shortage of material led to inadequate immunization and very weak antisera, also because the test method employed was densitometric, using the photron-reflectometer, and it proved to be impossible to obtain sufficiently clear test antigen solutions from the tissue extracts. Tran Van Ky *et al.* (1962) made antisera to *Biomphalaria glabrata*, *Planorbarius corneus*,

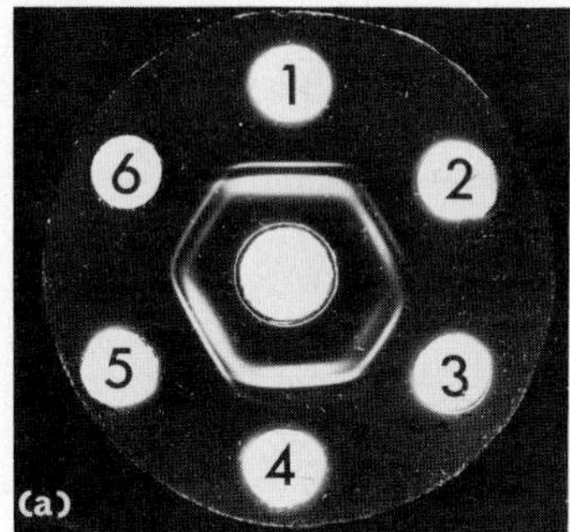

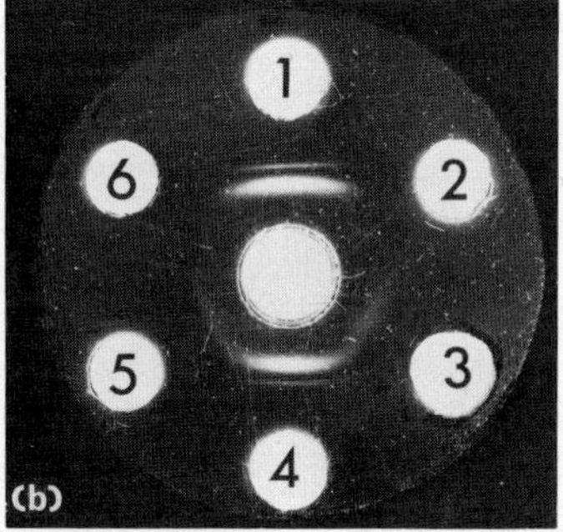

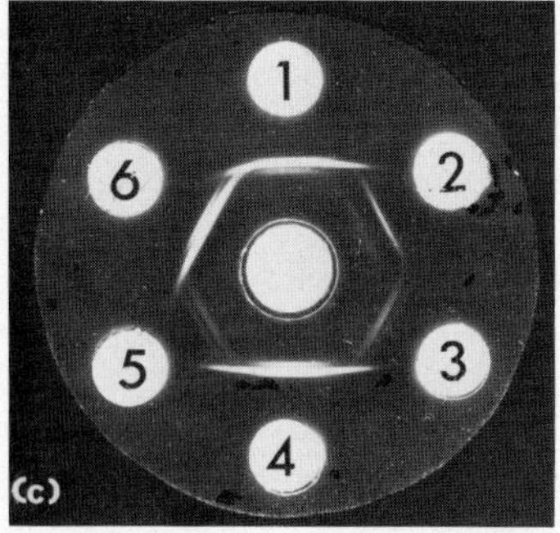

FIG. 5. Ouchterlony gel-diffusion plates showing the immunological relationships of the egg proteins of certain *Bulinus* species. (a) Antiserum to egg proteins of *B. wrighti* in centre well. Wells 1 and 4, egg proteins of *B. wrighti*; Well 2, egg proteins of *B. forskali*; Well 3, egg proteins of *B. natalensis* (*B. tropicus* group); Well 5, egg proteins of *B. truncatus*; Well 6, egg proteins of *B. globosus* (*B. africanus* group). (b) Arrangement as in (a) but antiserum in centre well absorbed by egg proteins of *B. forskali*. (c) Antiserum to egg proteins of *B. obtusispira* in centre well. (Wells 1 and 4 egg proteins of *B. obtusispira*; Wells 2, 3, 5 and 6 as in a).

In (a) the representatives of the four species groups of *Bulinus* are shown reacting with antiserum to *B. wrighti* and *B. truncatus* appears to give the strongest cross-reaction. (b) shows that after the antiserum is absorbed by *B. forskali* antigen the homologous reaction is still strong, the cross-reactions with the *africanus* and *forskali* groups are eliminated and the residual cross-reactions of the *tropicus* and *truncatus* groups are about equal. This result suggests that the affinities of *B. wrighti* are a little closer to the *tropicus*/*truncatus* complex than to the *forskali* group in which it was originally included. (c) shows unequivocaly the close relationship of *B. obtusispira* to the *africanus* group.

Lymnaea palustris and *L. stagnalis* using saline extracts of whole freeze-dried snails as antigens and giving subcutaneous injections with Freund's adjuvant to rabbits over a long period of time (fortnightly injections for 16 weeks). Using immunoelectrophoresis they reported about 15 precipitin lines in each homologous reaction and the heterologous cross reactions were said to accord with the existing systematic position of the species concerned.

Credit for the first production of a reasonably selective antiserum for taxonomic purposes undoubtedly goes to Morrill *et al.* (1964) who employed the egg albumen of *L. palustris* as their basic antigen. The purpose of their study was to compare the antigenic composition of various organs within the homologous species and their taxonomic interests were secondary. Using immunoelectrophoresis they were able to demonstrate the presence of 19 antigens in the egg albumen and all of these were also found in extracts of the albumen gland. Thirteen of these antigens were unique to the albumen gland and egg contents, three were found in all the organs tested and of the other three, one was present in both the prostate and seminal vesicle, one was found only in the prostate and the other only in the seminal vesicle. In heterologous tests against the egg proteins of *L. columella* and *L. stagnalis* Morrill *et al.* found nine antigens in common between *L. palustris* and *L. columella* and only six between *L. palustris* and *L. stagnalis*. Egg proteins were also adopted as the most useful antigens for work on planorbid snails by Wright and Klein (1967) who used Ouchterlony plate double diffusion methods as the test procedure. Comparisons between antisera to 'whole snail' (all organs except digestive gland and gonad), foot muscle, albumen gland and egg proteins of various species of *Biomphalaria* and *Bulinus* showed that the last two antigens produced much more selective antibodies capable of discrimination at species-group and species level. Egg proteins again have the obvious advantage that they may be obtained without damage to the snails producing them (see p. 362). The antigens in egg proteins were found to be shared fully by the albumen gland but scarcely any cross reaction was obtained with other adult organs and although newly hatched embryos also gave a strong reaction with antisera to egg proteins this was lost two days after emergence from the eggs. The Ouchterlony plate technique was used in preference to immunoelectrophoresis because the pH needed to obtain satisfactory separations of planorbid egg proteins (see p. 364) is too high for the formation of precipitin arcs. Laborious transfer procedures involving cutting the cellulose acetate strips on which egg proteins have been separated and placing these on agar diffusion slides have been tried (Wright *et al.*, 1967) but the results are somewhat unsatisfactory.

A. BLOOD PROTEINS

Michelson (1966a, b) used blood proteins of planorbids (*Biomphalaria glabrata*, *Helisoma anceps* and *Bulinus globosus*) as antigens and found that using interfacial ring-tests and a micro-ouchterlony diffusion methods his antisera were only able to discriminate at the generic level. Immunoelectrophoresis gave only five arcs in the homologous reaction of *Biomphalaria glabrata*, all of these corresponding to the slow moving benzidine-positive fractions identified as haemoglobins by disc electrophoresis (see p. 361). Absorption of the *B. glabrata* antiserum by blood of *Helisoma anceps* removed all but one of the precipitin arcs, the one remaining corresponding to the fourth fraction. These results tend to lend some weight to the arguments against the use of molluscan blood in taxonomic studies (see p. 361). Burch (1967, 1968) described a micro-ouchterlony absorption method using foot muscle tissue antigens. In a preliminary study on the polyploid complex of *Bulinus* he reported identity reactions between populations having the same chromosome number and non-identity between snails of different chromosome numbers. These results were confirmed in greater detail by Burch and Lindsay (1970), but only three antisera were used, one to a population of diploids, one to a tetraploid sample and the third to an octoploid population.

B. FOOT MUSCLE PROTEINS

The widest use of micro-ouchterlony absorption methods and immunoelectrophoresis has been made by Davis in his series of works on freshwater prosobranchs in the Far East. As in his electrophoretic studies the source of the protein used as antigen has always been foot muscle. In the first of these contributions Davis (1968a) drew attention to one of the most important technical points in the use of immunological methods, the variability in quality of antisera. His objective was to determine the precise relationships of *Oncomelania hupensis chiui*, a snail found in an isolated region of the northern part of Taiwan and originally described as *Tricula chiui*. Five antisera were prepared to *O. h. formosana*, the common form of *Oncomelania* on Taiwan and comparisons were made with other subspecies of *O. hupensis* from Japan and the Phillipines. One of the five antisera was capable of discriminating between these geographically isolated subspecies but despite absorption by heterologous antigens the other four were not able to do this. Even with the highly specific antiserum *O. h. chiui* proved to be indistinguishable from *O. h. formosana* thus providing support for the morphological and electrophoretic evidence (see p. 365) which suggested that it is no more than an isolated race of the more widely distributed species. This variation in quality of antisera is a common problem and one which leads many people to have doubts about the practicability of immuno-taxonomic work. The ratio of one good discriminating antiserum to four of lower specificity is by no means unusual and in cases where supplies of antigen are limited this can represent an enormous wastage, not only of material but also time. Davis and Suzuki (1971) point out that it may require several thousand individuals of one of the small *Oncomelania* species to yield enough foot muscle tissue to immunize a single rabbit. Similarly, in this laboratory, we have found that we need a packed volume of at least 5 ml of bulinid egg masses to prepare antigen for the injection of one rabbit. In some of the smaller species which do not breed easily this may represent at least six months collection from a laboratory colony. With this problem in mind Davis and Suzuki (1971) have described a method requiring small quantities of antigen innoculated intraperitoneally into mice together with Sarcoma 180 cells and Freund's adjuvant. Ascitic fluid with high antibody content, often comparable to the most discriminating rabbit antisera, can thus be produced in relatively short periods of time with small amounts of antigen. However, this technique has not been widely used and most of the work so far reported has been done with traditional rabbit antisera.

In both of Davis' studies on *Semisulcospira* (1968b, 1969) he was concerned with relationships at the specific level, in the first instance with the precise identity of *S. trachea* and in the second with the affinities of a population known to act as host to the lung-fluke *Paragonimus westermani.* In both cases the material under investigation proved to be no more than a local race of *S. libertina*, a species to which several antisera were available. On micro-ouchterlony tests reactions of identity were obtained between *S. libertina* and the other forms but with immunoelectrophoresis some unabsorbed antisera appeared to show minor differences between the homologous and heterologous antigens. However, absorption of the antisera failed to confirm these fine distinctions and the immunological results thus supported the findings based upon morphology and straight protein electrophoresis.

The electrophoretic data from Davis' (1971) study on *Brotia costula* established that two populations of this species from separate river drainage systems were virtually identical. Also, comparison of 12 key fractions in the foot muscle protein patterns of this species and *Semisulcospira libertina* showed that only three of the major peaks appeared to be common to both species (see p. 366). Micro-ouchterlony gel diffusion tests using both rabbit antisera and ascitic fluid from mice immunized with antigens to *S. libertina* showed from 4–7 precipitin lines in the homologous system. Absorption of the immune sera and ascitic fluids with antigens to *Brotia costula* still left 2–4 precipitin lines in the homologous reactions. Taking these lines as representing the part of the total homologous reaction which is unique to *S. libertina* and expressing their number as a percentage of the total for each of the eight immune fluids tested gives an average figure of 48%. The percentage of precipitin arcs peculiar to *S. libertina* on immunoelectrophoresis with unabsorbed antisera was 50–55%, a reasonably close figure to that established by gel diffusion absorption methods (Davis and Suzuki, 1971). The total of major electrophoretic peaks and Ouchterlony precipitin lines by which *B. costula differs* from *S. libertina* is 12 out of 18, and Davis suggests that, using these data, the degree of *relationship* between the two species is 34%. This level of similarity, Davis argues, indicates a closer affinity between the two families (Thiariidae and Pleuroceridae) to which these species belong than that which had previously been suggested on the basis of morphological data. One must have some misgivings about the validity of this index of relationship. In the first place only 12 out of a total of about 24 protein fractions were selected for comparison in the electrophoretic separations and their identity or non-identity with those of the other species was determined solely by a somewhat arbitrary standard of their relative mobility. Thus the three out of 12 'key' fractions common to the two species represent only 25% of the fractions selected for comparison and yet the immunoelectrophoretic results suggest that approximately 50% of the *total* foot muscle proteins are common to both. Further the same complex of unidentified proteins was used for both the electrophoretic and immuno-diffusion techniques and it is questionable as to whether the summation of these two sets of data is justified since they are merely different methods of comparing the same material. Davis himself, in discussing these results, draws attention to a further need for caution based upon experience with two species of *Semisulcospira*, *S. libertina* and *S. niponica*. Despite the fact that these are distinct species with haploid chromosome numbers of 18 and 12 respectively no differences could be found between them, either in the electrophoretic separations of their foot muscle proteins or in immunological tests using the same antigens. With these results, therefore, the index of relatedness between the two species (as applied to the *B. costula*/*S. libertina* system) would be 100%! The whole question of the quantification of degrees of affinity based upon the physical attributes of unidentified proteins is very much open to debate. Even in the field of isoenzyme studies, where it is possible to characterize more precisely the fractions separated, anomalous results are frequently encountered.

C. EGG PROTEINS

Recent applications of immunological techniques in the taxonomy of bulinid snails have been described by Wright (1971) and Brown and Wright (1972). In both of these contributions egg proteins were used as antigens in ouchterlony gel-diffusion plates, in the first instance to determine the species-group relationships of two species of uncertain affinities and in the second to attempt to differentiate between members of the polyploid series of *Bulinus* in Ethiopia. *B. obtusispira* was a little-known species from Madagascar usually considered to be no more than a form of the somewhat polymorphic *B. liratus*, which in turn is regarded simply as the Madagascan representative of the *B. tropicus* complex. However, discovery of snails identified as *B. obtusispira* carrying natural infections of *Schistosoma haematobium* called for a re-assessment of the situation because no member of the *B. tropicus* group is known to act as a host for this parasite. Morphological and cytological studies provided few distinctive characters, nor did electrophoresis of egg proteins. Chromatography of body surface mucus and isoenzyme electrophoresis (see p. 370) showed that *B. obtusispira* is distinct from *B. liratus* but neither technique gave

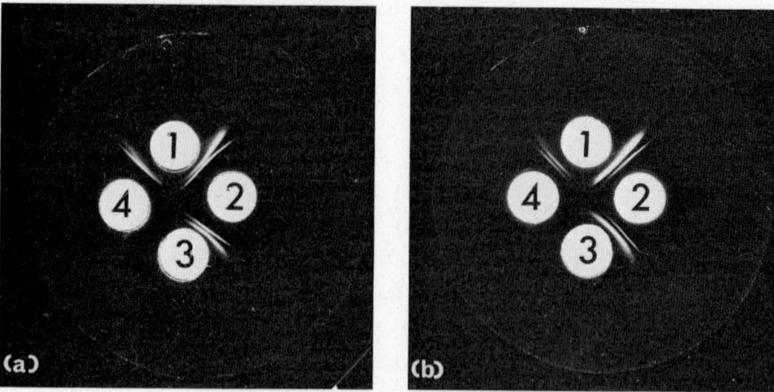

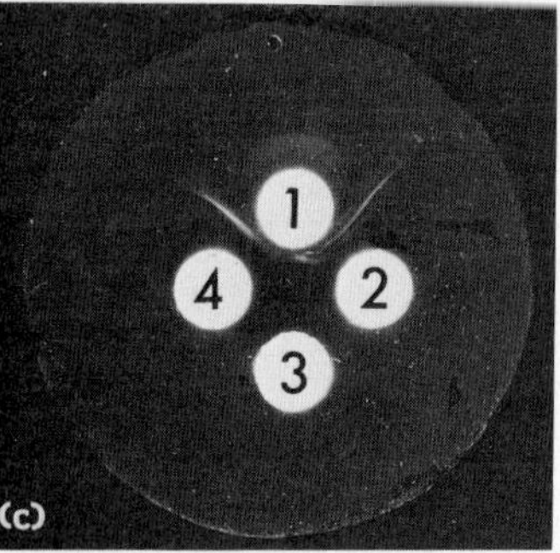

FIG. 6. Ouchterlony gel-diffusion plates using an absorption technique to show the relationships between pairs of species (based on the technique used by Davis 1968b, 1969 and 1971). (a) Well 1, antiserum to egg proteins of *B. wrighti*; Well 2, egg proteins of *B. wrighti*; Well 3, antiserum to *B. wrighti* absorbed by egg proteins of *B. scalaris*; Well 4, egg proteins of *B. scalaris*. (b) Well 1, antiserum to egg proteins of *B. wrighti*; Well 2, egg proteins of *B. wrighti*; Well 3, antiserum to *B. wrighti* absorbed by egg proteins of *B. forskali*; Well 4, egg proteins of *B. forskali*. (c) Well 1, antiserum to egg proteins of *B. scalaris*; Well 2, egg proteins of *B. scalaris*; Well 3, antiserum to *B. scalaris* absorbed by egg proteins of *B. forskali*; Well 4, egg proteins of *B. forskali*.

The precipitin lines between Wells 1 and 2 show the homologous reaction in the system, those between Wells 1 and 4 the heterologous reaction. The absence of precipitin lines between Wells 3 and 4 indicates that absorption of the antiserum in Well 3 by the heterologous antigen is complete and the presence of lines between Wells 2 and 3 gives an indication of the antigens *not* common to the two species. In (a) and (b) it is apparent that *B. wrighti* has substantial antigens not present in either *B. scalaris* or *B. forskali* and is therefore not closely related to either species. The antiserum to *B. scalaris* used in (c) is considerably weaker than that to *B. wrighti* used in (a) and (b). This is shown by the relatively poor homologous reaction between Wells 1 and 2. A very weak line between Wells 2 and 3 (scarcely visible in the photograph) shows that *B. scalaris* and *B. forskali* can be distinguished immunologically despite their much closer relationship to one another than either has to *B. wrighti*.

definite guidance as to its species-group affinities. Ouchterlony gel-diffusion tests using an antiserum to *B. tropicus* from the high veldt of South Africa emphasized that *B. obtusispira* has little in common with *B. tropicus* while tests with an antiserum to *B. obtusispira* showed clearly that the only strong cross-reactions are obtained with members of the *B. africanus* group. Further immunological tests using antisera to members of the *africanus* complex confirmed that while *B. obtusispira* is definitely correctly placed in this group it is a very distinctive species. The general conclusion is that *B. obtusispira* is something of a 'relic' which has survived in isolation on Madagascar in company with many other elements of the early African fauna. The affinities of *B. wrighti*, a species from South Arabia provided similar problems. Morphology indicated close similarity to *B. reticulatus*, an uncommon species occurring in isolated populations in East and Central Africa. *B. reticulus* was originally assigned to the *B. forskali* complex because of general similarities in the structure of the radular teeth. However, the principal distinction between *B. reticulatus* and *B. wrighti* is the form of the radular teeth which are quite exceptional in the Arabian species. Mucus chromatography showed *B. wrighti* to be lacking in the fluorescent greenish-blue band at high Rf which is characteristic for the *forskali* group (see p. 357). Electrophoresis of egg proteins gave an undistinguished pattern for *B. wrighti* but showed two complex patterns in *B. reticulatus*, samples from East Africa being quite distinct from those of southern Central Africa (see p. 364). Immunological tests using antisera to members of *africanus*, *forskali*, *tropicus* and *truncatus* groups showed that *B. wrighti* was consistently the odd one out in plates which included members of the homologous species complex. Finally, antisera to *B. wrighti* itself indicated a lack of very close affinity with members of any of the other groups, the *tropicus* complex being perhaps the closest. As a result *B. wrighti* and *B. reticulatus* were removed from the *forskali* group and united in a new separate group, the *reticulatus* group (Wright, 1971).

The immunological results of Burch and Lindsay's (1970) study of the polyploid series in Ethiopian *Bulinus* have already been mentioned (see p. 375). They found that members of populations having the same chromosome number gave reactions of identity using foot muscle antigens in a micro-ouchterlony system. Non-identity reactions were obtained when populations of different chromosome number were compared. Brown and Wright (1972) used both unabsorbed and absorbed antisera on normal ouchterlony plates and their findings proved to be more complex than those of Burch and Lindsay (1970). Ethiopian tetraploids proved to be identical with the tetraploid *B. truncatus* from North Africa and the Middle East. Octoploid populations from the Ethiopian highlands gave reactions closely similar to those of tetraploids when tested against unabsorbed tetraploid antisera but absorption of these antisera with some diploid antigens reduced the octoploid reactions much more severely than those of tetraploids. The early reactions of both tetraploids and octoploids to some unabsorbed diploid antisera were very similar but when these plates were developed for periods in excess of eight days strong spurs were formed by homologous diploid reactions against adjacent tetraploids but the spurs formed against heterologous octoploids were either much smaller or absent. These results suggested that octoploid populations have more antigens in common with some diploids than have tetraploids. Nevertheless, tetraploids tested against octoploid antiserum gave stronger cross reactions than did diploids suggesting that tetraploids have more antigens in common with octoploids than they have with diploids. Thus the octoploids appear to have more common antigens with diploids and tetraploids than either of these groups have with one another. The most interesting point raised by this study, both from the biological and the technical aspects, is the apparent heterogeneity of the diploids. So far no diploid antisera have been produced which are capable of discriminating between the diploid members of the *B. tropicus* complex but when diploid antigens are tested against tetraploid and octoploid antisera their reactions often differ. Using unabsorbed anti-octoploid serum spurs were formed both by the homologous octoploid and a tetraploid against one diploid population (from a stream near Haik village, Ethiopia) but not against the diploid from Lake Awasa, Ethiopia. Absorption of the same octoploid antiserum by the Haik diploid antigen eliminated cross reactions with several diploid populations but still resulted in a trace precipitin line from the Lake Awasa diploid. An unabsorbed antiserum to an Ethiopian population of *B. truncatus* formed spurs between the homologous antigen and those of a Haik diploid as well as *B. tropicus* from South Africa and the diploid *B. natalensis* from Mozambique. However, the spurs against *B. tropicus* were much more strongly marked than those against the Haik diploid and *B. natalensis* in turn formed spurs against *B. tropicus*. Absorption of this antiserum with antigens of the Haik diploid left a residual cross reaction with *B. natalensis* and after the third day of incubation a very faint precipitin line formed against the well containing *B. tropicus* antigen. From these results it would appear that these three diploid populations vary in the number of antigens which they have in common with this particular population of *B. truncatus*. Although tests using diploid antisera have so far failed to reveal differences of this kind it may be that wider use of heterologous (but closely related) antisera may help to sort out taxonomic problems in similar poorly differentiated complexes.

D. HETEROAGGLUTINATION

No account of immunological methods in molluscan taxonomy would be complete without mention of the indirect approaches derived from human blood-group studies. Various substances capable of agglutinating the erythrocytes of vertebrates have been reported in extracts of several snails (Boyd *et al.*, 1966) and some of them show considerable specificity for particular human blood-groups (Boyd and Brown, 1965; Prokop *et al.*, 1965). It is not the general distribution of these agglutinnis which is of particular interest here for their occurrence appears to be rather erratic and not closely associated with taxonomic relationships. What is interesting is their presence or absence in different populations of the same species and the possibility that they may serve as genetic markers. Thus Gilbertson and Etges (1967) found agglutinating activity for the red cells of hamsters, rabbits and man in the blood of *Biomphalaria sudanica* from Mwanza in Tanzania, for rabbit and human cells in two populations of *B. glabrata* from Salvador in Brazil and for man only in *B. glabrata* from Surinam. Three other populations of *B. glabrata* from Puerto Rico, Belo Horizonte in Brazil and Venezuela and samples of *B. straminea* from Brazil and *B. pfeifferi* from Liberia all failed to show agglutinating activity with any of the test cells. Gold *et al.* (1967) obtained agglutination of human red cells of groups A, B and O with extracts of the eggs of *Lymnaea stagnalis* from Bristol in south-west England but extracts of the whole snail failed to react. Lee-Potter (1969) found no agglutinating activity in extracts of *L. stagnalis* eggs from Sussex in south-east England but got a weak positive result with eggs from a Moroccan population. Lee-Potter also investigated the eggs of several other species of *Lymnaea* and *Bulinus* without finding agglutinins but extracts of eggs of a population of *B. globosus* from Zambia proved to be markedly haemolytic while those of a Ghanaian population of the same species showed no activity of any kind. Gold and Thompson (1969a) found that saline extracts of the albumen gland of *Helix aspersa* from the Bristol area were active in reverse passive haemagglutination tests with human A cells, although there was some variation in the activity of extracts from different individual snails. Material from Australian *H. aspersa* proved to be far less active and negative results were given by *H. pomatia*, *Helicella virgata*, *Cepaea nemoralis* and *Otala lactaea*. These results are interesting in that Boyd and Brown (1965) had previously reported strong specific activity in extracts of *O. lactaea* and Prokop *et al.* (1965) found very strong anti-A activity in extracts of the albumen gland of *H. pomatia* and *Cepaea hortensis* from Berlin. Gold and Thompson (1969b) found an extract of the albumen gland of one *H. aspersa* from Bristol that achieved complete haemolysis of human A, B and O cells while another extract from a different individual and diluted egg contents haemolysed only cells of group A. Again extracts from Australian *H. aspersa* were less active and gave only partial haemolysis of A cells.

So far these somewhat isolated reports do little more than draw attention to the existence of possible biochemical markers. In the future they may well serve as a fruitful link between those interested in the population genetics of both man and molluscs.

V. Summary

To summarize what is, in effect, little more than a summary is a difficult and perhaps unnecessary task. However, certain general considerations arising from this review suggest themselves.

The somewhat erratic and, in most cases, rather unsophisticated attempts which have been made to apply experimental methods to molluscan systematics reflect the present state of taxonomy in the phylum as a whole. This in turn is a reflection of the complex biology of the Mollusca. The reproductive patterns range from those of dioecious forms, some with internal and others with external fertilization, through hermaphrodites with varying potentials for cross- and self-fertilization to parthenogenetic species. Life-spans vary from periods of years to no more than a few months in different groups and the diversity of habitats colonized by molluscs results in an almost limitless variety of population types from marine species with pelagic larvae potentially capable of being carried throughout the geographical range of the species to self-fertilizing hermaphrodites in isolated temporary rainpools. The possible scope of different opportunities for genetic interchange, both between individuals and populations, makes generalizations about patterns of speciation within the Mollusca unrealistic. Add to this a long history of taxonomy based solely on features of the shell, a structure notoriously subject to modification by environmental influences and some of the reasons for the imperfections in molluscan classification become apparent. The demands of medical and veterinary parasitology which have provided such a stimulus to the development of biochemical and immunological methods in the taxonomy of gastropods have in turn contributed an element of confusion to the picture. Concerned less with determining the affinities of the snails than with defining their host-potential for a particular trematode, parasitologists have tended to seek for some precise character which may be linked with susceptibility to infection. No universal, key character determining susceptibility is likely to be discovered because such a concept fails to take proper account of genetic factors within the parasite which determine its infectivity to, and ability to survive in, a given population of snails. Thus there are populations of snails capable of acting as effective hosts to some strains of parasite but not others and there are strains of trematodes which are able to develop in a wide range of snail populations while other strains are more restricted in their host requirements. The only reliable guide to susceptibility on the part of any particular mollusc population is either successful experimental infection or the finding of naturally infected individuals. Determination of the affinities of the snail hosts must still be carried out by normal taxonomic procedures.

Of the range of techniques so far investigated there is no single method which can be designated as the most useful, each has its place depending upon the objective of the enquiry. There is, therefore, an advantage in having available a diversity of methods. Moreover some of the simple techniques can be applied in preliminary investigations before decisions are taken on the adoption of more complicated approaches requiring expensive equipment and specialized knowledge. Some techniques can provide results giving information at different taxonomic levels, for instance chromatography of mucus in *Lymnaea* has provided a method for detecting population differences in *L. peregra*, for differentiating between the closely related *L. peregra* and

L. auricularia and for showing a group relationship between these species and *L. natalensis*. Similarly cellulose acetate electrophoresis of planorbid egg proteins reveals differences at the population level but also provides an unequivocal character for distinguishing members of the *Bulinus truncatus* species group from the rest of the genus. Broadly speaking the methods of choice at different taxonomic levels are as follows: the individual—isoenzyme electrophoresis; the population—isoenzyme electrophoresis, general protein electrophoresis (foot muscle, egg albumen etc), mucus chromatography; the species—general protein electrophoresis, mucus chromatography, immunoelectrophoresis, ouchterlony gel immunodiffusion with absorbed antisera; the species group—general protein electrophoresis, mucus chromatography, gel immunodiffusion with or without absorption of antisera; above the species group level the only techniques of much value are those of quantitative immunology but, with the exception of Wilhelmi's (1944) study on general relationships of the phylum, these have not been pursued to any appreciable extent in the Mollusca.

In conclusion it cannot be too strongly stressed that the results of any of these experimental procedures are of little value in the absence of appropriate background information on the material studied. Also the material must be adequate both in respect to numbers and condition. Too often the conclusions drawn from elegant biochemical data are marred by the use of insufficient samples and questionable preliminary identification of the material. My earlier injunctions on this theme (Wright, 1959c; 1966b) have been reinforced by Davis and Lindsay (1967) whose concluding remarks bear repetition here 'Within a framework of precise anatomy and cytology, biophysical data have their useful place. Without this framework, however, one is caught in a morass of disjunct molecular populations'.

References and Bibliography

Ballantine, W. J. and Bradley, D. J. (1963). *Proc. malac. Soc. Lond.* **35**, 86–88.
Boyd, W. C. and Brown, R. (1965). *Nature, Lond.* **208**, 593.
Boyd, W. C., Brown, R. and Boyd, L. G. (1966). *J. Immunol.* **96**, 301.
Brown, D. S. and Wright, C. A. (1972). *J. Zool.* **167**, 97–132.
Burch, J. B. (1960). *Rep. Am. malac. Un. Pacific Div.* **27**, 15–16.
Burch, J. B. (1967). *Papua New Guin. sci. Soc. amn. Rep.* **18**, 29–36.
Burch, J. B. (1968). Symposium on Mollusca Part 1. 10–15. *Mar. biol. Ass. India Symp. Ser.* **3**.
Burch, J. B. and Lindsay, G. K. (1967). *Rep. Am. malac. Un.* **34**, 39–40.
Burch, J. B. and Lindsay, G. K. (1969). *Proc. Soc. exp. descr. Malacol.* **15**, 135.
Burch, J. B. and Lindsay, G. K. (1970). *Malac. Rev.* **3**, 1–18.
Cheng, T. C. (1964). *In* 'Taxonomic Biochemistry and Serology' (C. A. Leone, ed.), pp. 659–666. The Ronald Press Co., New York.
Coles, G. C. (1969a). *Comp. Biochem. Physiol.* **29**, 403–411.
Coles, G. C. (1969b). *Comp. Biochem. Physiol.* **31**, 1–14.

Coles, G. C. (1970). *Parasitology* **61**, 19–25.
Collyer, D. M. (1961). *J. mar. biol. Ass. U.K.* **41** (3), 683–693.
Davis, G. M. (1967). *Malacologia* **6** (1 & 2), 1–1v3.
Davis, G. M. (1968a). *Malacologia* **7** (1), 17–70.
Davis, G. M. (1968b). Biosystematic analysis of *Semisulcospira trachea* (Gastropoda: Pleuroceridae). *Symposium on Mollusca* Part 1, 16–35. *Mar. biol. Ass. India Symp. Ser.* **3**.
Davis, G. M. (1969). *Jap. J. Parasit*, **18** (1), 93–119.
Davis, G. M. (1971). *Proc. Acad. nat. Sci. Philad.* **123** (3), 53–86.
Davis, G. M. and Lindsay, G. (1964). *Rep. Am. malac. Un. Pacific Div.* **31**, 20–21.
Davis, G. M. and Lindsay, G. K. (1967). *Malacologia* **5** (2), 311–334.
Davis, G. M. and Suzuki, S. (1971). *Veliger* **13** (3), 207–225.
Davis, G. M. and Takada, T. (1969). *Expl. Parasit.* **25** (1–3), 193–201.
Dusanic, D. G. and Lewert, R. M. (1963). *J. inf. Diseases.* **112** (3), 243–246.
Friedl, F. E. (1961). *J. Parasit.* **47**, 773–776.
Gilbertson, D. E. and Etges, F. J. (1967). *Ann. trop. Med. Parasit.* **61**, 144–147.
Gilbertson, D. E., Etges, F. J. and Ogle, J. D. (1967). *J. Parasit.* **53** (3), 565–568.
Gold, E. R. and Thompson, T. E. (1969a). *Vox Sang.* **16**, 63–66.
Gold, E. R. and Thompson, T. E. (1969b). *Vox Sang.* **16**, 119–123.
Gold, E. R., Cann, G. B. and Thompson, T. E. (1967). *Vox Sang.* **12**, 461–464.
Goldberg, E. and Cather, J. N. (1963). *J. cellular comp. Physiol.* **16**, 31–38.
Grossu, A. V. and Tesio, C. (1971). *Atti Soc. ital. Sci. nat.* **112** (3), 289–300.
Hillman, R. E. (1964). *Syst. Zool.* **13** (1), 12–18.
Hiscock, I. D. (1949). *Aust. J. Sci.* **11** (6), 209.
Kirk, R. L., Main, A. R. and Beyer, F. G. (1954). *Biochem. J.* **57** (3), 440–442.
Lee-Potter, J. P. (1969). *Vox Sang.* **16**, 500–502.
Makino, K. (1934). *Zeitschr. f. Immunitätsforsch.*, **81**, 316–335.
Manwell, C. and Baker, C. M. A. (1968). *Comp. Biochem. Physiol.* **26**, 195–209.
Michejda, J. (1958). *Bull. Soc. Amis. Sci. Lett. Poznan* Ser. B. **14**, 341–344.
Michejda, J. and Turbanska, E. (1958). *Bull. Soc. Amis. Sci. Lett. Poznan* Ser. B. **14**, 359–365.
Michejda, J. and Urbanski, J. (1958). *Bull. Soc. Amis. Sci. Lett. Poznan* Ser. B. **14**, 345–358.
Michelson, E. H. (1966a). *J. Parasit.* **52** (3), 466–472.
Michelson, E. H. (1966b). *Ann. trop. Med. Parasit.* **60** (3), 280–287.
Morrill, J. B. (1963). *Acta Embryol. Morph. exp.* **6**, 393–443.
Morrill, J. B. (1964). *Acta Embryol. Morph. exp.* **7**, 131–142.
Morrill, J. B. and Norris, E. (1965). *Acta Embryol. Morph. exp.* **8**, 232–238.
Morrill, J. B., Norris, E. and Smith, S. D. (1964). *Acta Embryol. Morph. exp.* **7**, 155–166.
Norris, E. and Morrill, J. B. (1964). *Acta Embryol. Morph. exp.* **7**, 29–41.
Pace, G. L. and Lindsay, G. (1965). *Rep. Am. malac. Un. Pacific. Div.* **32**, 31–33.
Prokop, O., Rackwitz, A. and Schlesinger, D. (1965). *J. forens. Med.* **12**, 108–110.
Reid, R. G. B. (1968). *Comp. Biochem. Physiol.* **24**, 727–744.
Reid, R. G. B. and Dunnill, R. M. (1969). *Comp. Biochem. Physiol.* **29**, 601–610.
Richards, C. S. (1970). *Nature, Lond.* **227**, 806–810.
Rosenfield, A. and Sindermann, C. J. (1965). *Rep. Am. malac. Un. Pacific Div.* **32**, 8–9.
Rucker, J. B. (1965). *Can. J. Zool.* **43**, 351–355.
Stadnichenko, A. P. (1970). *Hydrobiol. J.* **6** (2), 106–109.
Stiglingh, I. and Van Eeden, J. A. (1970). *Wetenskap. Bydraes Potchefstroom University* Ser. B. **13**, 1–4.

Targett, G. A. T. (1962a). *Ann. trop. Med. Parasit.* **56** (1), 61–66.
Targett, G. A. T. (1962b). *Ann. trop. Med. Parasit.* **56** (2), 210–215.
Targett, G. A. T. (1962c). *J. Helminth.* **36** (1/2), 201–206.
Targett, G. A. T. (1963). *Expl. Parasit.* **14**, 143–151.
Tran van Ky, P., Rose, F and Laude, F. (1962). *C. r. hebd. Seanc. Acad. Sci., Paris* **255**, 366–7.
Wilhelmi, R. W. (1944). *Biol. Bull. mar. biol. Lab. Woods Hole.* **87**, 96–105.
Wium-Andersen, G. (1970). *Ophelia* **8**, 267–273.
Woods, K. R., Paulsen, E. C., Engle, R. L. and Pert, J. H. (1958). *Science, N.Y.* **127**, 519–520.
Wright, C. A. (1959a). *J. Linn. Soc. Lond.* **44**, 222.
Wright, C. A. (1959b). *Bull. serol. Mus. New Brunsw.* No. 21, 8.
Wright, C. A. (1959c). *Proc. 6th Intern. Congr. trop. Med. Mal., Lisbon 1958.* **2**, 38–42.
Wright, C. A. (1960). *Ann. trop. Med. Parasit.* **54** (1), 1–7.
Wright, C. A. (1962). *In* 'Bilharziasis', Ciba Found. Symp. (Wolstenholme and O'Connor, eds.), pp. 103–120.
Wright, C. A. (1964). *Proc. zool. Soc. Lond.* **142** (2), 371–378.
Wright, C. A. (1966a). *J. Helminth.* **40**, 403–412.
Wright, C. A. (1966b). *Int. Rev. gen. exp. Zool.* **2**, 1–42.
Wright, C. A. (1971). *Phil. Trans. Roy. Soc. Lond.* B. **260**, 299–313.
Wright, C. A. and File, S. K. (1968). *Comp. Biochem. Physiol.* **27**, 871–874.
Wright, C. A. and Klein, J. (1967). *J. Zool. Lond.* **151**, 489–495.
Wright, C. A. and Ross, G. C. (1959). *Trans. R. Soc. trop. Med. Hyg.* **53** (4), 308.
Wright, C. A. and Ross, G. C. (1963). *Ann. trop. Med. Parasit.* **57**, 47–51.
Wright, C. A. and Ross, G. C. (1965). *Bull. Wld. Hlth. Org.* **32**, 709–712.
Wright, C. A. and Ross, G. C. (1966). *Bull. Wld. Hlth. Org.* **35**, 727–731.
Wright, C. A., Harris, R. H. and Claugher, D. (1957). *Nature, Lond.* **180**, 1489.
Wright, C. A., File, S. K. and Ross, G. C. (1966). *Ann. trop. Med. Parasit.* **60**, 522–525.
Wright, C. A., Klein, J. and Eccles, D. H. (1967). *J. Zool. Lond.* **151**, 199–209.

7 Protozoa and Parasitic Helminths

T. K. R. BOURNS

Department of Zoology,
University of Western Ontario, London, Canada

I. Introduction

The editor of a recently published book on parasite taxonomy wrote in his introduction, 'The past few years have seen an upsurge of interest in systematics, particularly in the higher categories.—The fact that young persons are electing careers in the systematics of various groups is a healthy sign.—' In his conclusion 122 pages later, the same editor wrote, 'One of the major problems is the paucity of young persons electing careers in parasite systematics.' Now I have no desire to embarrass a colleague, yet this little slip is noteworthy not only for its humour but also because it draws attention to the fact that we do not know the state of affairs relating to taxonomy, especially immuno- and biochemical taxonomy, of groups like the protozoa and parasitic worms. The literature (Leone, 1968) shows quite clearly that little has been done in applying serological and biochemical methods to taxonomic problems of the protozoa and parasitic helminths, but more revealing still is the fact that a rather small percentage of even this literature concerns studies made in the last 25 years. It is true that many workers have analyzed worms and protozoa for enzymes, antigens, etc. (e.g. Tanner and Gregory, 1961; Baisden and Tromba, 1963; Huntley and Moreland, 1963; Kent, 1963; Fisher, 1964; Capron *et al.*, 1965; Damian, 1966; Ginger and Fairbairn, 1966; Greichus and Greichus, 1967; Sodeman, 1967) but clearly these have not been taxonomically-oriented studies.

Undoubtedly the fact that parasitism is involved has weighed heavily in this matter. The protozoologist and the parasitologist have had to be pragmatic; they have had compelling reasons for turning their attentions to the diagnostic problems of identification and differentiation as distinct from the more academic pursuit of assessing relative amounts of similarity and difference with a view to determining degrees of relationship. Perhaps this explains in part the paucity of published works, but there must be other reasons as well and there must be reasons for the men and women of these disciplines avoiding immuno- and biochemical taxonomy at a time when workers in other fields are embracing them ever more frequently.

As several writers have pointed out, most of the organisms under discussion are small, none has blood, and relatively few even have pseudocoelomic fluid. Consequently, extracts of entire organisms often must serve as the raw materials from which constituent antigens are derived. Such mixtures are far from ideal, yet, even coupled with the reminders of Desowitz (1966) that a variety of false results are likely to occur and mislead the unwary, this does not seem to be such an immense handicap. What special problems attend these organisms, then, that have discouraged our adventurers? We have known since the work of Canning (1929) that like other metazoa, the worms exhibit tissue- and organ-specific antigens, but special complications did become apparent when we were shown conclusively (Soulsby, 1963; Oliver-Gonzalez and Sala, 1963; Sadun *et al.*, 1965; Capron *et al.*, 1965; and others) that stage-specific substances exist as well. Given a life cycle as complex as that of a digenetic trematode then, with antigenically distinct eggs, cercariae, and adults, how do we select one life stage to be the species standard-bearer?

Moreover, fluids from such sources as hydatid cysts have been shown (Kagan and Norman, 1963) to contain host antigens, at times in more generous supply than parasite antigens. Further, endoparasites themselves may be serologically disguised by wearing host molecules on their surfaces (Smithers and Terry, 1967; Smithers *et al.*, 1969) or may, as a consequence of their particular evolutionary histories, have either come to possess antigens that mimic those of the host (Damian, 1962), or have lost kinds of molecules which had formerly contributed to the chemical disparity between host and parasite (Dineen, 1963a, b). Perhaps these phenomena have been sufficiently troublesome to disenchant the would-be immuno- or biochemical taxonomist.

Considered at the population level many of these organisms can be just as enigmatic. For example, as would be expected of creatures which commonly reproduce asexually and which, even among their sexual forms frequently permit self-fertilization so that mutants quickly find phenotypic expression at the population level (Smyth and Smyth, 1964), the protozoa and platyhelminths exist in myriads of eco-strains or eco-races. The taxonomist, therefore, is called upon to select not only the chemical substances but also the populations which are to represent the species.

Finally, the matter of getting together enough material to permit one to utilize the methods of 'molecular taxonomy' presents a greater problem to the protozoologist and parasitologist than it does to most other biologists. Goldman (1960), for example, has calculated that 100 culture tubes of *Entamoeba histolytica* would provide only 0·054 ml wet volume of organisms.

Having said that little work has been done and having presented as explanatory factors for this dearth, a series of hazards and difficulties inherent in the field of endeavor, I am compelled to ask whether or not this chapter is entitled to existence. The fact that I persevere rests partly upon the existence of a few works which are first rate, partly upon the message imparted by Crites (1969) that taxonomy makes other work possible while at the same time the fruit of the other work becomes the fabric of tomorrow's taxonomy, and finally upon the hope that it may be possible to bring the skimpiness of the literature into focus against a backdrop of promising methods and potentially fruitful approaches.

II. Methods

Given that the stuff of which organisms are made is basic to their natures (Boyden, 1953) and, consequently, that the study of this stuff may reasonably be expected to provide evidence for relationship, it becomes apparent that 'molecular taxonomy' has really only two major ways to proceed. (1) One may concentrate upon specifically-recognizable entities such as amino acids, sugars, fatty acids, etc., by 'calling the roll' as it were, listing those found to be present in a given species and comparing this list with those derived from other organisms, or (2) one may study polymorphic substances such as proteins by comparing the structure and/or the activity of homologous chemicals from two or more kinds of organisms.

Most of the 'roll call' studies have been made possible by the development of simple and inexpensive apparatus and by the publication of explicit instructions (see Bradshaw, 1966 for a general introduction; Crowle, 1961; Kagan and Pellegrino, 1961; Goldman, 1968; Clausen, 1969; and Campbell *et al.*, 1970, for serological methods; Wieme, 1965, and Gordon, 1969, for electrophoresis; and Stock and Rice, 1967; and Fischer, 1969, for chromatography). It has recently become possible for the most modestly prepared of us to become 'modern experimental biologists'. Indeed, we can even buy 'instant pH 8·4' in the form of a prepared buffering mixture to which we merely add water! I say this at the risk of sounding flippant because dangers lurk in the very simplicity of the whole affair. A few years ago many biologists shied away from serological and biochemical methods because they were expensive and because they sounded frighteningly sophisticated and difficult. Today just the reverse is true. Instead of seeming to be harder to practice than they really are they are now presented as being easier than they really are. Whereas we used to 'stay away in droves' because we were fearful of the 'black arts' of the chemist and immunologist, many of us summoned our courage and tried our hands at the new techniques of 'molecular taxonomy' but found that the 'simple' recipes didn't work for us and like brides weeping over burnt biscuits vowed never again to involve ourselves in chemistry, but to stick to Zoology where we belong. Each time this happens a scientist and his science loses something needlessly because it *is* possible to use serological and biochemical techniques as tools without becoming a serologist or a biochemist. One must, however, realize that each new biological system is precisely that—a *new* biological system, so the chemical or immunological technique may have to be modified slightly in order to 'work'. It is at this point that the neophyte requires personal help. The authors of his recipe books failed to anticipate his particular problem; the jump from recipe book to theoretical treatise is immense. He will flounder if left alone; he can make suitable modifications easily with the advice of one whose expertise lies in the appropriate direction. It follows, then, that one of the first steps to take when embarking upon these studies is to cultivate the acquaintance of an expert.

III. 'Roll Call' Methods

The methods used to execute what I term 'roll call' studies consist mainly of chromatography, electrophoresis, and immunoprecipitin tests (see the review by Wright, 1966) and histochemistry. Amino acids, either free or derived from hydrolyzed proteins, have been the substances most frequently studied by chromatography although peptides resulting from the 'finger-print' technique (Ingram, 1958) and sugars have also been employed (Table I). Whole proteins figure in most of the other studies although polysaccharides and lipids are also represented. (Table II.)

The entire battery of 'roll call' methods may, in one sense, be viewed as having restricted taxonomic value. If a study showed that species A and B share 20 amino acids, A and C 15, and B and C 13, one would suggest that A and B may be more similar to one another than either is to C. However, given a situation in which species A and B share substance *x*, A and C share substance *y* but not substance *x*, and B and C share substance *z* but not the others, and seeing clearly that there is no way of ranking the substances *x*, *y*, and *z*, it is obvious that the tests merely confirm the fact that A, B, and C differ one from the other. While such evidence may be of little value at most taxonomic levels it may be very useful at the extremes. At the population level, for example, it may be possible to determine whether or not hybridization has occurred. At the other extreme the universal presence or absence of certain substances may indicate broad relationships. Lyons (1964), for example, pointed out that the distribution of cystine sulphur in the hooks and sclerites of monogenetic flukes was consistent at the ordinal level and she was even able to suggest that its universal presence in larval hooks indicates affinity between the monogeneans and the cestodes.

It was pointed out above that ranking of chemical constituents is not possible. One cannot know if the presence or absence of one amino acid is more important or more fundamental than the presence or absence of another. To the classical taxonomist, then, this type of evidence may be of limited value. On the other hand if it were possible to execute a thorough inventory

Table I
Selected Reports Concerning Taxonomic Study of Helminths by Chromatography

Organisms	Constituents	Authors	Date
Anoplocephalids (3 spp)	Nitrogen, amino acids	Campbell	1958
Neoechiorhynchus (2 spp)	Peptides (finger-printing)	Dunagan	1961
Hemenolepis (3 spp, 3 strains)	Glucose, glycogen, protein, phospholipid, lipid	Huffman	1964
Corallobothrium (3 spp)	Amino acids	Timmons	1963
Ascarids (3 spp)	Amino acids	Krvavica *et al.*	1964
Cestodes (7 spp)	Amino acids	Goodchild and Dennis	1966
Strongyles (7 spp)	Amino acids	Herlich	1966
Taenia (3 spp)	Amino acids	Morseth	1966

Table II
Selected Reports Concerning Taxonomic Study of Helminths by Electrophoresis, Serology and Histochemistry

Organisms	Constituents and/or method	Author	Date
Helminths (7 spp)	Polysaccharides	Campbell	1937
Acanthocephala	Alk. Glycerophosphatase	Bullock	1958
Fasciola, Cotylophoron	Haemoglobin	Goil	1961
Cestodes	Histochemistry	Coil	1963
Cestodes	Histochemistry	Waitz and Schardein	1964
Parasites (review)	Haemoglobin	Lee and Smith	1965
Monogenea, Cestodes	Histochemistry, X-ray	Lyons	1964
Schistosomes (human)	Antigens	Capron *et al.*	1967

of the types of ultimate building blocks of which members of several species are made, surely we would have just the type of data that is conducive to analysis by the numerical taxonomist. The subject of numerical taxonomy (Sokal and Sneath, 1963, discussed in a parasitological context by Mettrick, 1964–5) has been largely avoided by protozoologists and parasitologists not, I think, because we are particularly cantankerous or reactionary but because the time has not been ripe. I would suggest that techniques and apparatus now or soon to be available, will enable us to utilize minute samples and to perform sufficient numbers of replications to enable us to study in great detail and to list completely the chemical building blocks of which protozoa and helminths are composed. When such studies are being planned, even though taxonomy may not be a prime motivating force, investigators should realize that their results will be splendid grist for the mill of numerical taxonomy and that they would be well advised to collect and record their data with this in mind.

IV. Comparative Serology

It should be apparent by this time that most biochemical methods and some immunological methods lead to the 'roll call' approach and thus, in my opinion, find limited application in present day taxonomy. Other immunological, or more correctly serological methods, on the other hand, have quite different potentialities. The reason is simply that instead of providing lists of attributes, these methods compare in an indirect way the amount of structural similarity between the macromolecules of a given physico-chemical type which were obtained from two or more kinds of organisms. The actual tests which have been used taxonomically have usually been variations of the precipitin system taking place in aqueous media (ring-test or turbidity test), or those in gels after single or double diffusion and sometimes preceded by electrophoretic fractionation, or as an envelope which forms around eggs, cercariae, the body openings of worms, or in some cases around entire worms. Somewhat less frequently used have been the agglutination tests (small organisms or antigen-coated particles of otherwise inert material), tests based upon immobilization of motile forms, and the use of tagged antibodies.

Bernheimer and Harrison (1940) studied four strains of *Paramecium multimicronucleatum* and one strain of both *P. aurelia* and *P. caudatum* using immobilization after 2 hours at 28° as their test criterion. These conditions produced a test of such delicacy that no inter-specific results at all were obtained and in some cases reactions were strain-specific, other strains of the same species producing negative results. Noguchi (1926) in a classical piece of work showed that the agglutination test, the complement fixation test, and fermentation tests were mutually supporting when applied to herpetomonads and leishmanias. Not only were the morphologically similar herpetomonads and culture-form leishmanias distinguished clearly from one another, and the morphologically dissimilar forms of one species taken from plant latex and the insect vector shown to be identical, but *L. infantum* was shown to be closely related to *L. donovani*.

A new 'twist' was put upon the quantitation of serological tests when Goldman (1960) designed a microfluorimeter which enabled him and his colleagues (Goldman *et al.*, 1960) to measure as average brightness per unit area, the fluorescence emanating from fluorescein-conjugated anti-*Entamoeba histolytica* when reacting with four strains of *E. histolytica*, or one strain each of *E. hartmanni*, *E. coli*, and *E. moshkovski*. Not only was *E. hartmanni* at least as unlike *E. histolytica* (strain 22) as were *E. coli* and *E. moshkovskii*, but the non-pathogenic Huff strain of *E. histolytica* was less like strain 22 than were the pathogenic strains K9 and M18.

The most ambitious study of helminths by serological means was undertaken by Wilhelmi (1940) who performed precipitin ring tests with saline solutions of fat-extracted lyophilized worms. Degrees of relationship were expressed in terms of their percentage of the homologous titre, the latter being based not upon nitrogen content but upon the amount (dry wt.) of antigen in solution. Wilhelmi was so enthused by his results that he proposed a tentative new definition for the helminth species—namely, 'a group of organisms the lipid-free antigen of which, when diluted to 1:4000 or more, yields a positive precipitin test within one hour with a rabbit antiserum produced by injecting 40 mg of dry-weight, lipid-free antigenic material and withdrawn ten to twelve days after the last of four intravenous injections administered every third day'. By his methods Wilhelmi showed that life stages of three trematodes and one cestode were not serologically distinct; he was able to place several trematode families and cestode genera with respect to one another, but he did not get cross-reactions between flukes and tapeworms.

More recently Vogel (1953) completed a study on helminths using methods very much like those of Wilhelmi, and Warren (1947) studied species and group specificity of filarioid nematodes. Abadie (1963), combining double diffusion precipitin tests in agar with paper chromatography and paper electrophoresis after absorption of antiserum with the polysaccharide antigen, studied the relative placement of four species of *Strongyloides*, and Le Flore (1961, 1965) was able to make suggestions about some trematode family relationships on the basis of his studies on the actions of 'immune' sera on cercariae.

V. The Future

It is perfectly clear that for good reasons, biochemical and serological methods have not been widely used in the taxonomic study of protozoa and parasitic helminths. I predict, however, that change is imminent, for recent technological advances have removed or reduced most of the former obstacles. Whereas it used to be difficult to gather enough material for study, and to obtain it free from contamination, it will not be long before we shall be able to harvest specimens of many species from *in vitro* cultures (Taylor and Baker, 1968). Moreover, we shall be able to create stockpiles of material by using the new apparatus and methods for freezing and for freeze-drying biologicals. In turn, the crude extracts which we derive from our organisms can now be fractionated easily, gently, and inexpensively, by gel chromatography. And finally, most analytical and serological methods have already been refined to the point where samples are measured in microliters.

If it is true that major gains in knowledge are made at times which are dictated by the conditions in ancillary disciplines, then surely we must expect significant advances very soon, for the stage is set.

References and Bibliography

Abadie, S. H. (1963). *Diss. Abstr.* **24** (6), 2623.
Baisden, L. A. and Tromba, F. G. (1963). *J. Parasit.* **49** (3), 375–379.
Bernheimer, A. W. and Harrison, J. A. (1940). *J. Immunol.* **39**, 73–83.

Boyden, A. (1953). *Syst. Zool.* **2**, 19–30.
Bradshaw, L. J. (1966). 'Introduction to Molecular Biological Techniques'. Prentice-Hall, Englewood Cliffs, N.J., U.S.A.
Bullock, W. L. (1958). *Expl. Parasit.* **7** (1), 51–68.
Campbell, D. H. (1937). *J. Parasit.* **23**, 348–353.
Campbell, D. H., Garvey, J. S., Cremer, N. E. and Sussdorf, D. H. (1970). 'Methods in Immunology', 2nd Edn. Benjamin, New York and Amsterdam.
Campbell, J. W. (1958). *Diss. Abstr.* **19** (5), 942–943.
Canning, G. A. (1929). *Am. J. Hyg.* **9**, 207–226.
Capron, A., Biguet, J., Rose, F. and Vernes, A. (1965). *Ann. Inst. Pasteur, Paris* **109** (5), 798–810.
Capron, A., Vernes, A., Biguet, J. and Tran Van Ky, P. (1967). *Ann. Soc. Belge. Med. trop.* **47** (2), 127–139.
Clausen, J. (1969). 'Immunochemical Techniques for the Identification and Estimation of Macromolecules'. North-Holland Publishing Co., Amsterdam and London.
Coil, W. H. (1963). *Proc. helminth. Soc. Wash.* **30** (1), 111–117
Crites, J. L. (1969). *In* 'Problems in Systematics of Parasites' (G. D. Schmidt, ed.), pp. 77–87. University Park Press, Baltimore and Manchester.
Crowle, A. J. (1961). 'Immunodiffusion'. Academic Press, New York and London.
Damian, R. T. (1962). *J. Parasit.* **48** (2, Sect. 2), 16.
Damian, R. T. (1966). *Expl. Parasit.* **18** (2), 255–265.
Dineen, J. K. (1963a). *Nature, Lond.* **197** (4864), 268–269.
Dineen, J. K. (1963b). *Nature, Lond.* **197** (4866), 471–472.
Desowitz, R. S. (1966). *Med. J. Malaya* **21** (1), 35–40.
Dunagan, T. T. (1961). *J. Parasit.* **47** (4, Sect. 2), 30.
Fischer, L. (1969). 'An Introduction to Gel Chromatography'. North-Holland Publishing Co., Amsterdam and London.
Fisher, J. S. (1964). *Diss. Abstr.* **25** (3), 2111.
Ginger, C. D. and Fairbairn, D. (1966). *J. Parasit.* **52** (6), 1086–1096.
Goil, M. M. (1961). *Z. Parasitenk.* **20** (6), 572–575.
Goldman, M. (1960). *Expl. Parasit.* **9**, 25–36.
Goldman, M. (1968). 'Fluorescent Antibody Methods'. Academic Press, New York and London.
Goldman, M., Carver, R. K. and Gleason, N. N. (1960). *Expl. Parasit.* **10**, 366–388.
Goodchild, C. G. and Dennis, E. S. (1966). *J. Parasit.* **52** (1), 60–62.
Gordon, A. H. (1969). 'Electrophoresis of Proteins in Polyacrylamide and Starch Gels'. North-Holland Publishing Co., Amsterdam and London.
Greichus, A. and Greichus, Y. A. (1967). *Expl. Parasit.* **21** (1), 47–52.
Herlich, H. (1966). *Proc. helminth. Soc. Wash.* **33** (1), 103–105.
Huffman, J. L. (1964). *Diss. Abstr.* **24** (11), 4878–4879.
Huntley, C. C. and Moreland, A. (1963). *Am. J. trop. Med. Hyg.* **12** (2), 204–208.
Ingram, V. M. (1958). *Biochim. Biophys. Acta* **28**, 539–545.
Kagan, I. G. and Norman, L. (1963). *Ann. N.Y. Acad. Sci.* **113** (Art. 1), 130–153.
Kagan, I. G. and Pellegrino, J. (1961). *Bull. World Health Org.* **25** (4/5), 611–674.
Kent, N. H. (1963). *Am. J. Hyg. Monogr. Ser.* No. 22, pp. 30–45.
Krvavica, S., Maloseja, Z. and Lui, A. (1964). *Vet. Arh.* **34** (7/8), 167–169. (Not seen).
Lee, D. L. and Smith, M. H. (1965). *Expl. Parasit.* **16** (3), 392–424.
LeFlore, W. B. (1961). *J. Parasit.* **47** (6), 899–904.
LeFlore, W. B. (1965). Ph.D. Thesis, Dept. of Biology, University of California.

Leone, C. A. (1968). *Bull. serol. Mus., New Brunsw.* No. 39, pp. 1–23.
Lyons, K. M. (1964). *Parasitology* **54** (4), 12P.
Mettrick, D. F. (1965). *Revta. Biol. Lisb. Yr.* 1964–65, **5** (1/2), 127–134.
Morseth, D. J. (1966). *Expl. Parasit.* **18** (3), 347–354.
Noguchi, H. (1926). *J. exp. Med.* **44**, 327–337.
Oliver-Gonzalez, J. and Sala, A. R. De. (1963). *Amer. J. trop. Med. Hyg.* **12** (4), 539–540.
Sadun, E. H., Schoenbechler, M. J. and Bentz, M. (1965). *Am. J. trop. Med. Hyg.* **14** (6), 977–995.
Smithers, S. R. and Terry, R. J. (1967). *Trans. Roy. Soc. trop. Med. Hyg.* **61**, 517–533.
Smithers, S. R., Terry, R. J. and Hockley, D. J. (1969). *Proc. Roy. Soc. (Lond.) B* **171**, 483–494.
Smyth, J. D. and Smyth, M. M. (1964). *Parasitology* **54** (3), 493–514.
Sodeman, W. A. (Jr.) (1967). *Am. J. trop. Med. Hyg.* **16** (5), 591–594.
Sokal, R. R. and Sneath, P. H. A. (1963). 'Principles of Numerical Taxonomy'. W. H. Freeman and Co., San Francisco.
Soulsby, E. J. L. (1963). *Am. J. Hyg. Monogr. Ser.* No. 22, pp. 47–57.
Stock, R. and Rice, C. B. F. (1967). 'Chromatographic Methods 2nd Edn'. Chapman and Hall, London.
Tandon, R. S. (1968). *Z. Parasitenk.* **30** (2), 149–151.
Tanner, C. E. and Gregory, J. (1961). *Can. J. Microbiol.* **7** (4), 473–481.
Taylor, A. E. R. and Baker, J. R. (1968). 'The Cultivation of Parasites *In Vitro*'. Blackwell, Oxford and Edinburgh.
Timmons, H. F. (1963). *Diss. Abstr.* **24** (6), 2630.
Vogel, H. (1953). *J. Immunol.* **70**, 503–506.
Waitz, J. A. and Schardein, J. L. (1964). *J. Parasit.* **50** (2), 271–277.
Warren, V. G. (1947). *Am. J. Hyg.* **45**, 299–304.
Wieme, R. J. (1965). 'Agar Gel Electrophoresis'. Elsevier, Amsterdam, London, New York.
Wilhelmi, R. W. (1940). *Biol. Bull. mar. biol. Lab., Woods Hole* **79**, 64–90.
Wright, C. A. (1966). *Int. Rev. gen. exp. Zool.* **2**, 1–42.

8 Microcomplement Fixation

A. B. CHAMPION, E. M. PRAGER, D. WACHTER and A. C. WILSON

Department of Biochemistry, University of California, Berkeley, California, U.S.A.

I. Introduction

The phenomenon of complement fixation has been known for over seventy years. Although various complement fixation techniques have long been used for detecting and measuring the reaction between antigen and antibody, the quantitative microcomplement fixation technique was introduced rather recently for studying reactions between soluble antigen and antibody at very low concentrations (Wasserman and Levine, 1961). This technique is particularly valuable for taxonomists and evolutionary biologists because it provides an extremely sensitive and economical way of detecting differences between related antigenic proteins. Furthermore, microcomplement fixation provides a way of estimating the approximate degree of amino acid-sequence difference between related proteins (Prager and Wilson, 1971a,b).

Although several descriptions of the microcomplement fixation technique are available (Fiset, 1964; Levine, 1967; Levine and Van Vunakis, 1967),

Abbreviations: Ab, antibody; Ag, antigen; As, antiserum; A_{413} and A_{541}, absorbance at 413 nm and 541 nm, respectively; C′, complement; MC′F, microcomplement fixation; SRBC, sensitized sheep red blood cells; ID, index of dissimilarity.

none deals specifically with the application of this technique to the measurement of cross reactions. We describe here how this technique is used in our laboratory to measure quantitatively the relatedness of proteins from different taxa.

II. Principle of the Method

Complement is a series of sequentially acting components found in vertebrate serum. Complement fixation techniques make use of two properties of complement. One is the ability of complement to bind irreversibly to antigen–antibody complexes. The second is the ability to lyse sensitized red blood cells. Experimentally, if complement is added to antigen (Ag) and antibody (Ab) in solution under suitable conditions, it will become fixed within the three-dimensional latticework of Ag–Ab complexes as they are formed (reaction 1). After an appropriate time of incubation, sensitized red blood cells (SRBC) are added, and any complement (C′) not fixed by the Ag–Ab complexes is available to lyse the cells (reaction 2).

1. $Ag + Ab + C' \longrightarrow AgAbC'$ aggregate + residual C′
2. Residual $C' + SRBC \longrightarrow$ Lysed cells

To determine the amount of lysis, unlysed cells are removed by centrifugation and the concentration of hemoglobin in the supernatant fluid is measured spectrophotometrically. Complement fixation is thus inversely proportional to the hemoglobin concentration.

III. Experimental Procedure

A. EQUIPMENT

1. Reaction tubes: 29 × 116 mm (40 ml) glass centrifuge tubes with a pour-spout; new tubes are washed thoroughly* and then conditioned by adding 6 ml of a solution containing 5 parts of diluent and one part of a 1/240 dilution† of C′ and incubating overnight at 0° to 4° C.
2. Pipettes: volumetric (1 ml); Ostwald-Folin (1 ml); measuring (0·1–10 ml); and serological (0·1–10 ml).
3. Automatic pipettes: Cornwall type, 2-ml and 5-ml automatic pipetting outfits, available from Becton, Dickinson and Co., Rutherford, N.J.
4. Metal racks for 24 reaction tubes, available from the Will Corporation, Rochester, N.Y.

* In this lab new tubes are cleaned by the following procedure. The tubes are soaked in cleaning solution, repeatedly rinsed in tap water, soaked in 1N HCl, repeatedly rinsed with tap water, and then with distilled water. After a day's experiment the tubes are washed by 3 to 4 rinses with distilled water.

† A $1/x$ dilution is used here to mean that one volume of reagent has been diluted to a final volume of x.

5. Centrifuge (unrefrigerated) capable of spinning 64 test tubes (13×100 mm); e.g. International Centrifuge size 2, Model 5, with a number 240 8-carrier head and eight 8-tube carriers.
6. Other standard equipment includes a large thermoregulated water bath, a refrigerator or cold room, plastic dishpans for chilling racks of reaction tubes, and a spectrophotometer.

B. PREPARATION OF REAGENTS

1. Diluents

Either isotris or isosatris buffers are used as diluents for all reagents in MC′F experiments. The only difference between the two is that isosatris contains bovine serum albumin (BSA). BSA (Fraction V) appears to have a stabilizing effect on some proteins (e.g., hemoglobin and lactic dehydrogenase) at high dilution. It is therefore advisable to try each diluent at the outset to determine whether BSA is needed. BSA should, of course, be omitted from experiments in which albumins are used as antigens. The first step in preparing these diluents is to make a stock solution containing NaCl (81·6 g) and Tris base (12·1 g). Adjust to pH 7·45 (at 25°C) with concentrated HCl (about 6·6 ml), and then bring to a final volume of 2·0 liters with distilled water. Isotris is prepared by combining the stock solution (400 ml) with $MgSO_4$ (0·15M, 6·66 ml), $CaCl_2$ (0·1M, 3·0 ml) and enough distilled water to make the final volume 2 liters. Isosatris is prepared in a similar way, except that BSA (2 g) is added; the pH must in this case be adjusted to pH 7·45 before bringing the solution to the final 2 liter volume.

2. Antiserum

MC′F experiments are generally done with antisera prepared in rabbits, although antisera prepared in other mammals (e.g. rat, man, and cow) can also be used. As discussed below, it is desirable in taxonomic work to use antisera prepared by immunizing rabbits for at least three months with a highly purified protein (Prager and Wilson, 1971a). We recommend that no MC′F experiments be done with antisera to protein mixtures.

Before use in MC′F experiments, antiserum is heated at 60°C for 20 minutes to inactivate rabbit complement. The treated serum is centrifuged at $30{,}000 \times g$ for 10 min, and the supernatant solution is collected and stored at −15°C. In this state rabbit antiserum is stable for decades.

To obtain optimal complement fixation curves it is generally necessary to dilute the antiserum by a factor of at least 1,000. For convenience a 1/100 dilution of the heat-treated antiserum is made and is used as a stock solution from which further dilutions are made. This stock solution is stored at −15°C.

3. *Antigen*

The various antigens are purified for immunization purposes by standard methods and stored at high concentration in the cold (5°C or −15°C). For MC′F experiments, however, impure antigen preparations can often be used provided that the antiserum was prepared against highly purified antigen (Wilson *et al.*, 1964; Sarich and Wilson, 1966). Maximum fixation of C′ generally occurs when the amount of antigen per tube is in the range of 20 to 100 ng (Levine and Van Vunakis, 1967). To ensure that this range is covered, a solution containing about 500 ng of antigen per ml is made and from it a geometric series of two-fold dilutions prepared, e.g. 500, 250, 125, 62·5, 31, and 16 ng/ml. Although six serial two-fold dilutions will provide a suitable range of antigen concentrations for most MC′F experiments, it is desirable with the lysozyme immune system to use seven serial 1·5-fold dilutions, while in the case of ovalbumin six serial 3-fold dilutions are required.

4. *Guinea Pig Complement*

Lyophilized complement (available from Hyland Laboratories, Costa Mesa, California) is stable for months at 4°C. Dissolved and diluted complement can deteriorate somewhat in a week at 4°C so that the concentration may have to be adjusted every few days in order to maintain a consistent rate of lysis. The stability of dissolved complement can be improved by storage at −15°C when not in regular use. Different batches of complement, and particularly complement from different suppliers, can vary considerably in both titer and stability.

The lyophilized powder is dissolved in 3 ml of distilled water. A typical dilution of freshly dissolved complement for experiments is 1/240. At this dilution lysis of sensitized red blood cells occurs in 25–35 min under standard conditions.

5. *Red Blood Cells*

Sheep blood, preserved in modified Alsever's solution and available from the Bennett Ranch Laboratory, Woodland, California, is stored at 4°C. At this temperature the red cells are usually stable for at least two weeks.

6. *Hemolysin*

Hemolysin, an antiserum obtained by immunizing rabbits with sheep red blood cells, is available from Baltimore Biological Laboratories, Baltimore, Maryland. This commercial preparation is a mixture of antiserum with an equal volume of glycerol and is stable indefinitely at −15°C.

7. *Sensitized Cells*

To prepare sensitized cells one must first mix an appropriate dilution of hemolysin and a suspension of washed sheep red blood cells. Prepare the dilute suspension of cells as follows. Centrifuge 5 ml of sheep blood in Alsever's

Table I
MC′F Protocol and Results

Tube	Diluent ml	Antiserum (1 ml) As	Antiserum (1 ml) Conc.	C′ ml	Antigen (1 ml) Ag	Antigen (1 ml) ng	Results A_{413}	Results % C′ Fixed
1	3·0	Anti-	1/4,000	1·0	Chicken	1,500	0·470	25
2		chicken			oval-	500	0·424	32
3		oval-			bumin	167	0·200	70
4		bumin				55·5	0·178	74
5		M19-11*				18·5	0·397	37
6			↓			6	0·550	11
7			1/4,600			1,500	0·603	4
8						500	0·568	10
9						167	0·267	59
10						55·5	0·258	61
11						18·5	0·445	30
12			↓			6	0·576	8
13			1/5,200			1,500	0·625	7
14						500	0·603	4
15						167	0·405	37
16						55·5	0·283	57
17						18·5	0·422	34
18			↓			6	0·550	13
19			1/5,800			1,500	0·618	3
20						500	0·625	2
21						167	0·431	33
22						55·5	0·352	46
23						18·5	0·461	28
24	↓		↓		↓	6	0·600	6
25	4·0		1/4,000		—	—	0·620	—
26			↓		—	—	0·613	—
27			1/4,600		—	—	0·610	—
28			↓		—	—	0·643	—
29			1/5,200		—	—	0·640	—
30			↓		—	—	0·615	—
31			1/5,800		—	—	0·638	—
32		↓	↓		—	—	0·632	—
33		—	—		Chicken oval-	1,500	0·642	—
34	↓	—	—		bumin	↓	0·641	—
35	5·0	—	—		—	—	0·634	—
36	↓	—	—	↓	—	—	0·647	—
37	4·0	—	—	2·0	—	—	0·750	—
38	↓	—	—	↓	—	—	0·742	—
39	6·0	—	—	—	—	—	0·020	—
40	↓	—	—	—	—	—	0·020	—

Conditions:

Incubation time at 4° C—16 hr; time for lysis of 2C′ control—16 min; incubation time at 35° C—34 min; C′ concentration—1/240; diluent—isotris buffer, pH 7·45; % lysis in As controls

$$= \frac{0{\cdot}626 - 0{\cdot}020}{0{\cdot}746 - 0{\cdot}020} \times 100 = 83{\cdot}4\%.$$

* M19-11 refers to antiserum obtained from rabbit M19 after 11 weeks of immunization.

solution at 3,000 rpm for 3 min. Decant the supernatant solution and suspend the cells in 5 ml of isosatris; repeat the washing cycle three more times. After the fourth centrifugation, dilute one ml of the cell pellet 1/18 with isosatris. Then lyse one ml of the dilute suspension by diluting to 25 ml with water. The A_{541} should be between 0·47 and 0·50; if it is not, adjust the concentration of the 1/18 dilution by addition of more washed cells or isosatris.

Sensitized cells are prepared by mixing the dilute suspension of cells with an equal volume of a 1/250 dilution of commercial hemolysin in isosatris. Add the hemolysin to the cell suspension slowly with continuous swirling. Incubate the mixture at 35° for 15 min and then dilute 1/10 with isosatris. Store the reagent (SRBC) at 4°C and use within 10 hr.

C. PROTOCOL FOR A TITRATION EXPERIMENT

We now describe how to set up a simple MC′F experiment in which an antiserum is tested at four concentrations against a series of six concentrations of the homologous antigen. This experiment involves the use of 40 reaction tubes, each of which receives 7 ml of reagents. The protocol is given in Table I, which shows that each experimental tube (1–24) receives the following five reagents: diluent (3 ml), diluted antiserum (1 ml), diluted complement (1 ml), diluted antigen (1 ml), and sensitized cells (1 ml). Tubes 25–40 are controls which lack antiserum, complement or antigen, the deficit being made up by additional diluent, as shown in Table I.

1. Procedure

Place the tubes in an ice-water bath. Add diluent to all tubes with a 5-ml automatic pipette. Add one ml of the desired antiserum concentration to the appropriate tubes with a volumetric or Ostwald-Folin pipette. Add one ml of a freshly prepared complement solution of proper dilution with a 2-ml automatic pipette to tubes 1–36, and 2 ml to each double complement (2C′) control; each cell control receives none. Swirl the tubes to rinse antiserum and C′ from the walls. Then add one ml of each antigen concentration to the appropriate tubes using an Ostwald-Folin pipette, blowing out the entire contents while swirling the reaction tube continuously. Cover the tubes with Saran wrap or aluminum foil and store them for 16 to 21 hours at 0° to 4°C. After this place the tubes in an ice-water bath, and add 1·0 ml of freshly made sensitized SRBC with a 2-ml automatic pipette. Swirl the tubes and place them in a 35°C water bath. Swirl the tubes again and every 5 min thereafter. Lysis, as seen by the clearing of the cloudy red suspensions, should occur in the 2C′ control tubes between 15 and 20 min. Lysis should begin to occur in the As, Ag and C′ controls between 30 and 35 min. When 80–85% of the cells in the control tubes have been lysed, remove all tubes and immediately place in an ice water bath. The optimal time for removal is readily learned through experience.

After the tubes have cooled, pour the contents into pre-chilled, numbered test tubes (13 × 100 mm) and immediately centrifuge for ten minutes to sediment the unlysed red cells. Some lysis occurs during the chilling of the tubes and subsequent centrifugation, and this fact has to be considered when removing the tubes from the 35°C water bath.

Gently remove the tubes from the centrifuge and place in a rack at room temperature. Carefully decant the upper 3 ml of the supernatant solution in each tube into a cuvette and determine the absorbance at 413 nm (A_{413}). With the pipettes recommended, cumulative pipetting errors should give absorbance values whose reproducibility is within ±5%.

To calculate the percent C′ fixed (y) in each tube, one normally uses the equation:

$$y = \frac{A_{413} \text{ (As control)} - A_{413} \text{ (experimental tube)}}{A_{413} \text{ (As control)} - A_{413} \text{ (cell control)}} \times 100$$

The antiserum control is usually taken as the upper limit of lysis. Usually the As, Ag and C′ controls are very similar, although the former two can vary with respect to the latter (see Section IVB).

The double complement controls (tubes 37 and 38) serve as an indicator of how long lysis at 35°C is going to take, as a standard to judge the extent of lysis in all tubes during incubation, and as a basis for calculation of the percent lysis in the other controls.

The single complement controls (tubes 35 and 36) serve as a standard to judge whether the antiserum controls (tubes 25–32) or antigen controls (tubes 33 and 34) are pro- or anticomplementary (see below).

D. RESULTS

Bell-shaped curves are obtained when the results, given in Table I, are plotted as % complement fixed versus antigen concentration (Fig. 1). The amount of antigen that produces maximum C′ fixation is approximately 100 ng. There is little or no C′ fixation in extreme antigen excess (tubes 1, 7, 13, and 19) or in extreme antibody excess (tubes 6, 12, 18, and 24).

When the peak height of the complement fixation curve is plotted versus the logarithm of the antiserum concentration (Fig. 2), a straight line is obtained with a slope of 200. The slope (m) is defined by the equation

$$Y = m \log X + b$$

where Y = % C′ fixed at peak, X = antiserum concentration and b = the Y intercept. Such a relationship has been observed for all immune systems studied to date (Allison and Kaplan, 1964; Reichlin, Hay and Levine, 1964; Sarich and Wilson, 1966; Arnheim and Wilson, 1967; Stanier *et al.*, 1970;

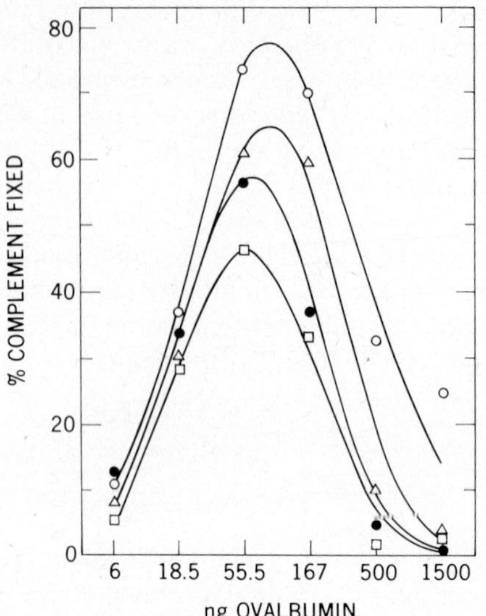

FIG. 1. The effect of antiserum concentration on complement fixation. The antiserum used was rabbit anti-chicken ovalbumin (M19–11). The concentrations were: ○—1/4,000; △—1/4,600; ●—1/5,200; □—1/5,800. The test antigen was chicken ovalbumin.

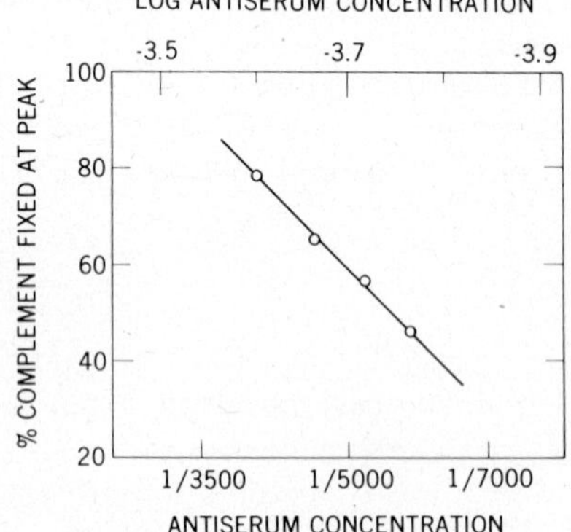

FIG. 2. The dependence of the peak height of the complement fixation curve on the concentration of antiserum. Each point represents the peak height of a complement fixation curve (obtained from Fig. 1) for a particular antiserum concentration.

Nonno *et al.*, 1970; Fink *et al.*, 1970). The slope varies from 100 to 380 depending on the immune system. (See Table II.)

The factor by which the antiserum must be diluted in order to give 75% fixation at the peak of the complement fixation curve is called the titer.* For the antiserum used in the experiment described here, the titer is 4,200 (Fig. 2).

* Some workers use 50% fixation instead of 75% fixation in defining the titer of an antiserum.

Table II
Dependence of Peak Height on Antiserum Concentration for Several Immune Systems

Protein	Source	Slope	Reference
F-I histones	laboratory rat (*Rattus*)	100	Bustin and Stollar (1973)
Muconolactone isomerase	*Pseudomonas putida*	190	Stanier *et al.* (1970)
Ferredoxin	*Clostridium acidi-urici*	195	Champion (1972)
Ovalbumin	chicken (*Gallus*)	200	Wachter and Wilson (unpublished results)
Lysozyme	chicken (*Gallus*)	205	Arnheim and Wilson (1967)
Serum albumin	alligator	220	Nakanishi (1971)
Muconate lactonizing enzyme	*Pseudomonas putida*	255	Stanier *et al.* (1970)
Serum albumin	chicken (*Gallus*)	265	Nakanishi (1971)
Glutamic dehydrogenase	chicken (*Gallus*)	265	Wilson (unpublished results)
Heart lactic dehydrogenase	chicken (*Gallus*)	290	Wilson (unpublished results)
Alkaline phosphatase	*Escherichia coli*	300	Cocks (1971)
Muscle triose phosphate dehydrogenase	lobster (*Homarus*)	300	Allison and Kaplan (1964)
Azurin	*Pseudomonas denitrificans*	310	Champion and Doudoroff (unpublished results)
Hemoglobin A_1	human (*Homo*)	310	Reichlin *et al.* (1964) Reichlin and Wilson (unpublished results)
Formyl-THF synthetase	*Clostridium cylindrosporum*	315	Champion (1972)

cont. p. 406

Table II—*cont.*

Protein	Source	Slope	Reference
Serum albumin	human (*Homo*)	360	Sarich and Wilson (1966)
Glycerol 3-phosphate dehydrogenase	bumble bee (*Bombus*)	370	Fink *et al.* (1970)
Serum albumin	ranid frogs	380	Wallace (1971)

This is a convenient measure of the concentration of complement-fixing antibodies in the antiserum.

E. MEASUREMENT OF IMMUNOLOGICAL RELATEDNESS

MC′F is most typically used in this laboratory to measure the degree of immunological difference between a reference protein from one species of organism and the corresponding protein from another species. Although in many immunological methods a fixed antiserum concentration is employed and cross reactions are expressed as a percentage of the reaction given by the reference protein, this is not appropriate for the MC′F method. Figure 3 shows, for example, that rhesus monkey albumin gave no C′ fixation with anti-human albumin at the anti-serum concentration used to demonstrate C′ fixation with human albumin. Rhesus albumin fixed C′ only when the anti-serum concentration was raised.

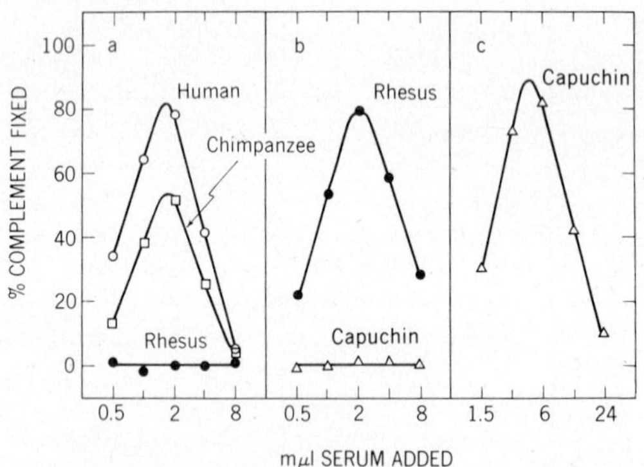

FIG. 3. The effect of antiserum concentration on complement fixation for primate serum albumins. The antiserum, rabbit anti-human albumin (6–15), was tested at three concentrations (a. 1/11,000; b. 1/4,600; c. 1/2,500). As the concentration of albumin in serum is about 40 mg/ml, the amount of albumin required for maximum fixation for human, chimpanzee, rhesus, and capuchin monkey was approximately 70, 70, 80 and 180 ng, respectively. This figure is taken from Sarich and Wilson (1966).

Consequently, in our studies of cross reactions, the antiserum is titrated first against the reference (homologous) protein and the slope determined. Then that same antiserum is titrated against the cross reacting (heterologous) protein. It is generally observed that the slopes given by heterologous proteins are identical to that of the homologous protein, as illustrated in Fig. 4.

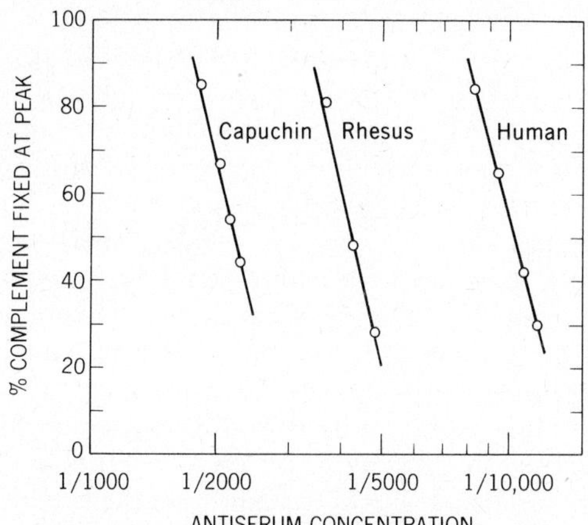

FIG. 4. The dependence of the peak height of the complement fixation curve on the concentration of human serum albumin antiserum. Each point represents the peak height of a complement fixation curve for a particular antiserum concentration. The slopes of the lines (see below) for human, rhesus and capuchin albumins were 360, 410 and 375, respectively. This figure is taken from Sarich and Wilson (1966).

The fact that the slopes are identical provides a basis for measuring the immunological relatedness between homologous and heterologous proteins. One measure of this relationship is called the *index of dissimilarity*. It is defined as the factor by which the antiserum concentration must be raised for a particular heterologous antigen to produce a C′ fixation curve equal in amplitude to that produced by the homologous antigen (Wilson *et al.*, 1964). In the example shown the indices of the serum albumins from chimpanzee, rhesus, and capuchin monkey, relative to human serum albumin, are 1·17, 2·38 and 4·64, respectively (Sarich and Wilson, 1966).

If there are many heterologous antigens to be tested, it is not necessary to determine the slope of the line for each heterologous antigen. Once it is established that the slopes of a few of the heterologous antigens are identical with that of the homologous antigen, the indices of dissimilarity (ID) for the remaining heterologous antigens can be calculated from the following equation:

$$\log \text{ID} = \frac{Y_H - Y_h}{m} + \log \frac{X_h}{X_H}$$

where m = slope,

$Y_H(Y_h)$ = percent complement fixed at the peak used with the homologous (heterologous) antigen,

$X_H(X_h)$ = antiserum concentration used with the homologous (heterologous) antigen.

The derivation of this equation is presented in the Appendix. Alternatively, the index of dissimilarity can be calculated graphically.

The results of cross reaction experiments may also be expressed as units of *immunological distance*. The term immunological distance is defined* as $100 \times \log$ ID. As pointed out below, it now appears probable that immunological distance is directly proportional to the degree of sequence difference among proteins. For this reason, it is desirable in publications to present immunological distance results.

When measuring reactions with heterologous antigens, it is imperative to include the homologous antigen (versus a single antiserum concentration) in the experiment since variations in reaction time and temperature can never be fully controlled, and also since antiserum titers may vary slightly with time.

As a rough estimate of the antiserum dilution to use for a heterologous antigen, the Ouchterlony test may be used to estimate qualitatively the strength of cross reactivity. It has been found that the extent of spur formation (or lack of it) is roughly correlated with the index of dissimilarity (Moore and Goodman, 1968; Stanier *et al.*, 1970; Gasser and Gasser, 1971; Gorman *et al.*, 1971).

IV. Practical Considerations

A. STANDARD EXPERIMENTAL CONDITIONS

At the outset, the operator should establish a set of standard reaction conditions and maintain them for any series of experiments. Practical reaction conditions used in this laboratory are the following: reaction time, 16–21 hr; reaction temperature, 0°–4°C; incubation time, 27–40 min; incubation temperature, 34°–37°C; A_{541} of diluted SRBC reagent, 0·47–0·50; percent lysis in controls, 80–85%.

1. *Reaction Time in the Cold*

The percentage of C′ fixed is dependent on reaction time. As the reaction time increases, the extent of Ag–Ab reaction increases and more complement is fixed. Hence, the titer of the antiserum is increased. If it is desired to alter the reaction time, small adjustments in antiserum concentration can be made.

* The term immunological distance was unfortunately used by Sarich and Wilson (1966) as a synonym for index of dissimilarity. We later discontinued this practice.

2. *Reaction Temperature in the Cold*

Temperature variation, even if only a few degrees centigrade, can have a significant effect on complement fixation; fixation occurs more slowly at lower temperatures. Therefore, maintain the tubes at a constant low temperature by, for example, placing them in an ice-water bath in a refrigerated location.

3. *Incubation Time at 35°C*

It is a matter of convenience to use a C′ concentration that results in an incubation time in the range of 27 to 40 min. When the incubation time is below 30 min (C′ concentration too high), lysis occurs more rapidly; consequently, judging the precise moment to remove the tubes is somewhat more difficult. Incubation times greater than 40 min (C′ concentration too low) are time consuming. If the incubation time is outside the above range, the complement concentration is adjusted accordingly.

4. *Incubation Temperature*

At 35°C, a 1/240 dilution of complement lyses sensitized red cells in a reasonable time and remains active for about 90 min.

5. *Concentration of SRBC*

Double complement controls giving A_{413} values of 0·60 to 0·85 are acceptable. Values outside this range reduce the sensitivity of the method.

6. *Percent Lysis in the Controls*

A range of 80 to 85% lysis in the controls is quite important for obtaining reproducible results. If the reaction tubes are taken out of the 35° bath too soon, less than about 80% lysis will occur; this results in lower absorbance values subject to greater experimental error. If the reaction tubes are left in the bath too long, lysis exceeds 85% in the controls, and the concentration of SRBC thereby becomes limiting. As a result, the rate of lysis in the controls is less than in the experimental tubes. The effect is a decrease in the %C′ fixed in the experimental tubes, with the result that the antiserum titer will appear to have dropped.

B. COMPLEMENTARITY

Theoretically, the antigen, antiserum and complement controls should give the same values in a given experiment, since none contains the complete Ag–Ab system which binds complement. Some antisera and some antigens, however, have the property or retarding or enhancing lysis in the presence of complement only. We refer to the retardation of lysis as *anti-complementarity*, and to the enhancement of lysis as *pro-complementarity*. Antiserum controls may be anti-complementary. This effect increases as the As concentration is increased, and can be a significant problem at the high As concentrations used for weak cross reactions.

Antigen controls can be either anti- or pro-complementary. The effect is greatest at the high Ag concentrations needed in studies of weak cross reactions.* These preparations may contain salts or other molecules that can affect the composition or concentration of the diluent buffer and thereby enhance or retard fixation. Pro-complementary behavior may also be due to molecules that absorb at 413 nm. These problems can usually be overcome by the following procedures.

1. Antiserum Controls

To compensate for the pro- or anti-complementary effect of an antiserum, all experimental tubes containing a particular As dilution are kept in the 35° bath until those antiserum controls exhibit 80 to 85% lysis. For a highly anti-complementary antiserum, this may require incubations of up to 90 min (see p. 409).

2. Antigen Controls

The problems of anti- or pro-complementarity of an antigen usually occur when impure preparations are used. If possible, the antigen should be further purified in lieu of attempting to correct for these effects. Alternatively, antigen controls have to be set up for each concentration for which there is a pro- or anti-complementary effect. If both antiserum and antigen are strongly anti- or pro-complementary, the experiment cannot be done.

3. Reaction Tubes

Occasionally a tube itself may be anti- (or pro-) complementary. This can occur either after the initial wash or after prolonged use. The problem is usually eliminated by soaking the tubes in hot water, followed by several rinses with hot water and then with distilled water. The tubes should then be checked with single complement controls (see above).

C. HEMOLYSIN TITRATION

Since different batches of commercial hemolysin vary in titer by a factor of up to about two, it may be desirable to determine the titer of each batch, rather than assuming that the titer is 1/250. The titration experiment is done with a series of reaction tubes, each containing 5 ml of diluent and one ml of a C′ dilution (1/240). Several preparations of sensitized cells are made, each with a different concentration of hemolysin. Each reaction tube receives one ml of sensitized red cells and the tubes are incubated at 35°C. The hemolysin concentration that produces 85% lysis in 35 min under standard conditions is then used for all future experiments.

* As seen in Fig. 3, it is necessary to increase the antigen concentration when measuring weak cross reactions.

V. Discussion

A. ANTISERUM VARIABILITY

Cross reactivity measurements in MC′F experiments are affected by several parameters of the immunological response of the animals being immunized. As shown in Fig. 5 the index of dissimilarity between chicken and duck A

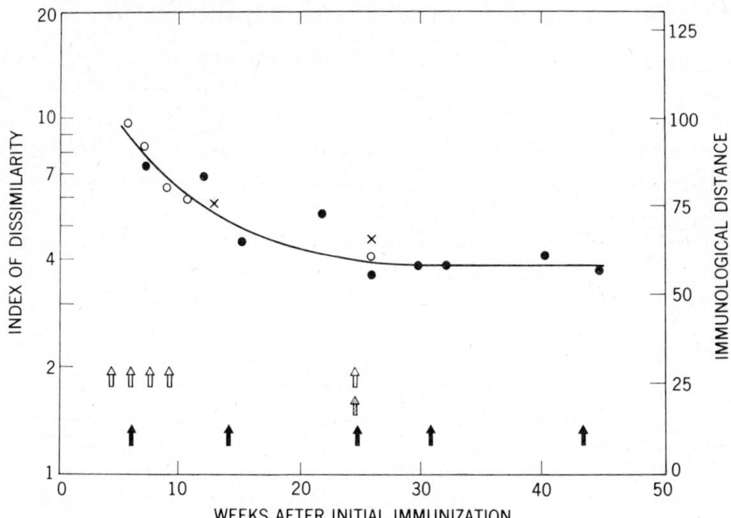

FIG. 5. Dependence of antiserum specificity on length of the immunization period. Three groups of 4 rabbits were immunized with pure chicken lysozyme, each group receiving a different pattern of injections. The arrows indicate the times at which booster injections were given (group I, solid arrows; group II, empty arrows; group III, hatched arrow). One week, or more, after each injection the rabbits were bled and titers determined with chicken lysozyme. The four antisera obtained from each group of rabbits were mixed in inverse proportion to their titers and the resulting pools were tested for reactivity with duck A lysozyme. The indices of dissimilarity are indicated by solid circles (group I), empty circles (group II) or crosses (group III). Regardless of the immunization schedule used, the ability of antisera to discriminate between chicken and duck lysozymes declined during the first three months of immunization and thereafter remained at a constant value of about 4. This figure is taken from Prager (1972).

lysozyme drops as the time course of immunization increases, and reaches a plateau between 20 and 25 weeks. This phenomenon has been observed with other immune systems (Cocks and Wilson, 1972).* Therefore, we recommend that an immunization period of three to four months be used. Furthermore, bleedings should be made seven days after the last intravenous injection (Prager and Wilson, 1971a; Cocks and Wilson, 1972). Individual rabbits also vary significantly in the specificity of the antiserum they produce (Prager and

* A possible explanation for the broadening of specificity is that more avid antibodies are produced with time. Thus, those determinants in the heterologous antigen which do not resemble homologous determinants sufficiently to react with early sera could react with later, more avid sera.

Wilson, 1971a); therefore, at least four rabbits should be immunized with the same antigen, and pools should be prepared by mixing the sera in inverse proportion to their titers, so that each rabbit contributes an equal number of complement fixing antibodies (Prager and Wilson, 1971a). By following these recommendations, the problems of antiserum variability can be minimized.

B. COMPARISON OF MC'F WITH OTHER TECHNIQUES

In order for a heterologous protein to fix complement or precipitate antibodies it must share several antigenic determinants with the homologous antigen (Kabat and Mayer, 1961). Many other immunological techniques, such as hemagglutination, neutralization and radio-immunoassays, require only a univalent antigen. Quantitative microcomplement fixation has two principal advantages over other immunological techniques that require a multivalent antigen, such as the quantitative precipitin method. MC'F uses extremely small amounts of material and is superior in its capacity to discriminate among closely related antigens (Wilson *et al.*, 1964; Prager and Wilson, 1971b). One

Table III
Reciprocal Tests

Protein	Number of reciprocal tests	Mean percentage deviation from average[a]	References
Mammalian serum albumins	300	6	Sarich (unpublished results)
Frog serum albumins	51	7	Wallace *et al.* (1973); Maxson (1973)
Mammalian transferrins	250	8	Sarich and Cronin (unpublished results)
Pseudomonad azurins	13	11	Champion and Doudoroff (unpublished results)
Enterobacterial alkaline phosphatases	6	16	Steffen, Cocks and Wilson (1972)
Bird lysozymes	21	17	Prager and Wilson (1971a)

[a] For each reciprocal test, the percentage deviation from the average immunological distance is calculated by means of the expression:

$$\frac{|A - B|}{A + B} \times 100$$

where A is the immunological distance between proteins P and Q, measured with antiserum to P; and B is the immunological distance between proteins P and Q, measured with antiserum to Q. The mean value of these percentage deviations is then calculated for all pairs of proteins.

uses, for homologous reactions, about 1,000-fold less antigen and antiserum; for weak heterologous reactions, one uses about 30-fold less antigen and antiserum in MC′F than in the precipitin technique. The precipitin technique, however, does have one technical advantage in that one can obtain a rapid measure of cross reactivity over a broad range of antigenic differences, since only one antiserum concentration is used.

An excellent correlation has been demonstrated for cross reactivity as measured by these two quantitative techniques (Sarich and Wilson, 1966; Prager and Wilson, 1971b), and both of these techniques are superior in discriminatory capacity over the qualitative Ouchterlony technique (Prager and Wilson, 1971b).

C. RECIPROCITY

For the quantitative evaluation of the MC′F method of measuring sequence resemblance between proteins, it is important to know how well the results of reciprocal tests agree. If immunological distance were strictly dependent on degree of sequence resemblance, the results of reciprocal tests should agree. Table III compares the results of reciprocal tests conducted with a number of immune systems. The albumin and transferrin values show good reciprocity compared to lysozyme and alkaline phosphatase. In general when working with a new system, reciprocal tests should be conducted.

D. SEQUENCE AND IMMUNOLOGY CORRELATION

As satisfactory reciprocity was observed in the above tests (Table III), it might be anticipated that the immunological distance between two proteins should be related to their sequence difference. Antigenic determinants of proteins are largely determined by the type and arrangement of the surface amino acids (Sela, 1969), and since most molecular evolution has been a surface phenomenon (Dickerson, 1971; Prager and Wilson, 1971b), one would expect that as sequence differences between related proteins increase, the number of antigenic sites in common would decrease. In addition, the structure of the antigenic sites in common would be expected to diverge, with a resultant lowering of the avidity of the antibodies for heterologous sites. Evidence for such decreases in both number and avidity of antigenic sites in common has been provided by Dietrich (1968) and Benjamin and Weigle (1971). Thus in order for a heterologous protein to produce an amount of complement fixation equivalent to the homologous protein, the concentration of the antiserum would have to be increased. Recently Prager and Wilson (1971a,b) have shown with bird egg-white lysozymes of known sequence that a correlation does exist between the immunological distance and percent sequence difference. Figure 6 summarizes the MC′F data for the lysozyme system. Human lysozyme, which differs in sequence by about 40% from bird lysozymes, does not cross react with bird lysozymes. A similar correlation exists for other proteins although the data are fragmentary. In all cases there is no cross reaction at sequence differences above 40% (Prager and Wilson, 1971b).

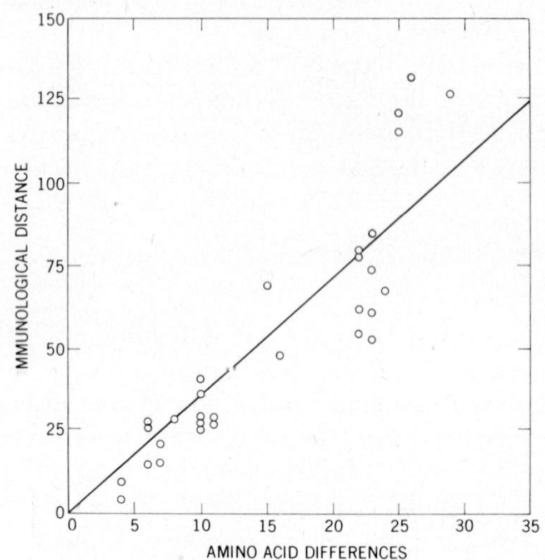

FIG. 6. Dependence of immunological distance on the number of amino acid sequence differences among bird lysozymes. The sequence data are from the following lysozymes: chicken (Canfield, 1963; Jollès *et al.*, 1963); Japanese quail (Kaneda *et al.*, 1969); turkey (La Rue and Speck, 1970); Duck II (Hermann and Jollès, 1970); Duck III (Hermann *et al.*, 1971); guinea fowl (Jollès *et al.*, 1972) and bobwhite quail (Prager *et al.*, 1972). Each lysozyme is a single polypeptide chain containing 129 amino acid residues. The immunological distance values are from Prager and Wilson (1971a and unpublished observations). The line can be described by the equation $y \approx 5x$, where y is an immunological distance units and x is the percent sequence difference. This figure is taken from Wilson and Prager (1974).

The immunological approach has a number of important advantages over conventional sequence determination. It is quick and easy to do, requires little material, and, if one dispenses with doing reciprocal tests (which would be acceptable when an approximate answer would suffice), permits a survey of a large number of related proteins without purification of all the antigens. There are, of course, disadvantages to this approach; internal substitutions may not be registered, the actual location of substitutions cannot be determined, only an approximate value for the number of substitutions can be determined, and distantly related proteins (differing in sequence by more than 30 to 40%) cannot be compared quantitatively.

In summary, if the proper precautions are taken in preparing the antisera and in executing the MC′F experiments, quantitative microcomplement fixation can be an extremely useful technique for obtaining an approximate measure of the degree of sequence resemblance between related proteins.

Acknowledgements

We thank Dr V. M. Sarich for helpful discussion. This work was supported in part by funds provided by the U.S. National Science Foundation (Grant GB-13119) and the U.S. National Institutes of Health (Grant GM-18578).

References and Bibliography

Allison, W. S. and Kaplan, N. O. (1964). *In* 'Taxonomic Biochemistry and Serology' (C. A. Leone, ed.), pp. 401–406. The Ronald Press Co., New York.

Arnheim, N. Jr. and Wilson, A. C. (1967). *J. biol. Chem.* **242**, 3951–3956.

Benjamin, D. C. and Weigle, W. O. (1971). *Immunochemistry* **8**, 1087–1097.

Bustin, M. and Stollar, B. D. (1973). *J. biol. Chem.* **248**, 3506–3510.

Canfield, R. E. (1963). *J. biol. Chem.* **238**, 2698–2707.

Champion, A. B. (1972). Comparative Studies on Clostridial Purine Fermenters. PhD thesis, University of California, Berkeley.

Cocks, G. T. (1971). Immunological Studies of Enterobacterial Alkaline Phosphatase. PhD thesis, University of California, Berkeley.

Cocks, G. T. and Wilson, A. C. (1972). *J. Bact.* **110**, 793–802.

Dickerson, R. E. (1971). *J. molec. Evolution* **1**, 26–45.

Dietrich, F. M. (1968). *Immunochemistry* **5**, 329–340.

Fink, S. C., Carlson, C. W., Gurusiddaiah, S. and Brosemer, R. W. (1970). *J. biol. Chem.* **245**, 6525–6532.

Fiset, P. (1964). *In* 'Techniques in Experimental Virology' (R. J. C. Harris, ed.), pp. 226–236. Academic Press, London and New York.

Gasser, F. and Gasser, C. (1971). *J. Bact.* **106**, 113–125.

Gorman, G. C., Wilson, A. C. and Nakanishi, M. (1971). *Syst. Zool.* **20**, 167–185.

Hermann, J. and Jollès, J. (1970). *Biochim. biophys. Acta* **200**, 178–179.

Hermann, J., Jollès, J. and Jollès, P. (1971). *Eur. J. Biochem.* **24**, 12–17.

Jollès, J., Jauregui-Adell, J., Bernier, I. and Jollès, P. (1963). *Biochim. biophys. Acta*, **78**, 668–689.

Jollès, J., Van Leemputten, E., Mouton, A., and Jollès, P. (1972). *Biochim. biophys. Acta*, **257**, 497–510.

Kabat, E. A. and Mayer, M. M. (1961). 'Experimental Immunochemistry' (2nd ed.), pp. 905, Charles C. Thomas, Publisher, Springfield, Ill.

Kaneda, M., Kato, I., Tominaga, N., Titani, K. and Narita, K. (1969). *J. Biochem., Tokyo* **66**, 747–749

La Rue, J. N. and Speck, J. C. Jr. (1970). *J. biol. Chem.* **245**, 1985–1991.

Levine, L. (1967). *In* 'Handbook of Experimental Immunology' (D. M. Weir, ed.), pp. 707–719. Blackwell Scientific Publications, Oxford and Edinburgh.

Levine, L. and Van Vunakis, H. (1967). *In* 'Methods in Enzymology', Vol XI, (C. H. W. Hirs, ed.), pp. 928–936. Academic Press, New York.

Maxson, L. R. (1973). A Molecular Approach to the Study of Hylid Evolution. PhD thesis, University of California, Berkeley.

Moore, G. W. and Goodman, M. (1968). *Bull. math. Biophys.* **30**, 279–289.

Nakanishi, M. (1971). Albumin Evolution in Birds and Crocodilians. Master's thesis, University of California, Berkeley.

Nonno, L., Herschman, H. and Levine, L. (1970). *Arch. Biochem. Biophys.* **136**, 361–367.

Prager, E. M. (1972). Immunological and Chemical Studies on Lysozyme. PhD thesis, University of California, Berkeley.

Prager, E. M. and Wilson, A. C. (1971a). *J. biol. Chem.* **246**, 5978–5989.

Prager, E. M. and Wilson, A. C. (1971b). *J. biol. Chem.* **246**, 7010–7017.
Prager, E. M., Arnheim, N., Mross, G. A. and Wilson, A. C. (1972). *J. biol. Chem.* **247**, 2905–2916.
Reichlin, M., Hay, M. and Levine, L. (1964). *Immunochemistry* **1**, 21–30.
Sarich, V. M. and Wilson, A. C. (1966). *Science, N.Y.* **154**, 1560–1566.
Sela, M. (1969). *Science, N.Y.* **166**, 1365–1374.
Stanier, R. Y., Wachter, D., Gasser, C. and Wilson, A. C. (1970). *J. Bact.* **102**, 351–367.
Steffen, D. L., Cocks, G. T. and Wilson, A. C. (1972). *J. Bact.* **110**, 803–808.
Wallace, D. G. (1971). A Comparison of Amphibian Albumins: Taxonomic and Evolutionary Significance. PhD thesis, University of California, Berkeley.
Wallace, D. G., King, M.-C. and Wilson, A. C. (1973). *Syst. Zool.* **22**, 1–13.
Wasserman, E. and Levine, L. (1961). *J. Immun.* **87**, 290–295.
Wilson, A. C. and Prager, E. M. (1974). *In* 'Lysozyme' (E. F. Osserman, R. E. Canfield and S. Beychokf), pp. 127–141. New York; Academic Press.
Wilson, A. C., Kaplan, N. O., Levine, L., Pesce, A., Reichlin, M. and Allison, W. S. (1964). *Fedn Proc. Fedn Am. Socs. exp. Biol.* **23**, 1258–1266.

Appendix

Derivation of the equation for the determination of the Index of Dissimilarity (ID): It has been seen (p. 403) that % complement fixed at peak is proportional to the log of the antiserum concentration, or:

$$(1) \quad Y = m \log X + b.$$

On the assumption that the slopes for both the heterologous (h) and the homologous (H) species are equal,

$$(2) \quad Y_H = m \log X_H + b_H \quad \text{or} \quad (2a) \quad b_H = Y_H - m \log X_H$$

and

$$(3) \quad Y_h = m \log X_h + b_h \quad \text{or} \quad (3a) \quad b_h = Y_h - m \log X_h.$$

Rearranging eqns. (2) and (3):

$$\log X_H = \frac{Y_H - b_H}{m} \quad \text{or} \quad (4) \quad X_H = 10^{[(Y_H - b_H)/m]}$$

$$\log X_h = \frac{Y_h - b_h}{m} \quad \text{or} \quad (5) \quad X_h = 10^{[(Y_h - b_h)/m]}.$$

Dividing eqn. (5) by eqn. (4), and rearranging

$$\frac{X_h}{X_H} = 10^{[(Y_h - Y_H + b_H - b_h)/m]}.$$

By definition:

$$\text{ID} = \frac{X_h}{X_H} \quad \text{when} \quad Y_h = Y_H.$$

Therefore:

$$(6) \quad \text{ID} = 10^{[(b_H - b_h)/m]}.$$

Substituting eqns. (2a) and (3a) into eqn. (6) and rearranging:

$$\text{ID} = 10^{[(Y_H - Y_h)/m + \log (X_h/X_H)]} \quad \text{or} \quad \log \text{ID} = \frac{Y_H - Y_h}{m} + \log \frac{X_h}{X_H}$$

Appendix: Techniques

I. Introduction

At the time when the compilation of this book was conceived it seemed not only possible but also highly desirable to include a section devoted to detailed descriptions of methods. Such a section would still be desirable but with the recent proliferation of technical advances, it is no longer practical within the confines of a single volume. Furthermore a wide range of specialist texts dealing with most of the basic techniques which are of use to taxonomists are now available. As a result this Appendix has been largely restricted to general comments and accounts of some of the methods particularly favoured by contributors to this volume.

Experience has shown that most taxonomic problems involve peculiarities requiring some modifications of basic methods. It is hoped that this Appendix will at least provide general guidance for those taxonomists interested in extending their investigations into areas which can supplement the results of comparative morphology.

II. Collection and Preparation of Materials

A. BLOOD

Collection should always be made from living or freshly killed specimens. With living animals some form of anaesthesia is often advisable, ether is generally useful for mammals but shrews may show poor recovery (Johnson, Ch. 1). Blood samples should always be handled with care to prevent haemolysis. Where blood is collected in a hypodermic syringe the needle should be removed before the sample is expelled gently into the collecting tube. For general information on bleeding methods and anaesthesia see the UFAW Handbook on the Care and Management of Laboratory Animals (Churchill Livingstone Publishers).

B. ANTICOAGULANTS

If blood is required solely for serum analysis then the sample may be allowed to clot and the serum drawn off. If the cells are also required (for haemoglobin studies etc) then anticoagulants should be used. To avoid dilution of the serum the collecting tubes may be coated with anticoagulant by adding the appropriate solution to them and evaporating to dryness before use (commercially coated tubes are available). Although trisodium citrate (3% solution in 0·28% saline) and sodium oxalate (1·4% in distilled water) may be used, heparin or ethylene diamine tetra-acetic acid (EDTA or sequestrene) are probably most useful. EDTA is used in 10% W/V solution and the dipotassium salt was found preferable for fish because of its rapid solubility (O'Rourke, Ch. 4).

C. STORAGE OF SERUM SAMPLES

Because of the possibility of differential deterioration of serum samples in storage they should be examined as soon after collection as possible. The addition of merthiolate (thiomersalate) to achieve a final dilution of 1/5,000 prevents bacterial degradation and this coupled with Seitz filtration and storage in sterile containers at 4° C usually gives a high standard of preservation. For long term storage freezing may be desirable but if regular samples are to be taken from the specimen repeated thawing and freezing are likely to have deleterious effects.

D. HAEMOGLOBIN

Blood samples treated with anticoagulants are centrifuged at low speed (about 1,000 rpm and not more than 2,000 rpm) for 5 min. The serum is drawn off for storage and the upper layer of white cells on the deposit is also removed. The red cells are resuspended in 5–10 times their volume of saline solution

(0·85–1·00%), centrifuged at 1,000 rpm for 5 min and the supernatant discarded. This washing procedure is repeated up to five times and after the final wash the cells are resuspended in twice their volume of distilled water and thoroughly shaken or stirred to cause lysis. The cell debris is removed by centrifugation at 4,000 rpm, the supernatant is removed and carbon monoxide bubbled through it for several seconds prior to storage by freezing at −75° C (Sibley *et al.*). At the lysis stage 1 volume of toluene may be added and after centrifugation the lysate drawn off from beneath the toluene layer and dialysed for 24 hr against potassium cyanide-phosphate buffer (pH 7·5; 0·65 g KCN, 7·1 g Na_2HPO_4 (anhydrous) in 1,000 ml, adjusted to pH 7·5 by addition of IN HCl). Subsequent storage may be in the refrigerator or frozen (Johnson, Ch. 1). It is important that direct comparisons of electrophoretic mobility of haemoglobins should be made only on samples prepared in the same way.

E. MUCUS

Body-surface mucus of both fish and snails may be scraped gently from the animal with a clean splinter of wood or similar implement. For delicate species of snail fine paint brushes may be used but must be thoroughly cleaned in 70% alcohol and water between individual samples. With snails it is advisable to discard the first sample since this is often contaminated with detritus. Too vigorous scraping will cause the snail to bleed, this will be obvious in species with pigmented blood such as planorbids but is not so readily detected in many others. The fresh mucus is applied directly to appropriate spots on chromatography paper and dried at room temperature. The papers may be stored in the dry state for several weeks or even longer periods until required for investigation (Wright, Ch. 6). From fish larger quantities of mucus can often be scraped off and collected in a bijou bottle. This is then homogenized for about 15 min and centrifuged at 5,000 rpm for 10 min. The clear supernatant is pipetted off and partially desalted by agitation for one hour with a cation exchange resin ('Zeo-Karb', H-form, 'Permutit' Ion exchange materials) and again centrifuged for 5 min at 1,5000 rpm. The supernatant is concentrated using a Colover concentrating cell with 30% w/w polyethylene glycol in distilled water for about 18 hr at 4° C and the concentrate taken up in small pieces of filter paper for subsequent electrophoresis in polyacrylamide gel (O'Rourke, Ch. 4).

F. TISSUES

Fish muscle tissue is best collected from the dorsal side of the fish and care should be taken that only white muscle is included in the sample (O'Rourke, Ch. 4).

Freshwater snails should be chilled on ice to immobilize them and the shells then gently cracked between the limbs of forceps. Dissection should be as rapid as possible; if tissues other than the digestive gland are required they should be removed first to avoid contamination. Tissues should be immediately transferred to chilled, labelled tubes for grinding and extraction (Wright, Ch. 6).

Thoracic or leg muscles of insects are excised over a petrie dish filled with ice and placed in pre-weighed, chilled grinder tubes (for tissue samples less than 0·08 mg use small, pointed grinders, for larger samples round-bottom grinders are adequate). Grinder tubes with samples are reweighed to determine the amount of material and two volumes (W/V) of 40% sucrose solution are added (this enables some quantitative standardization of subsequent tests when very small quantities of tissue are involved). Grinding is carried out in the cold with care to avoid aeration and extraction with intermittent grinding proceeds for at least 2 hr in the cold. Some enzymes appear to denature upon standing for more than 3–4 hr and there also appear to be generic or species-group tendencies in denaturation of certain systems when quick frozen tissue is used. After grinding, the extracts are centrifuged at 9,000 *g* for 30 min and the supernatant liquid is then decanted into a cold spot plate. Some extracts are reabsorbed if left in the tube in contact with the deposit (Stephen, Ch. 5).

G. EYE LENSES

Lenses are removed from the opened eye capsule with fine forceps and freed from any surrounding tissue. The lenses are then crushed in an approximately equal volume of phosphate buffer pH 7·3 and gently agitated before centrifuging at 3,000 rpm for 10 min. The supernatant is pipetted off ready for use and may be stored at 4° C for up to 1 week (O'Rourke, Ch. 4). Lenses from frozen fish are used by A. C. Smith (1972, *Comp. Biochem. Physiol.* **42**, 497–499), and extraction is carried out in a volume of 0·018 g% saline solution equal to seven times the wet weight of tissue for 24 hr at 5° C followed by centrifuging at 6,000 rpm for 60 min at 5° C. Lenses from birds may be collected into sufficient 2% phenoxyethanol solution to cover them and stored in the cold (Sibley *et al.*, Ch. 2).

H. EGG PROTEINS

Egg-white proteins of birds are obtained from the thin egg-white of fresh or slightly incubated eggs and may be stored at 4·0° C before use (Sibley *et al.*, Ch. 2).

Egg proteins of freshwater pulmonate snails are obtained from recently laid eggs (preferably less than 48 hr old). Egg masses are collected by gently scraping them from the substratum (polythene disks used as egg traps in aquaria if possible) with a piece of exposed photographic film. The egg masses are brushed clean with a fine paint brush, blotted on slightly damp filter paper and laid 'face down' on damp filter paper. Individual eggs are then punctured with a fine tungsten needle and the contents drawn up in a fine pasteur pipette. The proteins coagulate quickly if exposed to the air and should be used at once (Wright, Ch. 6).

III. Electrophoretic Techniques

Different electrophoretic techniques depend upon the supporting media on or in which separations are carried out and the choice of method is governed by the degree of resolution to be achieved, quantities of material available for examination, ease of manipulation and subsequent storage properties. All of the techniques are adequately described in standard textbooks and detailed methods of handling depend upon different manufacturers modifications of the apparatus. The traditional method of paper electrophoresis was widely used in the early days of experimental taxonomy but is seldom employed now because of the poor resolution of fractions and relatively large quantities of material required. Cellulose acetate is an excellent supporting medium similar in many ways to paper in its handling and storage characteristics but giving better resolution and requiring smaller quantities of material. Acrylamide gel gives perhaps the best resolution but more consistent results (and therefore more useful for long-term comparative work) are obtained with starch gel. Table I shows the methods of choice for different materials used by O'Rourke in fish studies.

A. CELLULOSE ACETATE METHOD FOR SNAIL EGG PROTEINS

Strips are floated on 0·1075M glycine-sodium hydroxide buffer pH 9·8 (240 ml of 0·2M glycine with 0·2M sodium chloride mixed with 50 ml of 0·203M sodium hydroxide and made up to 1 litre) care being taken to avoid trapping air beneath the strips. Buffer strength for use in the electrophoresis tank is 80% of that used for soaking strips, conveniently achieved by pouring off 200 ml of the original solution for soaking and making the remaining 800 ml up to 1 litre again as tank buffer. Standard electrophoresis tanks usually require modification to ensure that the air-space above the strips is adequately saturated with water vapour; this may be done by using plastic sponge material fixed to a perspex support sheet. Strips are connected to the tank buffer compartments by cellulose acetate wicks (soaked in tank buffer) placed along the bridge bars. Tank buffer and wicks are discarded after each run. Samples are applied to the strips *in situ* in the tank using a re-drawn Pasteur pipette with the tip smoothed in a low flame to avoid scratching the surface of the cellulose acetate. For delicate control of quantity the pipette is used with a mouth tube and uniformity of application point is assisted by a ruler supported across the tank and used to guide the pipette. Samples should be streaked on to the strips in the narrowest possible line and delay between the first and last applications avoided. Running time is 2 hr at 120 volts and 0·5 milliamps per strip (strip

Table I

Test material / Details of Technique	Supporting medium	Buffer system	pH	Running time	Current	Staining method	De-staining	Preservation	Scanning method
Haemoglobin	(Micro-) Starch gel	TRIS-HCl	8·2	60 min	3 mA p. slide	Amido-Black 10 B	Glacial Acetic Acid	Drying on Filter-paper	—
Haemoglobin	(Micro-) Agar gel	Phosphate	7·3	40 min	5 mA p. slide	Amido Black 10 B	Glacial Acetic Acid	Drying on Slide	—
Blood Serum	Poly-Acrylamide Gel	Veronal	8·6	120 min	30 mA p. big gel	Amido Black 10 B	Glacial Acetic Acid	Drying on Slide	Chromo-scan
Eye lenses	(Micro-) Agar gel	Phosphate	7·3	40 min	5 mA p. slide	Amido Black 10 B	Glacial Acetic Acid	Drying on Slide	Chromo-scan
Muscle Protein	Poly-Acrylamide Gel	Veronal	8·6	120 min	30 mA p. big gel	Amido Black 10 B	Glacial Acetic Acid	Drying on Slide	Chromo-scan
Body Mucus	Poly-Acrylamide Gel	Veronal	8·6	120 min	30 mA p. big gel	Amido Black 10 B	Glacial Acetic Acid	Drying on Slide	Chromo-scan

size 100 × 25 mm). Immediately after running the strips are removed from the tank and fixed in 5% trichloracetic acid for 15 min and stained in nigrosin (see p. 425) (G. C. Ross in Wright, Ch. 6).

B. ACRYLAMIDE SPLIT GEL METHOD FOR COMPARING IDENTICAL AMOUNTS OF DIFFERENT SAMPLES IN THE SAME TUBE

The basic procedure is essentially the same as that described by Davies, B. J. (1965, *Ann. N.Y. Acad. Sci.* **121**, 404–427). So that gels will slide out of tubes easily the tubes must be absolutely clean. After soaking in potassium dichromate solution they should be brushed with a solution of Alconox or Labtone, rinsed thoroughly in distilled water and given a final wash in a 1:200 solution of Pho-flo, dried in an oven and used immediately. The gel reagents should be mixed by swirling to avoid the formation of fine bubbles. Tubes are filled with 0·85 ml separating gel and a water layer, chemically polymerize, rinse twice, add 0·15 ml stacking gel, photopolymerize, rinse twice, protect from light. Dividers made from white plastic rulers 6″ long and 0·5 mm thick are inserted into the tubes. Before use the rulers are thoroughly scrubbed and soaked in buffer (pH 8·3) for a day, then dried and cut to fit the exact internal diameter of the tubes. The fit is right if a fine layer of plastic curls off as the divider is inserted, insertion must be in a straight continuous slide to avoid leaks. Samples of 5λ, 10λ or more are pipetted onto spot plates using Eppendorf microlitre pipettes. The sample is taken up in a 1 ml tuberculin syringe with a polyethylene tubing extension, sample gel is drawn up to 0·05 ml (a small crystal of Bromphenol Blue is added to the sample gel as a tag and to indicate the front in the subsequent run), the sample and sample gel are well mixed and carefully expelled into one side of the divided tube which is held in a slightly elevated horizontal position, care being taken to avoid the inclusion of small air bubbles. The tube is placed horizontally on a glass sheet over an inverted fluorescent desk lamp to polymerize. The procedure is repeated with a different sample for the other half of the tube and unmarked (without Bromphenol blue) sample gel is used in the mix. After polymerization is complete the gel is ready to run in the usual way. After electrophoresis the divider is removed with needle-nosed pliers, the gel is rimmed with a long, fine sewing needle in a metal holder and the extracted gels are placed in cold distilled water or buffer. If in inserting the divider it hits the top of the stacking gel with too much force the two sides will be uneven and the tube should be discarded. When the sample is placed in the tube it should be dropped straight in rather than being allowed to run down the tube wall or divider. The divider must be absolutely perpendicular to prevent migration slipping sideways. Tiny air bubbles at gel interfaces can be removed with the eye end of a fine sewing needle, being careful not to scratch the polymerized gel surface (Stephen).

Of the many variants of the starch gel method one which has proved to combine many desirable features is a thin-layer technique on microscope slides. A modification of the slide template used for agar gel electrophoresis (Shandon Scientific Company) in which the base is replaced by a slightly thicker, unperforated sheet of perspex is used to carry eight standard microscope slides. The slides are held in the template by a thin smear of grease. Starch (11·2 g) is mixed with 100 ml of appropriate buffer in a 500 ml side-arm flask containing a magnetic stirring bar. The mixture is heated with constant rotary swirling over a low gas until it clarifies and the stirring bar is clearly visible. The flask is transferred to a magnetic stirrer and stirred hard for 2 min, air is removed under vacuum for about 30 sec and the flask is returned to the stirrer and the gel is kept gently in motion until the temperature drops to 55° C. The gel is then poured onto the trays of slides (prewarmed to 50° C) and levelled immediately with a warm bevelled scraper giving a gel thickness of 1·5 mm. The gels are 'cured' by storage at 4° C for at least 16 hr before use. Sample slots are cut in the gels immediately before use with a ground down scalpel blade which must be kept sharp to avoid tearing the gel. After electrophoresis, staining development is carried out with the slides *in situ* in the tray. After development the gel is rinsed under running water and immersed under a 4% solution of glycerine in 5% acetic acid for fixation. The slides are then separated from one another with a scalpel and removed from the template tray to a shallow dish where a 7% solution of gelatine in 3% glycerol at 45° C is poured over them to a depth of 2 mm. The solution solidifies and is left to dehydrate at room temperature for about 72 hr. The slides are then cut out with a scalpel, trimmed and can be stored in a normal microscope slide cabinet indefinitely (G. C. Ross in Wright, Ch. 6).

IV. Localization and Identification of Proteins Separated by Electrophoresis

An excellent summary of some methods appropriate for starch gel electrophoresis is provided by H. C. Dessauer in Bulletin 36 of The Serological Museum, Rutgers University, New Brunswick.

For general staining of proteins in starch gel Amido black is widely used but in polyacrylamide gel Coomassie blue (Coomassie blue R250 Colab Product) is preferable. Stephen suggests the following modifications of the method described by Chramback *et al.* (1967, *Anal. Biochem.* **20**, 150–154). Gels are fixed for 12 hr in 12·5% trichloracetic acid then two identical samples are stained in 1% aqueous Coomassie blue (filtered and stored in refrigerator), one for 5 and one for 15 min. The gels are photographed about 36 hr later at the peak of differentiation and are subsequently stored in 7% acetic acid to prevent gel shrinkage. The reasons for these modifications are that Coomassie blue does not require destaining but if the gels are inadequately fixed (for periods of less than about 10 hr) the protein fractions continue to absorb dye and in storage the intensity of staining increases until fine bands become completely obscured in time. With two staining periods it is possible to visualize bands in proximity to the concentrated protein fractions in the lightly stained specimen and the longer staining period reveals fractions present in trace amounts.

For proteins separated on cellulose acetate the best stain so far tried is nigrosin. Many others, notably ponceau-S produce more rapid results but nigrosin gives consistently better resolution of more fractions. Immediately after running the strips are fixed in 5% trichloracetic acid for 15 min, rinsed in 5% acetic acid and stained in nigrosin solution, either 0·2% nigrosin in 20% acetic acid for quick staining or 0·02% nigrosin in 2% acetic acid overnight. Excess stain is removed by several rinses in 5% acetic acid and the strips are dried flat under pressure between sheets of filter paper (G. C. Ross in Wright, Ch. 6).

Despite the specificity of the chemical reactions involved in the localization of particular protein complexes, it is surprising to find that the physical conditions employed by different workers cover a considerable range. It is probable that the physical optima for the same enzyme systems in different animal groups do differ but there is little comparative work at this level to

Table II
Esterases

	Sibley *et al.* (Ch. 2)	Stephen (Ch. 5)	Ross (in Wright Ch. 6)
Buffer	0·1 M phosphate pH 6·0 100 ml	0·115 M tris-maleate pH 5·6 100 ml	0·02 M tris-HCl pH 7·0 100 ml
1-Napthyl acetate	20 mg	40 mg	40 mg
Diazonium salt	Fast garnett GBC 50 mg	Fast Blue 2R 70 mg	Fast Blue 2R 50 mg
		Lactic Dehydrogenases	
Buffer	0·09 M tris-HCl pH 7·1 100 ml	0·09 M tris-maleate pH 8·3 100 ml	0·05 tris-HCl pH 7·0 100 ml
Containing sodium lactate	0·1 M	0·043 M	0·1 M
DPN	30 mg	30 mg	40 mg
Nitro blue tetrazolium	20 mg	80 mg	25 mg
Phenazine metho-sulphate	2 mg	14·3 mg	2 mg

provide the necessary evidence. It is equally probable that the degree of precision accorded to these techniques by some workers is not wholly justified although for purposes of direct comparison it is essential that the methods used should be standardized. As examples of the variations encountered Table II gives the incubation media for acetyl esterases and lactic dehydrogenases used by three of the contributors to this volume. The media have been adjusted to equivalent volumes for comparative purposes.

In these examples Sibley *et al.* are working with avian materials in starch gel, Stephen with insect proteins separated in polyacrylamide gel and Ross is investigating enzyme systems in freshwater snails, trematodes and reptiles using a thin-layer starch technique. Reference to the literature in which specific protein complexes have been studied in a taxonomic context will reveal further variations, although the precise nature of the discrepancies is often difficult to detect without reducing the figures given to equivalent volumes. To single out any particular methods for general application to all animal groups would be unwise in a field which is continuing to develop rapidly. There exist several texts which describe techniques for localizing specific substances separated by electrophoresis (e.g. Smith, I. 1968, 'Chromatographic and Electrophoretic Techniques', Vol 2. 'Zone Electrophoresis', 2nd ed. Heinemann Medical Books; and Brewer, G., 1970. 'An introduction to Isozyme Techniques', Academic Press, New York). Those interested in pursuing this approach to taxonomy would be well advised to follow the methods set out in such texts and subsequently to introduce modifications appropriate to the materials with which they are working.

V. Chromatographic Techniques

Routine methods of paper chromatography, ascending or descending and two-dimensional have been used in several taxonomic studies and are all adequately described in appropriate textbooks. Continuous flow ion-exchange chromatography and two-dimensional chromatography and electrophoresis on thin-layer plates are referred to later in specific studies of tryptic peptides of purified proteins.

Horizontal circular chromatography using dishes designed by Kawerau and manufactured by the Shandon Scientific Company Ltd has proved to be particularly useful for the study of fluorescent substances in mucus. Specially cut filter paper disks 25 cm diameter are available in several grades but Whatman No. 3 is to be recommended. Mucus from separate individuals is spotted onto the paper near the central apex of each of five segments and allowed to dry. Disks can be kept in this condition for several months before development if required. When disks are placed in the dishes for development care must be taken to ensure that the centre remains in contact with the capillary feeding solvent from the bottom of the dish, this can be assisted by placing a drop of solvent on the centre with a pipette. For studies on snail mucus, the solvent most frequently used is butanol/acetic acid/water (100:22:50) (Wright, Ch. 6) and for fish butanol/acetic acid/water (120:30:50), phenol water (500 g phenol dissolved in 125 ml distilled water) or phenol ammonia (1 ml ammonium hydroxyde 0·880 in 200 ml of the phenol water solvent) (O'Rourke, Ch. 4). A running time of about 6 hr is usually satisfactory. After development the disks are examined under ultra-violet light (3650° A). For fish mucus the fluorescent patterns are examined both before and after drying but for snail mucus only after drying. For certain groups of snails it is necessary to expose the dried chromatograms to ammonia fumes to enhance the fluorescent pattern and excite fluorescence in substances which may not otherwise be revealed.

Colour photographs of fluorescent patterns may be made for record purposes using high speed Ektachrome daylight film (EH 135) with an ultra-violet filter and exposures of 4–6 sec depending on the brightness of the pattern. It is important to obtain photographs of ammonia enhanced patterns as soon as possible after fuming because they tend to fade rapidly. After development and particularly after fuming changes may occur in the fluorescent patterns and old chromatograms should not be compared with freshly developed ones. After recording fluorescent patterns the chromatograms may be sprayed with ninhydrin (0·3% solution in acetone W/V) and rapidly dried out at 80° C to reveal the presence of ninhydrin positive materials. These patterns can be stabilized for up to a month by spraying with copper nitrate (1 ml saturated copper nitrate solution in 100 ml ethanol and 0·2 ml of 10% nitric acid) (O'Rourke, Ch. 4).

VI. Immunological Techniques

The production of specific antisera for use in immunotaxonomy is fundamental to the application of most techniques and yet it remains a somewhat haphazard process. Rabbits are usually the animals of choice for production of antisera although other species have been used for particular purposes. Individual rabbits vary in their response to antigens and even litter-mate animals subjected to identical injection schedules with the same antigen mixture may give antisera with quite different characteristics. Standard texts on immunological procedures (e.g. D. H. Campbell *et al.*, 1964, 'Methods in Immunology', W. A. Benjamin Inc., New York and Amsterdam; and C. A. Williams and M. W. Chase, 1967, 'Methods in Immunology and Immunochemistry', Academic Press, London and New York) recognise this problem and recommend the use of up to 10 rabbits with any particular antigen system in order to ensure at least one or two good antisera. This may be possible when the source of antigen is some relatively large or easily cultivated organism and the supply is plentiful but it is rarely realistic with the limited resources available to many taxonomists of invertebrate groups; variations in injection sites and schedules and the use of adjuvants may help to improve the standard of antisera produced. It is also possible in some cases to find particular strains of rabbit which appear to react to some types of antigen better than others. As a rule short injection schedules yield more specific antisera and longer schedules produce a broader spectrum of antibodies. Before undertaking immunotaxonomic investigations it is advisable to study appropriate texts in order to avoid some of the more usual errors such as unecessary repetition of adjuvant injections or loss of rabbits by anaphylaxis due to inappropriate timing of intravenous injections.

When a general grounding in the methods is obtained it is necessary for each individual to work out his own approach best suited to the antigens with which he is working and be prepared for frequent disappointments. Rather than recount a series of different injection schedules with conflicting information which might be confusing a single example is given here, not necessarily as a model but as an illustration of a method which has given reasonably satisfactory results over a period of several years.

The antigens used have been planorbid snail egg proteins and the objective has been to produce antisera capable of discrimination at the specific level by agar gel double diffusion precipitin techniques (ouchterlony method). Freshly laid snail egg masses are collected, rinsed in a 1/5,000 solution of merthiolate (thiomersalate) in distilled water, blotted on filter paper and frozen at $-25°$ C until a sufficient quantity (usually about 5 ml packed volume) has been accumulated. The eggs are then thawed and emulsified with about half their volume of 0·85% saline solution in a glass tissue grinder. The resultant gelatinous mixture is centrifuged for an hour at about 70,000 g (slower speeds usually fail to throw down the egg-capsule debris), the supernatant fluid is drawn off and sterilized by passage through membrane filters under pressure (Swinnex filter units, pore size 0·22 μm, Millepore) into sterile phials and rubber-capped. This constitutes the injection antigen and it usually contains about 0·2% protein. Experience has shown that the most useful strain of rabbit for these antigens is the half-lop and young adults are used, irrespective of sex. Four intravenous injections are given in the marginal ear vein, 0·25, 0·5, 1·0 and 1·0 ml, two-day intervals between each of the first three injections and three days between the last two. The total volume of antigen solution (2·75 ml) contains usually about 5–6 mg of protein. Since this often represents the results of several months' collection of eggs there is rarely sufficient available for further injections. Under these circumstances even if trial bleedings show a poor antibody response there is nothing that can be done about it and therefore trial bleedings are not usually carried out. Rabbits are killed and bled-out immediately by jugular section ten days after the final injection. The blood is collected in large glass centrifuge tubes (between 80–100 ml is a reasonable yield per rabbit), and the clot is detached at the meniscus by careful ringing with a thin glass rod. Clot formation is allowed to continue at room temperature and the separated serum is decanted, centrifuged to remove free cells, Seitz filtered after the addition of 1% merthiolate to achieve a final dilution of 1/5,000, and stored in sterile bottles in the refrigerator.

These antisera vary in quality and the results from one cannot be compared directly with another. A check on the last 20 antisera made following this schedule classified 3 as very good, 7 good, 7 fair and 3 useless; very good indicates an antiserum capable of distinguishing between isolated populations of the same species using an absorption technique, good antisera discriminate between related species, fair are only of use at the generic level and useless antisera (in this context) are those that give either a single nonspecific precipitin line or none at all. On occasion when certain antigens have been relatively abundant several antisera have been made at the same time, some following the schedule described here and others using Freund's adjuvant (both complete and incomplete). No improvement in antibody response was obtained with adjuvant and there was a tendency for the precipitin lines produced by these antisera to be less clearly defined. The same injection schedule has been used for the preparation of antibodies to both mammalian sera and eye-lens proteins of freshwater fish but the results of these attempts have been less successful than those with snail egg proteins (Wright).

Of the immunological techniques available to taxonomists the choice will usually depend upon the objective of the study and the materials available. Traditional ring-test methods using serial dilutions of antigen to determine the titre point beyond which a given antigen no longer produces a ring of precipitate at an antiserum-antigen interface are still used, usually in a modified form in capillary tubes which are less demanding in terms of materials. These interfacial reactions are useful to give a quantitative value to the relationships between organisms but they demand a fairly accurate knowledge of antigen concentrations in the test solutions. More detailed quantitative information can be obtained from the nephelometric methods developed particularly by Alan Boyden and his co-workers at Rutgers University. These methods depend upon measuring the amount of suspended precipitate in serial dilutions of antigen mixed with constant volumes of antiserum. Where antigen is present in excess the precipitate is re-dissolved and when the antigen is too dilute no precipitate is formed. As a result, if an adequate range of antigen dilutions are considered and the amounts of precipitate formed are plotted against antigen concentration, a symmetrical distribution curve is obtained with a peak at the point of optimum proportions of antigen and antibody. Antigens of related species titrated in the same way against the same antiserum will produce similar curves but of lower amplitude and the area beneath each curve (treated as a percentage of the area enclosed by the homologous curve) is indicative of the relationship between the other species and the homologous species. By using a range of antisera a three-dimensional picture of relationships within a group may be built up. These techniques are extravagant in their use of both antisera and antigens and for this reason their application is limited to situations where both are in abundant supply. If the antigen used is a tissue extract it may be difficult to obtain sufficiently clear test solutions. Although a blank run on both antigen and antiserum is normally included in order to make correction for the natural turbidity of either solution experience has shown that in some cases the correction factor is so large that the lower readings in less closely related forms are not reliable. A further problem arises where the antigen solutions are complex mixtures for in these cases the curves of antigen concentration plotted against turbidity tend to be polymodal with separate peaks for different antigens and such curves are less easy to interpret. Another complication which may arise in such cases derives from the fact that not all of the components in a complex mixture are equally antigenic to the animal in which antiserum is prepared. Sibley *et al.* (p. 109) have commented that the majority of immune responses by rabbits to bird egg-white proteins appear to be to minor components, e.g. *not* ovalbumin, conalbumin, lysozyme and ovomucoid which together make up 92% of the total protein. Thus although the antigen dilution series will usually be based upon a total protein estimation for the raw antigen, in the case of bird egg-white it is possible that not more than 8% of that total will have been involved in appreciable antibody production.

Of perhaps wider application to taxonomic problems are the gel-diffusion techniques in which antigen and antibody diffuse through agar gel and form discrete lines of precipitate where they meet. Although less readily quantified than the previous methods they do allow some analysis of the number of antigen/antibody systems involved in a complex reaction because each antigen forms a separate line of precipitate although these are often so close together as to be impossible to resolve. The one-dimensional method of Oudin involves the layering of antigen solutions onto the surface of tubes of agar gel into which antiserum has been incorporated. This involves the use of relatively large amounts of antigen and antiserum and is rather less versatile than the two-dimensional technique of Ouchterlony in which symmetrical patterns of wells are cut into agar plates, usually in petri dishes. Antigens and antisera are placed in appropriate wells and precipitin lines form in the gel between the wells. Assessments of affinity between species on the basis of the number of lines formed may be confusing because in a strong homologous reaction the lines may coalesce into a few broad bands while heterologous antigens may form narrower, more discreet lines which can appear to be more numerous. If a particular line in one system forms a spur against the comparable line from an adjacent antigen well it is an indication that the first antigen is more closely related to the homologous (antiserum-producing) antigen than is the second (see Fig. 5, Ch. 6). Further refinements involve the use of antisera absorbed by one of the heterologous test antigens. This is achieved by placing the absorbing antigen in the antiserum well and allowing it to diffuse slightly into the agar before the antiserum is added. This will precipitate all of the antibodies capable of reacting with that particular antigen system and the occurrence of precipitin lines in the agar between the absorbed antiserum well and those of other antigen systems will indicate the presence of antigens common to the homologous but lacking in the absorbing system (see Figs 5 and 6, Ch. 6). The rapidity with which precipitin lines appear can also be an indication of the affinities of the species under examination for the homologous system normally shows lines developing well in advance of the heterologous. The age of the agar can also be important in this respect, development usually occurring more slowly in freshly-poured plates than in those which have been stored in the refrigerator for a few days prior to use.

More accurate estimates of the numbers of antigens common to different organisms can be made by the use of immunoelectrophoresis. The test antigens are separated by electrophoresis in agar gel and antiserum is placed in trenches cut between the lines of separated antigens. After appropriate incubation precipitin arcs form between the separated antigens and the antiserum trenches. The ease with which the precipitin arcs can be counted depends largely on the extent to which the antigen fractions have been separated but even where areas of overlap occur it is usually possible to detect the ends of the arcs. The number of arcs do not, of course, necessarily represent the total number of antigens present even in the homologous system but only those to which antibodies have been formed in the particular antiserum used and it is therefore not proper to treat these numbers as absolute quantities for comparative purposes. Technical details for the performance of most gel-diffusion methods can be found in Crowle, A. J. (1961), *Immunodiffusion*, Academic Press, and many of them can be carried out on a semi-microscale on cellulose acetate membrane instead of in agar gel.

The most refined of the immunological techniques at present employed in taxonomic studies is that of microcomplement fixation described in detail by Champion *et al.* (Ch. 8).

VII. Peptide Analysis of Purified Proteins

Electrophoretic and immunological methods allow comparisons of certain physical characteristics of protein molecules. More precise taxonomic information can be obtained by comparing the peptides of homologous proteins from related species. The following methods used by Sibley *et al.* for ovalbumin and haemoglobin provide good examples of this approach.

> 'Ovalbumin was partially isolated from the other egg white proteins by salting-out with ammonium sulfate and then purifying by ion-exchange column chromatography. The ovalbumin crystals obtained in the salting-out step were dissolved in distilled water, the salts removed by dialysis against distilled water and Amberlite MB-3, and the ovalbumin solution dialyzed against the starting ion-exchange buffer. That buffer was composed of 0·02M glycine, 0·017M K_2HPO_4, 0·029M KH_2PO_4 and adjusted to pH 6·0 with 0·1N HCl. After the final period of dialysis, the sample was applied to a diethylaminoethyl (DEAE)-cellulose column. Ovalbumin and the protein contaminants, ovotransferrin and ovomucoid, were eluted

from the column by a gradual pH gradient that dropped from 6·0 to 3·5 over a volume of 450 ml. The gradient was produced and maintained by a Buchler Varigrad mixer. A Technicon auto-analyzer proportioning pump was used to maintain a constant flow rate during the elution, and the eluate was monitored and collected by a Gilson Medical Electronics fraction collector. The UV absorption curve (at 280 mμ) was recorded by a Texas Instruments Recti/riter.

The fractions included in the ovalbumin portion of the optical density curve were pooled, dialyzed against distilled water and Amberlite MB-3, lyophilized and stored in rubber-capped serum bottles at −60° C if not used immediately. An aliquot of the chromatographic fractions was examined by starch-gel electrophoresis using a discontinuous buffer system to monitor the progress of ovalbumin purification. The ovalbumin was considered to be free from contamination if the electrophoretic pattern did not contain other protein fractions. If contaminants were present, the sample was rechromatographed on a DEAE-cellulose column. It is possible that immunological techniques might have demonstrated the presence of other protein components that were present in amounts too small to be detected by the Amido Black 10B protein stain. However, the tryptic peptides of such trace proteins would not be detectable on a peptide map because of their low concentrations.

The final procedures involved the digestion of the purified ovalbumins with trypsin and the comparison of these tryptic digests by two-dimensional thin-layer chromatography and electrophoresis. Prior to each digestion, the ovalbumins were reduced with sodium borohydride and alkylated with iodacetamide. Two-times crystallized trypsin (Worthington Biochemical Corporation, Freehold, New Jersey) was used to digest each sample of reduced and alkylated ovalbumin. The ratio of trypsin to ovalbumin was 1:100, and the digestion, performed under a nitrogen atmosphere in a semi-automatic pH stat at pH 8·0 and 38° C, was complete at the end of 2·5 hours. The sample was acidified to pH 2·0 and then lyophilized.

Two-dimensional thin-layer chromatography and electrophoresis on Silica Gel C was used to produce the peptide maps. To insure that separation distances were equivalent from one thin-layer plate to another, 5 μl of Gelman RBY dye were spotted onto the corner of each thin-layer plate prior to electrophoresis. A pyridine, acetic acid, and water buffer (1:10:289) was used for the electrophoretic separation which was effected by a constant voltage of 40 volts/cm for 30 minutes. After electrophoresis, the plates were dried in a forced draft oven preheated to 105° C. The peptides were chromatographed in the second dimension using a solvent composed of *n*-butanol, acetic acid, and water (13:2:5). After the primary solvent front had migrated 12 cm from the point of sample application, the thin-layer plate was placed in a hood and partially dried before spraying with a ninhydrin reagent (0·4 per cent ninhydrin in ethanol, water, and collidine, 20:5:1). Maximum color development was obtained by placing the sprayed plate in the oven for 10 minutes. When the ninhydrin spots were fully developed, the plate was cooled and placed on a light box where the spots could be outlined by dotting their margins with a pin-tipped stylus. The plate was next sprayed with a modified Ehrlich's reagent (1 per cent *p*-dimethylaminobenzaldehyde in ethanol, acetone, and HCl, 30:15:2). This combination reduced the flaking of the Silica Gel G without decreasing the intensity of the purple color produced by the reaction of the reagent with tryptophan. Those peptides that contained tryptophan were marked and the peptide pattern recorded by using the plate itself as a negative to produce a contact print of the peptide map. Each ovalbumin sample was digested two or more times and at least three peptide maps were prepared from each digest.

Tryptic digestion of hemoglobin. Hemoglobin samples to be digested were shaken with an equal volume of toluene and then centrifuged at 10,000 rpm for 20 min at 4° C. This was repeated until the toluene layer was clear. The hemoglobin was then lyophilized. Heme-free globin was prepared as follows. Fifty mg of the lyophilized hemoglobin were dissolved in 5 ml of distilled water and placed in a 50 ml separatory funnel fitted with a rubber stopper. Two syringe needles (20 gauge) and a glass tube (3 mm o.d.) were passed through the stopper. The glass tube reached to the bottom of the funnel. Carbon monoxide was bubbled through the hemoglobin solution for 5 min followed by nitrogen for the remainder of the time required to remove the heme. One ml of 0·1 N HCl was added through one of the syringe needles; the other served as a vent. The solution was mixed by shaking the funnel gently and then cooled in a bath of crushed ice and water. When the solution was cold, 10 ml of chilled methylethylketone saturated with water was added through a syringe needle. The rubber stopper was replaced with a glass stopper and the contents of the funnel were shaken. After the phases were allowed to separate, the lower aqueous phase containing the globin was drained into a dialysis bag and dialyzed against water and Amberlite MB-3 at 4° C until the odor of methylethylketone could not be detected. This required 4–5 changes of distilled water, one liter each. The dialyzed globin sample was placed in the digestion chamber of a pH-stat (Radiometer Automatic Titrator). The sample was warmed to 37° C under a nitrogen atmosphere and 0·02M ammonium hydroxide was added until the pH held constant at 8·5. One-half ml of distilled water containing 0·2 mg of 2 × crystallized trypsin (Worthington Biochemical Corp.) was added to the sample and digestion was allowed to continue until the pH remained constant without the addition of base. The time required to achieve this varied from species to species. Seventeen hours was sufficient time for the *Phoenicopterus*, *Ardea*, and *Anas* samples, but the *Branta bernicla* sample was digested for 24 hr, the *Gavia stellata* sample for 40 hr, and the *Larus argentatus* sample for 44 hr. At the end of digestion, each sample was acidified to pH 3·2 by the addition of 0·1N HCl and lyophilized.

Ion-exchange column chromatography of tryptic peptides of hemoglobin. A Technicon 'Auto-Analyzer' set up for peptide analysis was used to characterize the trypic peptides of the hemoglobins of nine species. The column was packed to 100 mm with Technicon 'Chromobead' type 'P' resin and equilibrated with the starting buffer. Thirteen mg of each hemoglobin digest were dissolved in 0·5 ml of the starting buffer and applied to the column. Peptides were eluted by a gradient of increasing pH and ionic strength using a Buchler Varigrad with the following buffers in the respective chambers: chambers 1 and 2 (90 ml each), 0·10M critic acid-H_2O, 0·212M NaOH, adjusted to pH 3·2 with 6·0N HCl; chambers 3 and 4 (90 ml each), 0·132M citric acid-H_2O, 0·297M NaOH, 0·143M sodium acetate-$3H_2O$, adjusted to pH 4·60 with 6·0N HCl; chamber 5 (88 ml), 0·224M citric acid-H_2O, 0·527M NaOH, 0·436M sodium acetate-$3H_2O$ adjusted to pH 5·0 with 6·0N HCl; chamber 6 (87 ml), 0·428M citric acid-H_2O, 1·0M NaOH, 0·855M sodium acetate-$3H_2O$ adjusted to pH 5·1 with 0·87M acetic acid; chamber 7 (87 ml), 0·320M citric acid-H_2O, 0·75M NaOH, 1·145M sodium acetate-$3H_2O$, adjusted to pH 5·25 with 0·87M acetic acid; chambers 8 and 9 (87 ml each), 2·0M sodium acetate-$3H_2O$ adjusted to pH 6·8 with 0·87M acetic acid. The column was heated to 55° C during the elution of the peptides and the flow rate was 35 ml/hr. The Technicon peptide manifold was modified in several ways. The eluate stream was split as it emerged from the column and 75 per cent of the eluate was combined with a stream containing 4·87M NaOH. This was passed through the Teflon coil of a Technicon hydrolysis bath and heated to 95° C for approximately 2 hr. After alkaline hydrolysis the stream was split in half and each resulting stream was buffered to pH 5·5 with water, glacial acetic acid, and 4·0N sodium acetate-$3H_2O$ pH 5·1 (10:10:3). Each stream then was combined with ninhydrin reagent and passed through the standard length glass coil of a Technicon heating bath. The color developed during the reaction of the ninhydrin with the peptides was monitored continuously at 570 mμ in a Technicon colorimeter equipped with a cuvet of standard path length (Technicon tubular flow cell).

The remaining 25 per cent of the eluate stream from the column was shunted to a manifold designed to assay for the amino acid histidine (Pauly reaction). The sample stream was sequentially combined with streams containing 2·0M $Na_2 3O_3$, 0·029M sulfanilic acid in 0·5N HCl, and 0·057M $NaNO_2$. The extent of this reaction was monitored at 505 mμ in a Technicon colorimeter equipped with a tubular flow cell of standard path length.

Thin-layer electrophoresis (TLE) of tryptic peptides. A portion of each digest (0·5 mg) was dissolved in 0·01 ml of 0·1N $NaHCO_3$ or distilled water and applied 3 cm from the edge of a glass plate precoated with 250 μ of Silica Gel H without UV indicator (E. Merck, Darmstadt). After heating to 110° C for 10 min, the plate was sprayed with pyridine (redistilled Mallinchrodt AR grade), acetic acid, and water (25:1:225) and placed on a Savent FP-18 flat plate electrophoresis apparatus in a cold room. Water cooled to 2° C was circulated through the cooling block. The tryptic peptides of hemoglobin were separated for 2 hr at 30 v/cm. The tryptic peptides of ovalbumin were separated for 90 min at 40 v/cm. The TLE plates were dried in a forced draft oven heated to 105° C and then sprayed with 0·4 per cent ninhydrin in n-butanol. The rate of the reaction of the peptides and the ninhydrin was increased by heating the plates in the oven at 105° C for 5 min.'

Author Index

Numbers in *italics* indicate pages where references are listed at the end of chapters.

B

C

I

J

N

Y

Z

Subject Index

Abbreviations used in compilation of Subject Index

AA	amino acid
AAT	aspartate amino transferase
DH	dehydrogenase
DNA	deoxyribonucleic acid
EP	electrophoresis
GAPD	glyceraldehyde phosphate dehydrogenase
GDH	glycerophosphate dehydrogenase
G6PD	glucose 6 phosphate dehydrogenase
Hb	Hemoglobin
ICDH	isocitrate dehydrogenase
ID	immunological difference
IEP	immunoelectrophoresis
LAP	leucine amino peptidase
LDH	lactate dehydrogenase
MCF(T)	microcomplement fixation (test)
MDH	malate dehydrogenase
PGD	phosphogluconate dehydrogenase
RNA	ribonucleic acid
SCM	S carboxymethyl

F

G

H

I

J

K

L

M

P

T

Z